The Organic Chemistry of Drug Design and Drug Action

Second Edition

To Mom and the memory of Dad, for their love, their humor, their ethics, their inspiration, but also for their genes

The Organic Chemistry of Drug Design and Drug Action

Second Edition

Richard B. Silverman
Northwestern University
Department of Chemistry
Evanston, Illinois

ELSEVIER
ACADEMIC
PRESS

Amsterdam • Boston • Heidelberg • London • New York • Oxford
Paris • San Diego • San Francisco • Singapore • Sydney • Tokyo

Senior Publishing Editor   Jeremy Hayhurst
Project Manager   Simon Crump
Editorial Assistant   Desiree Marr
Marketing Manager   Linda Beattie
Cover Design   Eric DeCicco
Composition   Newgen Imaging Systems (P) Ltd
Printer   Maple-Vail Book Manufacturing Group

Elsevier Academic Press
200 Wheeler Road, Burlington, MA 01803, USA
525 B Street, Suite 1900, San Diego, California 92101-4495, USA
84 Theobald's Road, London WC1X 8RR, UK

This book is printed on acid-free paper. ∞

Library of Congress Cataloging-in-Publication Data
Application submitted

British Library Cataloguing in Publication Data
A catalogue record for this book is available from the British Library

ISBN-13: 978-0-12-643732-4
ISBN-10: 0-12-643732-7

For all information on all Academic Press publications
visit our Web site at www.academicpressbooks.com

Printed in the United States of America
08 9 8 7 6

Contents

Color plate section between pages 172 and 173

Preface to the First Edition

From 1985 to 1989 I taught a one-semester course in medicinal chemistry to senior undergraduates and first-year graduate students majoring in chemistry or biochemistry. Unlike standard medicinal chemistry courses that are generally organized by classes of drugs, giving descriptions of their biological and pharmacological effects, I thought there was a need to teach a course based on the organic chemical aspects of medicinal chemistry. It was apparent then, and still is the case now, that there is no text that concentrates exclusively on the organic chemistry of drug design, drug development, and drug action. This book has evolved to fill that important gap. Consequently, if the reader is interested in learning about a specific class of drugs, its biochemistry, pharmacology, and physiology, he or she is advised to look elsewhere for that information. Organic chemical principles and reactions vital to drug design and drug action are the emphasis of this text with the use of clinically important drugs as examples. Usually only one or just a few representative examples of drugs that exemplify the particular principle are given; no attempt has been made to be comprehensive in any area. When more than one example is given, it generally is to demonstrate different chemistry. It is assumed that the reader has taken a one-year course in organic chemistry that included amino acids, proteins, and carbohydrates and is familiar with organic structures and basic organic reaction mechanisms. Only the chemistry and biochemistry background information pertinent to the understanding of the material in this text is discussed. Related, but irrelevant, background topics are briefly discussed or are referenced in the general readings section at the end of each chapter. Depending on the degree of in-depthness that is desired, this text could be used for a one-semester or a full-year course. The references cited can be ignored in a shorter course or can be assigned for more detailed discussion in an intense or full-year course. Also, not all sections need to be covered, particularly when multiple examples of a particular principle are described. The instructor can select those examples that may be of most interest to the class. It was the intent in writing this book that the reader, whether a student or a scientist interested in entering the field of medicinal chemistry, would learn to take a rational physical organic chemical approach to drug design and drug development and to appreciate the chemistry of drug action. This knowledge is of utmost importance for the understanding of how drugs function at the molecular level. The principles are the same regardless of the particular receptor or enzyme involved. Once the fundamentals of drug design and drug action are understood, these concepts can be applied to the understanding of the many classes of drugs that are described in classical medicinal chemistry texts. This basic understanding can be the foundation for the future elucidation of drug action or the rational discovery of new drugs that utilize organic chemical phenomena.

Richard B. Silverman
Evanston, Illinois
April 1991

Preface to the Second Edition

In the 12 years since the first edition was written, certain new approaches in medicinal chemistry have appeared or have become commonly utilized. The basic philosophy of this textbook has not changed, that is, to emphasize general principles of drug design and drug action from an organic chemical perspective rather than from the perspective of specific classes of drugs. Several new sections were added (in addition to numerous new approaches, methodologies, and updates of examples and references), especially in the areas of lead discovery and modification (Chapter 2). New screening approaches, including high-throughput screening, are discussed as are the concepts of privileged structures and drug-likeness. Combinatorial chemistry, which was in its infancy during the writing of the first edition, evolved, became a separate branch of medicinal chemistry, then started to wane in importance during the 21st century. Combinatorial chemistry groups, prevalent in almost all pharmaceutical industry at the end of the 20th century, began to be dissolved, and a gradual return to traditional medicinal chemistry has been seen. Nonetheless, combinatorial chemistry journals have sprung up to serve as the conduit for dissemination of new approaches in this area, and this along with parallel synthesis are important approaches that have been added to this edition. New sections on SAR by NMR and SAR by MS have also been added. Peptidomimetic approaches are discussed in detail. The principles of structure modification to increase oral bioavailability and effects on pharmacokinetics are presented, including log P software and "rule of five" and related ideas in drug discovery. The fundamentals of molecular modeling and 3D-QSAR also are expanded. The concepts of inverse agonism, inverse antagonism, racemic switches, and the two-state model of receptor activation are introduced in Chapter 3. In Chapter 5 efflux pumps, COX-2 inhibitors, and dual-acting drugs are discussed; a case history of the discovery of the AIDS drug ritonavir is used to exemplify the concepts of drug discovery of reversible enzyme inhibitors. Discussions of DNA structure and function, topoisomerases, and additional examples of DNA-interactive agents, including metabolically activated agents, are new or revised sections in Chapter 6. The newer emphasis on the use of HPLC/MS/MS in drug metabolism is discussed in Chapter 7 along with the concepts of fatty acid and cholesterol conjugation and antedrugs. In Chapter 8 a section on enzyme-prodrug therapies (ADEPT, GDEPT, VDEPT) has been added as well as a case history of the discovery of omeprazole. Other changes include the use of both generic names and trade names, with generic names given with their chemical structure, and the inclusion of problem sets and solutions for each chapter.

The first edition of this text was written primarily for upperclass undergraduate and first-year graduate students interested in the general field of drug design and drug action. During the last decade it has become quite evident that there is a large population, particularly of synthetic organic chemists, who enter the pharmaceutical industry with little or no knowledge of medicinal chemistry and who want to learn the application of their skills to the process of drug discovery. The first edition of this text provided an introduction to the field for both students and practitioners, but the latter group has more specific interests in how to accelerate the drug discovery process. For the student readers, the basic principles described in the second edition are sufficient for the purpose of teaching the general process of how drugs are discovered and how they function. Among the basic principles, however, I have now interspersed many more specifics that go beyond the basics and may be more directly related to procedures and applications useful to those in the pharmaceutical industry. For example, in Chapter 2 it is stated that "Ajay and coworkers proposed that *drug-likeness* is a possible inherent property of some molecules,[1] and this property could determine which molecules should be selected for screening." The basic principle is that some molecules seem to have scaffolds found in many drugs and should be initially selected for testing. But following that inital statement is added more specifics: "They used a set of one- and two-dimensional parameters in their computation and were able to predict correctly over 90% of the compounds in the Comprehensive Medicinal Chemistry (CMC) database.[2] Another computational approach to differentiate drug-like and nondrug-like molecules using a scoring scheme was developed,[3] which was able to classify correctly 83% of the compounds in the Available Chemicals Directory (ACD)[4] and 77% of the compounds in the World Drug Index (WDI).[5] A variety of other approaches have been taken to identify drug-like molecules."[6] I believe that the student readership does not need to clutter its collective brain with these latter specifics, but should understand the basic principles and approaches; however, for those who aspire to become part of the pharmaceutical research field, they might want to be aware of these specifics and possibly look up the references that are cited (the instructor for a course who believes certain specifics are important may assign the references as readings).

For concepts peripheral to drug design and drug action, I will give only a reference to a review of that topic in case the reader wants to learn more about it. If the instructor believes that a particular concept that is not discussed in detail should have more exposure to the class, further reading can be assigned.

To minimize errors in reference numbers, several references are cited more than once under different endnote numbers. Also, although multiple ideas may come from a single reference, the reference is only cited once; if you want to know the origin of discussions in the text, look in the closest reference, either the one preceding the discussion or just following it. Because my expertise extends only in the areas related to enzymes and the design of enzyme inhibitors,

---

[1] Ajay; Walters, W. P.; Murcko, M. A. *J. Med. Chem.* **1998**, *41*, 3314.

[2] This is an electronic database of Volume 6 of *Comprehensive Medicinal Chemistry* (Pergamon Press) available from MDL Information systems, Inc., San Leandro, CA 94577.

[3] Sadowski, J.; Kubinyi, H. *J. Med. Chem.* **1998**, *41*, 3325.

[4] The ACD is available from MDL Information systems, Inc., San Leandro, CA, and contains specialty and bulk commercially available chemicals.

[5] The WDI is from Derwent Information.

[6] (a) Walters, W. P.; Stahl, M. T.; Murcko, M. A. *Drug Discovery Today* **1998**, *3*, 160. (b) Walters, W. P.; Ajay; Murcko, M. A. *Curr. Opin. Chem. Biol.* **1999**, *3*, 384. (c) Teague, S. J.; Davis, A. M.; Leeson, P. D.; Oprea, T. *Angew. Chem. Int. Ed. Engl.* **1999**, *38*, 3743. (d) Oprea, T. I. J. *Comput.-Aided Mol. Des.* **2000**, *14*, 251. (e) Gillet, V. J.; Willett, P. L.; Bradshaw, J. *J. Chem. Inf. Comput. Sci.* **1998**, *38*, 165. (f) Wagener, M.; vanGeerestein, V. J. *J. Chem. Inf. Comput. Sci.* **2000**, *40*, 280. (g) Ghose, A. K.; Viswanadhan, V.N.; Wendoloski, J. J. *J. Comb. Chem.* **1999**, *1*, 55. (h) Xu, J.; Stevenson, J. *J. Chem. Inf. Comput. Sci.* **2000**, *40*, 1177. (i) Muegge, I.; Heald, S. L.; Brittelli, D. *J. Med. Chem.* **2001**, *44*, 1841. (j) Anzali, S.; Barnickel, G.; Cezanne, B.; Krug, M.; Filimonov, D.; Poroikiv, V. *J. Med. Chem.* **2001**, *44*, 2432. (k) Brstle, M.; Beck, B.; Schindler, T.; King, W.; Mitchell, T.; Clark, T. *J. Med. Chem.* **2002**, *45*, 3345.

I want to thank numerous experts who read parts or whole chapters and gave me feedback for modification. These include (in alphabetical order) Shuet-Hing Lee Chiu, Young-Tae Chang, William A. Denny, Perry A. Frey, Richard Friary, Kent S. Gates, Laurence H. Hurley, Haitao Ji, Theodore R. Johnson, Yvonne C. Martin, Ashim K. Mitra, Shahriar Mobashery, Sidney D. Nelson, Daniel H. Rich, Philippa Solomon, Richard Wolfenden, and Jian Yu. Your input is greatly appreciated. I also greatly appreciate the assistance of my two stellar program assistants, Andrea Massari and Clark Carruth, over the course of writing this book, as well as the editorial staff (headed by Jeremy Hayhurst) of Elsevier/Academic Press.

Richard B. Silverman
still in Evanston, Illinois
May 2003

Chapter 1

Introduction

1.1 Medicinal Chemistry Folklore

Medicinal chemistry is the science that deals with the discovery and design of new therapeutic chemicals and their development into useful medicines. *Medicines* are substances used to treat diseases. *Drugs* are molecules used as medicines or as components in medicines to diagnose, cure, mitigate, treat, or prevent disease.[1] Medicinal chemistry may involve isolation of compounds from Nature or the synthesis of new molecules, investigations of the relationships between the structure of natural and/or synthetic compounds and their biological activities, elucidations of their interactions with receptors of various kinds, including enzymes and DNA, the determination of their absorption, transport, and distribution properties, and studies of the metabolic transformations of these chemicals into other chemicals and their excretion. More recently, *genomics*, the investigations of an organism's *genome* (all of the organism's genes) to identify important target genes and gene products (that is, proteins expressed by the genes), and *proteomics*, the investigations of new proteins in the organism's *proteome* (all of the proteins expressed by the genome)[2] to determine their structure and/or function often by comparison with known proteins, have become increasingly important approaches to identify new drug targets.

Medicinal chemistry, in its crudest sense, has been practiced for several thousand years. Man has searched for cures for illnesses by chewing herbs, berries, roots, and barks. Some of these early clinical trials were quite successful; however, not until the last 100–150 years has knowledge of the active constituents of these natural sources been known. The earliest written records of the Chinese, Indian, South American, and Mediterranean cultures described the therapeutic effects of various plant concoctions.[3–5] A Chinese health science anthology called *Nei Ching* is thought to have been written by the Yellow Emperor in the 13th century B.C., although some believe that it was backdated by the 3rd-century compilers.[6] The Assyrians described on 660 clay tablets 1000 medicinal plants used from 1900–400 B.C.

Two of the earliest medicines were described about 5100 years ago by the Chinese Emperor Shen Nung in his book of herbs called *Pen Ts'ao*.[7] One of these is *Ch'ang Shan*, the root

Dichroa febrifuga, which was prescribed for fevers. This plant contains alkaloids that are used even today in the treatment of malaria. Another plant called *Ma Huang* (now known as *Ephedra sinica*) contains ephedrine, a drug that raises the blood pressure and relieves bronchial spasms; this herb was used as a heart stimulant, a diaphoretic agent (perspiration producer), and for treatment of asthma, hay fever, and nasal and chest congestion. It also is used today (unadvisably) by some body builders and endurance athletes because it promotes themogenesis (the burning of fat) by release of fatty acids from stored fat cells, leading to quicker conversion of the fat into energy. Ephedra also tends to increase the contractile strength of muscle fibers, which allows body builders to work harder with heavier weights.

Theophrastus in the 3rd-century B.C. mentioned opium poppy juice as an analgesic agent, and in the 10th-century A.D. Rhazes (Persia) introduced opium pills for coughs, mental disorders, aches, and pains. The opium poppy, *Papaver somniferum*, contains morphine, a potent analgesic agent and codeine, which is prescribed today as a cough suppressant. The East Asians and the Greeks used henbane, which contains scopolamine (truth serum) as a sleep inducer. Inca mail runners and silver miners in the high Andean mountains chewed coca leaves (cocaine) as a stimulant and euphoric. The antihypertensive drug reserpine was extracted by ancient Hindus from the snakelike root of the *Rauwolfia serpentina* plant and was used to treat hypertension, insomnia, and insanity. Alexander of Tralles in the 6th-century A.D. recommended the autumn crocus (*Colchicum autumnale*) for relief of pain of the joints, and it was used by Avrienna (11th-century Persia) and by Baron Anton von Störck (1763) for the treatment of gout. Benjamin Franklin heard about this medicine and brought it to America. The active principle in this plant is the alkaloid colchicine, which is used today to treat gout.

In 1633 a monk named Calancha, who accompanied the Spanish Conquistadors to Central and South America, introduced one of the greatest herbal medicines to Europe on his return. The South American Indians would extract the cinchona bark and use it for chills and fevers; the Europeans used it for the same and for malaria. In 1820 the active constituent was isolated and later determined to be quinine, an antimalarial drug.

Modern therapeutics is considered to have begun with an extract of the foxglove plant, which was cited by Welsh physicians in 1250, named by Fuchsius in 1542, and introduced for the treatment of dropsy (now called congestive heart failure) in 1785 by Withering.[3,8] The active constituents are secondary glycosides from *Digitalis purpurea* (the foxglove plant) and *Digitalis lanata*, namely, digitoxin and digoxin, respectively, both important drugs for the treatment of congestive heart failure. Today, digitalis, which refers to all of the cardiac glycosides, is still manufactured by extraction of foxglove and related plants.

1.2 Discovery of New Drugs

Nature is still an excellent source of new drugs or, more commonly, of precursors to drugs. Of the 20 leading drugs in 1999, 9 of them were derived from natural products.[9] Almost 40% of the 520 new drugs approved for the drug market between 1983 and 1994 were natural products or derived from natural products. Greater than 60% of the anticancer and anti-infective agents that are on the market or in clinical trials are of natural product origin or derived from natural products.[10] This may be a result of the inherent nature of these secondary metabolites to act in defense of their producing organisms; for instance, a fungal natural product might be produced against bacteria or other fungi or against cell replication of foreign organisms.[11]

Typically, when a natural product is found to be active, it is chemically modified to improve its properties. As a result of advances made in synthesis and separation methods and in

Figure 1.1 ► "That's Dr Arnold Moore. He's conducting an experiment to test the theory that most great scientific discoveries were hit on by accident."
Drawing by Hoff; © *1957*
The New Yorker Magazine, Inc.

biochemical techniques since the late 1940s, the early random approach to drug discovery (Figure 1.1) was supplanted by a more rational approach, namely, one that involves the element of design. A discussion of how drugs are discovered and chemically modified to improve or change their medicinal properties is presented in Chapter 2. As we will see, the random approach still is important!

1.3 General References

The following references are excellent sources of material for this entire book.

Journals

Annual Reports in Medicinal Chemistry; Academic Press, San Diego
Annual Review of Biochemistry
Annual Review of Medicinal Chemistry
Biochemical Pharmacology
Bioorganic and Medicinal Chemistry
Bioorganic and Medicinal Chemistry Letters
Chemistry & Biology
Current Drug Metabolism
Current Drug Targets
Current Genomics
Current Medicinal Chemistry
Current Opinion in Chemical Biology
Current Opinion in Drug Discovery and Development
Current Opinion in Investigational Drugs
Current Opinion in Therapeutic Patents
Current Pharmaceutical Biotechnology
Current Pharmaceutical Design
Current Protein & Peptide Science
Drug Design and Discovery
Drug Development Research
Drug Discovery and Development
Drug Discovery Today
Drug News and Perspectives
Drugs
Drugs of the Future
Drugs of Today
Drugs Under Experimental and Clinical Research

Emerging Drugs   *Modern Drug Discovery*
Emerging Therapeutic Targets   *Modern Pharmaceutical Design*
European Journal of Medicinal Chemistry   *Molecular Pharmacology*
Expert Opinion on Investigational Drugs   *Nature Medicine*
Expert Opinion on Pharmacotherapy   *Perspectives in Drug Discovery and Design*
Expert Opinion on Therapeutic Patents   *Progress in Drug Research*
Expert Opinion on Therapeutic Targets   *Progress in Medicinal Chemistry*
Journal of Medicinal Chemistry   *Trends in Biochemical Sciences*
Medicinal Research Reviews   *Trends in Pharmacological Sciences*
Mini Reviews in Medicinal Chemistry

Books

Advances in Medicinal Chemistry; Elsevier, New York; (continuing series).

Albert, A. *Selective Toxicity*, 7th ed., Chapman and Hall, London, 1985.

Arins, E. J. (Ed.) *Drug Design*, Academic, New York, 1971–1980, Vols. 1–10.

Borchardt, R. T.; Freidinger, R. M.; Sawyer, T. K. *Integration of Pharmaceutical Discovery and Development: Case Histories*, Plenum Press, New York, 1998.

Burger, A. *A Guide to the Chemical Basis of Drug Design*, Wiley, New York, 1983.

Hansch, C.; Emmett, J. C.; Kennewell, P. D.; Ramsden, C. A.; Sammes, P. G.; Taylor, J. B. (Eds.) *Comprehensive Medicinal Chemistry*, Pergamon Press, Oxford, 1990, Vols. 1–6.

Hardman, J. G.; Limbird, L. E.; Gilman, A. G. (Eds.) *Goodman and Gilman's The Pharmacological Basis of Therapeutics*, 10th ed., McGraw-Hill, New York, **2001**.

Lednicer, D. *Chronicles of Drug Discovery*, ACS, Washington, DC, **1993**.

Lednicer, D. *Strategy for Organic Drug Synthesis and Design*, Wiley, New York, **1998**.

Lednicer, D. Mitscher, L. A. *The Organic Chemistry of Drug Synthesis*, Wiley, New York, **1996**, five-volume set.

Lunn, G.; Schmuff, N. *HPLC Methods for Pharmaceutical Analysis*, Wiley-VCH, New York, **1997**.

Methods and Principles in Medicinal Chemistry, VCH, Weinheim. (continuing series).

O'Neil, M. J.; Smith, A. (Eds.) *The Merck Index*, 13th ed., Merck & Co., Rahway, NJ, **2001**.

Wolff, M. E. (Ed.) *Burger's Medicinal Chemistry and Drug Discovery*, John Wiley & Sons, New York, **1995–1997**, Vols. 1–6.

1.4 References

1. *Webster's Ninth New Collegiate Dictionary*, Merriam-Webster, Springfield, MA, 1987.

2. Wilkins, M. R.; Williams, K. L.; Appel, R. D.; Hochstrasser, D. F. (Eds.) *Proteome Research: New Frontiers in Functional Genomics*, Springer-Verlag, Berlin, **1997**.

3. Bauer, W. W. *Potions, Remedies and Old Wives' Tales*, Doubleday, New York, 1969.

4. Withering, W. *An Account of the Foxglove and Some of Its Medicinal Uses: With Practical Remarks on Dropsy and Other Diseases*, Robinson, C. G. J., London, 1785; reprinted in *Med. Class.* **1937**, *2*, 305.

5. Sneader, W. *Drug Discovery: The Evolution of Modern Medicines*, Wiley, Chichester, 1985.

6. Nakanishi, K. In *Comprehensive Natural Products Chemistry*, Barton, D.; Nakanishi, K. (Eds.), Elsevier, Amsterdam and New York, 1999, Vol. 1, pp. xxiii–xl.

7. Chen, K. K. *J. Am. Pharm. Assoc.* **1925**, *14*, 189.

8. Burger, A. In *Burger's Medicinal Chemistry*, 4th ed., Wolff, M. E. (Ed.), Wiley, New York, **1980**, Part I, Chap. 1.

9. Harvey, A. *Drug Discovery Today* **2000**, *5*, 294.

10. Cragg, G. M.; Newman, D. J.; Snader, K. M. *J. Nat. Prod.* **1997**, *60*, 52.

11. Hung, D. T.; Jamison, T. F.; Schrieber, S. L. *Chem. Biol.* **1996**, *3*, 623.

Chapter 2

Drug Discovery, Design, and Development

2.1 Drug Discovery

Drug discovery is a very time-consuming and expensive process. Estimates of the average time required to bring a drug to the market range from 12–15 years at an average cost of about $800 million. For approximately every 10,000 compounds that are evaluated in animal studies, 10 will make it to human clinical trials in order to get 1 compound on the market. The clinical trials consist of three phases prior to drug approval: phase I (generally a few months to a year and a half) evaluates the safety, tolerability (dosage levels and side effects), pharmacokinetic properties, and pharmacological effects in 20–100 healthy volunteers; phase II (about 1–3 years) assesses the effectiveness of the drug, determines side effects and other safety aspects, and clarifies the dosing regimen in a few hundred diseased patients; and phase III (about 2–6 years) is a larger trial with several thousand patients in clinics and hospitals that establishes the efficacy of the drug and monitors adverse reactions from long-term use. Once the new drug application (NDA) is submitted to the Food and Drug Administration (FDA), it can be several months to several years before it is approved for commercial use. Phase IV studies are considered to be the results found with a drug that has already been allowed onto the drug market and is in general use. Drug candidates (or *new chemical entities*, NCE, as they are often called) that fail late in this process result in huge, unrecovered financial losses for the company. This is why the cost to purchase a drug is so high. It is not that it costs that much to manufacture that one drug, but that the profits are needed to pay for all of the drugs that fail to make it to market after large sums of research funds have already been expended.

In general, drugs are not discovered. What is more likely discovered is known as a *lead* compound. The lead is a prototype compound that has a number of attractive characteristics, such as the desired biological or pharmacological activity, but may have other undesirable characteristics, for example, high toxicity, other biological activities, absorption difficulties, insolubility, or metabolism problems. The structure of the lead compound is modified by synthesis to amplify the desired activity and to minimize or eliminate the unwanted properties to a point where a *drug candidate*, a compound worthy of extensive biological, pharmacological, and animal studies, is identified; then a *clinical drug*, a compound ready for clinical trials, is developed. Prior to an elaboration of approaches to lead discovery and lead modification, two common drugs discovered without a lead are discussed.

2.1.A Drug Discovery without a Lead

A.1 Penicillins

In 1928 Alexander Fleming noticed a green mold growing in a culture of *Staphylococcus aureus*, and where the two had converged, the bacteria were lysed.[1] This led to the discovery of penicillin, which was produced by the mold. Actually, Fleming was not the first to make this observation; John Burdon-Sanderson had done so in 1870, ironically also at St. Mary's Hospital in London, the same institution where Fleming made the rediscovery![2] Joseph Lister had treated a wounded patient with *Penicillium*, the organism later found to be the producer of penicillin (although the strains discovered earlier than Fleming's strain did not produce penicillin, but, rather, another antibiotic, mycophenolic acid). After Fleming observed this phenomenon, he tried many times to repeat it without success; it was his colleague, Dr. Ronald Hare,[3,4] who was able to reproduce the observation. It only occurred the first time because a combination of unlikely events all took place simultaneously. Hare found that very special conditions were required to produce the phenomenon initially observed by Fleming. The culture dish inoculated by Fleming must have become accidentally and simultaneously contaminated with the mold spore. Instead of placing the dish in the refrigerator or incubator when he went on vacation as is normally done, Fleming inadvertently left it on his lab bench. When he returned the following month, he noticed the lysed bacteria. Ordinarily, penicillin does not lyse these bacteria; it prevents them from developing, but it has no effect if added after the bacteria have developed. However, while Fleming was on vacation (July to August) the weather was unseasonably cold, and this provided the particular temperature required for the mold and the staphylococci to grow slowly and produce the lysis. Another extraordinary circumstance was that the particular strain of the mold on Fleming's culture was a relatively good penicillin producer, although most strains of that mold (*Penicillium*) produce no penicillin at all. The mold presumably came from the laboratory just below Fleming's where research on molds was going on at that time.

Although Fleming suggested that penicillin could be useful as a topical antiseptic, he was not successful in producing penicillin in a form suitable to treat infections. Nothing more was done until Sir Howard Florey at Oxford University reinvestigated the possibility of producing penicillin in a useful form. In 1940 he succeeded in producing penicillin that could be administered topically and systemically,[5] but the full extent of the value of penicillin was not revealed until the late 1940s.[6] Two reasons for the delay in the universal utilization of penicillin were the emergence of the sulfonamide antibacterials (sulfa drugs, **2.1**; see Chapter 5, Section 5.4.B.2, p. 254) in 1935 and the outbreak of World War II.

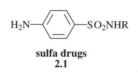

sulfa drugs
2.1

No studies related to the pharmacology, production, and clinical application of penicillin were permitted until after the war to prevent the Germans from having access to this wonder drug. Allied scientists who were interrogating German scientists involved in chemotherapeutic research were told that the Germans thought the initial report of penicillin was made just for commercial reasons to compete with the sulfa drugs. They did not take the report seriously.

The original mold was *Penicillium notatum*, a strain that gave a relatively low yield of penicillin. It was replaced by *Penicillium chysogenum*,[7] which had been cultured from a mold growing on a grapefruit in a market in Peoria, Illinois!

For many years debate raged regarding the actual structure of penicillin (**2.2**),[8] but the correct structure was elucidated in 1944 with an X-ray crystal structure by Dorothy Crowfoot Hodgkin (Oxford); the crystal structure was not actually published until 1949.[9] Several different penicillin analogs (R group varied) were isolated early on; only two of these early analogs (**2.2**, R = PhOCH$_2$, penicillin V; and **2.2**, R = CH$_2$Ph, penicillin G) are still in use today.

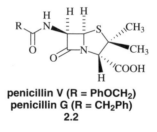

penicillin V (R = PhOCH$_2$)
penicillin G (R = CH$_2$Ph)
2.2

A.2   Librium

The first benzodiazepine tranquilizer drug, chlordiazepoxide HCl [7-chloro-2-(methylamino)-5-phenyl-3H-1,4-benzodiazepine 4-oxide; **2.3**; Librium], was discovered serendipitously.[10] Dr. Leo Sternbach at Roche was involved in a program to synthesize a new class of tranquilizer drugs. He originally set out to prepare a series of benzheptoxdiazines (**2.4**), but when R^1 was CH$_2$NR$_2$ and R^2 was C$_6$H$_5$, it was found that the actual structure was that of a quinazoline 3-oxide (**2.5**).

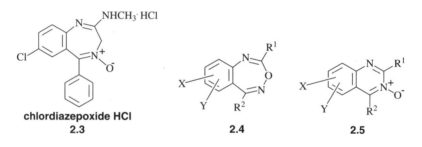

chlordiazepoxide HCl   **2.4**   **2.5**
2.3

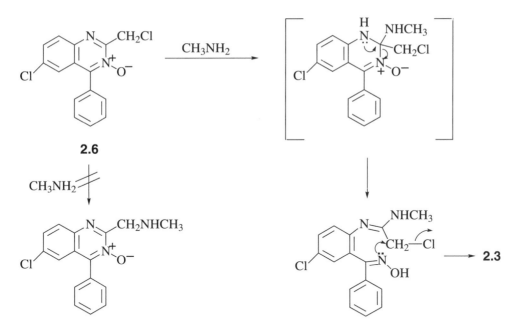

Scheme 2.1 ► Mechanism for formation of Librium.

However, none of these compounds gave any interesting pharmacological results. The program was abandoned in 1955 in order for Sternbach to work on a different project. In 1957 during a general laboratory cleanup a vial containing what was thought to be **2.5** (X = 7-Cl, $R^1 = CH_2NHCH_3$, $R^2 = C_6H_5$) was found and, as a last effort, was submitted for pharmacological testing. Unlike all of the other compounds submitted, this one gave very promising results in six different tests used for preliminary screening of tranquilizers. Further investigation revealed that this compound was not a quinazoline 3-oxide, but was instead the benzodiazepine 4-oxide, **2.3**, presumably produced in an unexpected reaction of the corresponding chloromethyl quinazoline 3-oxide (**2.6**) with methylamine (Scheme 2.1). If this compound had not been found in the laboratory cleanup, all of the negative pharmacological results would have been reported for the quinazoline 3-oxide class of compounds, and benzodiazepine 4-oxides may not have been discovered for many years to come.

The examples of drug discovery without a lead are relatively few in number. The typical occurrence is that a lead compound is identified, and its structure is modified to give, eventually, the compound that goes to the clinic.

2.1.B Lead Discovery

Penicillin V and Librium are, indeed, two important drugs that were discovered without a lead. However, once they were identified, they then became lead compounds for second-generation analogs. A myriad of penicillin-derived antibacterials have been synthesized as a result of the structure elucidation of the earliest penicillins. Diazepam (**2.7**, Valium) was synthesized at Roche even before Librium was introduced on to the market; this drug was derived from the lead compound Librium and is almost 10 times more potent than the lead.

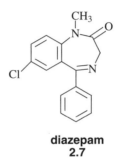

diazepam
2.7

The initial difficulty arises in the discovery of the lead compound. Several approaches can be taken to identify a lead. The first requirement for all of the approaches is to have a means to assay compounds for a particular biological activity, so that researchers can tell when a compound is active. A *bioassay* (or *screen*) is a means of determining in a biological system, relative to a control compound, if a compound has the desired activity, and, if so, what the relative potency of the compound is. Note the distinction between the terms *activity* and *potency*. *Activity* is the particular biological or pharmacological effect (for example, antibacterial activity or anticonvulsant activity); *potency* is the strength of that effect.

Some screens are *in vitro* tests, for example, the inhibition of an enzyme or antagonism of a receptor; others are *in vivo* tests, for example, the ability of the compound to prevent an induced seizure in a mouse. In general, the *in vitro* tests are quicker and less expensive. Currently, *high-throughput screens* (HTS),[11] very rapid and sensitive *in vitro* screens initially developed about 1989–1991, that now can be carried out robotically in 1536- or 3456-well titer plates on small (submicrogram) amounts of compound (dissolved in submicroliter volumes) are becoming universally used. With these ultra-high-throughput screening approaches, it is possible to screen 100,000 compounds in a day! As we will see below, combinatorial chemistry (see Section 2.2.E.5, p. 34) can supply huge numbers of compounds in a short period of time, which, theoretically, should provide an increased number of *hits*, i.e., compounds that elicit a predetermined level of activity in the bioassay and, therefore, provide more leads. According to Drews,[12] the number of compounds assayed in a large pharmaceutical company in the early 1990s was about 200,000 a year; that number rose to 5–6 million during the mid-1990s, and by the end of the 1990s it was >50 million! However, the increase in the assay rate did not result in a commensurate increase in research productivity, as measured by new compounds entering the market. Of course, it can take 12–15 years for a drug to reach the market, so productivity in the early part of the 21st century should provide a more accurate ruler for success of drug discovery changes made at the end of the 20th century. Currently, HTS appears to have resulted in an increase in the number of hits, but this may be because more lipophilic compounds, which may have more drug-like properties (see Section 2.2.F.2, p. 53), can be tested by dissolving them in dimethylsulfoxide (DMSO) rather than in water. Nonetheless, it is not yet clear if this increase in hit rate is translating into a much greater number of leads and development compounds.[13]

An exciting approach for screening compounds that might interact with an enzyme in a metabolic pathway was demonstrated by Wong, Pompliano, and coworkers for the discovery of lead compounds that block bacterial cell wall biosynthesis (as potential antibacterial agents).[14] Conditions were found to reconstitute all six enzymes in the cell wall biosynthetic pathway so that incubation with the substrate for the first enzyme leads to the formation of the product of the last enzyme in the pathway. Then by screening compounds and looking for the buildup of an intermediate, it is possible to identify not only compounds that block the pathway (and prevent the formation of the bacterial cell wall), but to determine which enzyme is blocked (the buildup of an intermediate means that the enzyme that acted on that intermediate was blocked).

Compound screening also can be carried out by electrospray ionization mass spectrometry[15] (the technique for which John Fenn received the Nobel prize in 2002) and by NMR spectrometry.[16] Tightly bound noncovalent complexes of compounds with a macromolecule (such as a receptor or enzyme) can be observed in the mass spectrum. The affinity of the *ligand* (a small molecule that binds to a receptor) can be measured by varying the collision energy and determining at what energy the complex dissociates. This method also can be used to screen mixtures (a library) of compounds, provided they have different molecular masses and/or charges, so the m/z for each complex with the biomolecule can be separated in the mass spectrometer. By varying the collision energy, it is possible to determine which test molecules bind to the biomolecule best. The ^{1}H NMR method exploits changes in either relaxation rates or diffusion rates of small molecules when they bind to a macromolecule. This method also can be used to screen mixtures of compounds to determine the ones that bind best.

Once the screen is developed, a variety of approaches can be taken to obtain a lead. As we will see below, the typical lead compound for a receptor or enzyme is the natural ligand for the receptor or substrate for the enzyme. Another good source of lead compounds is marketed drugs.[17] In this case the target will generally be well established, and the lead structure will be known to bind well to the target and to have good absorption properties. The main stumbling block to the use of marketed drugs as leads may be patent issues for commercialization. If the target macromolecule is not known or if no new leads have come from a marketed drug, other approaches can taken.

B.1 Random Screening

In the absence of known drugs and other compounds with desired activity, a random screen is a valuable approach. *Random screening* involves no intellectualization; all compounds are tested in the bioassay without regard to their structures. Prior to 1935 (the discovery of sulfa drugs), this was essentially the only approach; today this method is still an important approach to discover drugs or leads, particularly because it is now possible to screen such huge numbers of compounds rapidly with HTSs. This is the lead discovery method of choice when nothing is known about the receptor target.

The two major classes of materials screened are synthetic chemicals and natural products (microbial, plant, and marine). An example of a random screen of synthetic and natural compounds was the "war on cancer" declared by Congress and the National Cancer Institute in the early 1970s. Any new compound submitted was screened in a mouse tumor bioassay. Few new anticancer drugs resulted from that screen, but many known anticancer drugs also did not show activity in the screen used, so a new set of screens was devised that gave more consistent results. In the 1940s and 1950s, a random screen of soil samples by various pharmaceutical companies in search of new antibiotics was undertaken. However, in this case, not only were numerous leads uncovered, but two important antibiotics, streptomycin and the tetracyclines, were found. Screening of microbial broths, particular strains of *Streptomyces*, was a common random screen methodology prior to 1980.

B.2   Nonrandom (or Targeted or Focused) Screening

Nonrandom screening, also called *targeted* or *focused screening*, is a more narrow approach than is random screening. In this case, compounds having a vague resemblance to weakly active compounds uncovered in a random screen, or compounds containing different functional groups than leads, may be tested selectively. By the late 1970s, the National Cancer Institute's random screen was modified to a nonrandom screen because of budgetary and man-power restrictions. Also, the single tumor screen was changed to a variety of tumor screens because it was realized that cancer is not just a single disease.

B.3   Drug Metabolism Studies

During drug metabolism studies (see Chapter 7) *metabolites* (drug degradation products gen-erated *in vivo*) that are isolated are screened to determine if the activity observed is derived from the drug candidate or from a metabolite. For example, the anti-inflammatory drug sulin-dac (**2.8**, Clinoril) is not the active agent; the metabolic reduction product, **2.9**, is responsible for the activity.[18]

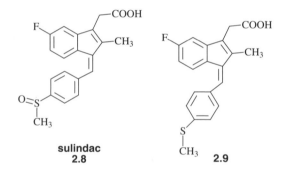

sulindac
2.8

2.9

The nonsedating antihistamine terfenadine hydrochloride (**2.10**, Seldane) was found to cause an abnormal heart rhythm in some users who also were taking certain antifungal agents, which were found to block the enzyme that metabolizes terfenadine. This caused a buildup of terfenadine, which led to the abnormal heart rhythms. However, a metabolite of terfenadine, fexofenadine hydrochloride (**2.11**, Allegra), was also found to be a nonsedating antihistamine, but it can be metabolized even in the presence of antifungal agents. This, then, is a safer drug. Metabolites can be screened for other activities as well.

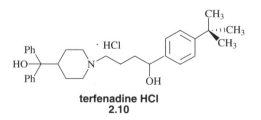

terfenadine HCl
2.10

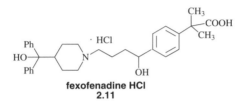

fexofenadine HCl
2.11

B.4 Clinical Observations

Sometimes a drug candidate during clinical trials will exhibit more than one pharmacological activity; that is, it may produce a side effect. This compound, then, can be used as a lead (or, with luck, as a drug) for the secondary activity. In 1947 an antihistamine, dimenhydrinate (**2.12**, Dramamine) was tested at the allergy clinic at Johns Hopkins University and was found also to be effective in relieving a patient who suffered from car sickness; a further study proved its effectiveness in the treatment of seasickness[19] and airsickness.[20] It then became the most widely used drug for the treatment of all forms of motion sickness.

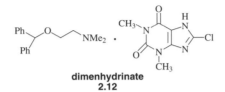

dimenhydrinate
2.12

There are other popular examples of drugs derived from clinical observations. Bupropion hydrochloride (**2.13**), an antidepressant drug (Wellbutrin), was found to help patients stop smoking and is now the first drug marketed as a smoking cessation aid (Zyban). The impotence drug sildenafil citrate (**2.14**; Viagra) was designed for the treatment of angina and hypertension by blocking the enzyme phosphodiesterase-5, which hydrolyzes cyclic guanosine monophosphate (cGMP), a vasodilator that allows increased blood flow.[21] In 1991 sildenafil went into phase I clinical trials for angina. In phase II clinical trials, it was not as effective against angina as Pfizer had hoped, so it went back to phase I clinical trials to see how high of a dose could be tolerated. It was during that clinical trial that the volunteers reported increased erectile function. Given the weak activity against angina, it was an easy decision to try to determine its effectiveness as the first treatment for erectile dysfunction. Sildenafil works by the mechanism for which it was designed as an antianginal drug, except it inhibits the phosphodiesterase in the penis (phosphodiesterase-5) instead of the heart (Figure 2.1). Sexual stimulation causes release of nitric oxide in the penis.

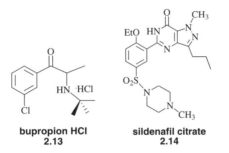

bupropion HCl   **sildenafil citrate**
2.13   **2.14**

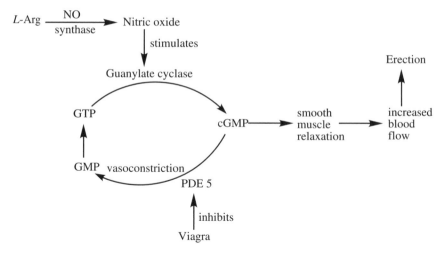

Figure 2.1 ▶ Mechanism of action of sildenafil (Viagra)

Nitric oxide is a second messenger molecule that stimulates the enzyme guanylate cyclase, which converts guanosine triphosphate to cGMP. The vasodilator cGMP relaxes the smooth muscle in the *corpus cavernosum*, allowing blood to flow into the penis, thereby producing an erection. However, phosphodiesterase-5 (PDE 5) hydrolyzes the cGMP, which causes vaso-constriction and the outflow of blood from the penis. Sildenafil inhibits this phosphodiesterase, preventing the hydrolysis of cGMP and prolonging the vasodilation effect.

B.5 Rational Approaches to Lead Discovery

None of the above approaches to lead discovery involves a major rational component. The lead is just found by screening techniques, as a by-product of drug metabolism studies, or from clinical investigations. Is it possible to *design* a compound having a particular activity? Rational approaches to drug design now have become the major routes to lead discovery. The first step is to identify the cause for the disease state. Many diseases, or at least the symptoms of diseases, arise from an imbalance (either excess or deficiency) of particular chemicals in the body, from the invasion of a foreign organism, or from aberrant cell growth. As will be discussed in later chapters, the effects of the imbalance can be corrected by antagonism or agonism of a receptor (see Chapter 3) or by inhibition of a particular enzyme (see Chapter 5); foreign organism enzyme inhibition or interference with DNA biosynthesis or function are important approaches to treat diseases arising from microorganisms and aberrant cell growth (see Chapter 6).

Once the relevant biochemical system is identified, initial lead compounds then become the natural receptor ligands or enzyme substrates. For example, lead compounds for the contraceptives (+)-norgestrel (**2.15**, Ovral) and 17α-ethynylestradiol (**2.16**, Activella) were the steroidal hormones progesterone (**2.17**) and 17β-estradiol (**2.18**). Whereas the steroid hormones **2.17** and **2.18** show weak and short-lasting effects, the oral contraceptives **2.15** and **2.16** exert strong progestational activity of long duration.

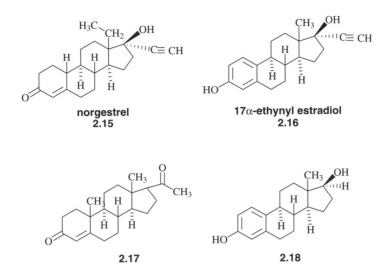

norgestrel
2.15

17α-ethynyl estradiol
2.16

2.17

2.18

At Merck it was believed that serotonin (**2.19**) was a possible mediator of inflammation. Consequently, serotonin was used as a lead for anti-inflammatory agents, and from this lead the anti-inflammatory drug indomethacin (**2.20**, Indocin) was developed.[22]

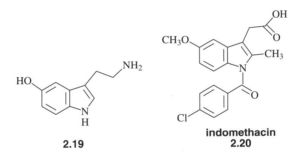

2.19

indomethacin
2.20

The rational approaches are directed at lead discovery. It is not possible, with much accuracy, to foretell toxicity and side effects, anticipate transport characteristics, or predict the metabolic fate of a drug. Once a lead is identified, its structure can be modified until an effective drug is obtained.

2.2   Lead Modification: Drug Design and Development

Once your lead compound is in hand, how do you know what to modify in order to improve the desired pharmacological properties?

2.2.A   Identification of the Active Part: The Pharmacophore

Interactions of drugs with receptors, known as *pharmacodynamics*, are very specific (see Chapter 3). Therefore, only a small part of the lead compound may be involved in the appropriate receptor interactions. The relevant groups on a molecule that interact with a receptor and are responsible for the activity are collectively known as the *pharmacophore*. The other atoms in the lead molecule, sometimes referred to as the *auxophore*, may be extraneous.

Some of the atoms, of course, are essential to maintain the integrity of the molecule and hold the pharmacophoric groups in their appropriate positions. Some of these extraneous atoms, however, may be interfering with the binding of the pharmacophore, and those atoms need to be excised from the lead compound. Other atoms in the auxophore may be dangling in space within the receptor and are neither binding to the receptor nor preventing the pharmacophoric atoms from binding. Although these atoms appear to be innocuous, it is important to know which atoms these are, because these are the ones that can be modified without loss of potency. As we will see later, there are other aspects to lead modification that are as important as increasing binding to the target receptor, such as *pharmacokinetics* (absorption, distribution, metabolism, and excretion or ADME). Modification of the atoms that are not interfering with binding could be very important to solving pharmacokinetics problems.

By determining which are the pharmacophoric groups and which are the auxophoric groups on your lead compound, and of the auxophoric groups, which are interfering with lead compound binding and which are not detrimental to binding, you will know which groups must be excised and which you can retain or modify as needed. One approach in lead modification to help make this determination is to cut away sections of the lead molecule and measure the effects of those modifications on potency. Consider this artificial example of how this might be done. Assume that the addictive analgesics morphine (**2.21**, R = R' = H), codeine (**2.21**, R = CH$_3$, R' = H), and heroin (**2.21**, R = R' = COCH$_3$) are the lead compounds, and we want to know which groups are pharmacophoric and which are auxophoric.

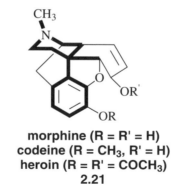

morphine (R = R' = H)
codeine (R = CH$_3$, R' = H)
heroin (R = R' = COCH$_3$)
2.21

The morphine family of analgesics binds to the μ opioid receptors. The pharmacophore is known and is shown as the darkened part in **2.21**. A decrease in potency on removal of a group will suggest that it may have been pharmacophoric, an increase in potency means it was auxophoric and interfering with proper binding, and essentially no change in potency will mean that it is auxophoric but not interfering with binding.

Let's start by excising the dihydrofuran oxygen atom, which is not in the pharmacophore. This may not seem to be sensible because that atom connects the cyclohexene ring to the benzene ring; its removal will result in a change in the conformation of the cyclohexene ring and an increase in the degrees of freedom of the molecule. Excision of the dihydrofuran oxygen gives morphinan (**2.22**, R = H)[23]; the hydroxy analog, levorphanol[24] (**2.22**, R = OH, Levo-Dromoran), is three to four times *more* potent than morphine as an analgesic, but it retains the addictive properties (note that in **2.22** the cyclohexene ring conformation has not been changed for ease of comparison with **2.21**; surely, a lower energy conformer will be favored).

levorphanol (R = OH)
2.22

Possibly, the additional conformational mobility allowed the molecule to approximate its *bioactive conformation* (the conformation that most effectively binds to a receptor). Removal of half of the cyclohexene ring (also not in the pharmacophore), leaving only methyl substituents, gives benzomorphan (**2.23**, R = CH₃).[25] This compound shows some separation of analgesic and addictive effects; pentazocine hydrochloride (**2.23**, R = CH₂CH=C(CH₃)₂; component of Talwin) is less potent than morphine (about as potent as codeine), but has a much lower addiction liability. Remember, your goal is both to increase potency and decrease adverse effects, such as addictive properties. Although this analog is not more potent than morphine, it is less addicting. Cutting away the methylene group of the cyclohexane fused ring (**2.24**) also, surprisingly, has little effect on the analgesic activity in animal tests.

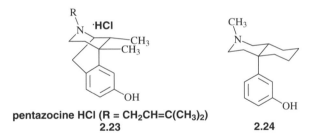

pentazocine HCl (R = CH₂CH=C(CH₃)₂)   **2.24**
2.23

Again, this excision removes the rigidity of the parent structure. Removal of all fused rings, for example, in the case of meperidine (**2.25**, Demerol), gives an analgesic still possessing 10–12% of the overall potency of morphine.[26] Although the potency is lower, it certainly will be much easier to synthesize analogs of meperidine than of morphine. Even acyclic analogs are active. Dextropropoxyphene (**2.26**, Darvon; again note the side chain is left in a conformation to resemble the structure of morphine) is one-half to two-thirds as potent as codeine. Both morphine and dextropropoxyphene bind to the μ opioid receptor, so the activity of dextropropoxyphene can be ascribed to the fact that it can assume a conformation related to that of the morphine pharmacophore. I don't think any of us, seeing dextropropoxyphene written in a more energetically favorable conformation, would ever make the connection between this structure and that of morphine. By cutting pieces off of the lead compound, it gives you new perspectives on possible active structures, which should open up completely new scaffolds to consider in synthesis. Another acyclic analog is methadone (**2.27**, Methadose) which is as potent an analgesic as morphine; the (−)-isomer is used in the treatment of opioid abstinence syndromes in heroin abusers because it is eliminated from the body slower than morphine, allowing the body to adapt to the falling levels of drug gradually.

meperidine
2.25

dextropropoxyphene
2.26

methadone
2.27

What if every cut in the lead produces a compound with lower potency? Then, either every excision is removing part of the pharmacophore or each cut causes a conformational change that gives a structure *less* similar to the bioactive conformation. The latter possibility is particularly relevant to such a rigid structure as in morphine. In this case, groups need to be added to the lead structure to *increase* the pharmacophore. For example, oripavine derivatives such as etorphine (**2.28**, R = CH_3, R' = C_3H_7; Immobilon), which has a two-carbon bridge and substituent not in morphine, is 3200 times more potent than morphine[27] and is used in veterinary medicine to immobilize large animals. The related analog, buprenorphine (**2.28**, R = CH_2—◁, R' = *t*-Bu, double bond reduced; Buprenex) is 10–20 times more potent than morphine and has a very low level of dependence liability.

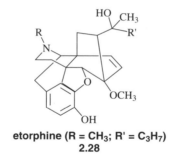

etorphine (R = CH_3; R' = C_3H_7)
2.28

Apparently, the additional rigidity of the oripavine derivatives increases the appropriate receptor interactions (see Chapter 3).

The activity and potency of a molecule are related to the interactions of the pharmacophoric groups with groups on the receptor (see Chapter 3, Section 3.2.B). The binding constants of 200 drugs and potent enzyme inhibitors were used by Andrews and coworkers[28] to calculate the average binding energies of common functional groups; these energies can be used to determine how well a new molecule binds to its receptor. If the test molecule has a measured binding energy that is lower than the calculated average value, it suggests that the molecule contains groups that do not interact with the receptor (are not in the pharmacophore). These groups, then, could be excised without loss of potency, giving a simplified lead for further structural modification. This *Andrews analysis* was carried out on a highly substituted lead compound, leading to a more simple analog structure that was modified to give molecules with enhanced potency.[29] If the test compound has a binding energy greater than the calculated average value, then the molecule may bind differently than suspected, leading to enhanced binding interactions. This indicates that manipulation of functional groups is an important lead modification approach.

2.2.B Functional Group Modification

The importance of functional group modification is demonstrated by **2.29**. The antibacterial agent, carbutamide (**2.29**, R = NH$_2$), was found to have an antidiabetic side effect; however, it could not be used as an antidiabetic drug because of its antibacterial activity, which could lead to bacterial resistance (see Chapter 5, Section 5.2, p. 231). The amino group of carbutamide was replaced by a methyl group to give tolbutamide (**2.29**, R = CH$_3$; Orinase) and in so doing the antibacterial activity was eliminated from the antidiabetic activity.

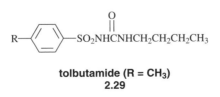

tolbutamide (R = CH$_3$)
2.29

In some cases, an experienced medicinal chemist knows what functional group will elicit a particular effect. Chlorothiazide (**2.30**, Aldocor) is an antihypertensive agent that has a strong diuretic effect as well. It was known from sulfanilamide work that the sulfonamide side chain can give diuretic (increased urine excretion) activity (see Section 2.2.C below). Consequently, diazoxide (**2.31**, Hyperstat) was prepared as an antihypertensive drug without diuretic activity.

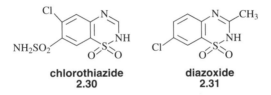

chlorothiazide
2.30

diazoxide
2.31

Obviously, a relationship exists between the molecular structure of a compound and its activity. This phenomenon was first realized about 135 years ago.

2.2.C Structure–Activity Relationships

In 1868 Crum-Brown and Fraser,[30] suspecting that the quaternary ammonium character of curare, a potent poison known since the 16th century that was used on arrowheads, may be responsible for its muscular paralytic properties (it blocks the action of the excitatory neurotransmitter acetylcholine on muscle receptors), examined the neuromuscular blocking effects of a variety of simple quaternary ammonium salts and quaternized alkaloids in animals. From these studies they concluded that the physiological action of a molecule was a function of its chemical constitution. Shortly thereafter, Richardson[31] noted that the hypnotic activity of aliphatic alcohols was a function of their molecular weight. These observations are the basis for future *structure–activity relationships* (SARs).

Drugs can be classified as being structurally specific or structurally nonspecific. *Structurally specific drugs*, which most drugs are, act at specific sites, such as a receptor or an enzyme. Their activity and potency are very susceptible to small changes in chemical structure; molecules with similar biological activities tend to have common structural features. *Structurally nonspecific drugs* have no specific site of action and usually have lower potency. Similar biological activities may occur with a variety of structures. Examples of these drugs are gaseous anesthetics, sedatives and hypnotics, and many antiseptics and disinfectants.

Even though only a part of the molecule may be associated with its activity, a multitude of molecular modifications could be made. The hallmark of SAR studies is the synthesis of as many analogs as possible of the lead and their testing to determine the effect of structure on activity (or potency). Once enough analogs are prepared and sufficient data accumulated, conclusions can be made regarding structure–activity relationships. Unfortunately, ease of synthesis, rather than cogent rationales, is often the guiding force behind the choice of analogs made.

An excellent example of this approach came from the development of the sulfonamide antibacterial agents (sulfa drugs). After a number of analogs of the lead compound sulfanilamide (**2.1**, R = H; AVC) were prepared, clinical trials determined that compounds of this general structure exhibited diuretic and antidiabetic activities as well as antimicrobial activity. Compounds with each type of activity eventually were shown to possess certain structural features in common. On the basis of the biological results of greater than 10,000 compounds, several SAR generalizations were made.[32] Antimicrobial agents have structure **2.32** (R = SO$_2$NHR′ or SO$_3$H).

2.32

In **2.32**, (1) the amino and sulfonyl groups on the benzene ring should be para; (2) the anilino amino group may be unsubstituted (as shown) or may have a substituent that is removed *in vivo*; (3) replacement of the benzene ring by other ring systems, or the introduction of additional substituents on it, decreases the potency or abolishes the activity; (4) R may be any of the alternatives shown below, but the potency is reduced in most cases; (5) N′-monosubstitution

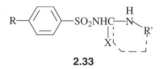

(R = SO$_2$NHR′) results in more potent compounds, and the potency increases with heteroaromatic substitution; and (6) N′-disubstitution (R = SO$_2$NR′$_2$), in general, leads to inactive compounds.

Antidiabetic agents are compounds with structure **2.33**, where X may be O, S, or N incorporated into a heteroaromatic structure such as a thiadiazole or a pyrimidine or in an acyclic structure such as a urea or thiourea. In the case of ureas, the N-2 should carry as a substituent a chain of at least two carbon atoms.[33]

R—⟨benzene⟩—SO$_2$NHC—N(H)—R′, X

2.33

Sulfonamide diuretics are of two general structural types: hydrochlorthiazides (**2.34**) and the high ceiling type[34] (**2.35**). The former compounds have 1,3-disulfamyl groups on the benzene ring and R^2 is an electronegative group such as Cl, CF$_3$, or NHR.

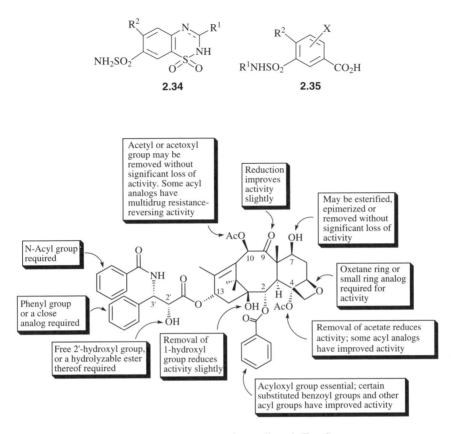

Figure 2.2 ▶ SAR for paclitaxel (Taxol)

The high ceiling compounds contain 1-sulfamyl-3-carboxy groups. Substituent R^2 is Cl, Ph, or PhZ, where Z may be O, S, CO, or NH and X can be at position 2 or 3 and is normally NHR, OR, or SR.[35]

A more recent example of a SAR is that of the natural product anticancer drug paclitaxel (**2.36**, Taxol), which was the first anticancer compound found to act by promoting the assembly of tubulin into microtubules, thereby blocking mitosis.[36] After a large number of modifications were introduced,[37] many SAR (actually, structure–*potency* relationship) conclusions could be made (Figure 2.2). A common way to track the structural changes is with the use of *molecular activity maps*, structural drawings of a lead compound annotated to show where in the molecule specific structural changes affect activity or potency measured in a single bioassay.

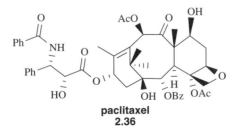

paclitaxel
2.36

For large lead molecules that have been extensively modified, these maps concisely summarize a huge number of facts relating structures with their activities and potencies. Effects could include the abolishment of an activity, unexpected toxicity, or large changes in potency. These maps may depict the results of a long-lasting drug discovery effort involving numerous chemists (and the biologists who do the screens). Their main virtues are that they can prevent your coworkers from synthesizing analogs that have already been made and tested, and they may direct the chemists' creativity to unexplored regions of the lead compound, yielding novel structural changes.

The above examples provide strong evidence to support the notion that a correlation does exist between structure and activity, but does each structure interact with only one receptor and lead to only one activity?

2.2.D   Privileged Structures and Drug-Like Molecules

Evans and coworkers first introduced the term *privileged structures* for certain molecular scaffolds that appear to be capable of binding to multiple receptor targets, and, consequently, with appropriate structure modifications, could exhibit multiple activities.[38] The Merck group used benzodiazepines (**2.37**) as the example of this phenomenon, which was earlier mentioned by Ariëns and coworkers without referring to them as privileged structures.[39] A number of other privileged structures are known.[40]

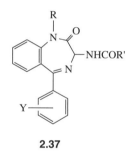

2.37

The commonality of molecular features in a variety of drugs was apparent by the revelation that only 32 scaffolds describe half of all known drugs.[41] Likewise, a small number of moieties account for a large majority of the side chains found in drugs.[42] The average number of side chains per molecule is four. If the carbonyl side chain is ignored, then 73% of the side chains in drugs are from the top 20 most common side chains. On the basis of what is known about privileged structures and common scaffold structure, Ajay and coworkers proposed that *drug-likeness* is a possible inherent property of some molecules,[43] and this property could determine which molecules should be selected for screening. They used a set of one- and two-dimensional parameters in their computation and were able to predict correctly more than 90% of the compounds in the Comprehensive Medicinal Chemistry (CMC) database.[44] Another computational approach to differentiate drug-like and nondrug-like molecules using a scoring scheme was developed[45] that was able to classify correctly 83% of the compounds in the Available Chemicals Directory (ACD)[46] and 77% of the compounds in the World Drug Index (WDI).[47] A variety of other approaches have been taken to identify drug-like molecules.[48]

There also are many nondrug-like molecules that show up as active compounds in screens, but later are demonstrated to be *false positives* (inactive compounds that appear to be active in the screen). These compounds initially appear to interact with a variety of receptors and are known as *promiscuous binders* (*promiscuous antagonists* if they antagonize many receptors or *promiscuous inhibitors* if they inhibit many enzymes). However, they are nondrug-like because they act at a site different from that of the natural ligand or substrate, show little relationship between structure and activity, and have poor selectivity for a specific receptor or enzyme. As a result, much time is wasted following up on the activity of these compounds, later to find out that they are inactive. Shoichet and coworkers showed that many of these compounds are actually aggregates of molecules, and it is the aggregate that produces the false-positive activity.[49] If the conditions of the screen are changed, these aggregates dissociate, and the individual molecules can be shown to be inactive.

2.2.E Structure Modifications to Increase Potency and the Therapeutic Index

How do you know what molecular modifications to make in order to fine tune the lead compound? The preceding section makes clear that structure modifications are the keys to activity and potency manipulations. After years of structure–activity relationship studies, various standard molecular modification approaches have been developed for the systematic improvement of the *therapeutic index* (also called the *therapeutic ratio*), which is a measure of the safety of a drug as determined from the ratio of the concentration of a drug that gives undesirable effects to that which gives desirable effects. The therapeutic index can be determined by any method that measures undesirable and desirable drug effects, but often it is taken as the dose-limiting toxicity versus the desirable pharmacological effect in an animal model (preferably humans). For example, the therapeutic index could be the ratio of the LD_{50} (the lethal dose for 50% of the test animals) to the therapeutic ED_{50} (the effective dose that produces the maximum therapeutic effect in 50% of the test animals); a toxic ED_{50} (the dose that produces toxicity in 50% of the test animals) may substitute for the LD_{50}. The larger the therapeutic index, the greater the margin of safety of the compound. In other words, you would like to have to administer gram quantities of the drug before any undesirable effects are observed, but administer only milligrams of the drug to attain the desirable effects. There is no specific minimum value for a therapeutic index that must be attained before a drug can be approved; it depends on the disease that is being treated and whether other therapies are already available. A low therapeutic index is tolerable for lethal diseases, such as cancer or AIDS (maybe even as low as 1–5), especially if no other treatment is available or if the side effect is minor compared with the treatment benefit. For less threatening diseases, therapeutic indices on the order of 10–100 may be reasonable. As an example, the therapeutic index for the antitumor agent chlorambucil (**2.38**, Leukeran) is 23.[50] The Merck Index is a good source for obtaining LD_{50} data for drugs in animals.

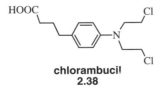

chlorambucil
2.38

A number of structural modification methodologies follow.

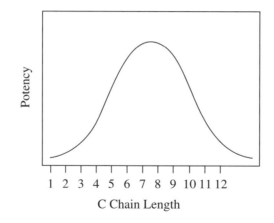

Figure 2.3 ▶   General effect of carbon chain length on drug potency

E.1   Homologation

A *homologous series* is a group of compounds that differ by a constant unit, generally a CH_2 group. As will become more apparent in Section 2.2.F.2, p. 53, biological properties of homologous compounds show regularities of increase and decrease. For many series of compounds, lengthening of a saturated carbon side chain from one (methyl) to five to nine atoms (pentyl to nonyl) produces an increase in pharmacological effects; further lengthening results in a sudden decrease in potency (Figure 2.3). In Section 2.2.F.2.b, p. 55, I show that this phenomenon corresponds to increased lipophilicity of the molecule to permit penetration into cell membranes until its lowered water solubility becomes problematic in its transport through aqueous media. In the case of aliphatic amines, another problem is micelle formation, which begins at about C_{12}. This effectively removes the compound from potential interaction with the appropriate receptors. One of, if not the, earliest examples of this potency versus chain length phenomenon was reported by Richardson,[51] who was investigating the hypnotic activity of alcohols. The maximum effect occurred for 1-hexanol to 1-octanol; then the potency declined on chain lengthening until no activity was observed for hexadecanol.

A study by Dohme *et al.*[52] on 4-alkyl-substituted resorcinol derivatives showed that the peak antibacterial activity occurred with 4-*n*-hexylresorcinol (see Table 2.1), a compound now used as a topical anesthetic in a variety of throat lozenges. Funcke *et al.*[53] found that the peak spasmolytic activity of a series of mandelate esters occurred with the *n*-nonyl ester (see Table 2.1).

E.2   Chain Branching

When a simple lipophilic relationship is important, as described above, then chain branching lowers the potency of a compound because a branched alkyl chain is less lipophilic than the corresponding straight alkyl chain as a result of larger molar volumes and shapes of branched compounds. This phenomenon is exemplified by the lower potency of the compounds having isoalkyl chains in Table 2.1. In this case, pharmacokinetics would be the overriding factor for potency. However, another explanation for lower potency with branching could be pharmacodynamics; chain branching may interfere with receptor binding. For example, phenethylamine ($PhCH_2CH_2NH_2$) is an excellent substrate for monoamine oxidase, but

TABLE 2.1 ▶ Effect of Chain Length on Potency. Antibacterial activity of 4-*n*-alkylresorcinols and spasmolytic activity of mandelate esters

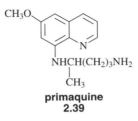

R	Phenol coefficient	% Spasmolytic activity[a]
methyl	—	0.3
ehtyl	—	0.7
n-propyl	5	2.4
n-butyl	22	9.8
n-pentyl	33	28
n-hexyl	51	35
n-heptyl	30	51
n-octyl	0	130
n-nonyl	0	190
n-decyl	0	37
n-undecyl	0	22
i-propyl	—	0.9
i-butyl	15.2	8.3
i-amyl	23.8	28
i-hexyl	27	—

[a] Relative to 3,3,5-trimethylcyclohexanol, set at 100%.

α-methylphenethylamine (amphetamine) is a poor substrate. Primary amines often are more potent than secondary amines, which are more potent than tertiary amines. For example, the antimalarial drug primaquine phosphate (**2.39**, Primaquine) is much more potent than its secondary or tertiary amine homologs.

CH₃O— [quinoline structure]

NHCH(CH₂)₃NH₂
|
CH₃
primaquine
2.39

Major pharmacological changes can occur with chain branching or homologation. Consider the 10-aminoalkylphenothiazines (**2.40**, X = H). When R is $CH_2CH(CH_3)N(CH_3)_2$ (promethazine HCl; Phenergan), antispasmodic and antihistaminic activities predominate. However, the straight-chain analog **2.40** with R being $CH_2CH_2CH_2N(CH_3)_2$ (promazine) has greatly reduced antispasmodic and antihistaminic activities, but sedative and tranquilizing activities are greatly enhanced. In the case of the branched chain analog **2.40** with R equal to $CH_2CH(CH_3)CH_2N(CH_3)_2$ (trimeprazine) (next larger branched-chain homolog), the tranquilizing activity is reduced and antipruritic (anti-itch) activity increases. This indicates that multiple receptors are involved, and branching or homologation can cause the molecule to bind more or less well to the receptors responsible for antispasmodic activity, antihistamine activity, tranquilizing activity, or antipruritic activity.

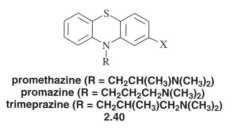

promethazine (R = CH₂CH(CH₃)N(CH₃)₂)
promazine (R = CH₂CH₂CH₂N(CH₃)₂)
trimeprazine (R = CH₂CH(CH₃)CH₂N(CH₃)₂)
2.40

E.3   Ring-Chain Transformations

Another modification that can be made is the transformation of alkyl substituents into cyclic analogs, which often does not affect potency greatly. Consider the promazines again (**2.40**). Chlorpromazine ($X = Cl$, $R = CH_2CH_2CH_2N(CH_3)_2$; Thorazine) and **2.40** ($X = Cl$, R = CH₂CH₂CH₂N⎯⟨⟩) are equivalent as tranquilizers in animal tests. The branched methyl group and one of the dimethylamino methyl groups of trimeprazine (**2.41**, Vallergan or Temaril) could be connected to give methdilazine (**2.42**, Dilosyn), which have similar antipruritic activity in man.

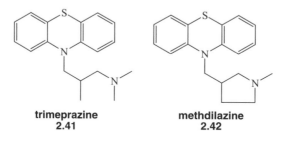

trimeprazine   **methdilazine**
2.41   **2.42**

However, a ring-chain transformation could have an important pharmacokinetic effect, such as to increase lipophilicity or decrease metabolism, which could make the drug more effective *in vivo*. Also, by connecting substituents into a ring, pharmacodynamic properties could be enhanced by constraining the groups into a particularly favorable conformation. Of course, it also could constrain the molecule into an unfavorable conformation, and potency could drop!

Different activities can result from a ring-chain transformation as well. For example, if the dimethylamino group of the tranquilizer chlorpromazine is substituted by a methylpiperazine ring (**2.40**, $X = Cl$, R = CH₂CH₂CH₂N⎯⟨⟩NCH₃; prochlorperazine), antiemetic (prevents nausea

TABLE 2.2 ▶ Classical Isosteres

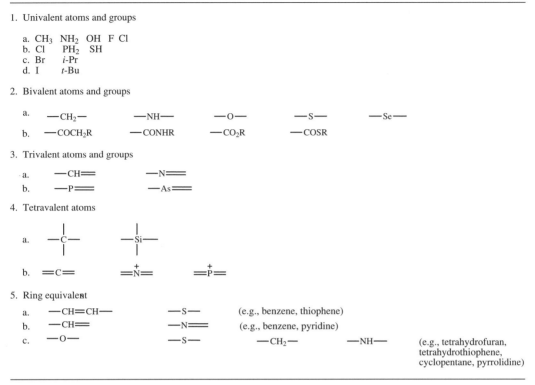

1. Univalent atoms and groups

    a. CH_3 NH_2 OH F Cl
    b. Cl PH_2 SH
    c. Br   *i*-Pr
    d. I   *t*-Bu

2. Bivalent atoms and groups

    a. —CH_2—   —NH—   —O—   —S—   —Se—
    b. —$COCH_2R$ —CONHR   —CO_2R —COSR

3. Trivalent atoms and groups

    a. —CH═   —N═
    b. —P═   —As═

4. Tetravalent atoms

    a. —C—   —Si—
    b. ═C═   ═N⁺═   ═P⁺═

5. Ring equivalent

    a. —CH═CH—   —S—   (e.g., benzene, thiophene)
    b. —CH═   —N═   (e.g., benzene, pyridine)
    c. —O—   —S—   —CH_2—   —NH—   (e.g., tetrahydrofuran, tetrahydrothiophene, cyclopentane, pyrrolidine)

and vomiting) activity is greatly enhanced. In this case, however, an additional methylamino group is also added, which may have an effect in changing the activity.

E.4 Bioisosterism

Bioisosteres are substituents or groups that have chemical or physical similarities, and which produce broadly similar biological properties.[54] Bioisosterism is an important lead modification approach that has been shown to be useful to attenuate toxicity or to modify the activity of a lead, and may have a significant role in the alteration of pharmakinetics of a lead. There are classical isosteres[55,56] and nonclassical isosteres.[57] In 1925 Grimm[58] formulated the *hydride displacement law* to describe similarities between groups that have the same number of valence electrons, but may have a different number of atoms. Erlenmeyer[59] later redefined isosteres as atoms, ions, or molecules in which the peripheral layers of electrons can be considered to be identical. These two definitions describe *classical isosteres*; examples are shown in Table 2.2. *Nonclassical bioisosteres* do not have the same number of atoms and do not fit the steric and electronic rules of the classical isosteres, but do produce similar biological activity. Examples of these are shown in Table 2.3. Ring-chain transformations also can be considered to be isosteric interchanges. There are hundreds of examples of compounds that differ by a bioisosteric interchange;[60] some examples are shown in Table 2.4.

TABLE 2.3 ► Nonclassical Isosteres

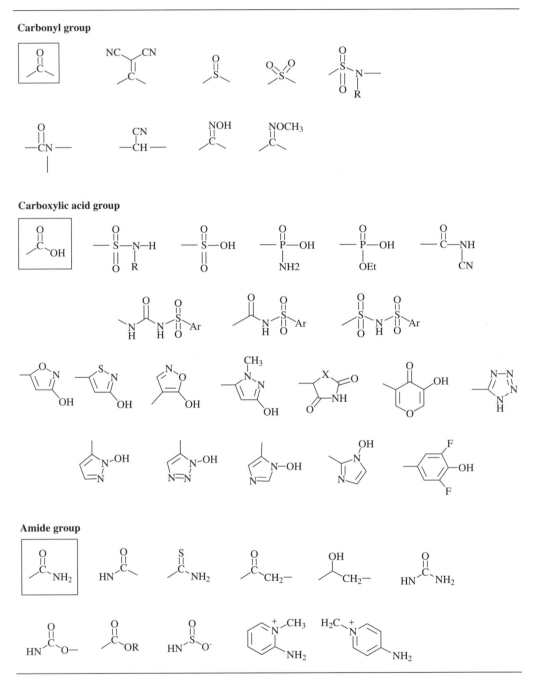

TABLE 2.3 ► *Continued*

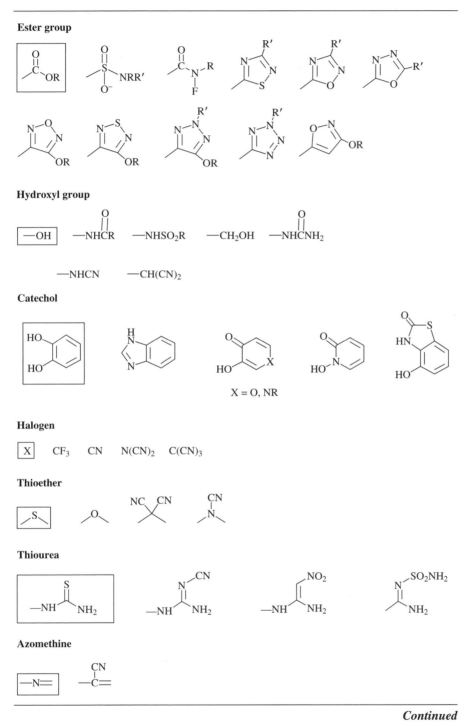

Ester group

Hydroxyl group

Catechol

X = O, NR

Halogen

Thioether

Thiourea

Azomethine

Continued

TABLE 2.3 ► *Continued*

Pyridine

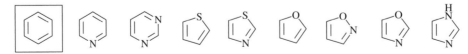

Benzene

Ring equivalents

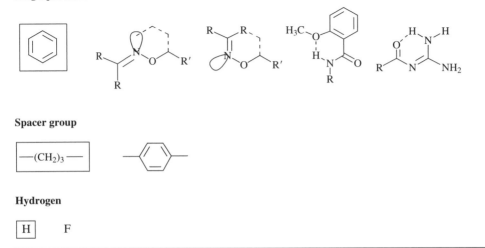

Spacer group

—(CH₂)₃—

Hydrogen

H   F

Bioisosterism, however, also can lead to changes in activity or potency. For example, if the sulfur atom of the phenothiazine neuroleptic drugs (**2.40**) is replaced by −CH=CH− or −CH₂CH₂− bioisosteres, then dibenzazepine antidepressant drugs (**2.43**) result. A change in enzyme affinity also can be observed. For example, when the thiazolone ring in a series of anti-inflammatory compounds selective for the cyclooxygenase-2 isozyme over the cyclooxygenase-1 isozyme (see Chapter 5, Section 5.5.B.2.b, p. 280) was substituted by an oxazolone ring (i.e., the S was replaced by O), the selectivity for the two isozymes reversed.[61]

2.43

It is actually quite surprising that bioisosterism should be such a successful approach to lead modification. Perusal of Table 2.2, and especially of Table 2.3, makes it clear that in making a bioisosteric replacement, one or more of the following parameters will

TABLE 2.4 ► Examples of Bioisosteric Analogs

1. **Neuroleptics (antipsychotics)**

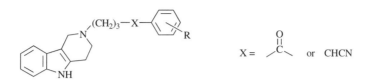

$$X = \overset{O}{\underset{||}{C}} \quad \text{or} \quad CHCN$$

2. **Anti-inflammatory agents**

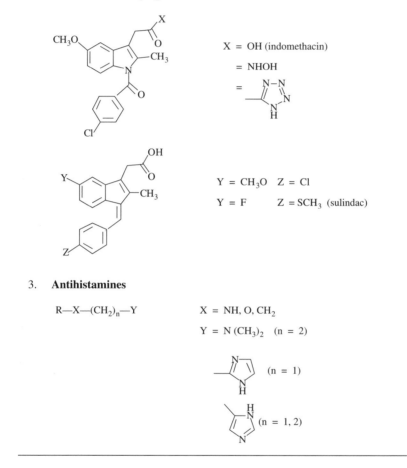

X = OH (indomethacin)

= NHOH

= (tetrazole)

Y = CH$_3$O   Z = Cl

Y = F   Z = SCH$_3$ (sulindac)

3. **Antihistamines**

R—X—(CH$_2$)$_n$—Y

X = NH, O, CH$_2$

Y = N (CH$_3$)$_2$ (n = 2)

(n = 1)

(n = 1, 2)

change: size, shape, electronic distribution, lipid solubility, water solubility, pK_a, chemical reactivity, and hydrogen bonding. Because a drug must get to the site of action, then interact with it (see Chapter 3), bioisosteric modifications made to a molecule may have one or more of the following effects:

1. *Structural.* If the moiety that is replaced by a bioisostere has a structural role in holding other functionalities in a particular geometry, then size, shape, and hydrogen bonding will be important.

2. *Receptor interactions*. If the moiety replaced is involved in a specific interaction with a receptor or enzyme, then all of the parameters except lipid and water solubility will be important.

3. *Pharmacokinetics*. If the moiety replaced is necessary for absorption, transport, and excretion of the compound, then lipophilicity, hydrophilicity, pK_a, and hydrogen bonding will be important.

4. *Metabolism*. If the moiety replaced is involved in blocking or aiding metabolism, then the chemical reactivity will be important.

It is because of these subtle changes that bioisosterism is effective. This approach allows the medicinal chemist to tinker with only some of the parameters in order to augment the potency, selectivity, and duration of action and to reduce toxicity. Multiple alterations may be necessary to counterbalance effects. For example, if modification of a functionality involved in binding also decreases the lipophilicity of the molecule, thereby reducing its ability to penetrate cell walls and cross other membranes, the molecule can be modified at a different site with a lipophilic group to increase absorption. But where can these bioisosteric replacements be made? A pharmacophore study (Section 2.2.A, p. 17) could have identified those groups on the lead that could be modified without an effect on receptor binding (the scissions that led to little change in potency). Those are the positions that can be safely modified. Modifications of this sort, however, may change the overall molecular shape and result in another activity.

E.5 Combinatorial Chemistry

a. General Aspects

Combinatorial chemistry involves the synthesis or biosynthesis of *chemical libraries* (a family of compounds having a certain base chemical structure) of molecules for the purpose of biological screening, particularly for lead discovery or lead modification.[62] Typically, these chemical libraries are prepared in a systematic and repetitive way by covalent assembly of *building blocks* (various reactant molecules that build up parts of the overall structure) to give a diverse array of molecules with a common *scaffold* (the parent structure in the family of compounds). The advantage of this methodology is that it is carried out on a solid (polymeric) support, so that isolation and purification of the product of each reaction can be performed by simple filtration and washing (with a variety of solvents) of the polymeric support to which the building blocks have been attached. Because of the insolubility of the polymer, everything not attached to the polymer is removed, which allows the use of excess reagents to drive the synthetic reactions. The disadvantages of this methodology are the difficulty in scaling up the reactions and the sluggishness of reactions. An alternative strategy (*covalent scavenger technology*) is to carry out the reactions in solution with excess reagent, which is then scavenged with a polymeric-supported scavenger after the reaction is completed (Figure 2.4). In this approach, filtration removes the excess reagent attached to the scavenger polymer, leaving the product in solution.[63] Another approach is to use polymer-supported reagents with solution reactions. To avoid problems of heterogeneous polymer reactions, soluble polyethylene glycol polymers can be used.[64]

The main differences among the various combinatorial approaches are the solid support used, the methods for assembling the building blocks, the state (immobilized or in solution) and numbers (a fraction of the total library or individual entities) in which the

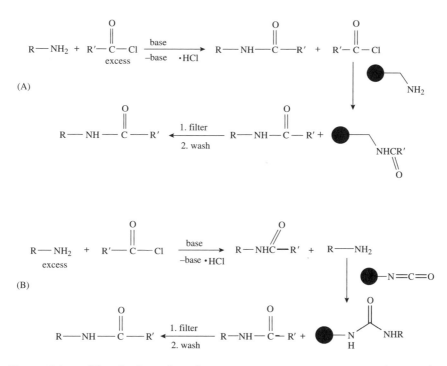

Figure 2.4 ▶ Use of polymer-bound reagents to scavenge excess reactants in a reaction

libraries are screened, and the manner in which the structures of active compounds are determined.

The number of possible different compounds in a library (N) is determined by the number of building blocks used in each step (b) and the number of synthetic steps (x). If an equal number of building blocks is used in each synthetic step, then Equation 2.1 holds. If the number of building blocks in each step is varied (e.g., b, c, and d for a three-step synthesis), then Equation 2.2 is relevant.

$$N = b^x \tag{2.1}$$

$$N = bcd \tag{2.2}$$

A combinatorial library of all of the pentapeptides that comprise the 20 commonly encoded amino acids would be $N = 20^5$ or 3.2 million different peptides. Combinational chemistry originally was used to make peptide libraries, but now is most commonly employed for the synthesis of large arrays of diverse nonpeptidic small molecules. It is only because of the discovery and development of HTS techniques that combinatorial chemistry was able to thrive. Unless there were a method to test 3 million new compounds generated in a library in a short period of time, there would be no advantage to being able to make that many compounds. Theoretical estimates conclude that there are up to 10^{180} possible drug-like molecules with molecular weights below 800, but this amount exceeds the mass of the universe. The much lower estimate of 10^{60} molecules[65] would still take over 10^{51} years to synthesize, even if a million compounds a day were prepared![66] Therefore, even combinatorial methods are inconsequential relative to the total number of compounds possible, but the belief is that it can approach the theoretical value of compounds quicker than by conventional synthetic methods. However, the diversity of molecules that can be attained by combinatorial chemistry may not equal conventional synthetic approaches. The beginnings of combinatorial chemistry are attributed to Furka[67] with applications in peptide synthesis by Geysen and coworkers[68] and Houghten.[69] These initial efforts in peptide library synthesis were followed by synthesis of peptoids by Zuckermann and coworkers[70] and small molecule nonpeptide libraries by Ellman and coworkers[71] and Terrett and coworkers.[72]

b.   Split Synthesis: Peptide Libraries

The initial approach, known as a *split synthesis* (also called *mix and split, split and pool*, or the *divide, couple, recombine* method), is the most common general lead discovery approach for making large libraries (10^4–10^6 compounds) that are assayed as library mixtures.[73] The result of a split synthesis is a collection of polymer beads, each containing one library member, i.e., one bead, one compound. The library contains every possible combination of every building block. The serious limitations are that it is applicable only to the synthesis of sequenceable oligomers and each bead carries only about 100–500 pmol of product, which makes structure determination difficult or impossible. For simple compounds mass spectrometric methods may be used,[74] but this is not applicable if the library contains many thousands or millions of members that may not be pure or are isomeric with other library members. In that case, encoding methods (see next section) need to be utilized.

Below is an example of how the split synthesis approach would be applied to a small (27-member) library of all possible tripeptides of three amino acids. This method can be extrapolated to any size library. A homogeneous mixture of all of the tripeptides of His, Val, and Ser ($3^3 = 27$) could be synthesized on a Merrifield resin as shown in Scheme 2.2. Note that a Merrifield synthesis starts at the C terminus and builds to the N terminus. The homogenization step is very important to ensure that each tube contains the same mixture of resin-bound compounds.

What if you want to determine the most active peptide for binding to a particular receptor? Houghten and coworkers[75] prepared a combinatorial library of more than 52 million *L*-[76] and *D*-hexapeptides[77] to identify the best hexapeptide antagonist of the μ opioid receptor (the receptor to which morphine and endorphins bind). The process shown in Scheme 2.2 was carried out, but starting with 20 separate tubes containing methylbenzhydrylamine (MBHA) polystyrene as the resin. (This resin produces peptide amides when peptides are cleaved from it.) A combinatorial library of pentapeptides containing the 20 standard amino acids was constructed on the MBHA resin, homogenized, then separated into 20 tubes. To each of the 20 different tubes was added a different *N*-acetylamino acid, so that in each tube there was a combinatorial library of all possible resin-linked *N*-acetylhexapeptides having the same N terminus; each tube contained all of the *N*-acetylhexapeptides starting with a different N-terminal amino acid (Figure 2.5A). For example, the first tube may contain all resin-linked *N*-acetylhexapeptides that have *N*-acetylalanine at the N terminus (this resin synthesis, as with the Merrifield synthesis, builds the peptide from the C terminus back to the N terminus), the second tube could contain all resin-linked *N*-acetylhexapeptides that start with *N*-acetylcysteine, and so forth. All of the *N*-acetylpeptides can be cleaved from the resin to give *N*-acetylhexapeptide amides. An aliquot from each of the 20 tubes is removed and assayed. The most potent aliquot indicates which amino acid is best at the N terminus (in this case *N*-acetylArg was found to be best). Then this process is repeated, except in the next iteration a combinatorial library of MBHA-bound tetrapeptides is made, is split into 20 tubes, a different amino acid is coupled in each tube at the next-to-N-terminal position, then each tube is N-terminal capped with *N*-acetylArg, because that was shown in the previous assay

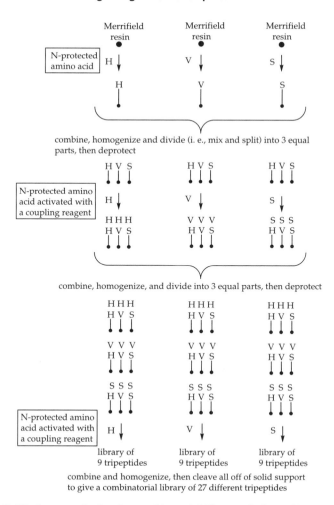

Scheme 2.2 ▶ Solid-phase synthesis of a combinatorial library of all possible tripeptides of histidine, valine, and serine

to be best (Figure 2.5B). Again, an aliquot from each tube, after cleavage from the resin, is assayed, and the best amino acid at the penultimate position is determined. This process is repeated until there are 20 tubes, each containing only one N-acetylhexapeptide, and the best of the 20 is determined by assay. Following this procedure, the most potent antagonist for the μ opioid receptor found up to that time was identified (Ac-Arg-Phe-Met-Trp-Met-Thr-NH$_2$). The entire process took about 2 months! Can you imagine how long this would take to carry out one compound at a time?

This methodology sounds foolproof, but it is not. Sometimes, the most potent aliquot is less potent than what was observed in the assays from the previous iteration. How is that possible, because the previous iteration had to have contained all of the peptides in the next iteration (and then some)? This is a common phenomenon with assaying multiple compounds simultaneously, particularly peptides. One explanation is peptide–peptide interactions. In the next iteration there are fewer peptides in the tube than in the previous assay; the necessary peptide–peptide interaction may be lost in the later assay. Possibly the active component is really two or more interacting peptides and one is removed in the next iteration. Or maybe

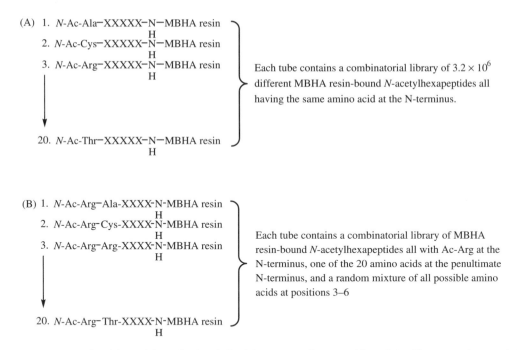

Figure 2.5 ▶ Combinatorial synthesis of all of the *N*-acetylhexapeptides of the 20 commonly encoded amino acids

there is a conformational difference in the active peptide as the number of peptides diminishes. Peptides generally do not make useful drugs, as discussed in Section 2.2.E.7, p. 47.

c. Encoding Combinatorial Libraries

Before turning our attention to the more important nonpeptide libraries, let's consider an alternative approach to the identification of the most potent analog in a combinatorial library other than repeated iterations until one compound remains. A more rapid approach would be to test the entire library at once and identify the active component of the library directly. As mentioned above, with large libraries of complex molecules it is not readily possible to determine the structure of the active component. In that case, encoding methods are needed.[78] This is similar to the way in which proteins are often sequenced in biology; the protein is not sequenced, but the gene that encodes the protein is.[79] Although the structure of the actual compound may not be directly elucidated, certain tag molecules that encode the structure may be determined.[80] One important approach that involves the attachment of unique arrays of readily analyzable, chemically inert, small molecule tags to each bead in a split synthesis was reported by Still and coworkers.[81] Ideal encoding tags must survive organic synthesis conditions, not interfere with screening assays, be readily decoded without ambiguity, encode large numbers of compounds, and the test compound and the encoding tag must be able to be packed into a very small volume. In the Still method, groups of tags are attached to a bead at each combinatorial step in a split synthesis. The tags create a record of the building blocks used in that step. At the end of the synthesis, the tags are removed and analyzed, which decodes the structure of the compound attached to that bead. As depicted in Scheme 2.3, one or more readily cleavable tag molecules (TagsX) are attached to about 1% of the polymer bead sites (about 1 pmol/bead), and these encode building block 1 (BB1). Then

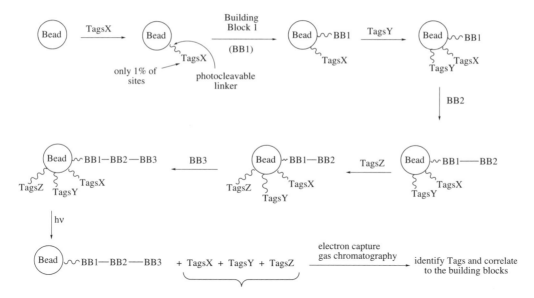

Scheme 2.3 ▶ Still methodology for encoding combinatorial peptide libraries on a polymer bead

one or more other cleavable tags (TagsY) are attached to encode building block 2 (BB2), followed by TagsZ to encode BB3, and so forth. Although a different tag could be used for each building block, it is more efficient to use mixtures of tags because mixtures of N different tags can represent 2^N different syntheses; with just 10 tags, 1024 (2^{10}) syntheses can be performed.

The tags need to be chemically inert and reliably analyzed on the femtomolar scale. Two examples of tags are **2.44** and **2.45**. The thirty **2.44** tags are photocleavable, and the forty **2.45** tags are oxidatively (ceric ammonium nitrate) cleaved. The released tags are analyzed by capillary gas chromatography using electron capture detection; all of the tags have different retention times. Other sensitive detection methods are fluorescence-based HPLC[82] and GC mass spectrometry.[83]

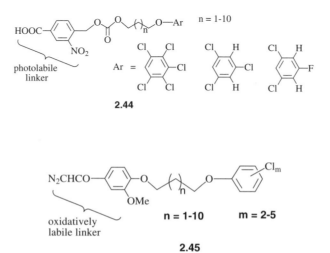

If you are using the split synthesis method, how will you know which of the large number of polymer beads has an active compound? One approach[84] is to chemically attach a commercial dye to the target receptor. The assay is run with the library still attached to each polymer bead. If a compound binds to the dye-labeled receptor, then the bead to which the compound is attached will take on the color of the dye, and the colored beads can be removed manually and decoded. The intensity of the color in the bead is an indication of the tightness of binding of the compound to the receptor.

Let's go through an example, which may make this process clearer (or maybe not). A combinatorial library of all of the hexapeptides of serine, leucine, lysine, isoleucine, and glutamate (5^6- or 15,625-member library) is prepared with the appropriate tag molecules at each iteration. To do this you need 6×3 (3-bit binary code) or 18 different tag molecules. Tags 1–3 are only used to define building block 1, tags 4–6 are for the second building block, tags 7–9 for the third, tags 10–12 for the fourth, tags 13–15 for the fifth, and tags 16–18 for the sixth building block. Arbitrarily assign a 3-bit binary code to each amino acid building block, for example, 001 = Ser, 010 = Leu, 011 = Lys, 100 = Ile, and 110 = Glu. If a tag is used, it represents binary bit 1; if no tag is used, it means binary bit 0. Because peptide syntheses are typically done on a resin to which the C-terminal amino acid is attached, and then coupling occurs back to the N terminus, tags 16–18 encode the N-terminal amino acid, and tags 1–3 encode the C-terminal amino acid. Let's say an active bead is identified, and you want to know what the hexapeptide structure is. The tags are removed from the bead (by photolysis or oxidation, depending on which linker was used), the carboxylic acid produced by cleavage is trimethylsilylated to make it more volatile, and the electron capture GC is run. In this example, let's say that tags 1, 2, 5, 6, 8, 9, 12, 14, and 16 were detected. These tags can be decoded to identify the hexapeptide as shown in Figure 2.6.

Another encoding process segregates the test compound from the coding tag molecule by attaching the test compound to the exterior of the bead and the coding tag molecule to the interior of the bead.[85] This prevents the tag molecules from interfering with binding of the test compound to the target receptor. A polymer bead system was developed in which only the surface of the bead is exposed to an organic solvent that contains the organic-soluble derivatizing reagent; this allows a nonpeptidic test compound to be constructed while the interior of the bead remains in water without derivatizing reagent. The coding molecule is a

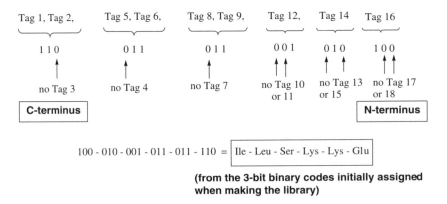

Figure 2.6 ► An example of encoding the structure of a peptide bound to a polymer bead by the Still methodology

peptide, which can be synthesized in the aqueous medium of the interior of the bead. Once the active bead is identified by a colorimetric assay,[86] the tag peptide molecule in the interior of that bead is sequenced by Edman degradation or by mass spectrometry,[87] which encodes the structure of the test compound attached to the exterior of the bead.

Other interesting alternatives to molecular tag encoding are encoding by radio-frequency tags[88] and encoding with a polymeric matrix having unusual shapes.[89] Encoding methods may some day be displaced by mass spectral analyses, such as imaging time-of-flight secondary ion mass spectrometry (TOF-SIMS).[90] Many hundreds of beads can be assayed in a single measurement at the rate of about 10 beads/sec, although there is a problem with fragmentation of compounds, leading to a complicated analysis. Mass accuracy is about ± 0.01 amu, so in the 3.2×10^6-member library of pentapeptides from 20 amino acids, all peptides, except those containing leucine and isoleucine (same molecular weight), can be separated. By incorporation of a ^{15}N label into either leucine or isoleucine, even those peptides can be differentiated.

d. Nonpeptide Libraries

As discussed later in Section 2.2.E.7, p. 47, peptides do not make very useful drugs, especially if an orally active drug is sought. The same techniques described above for the synthesis of peptide libraries could be utilized to prepare nonpeptide libraries; however, there is an important difference between the chemistry with peptides versus nonpeptides, namely, reactivity. In a typical peptide coupling reaction the carbodiimide-activated N-protected amino acids are all about the same in reactivity with the different amino acids in the growing peptide chain. Because of that, the split synthesis method works well. However, with nonamino acid reagents, such as different acid chlorides, the structure of the acid chloride will affect the rates of reaction with different nucleophiles. That could lead to mixtures in which some of the components have reacted and others have not. For each reaction the conditions have to be worked out to be sure complete reaction has occurred.

Over the years it has been recognized that when large numbers of nonpeptide analogs are screened simultaneously, many *false negatives* (an active compound that does not produce a *hit*, i.e., a compound that shows a predetermined level of activity in the assay) and *false positives* (an inactive compound that gives a hit) are observed. A false positive may arise from an impurity in the sample tested or as a result of a complex between more than one compound. False positives are a waste of time, but false negatives mean that potential drugs (or at least lead compounds) are being overlooked. It is typical for pharmaceutical companies to carry out single entity screens to avoid these problems. Because of this, individual compounds, rather than mixtures, are synthesized. Nonetheless, synthesis on a solid support allows the synthesis of large numbers of individual compounds rapidly and robotically. The reactions are carried out individually in separate microtubes containing the polymeric support. This method is referred to as *parallel synthesis* rather than combinatorial synthesis because the library of compounds (in the range of $50-10^4$ compounds in amounts of 1–50 mg) is synthesized in parallel without combining any of the tubes. One strategy that can be used for potentially more effective libraries is to select privileged structures as the scaffold. Another strategy is to design a scaffold based on an important molecular recognition motif in the target receptor. The libraries should incorporate different sets of (commercially available) building blocks to provide a large number of diverse structures, and they should contain as much functionality as possible as recognition elements. Molecular diversity, however, is difficult to determine; Dixon and Villar have found that a protein can bind a set of structurally diverse molecules with

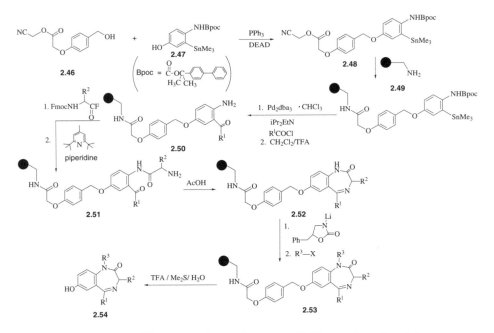

Scheme 2.4 ▶ Solid-phase synthesis of a nonpeptide library of privileged structures

similar potent binding affinities, but analogs closely related to these compounds can exhibit very weak binding.[91] Parallel synthesis can generate many more compounds than can be synthesized traditionally, and the cost per compound is much lower.[92]

An example of a nonpeptide library of a privileged structure (benzodiazepines) is shown in Scheme 2.4.[93] Note that the first piece of the benzodiazepine (**2.47**) is not attached directly to the polymer (aminomethylpolystyrene, **2.49**), but is attached to **2.46** instead to give **2.48** in a Mitsunobu coupling (PPh$_3$/diethylazodicarboxylate or DEAD). If **2.47** were attached directly to the polymer, then steric hindrance by the polymer to the first chemical reaction may result, i.e., the Stille coupling of an acid chloride to the aryl stannane in the presence of palladium to give **2.50**. To avoid that problem, a spacer group is typically attached to the polymer which moves the first reactant away from the polymer so that steric hindrance is not a problem. Compound **2.46** serves as the spacer, which will be removed at the end of the synthesis. In this solid-phase synthesis, three *diversity elements* can be varied: R^1, R^2, and R^3. The Stille coupling can be carried out with as many acid chlorides as are available (to vary R^1). Each of those products (**2.50**) can be coupled to the same Fmoc-protected amino acid fluoride in the next step or each of the **2.50** products can be treated with different Fmoc-protected amino acid fluorides, so that a wide variety of R^2 groups can be incorporated. Acetic acid causes autocyclization of **2.51** to **2.52**. Again, each **2.52** can be treated with base and the same alkyl halide (R^{3}X) or each of **2.52** can be treated with a different alkyl halide to give a library of **2.53**; cleavage from the spacer and polymer resin gives the library of benzodiazepines (**2.54**). If 25 different acid chlorides were used in the Still coupling, and each of those products was treated with 25 different Fmoc-protected amino acid fluorides, then each of those compounds alkylated with 25 different alkyl halides, there would be a library of 25 × 25 × 25 or 15,625 different benzodiazepines using only 75 different building blocks.

Despite the potential of combinatorial library (or parallel) synthesis, natural products seem to provide greater structural diversity than standard combinatorial chemistry. About 40% of the chemical scaffolds in the Dictionary of Natural Products and the Bioactive Natural Products Database (a total of more than 100,000 compounds) have not been synthesized.[94] Furthermore, natural products that are biologically active in assays generally have drug-like properties, i.e., are capable of being absorbed and metabolized.[95a] Isolation of compounds from natural sources and from combinatorial approaches, however, should be complementary. Generally the excessive time taken to isolate and characterize bioactive compounds from natural product extracts is a disadvantage of the method, but the reward is greater molecular diversity,[95b] which gives the greatest opportunity to identify a variety of scaffolds for screening. To attain a wide diversity of chemical structures for screening purposes, computational chemists often reject compounds that are similar in structure, believing that similar compounds would have similar biological activities. In general, structurally similar compounds have similar biological activity. However, the biological similarity may not be very strong; it has been shown that only 30% of compounds considered to be 85% structurally similar to an active compound will themselves have the same activity.[96] Adding just one methylene group to a 4-hydroxypiperidine analog changed it from a poor binder of the chemokine receptor CCR1 into a potent binder.[97] This may be because similar compounds do not necessarily bind to the target receptor the same way.

Construction of chemical libraries based on natural product hits is a sensible compromise approach. For example, Nicolaou and coworkers developed a solid-phase method for the preparation of large natural product-like combinatorial libraries based on the privileged structure 2,2-dimethylbenzopyran,[98] a scaffold found in many natural products. The general methodology is shown in Scheme 2.5. In this example the 2,2-dimethylbenzopyran scaffold (**2.57**) is generated by reaction of the starting *o*-prenyl phenol (**2.55**) with a polystyrene-based selenenyl bromide resin. A variety of reactions are possible to elaborate the side chains to further enhance the library (**2.56**). The final products can be released from the polymeric support by oxidation/elimination to **2.57**. A library of more than 10,000 analogs was readily prepared by this approach.

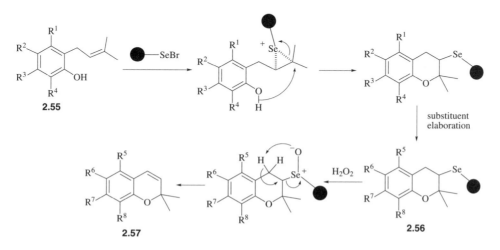

Scheme 2.5 ▶ Solid-phase synthesis of a natural product-like combinatorial library

E.6   SAR by NMR/SAR by MS

Fesik and coworkers at Abbott Laboratories developed a NMR-based approach to screen libraries of small organic molecules and to identify and optimize high-affinity *ligands* (compounds that bind to receptors) for proteins.[99] This approach, termed *SAR by NMR*, was initially used to discover compounds with nanomolar affinities (highly potent; see Chapter 3, Sections 3.2.A and 3.2.C) for the immunosuppressant FK506 binding protein by tethering two molecules with micromolar affinities (low potency). The first step of the process (Figure 2.7) involves screening a library of small compounds, 10 at a time, by observation of the amide ^{15}N-chemical shift in the heteronuclear single quantum coherence (HSQC) NMR spectrum. Once a lead is identified, a library of analogs is screened to identify compounds with optimal binding at that site. Then a second library of compounds is screened to find a compound that binds at a nearby site, and again this compound is optimized by screening a library of related compounds. Based on the NMR spectrum of the ternary complex of the protein and the two bound ligands, the location and orientation of these ligands are determined, and compounds are synthesized in which the two ligands are covalently attached. Although each individual ligand may be a relatively weak binder, when the two are attached, the binding affinity increases dramatically. This is because the free energy of binding becomes the sum of three free energies: the two ligands and the linker; the binding affinity is the multiplier of the three binding affinities. There is a gain of about a factor of 100 in binding affinity by freezing out one bond rotation. Therefore, it is not necessary to optimize the lead much, because ligands with micromolar or even millimolar affinities can attain nanomolar affinities when linked.

An example of this is the identification of the first potent inhibitor of the enzyme stromelysin, a *matrix metalloprotease* (a family of zinc-containing hydrolytic enzymes responsible for degradation of extracellular matrix components such as collagen and proteoglycans in normal tissue remodeling and in many disease states such as arthritis, osteoporosis, and cancer),[100] as a potential antitumor agent.[101] Matrix metalloproteases are generally inhibited by compounds that contain a hydroxamate moiety to bind to the zinc ion. A library of hydroxamates was screened, and acetohydroxamic acid (**2.58**) was identified with a K_d of 17 mM (very poor binding affinity). A focused screen of hydrophobic compounds was carried out in the presence of saturating (excess) amounts of acetohydroxamic acid, and biphenyl analogs were identified; optimization led to **2.59** with a K_d of 20 μM. From the NMR spectrum, the best site for a linker was expected to be between the methyl of acetohydroxamic acid and the hydroxyl group of **2.59**. Consequently, alkyl linkers of varying chain length were

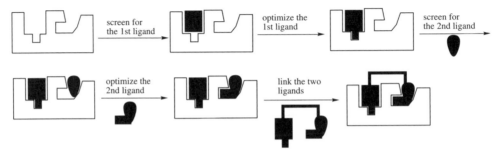

Figure 2.7 ▶ SAR by NMR methodology

tried, and the best was a one-carbon linker, giving **2.60** having a K_d of 15 nM! The ΔG for **2.59** is –2.4 kcal/mol, for **2.60** is –4.8 kcal/mol, and for the linker is –2.6 kcal/mol; the total, therefore, is –9.8 kcal/mol. It took about 6 months to identify this inhibitor; prior to this study 115,000 compounds had been screened with no leads.

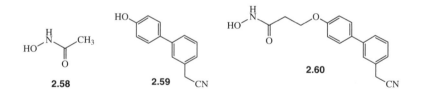

2.58   **2.59**   **2.60**

Sounds simple, doesn't it? But let's think about what is involved. The method requires screening compounds and observing a specific ^{15}N-amide chemical shift for binding. Where did the ^{15}N come from? This had to be incorporated into the protein because natural abundance ^{15}N is not sufficiently high in concentration for detection. To incorporate ^{15}N, it is necessary to be able to express the protein in a microorganism, then grow the microorganism on ^{15}NH$_4$Cl as its sole nitrogen source; that gives the protein with all ^{15}N-containing amino acids. To perform the NMR experiments, large amounts of soluble ($>100\,\mu$M) protein (>200 mg per spectrum) are needed; therefore, an efficient overexpression system for the protein is needed. Then the protein has to be purified, and its complete structure determined by three- and four-dimensional NMR techniques, so that the position of every amino acid residue in the protein is known (which is needed to determine when the two ligands have bound in nearby sites). This means that the protein target should have a mass less than about 40 kDa (the current limit for rapid protein NMR spectra, although spectra of larger proteins is possible).[102] Although it appears that this is a highly specialized technique, it is used widely because molecular biology and protein chemistry techniques are well developed, making overexpression of proteins in microorganisms and their purification routine.[103] Newer NMR instrumentation and methods also have made structure determination plausible. If the structure can be determined, SAR by NMR provides a technique to screen by automation about 1000 compounds a day and identify, relatively rapidly, potent protein binders.[104] Even covalent binders can be identified by high-throughput NMR-based screens.[105] In drug discovery programs, it is often not too difficult to find compounds that bind to proteins in the micromolar range; what becomes time consuming is increasing the potency of the lead into the low nanomolar range. SAR by NMR may shorten that time.

Ellman and coworkers have developed a combinatorial lead optimization approach using the basic principles of SAR by NMR, except without the use of NMR and without needing any structural or mechanistic information about the target protein![106] First, a diverse library of compounds is synthesized in which each molecule incorporates a common chemical linkage group (Figure 2.8). Next, the library is screened to identify any members that show even weak binding to the target. Third, a new library is constructed containing all combinations of any two of the active compounds linked to each other by the common chemical linkage group through a set of flexible linkers. Then this combinatorial library is screened to identify the most potent analog. The method depends on two analogs binding in nearby sites (although it is not known which two will bind or where the sites are) and finding the appropriate linker size combinatorially so the linked active compounds take advantage of the additive free-energy gain of the three elements, the two compounds and the linker. This approach was used to identify a potent (IC$_{50}$ = 64 nM) and selective inhibitor of one subtype of tyrosine kinase.

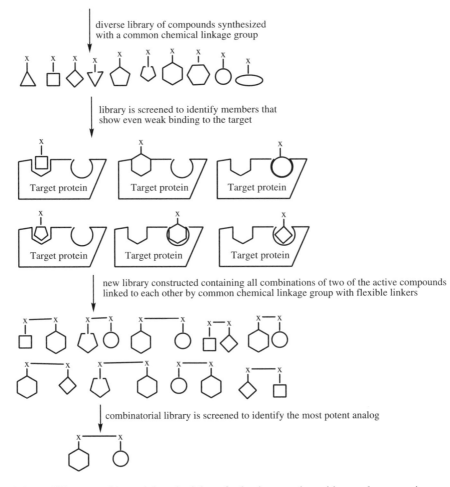

Figure 2.8 ▶ Ellman combinatorial methodology for lead generation with an unknown or impure receptor or enzyme

A complementary method to SAR by NMR is *SAR by MS*.[107] This is a high-throughput mass spectrometry-based screen that quantifies the binding affinity, stoichiometry, and specificity over a wide range of ligand binding energies. A set of diverse compounds is screened by mass spectrometry to identify those that bind to the receptor. Competition experiments are used to identify the ones that bind to the same site and those that do not. If two compounds bind at different binding sites, then a ternary complex of the two molecules plus the receptor is detected in the mass spectrum. If the two compounds bind at the same site, the tighter binding molecule displaces the other from the binding site, and only a binary complex is detected. By varying the substituent size on various classes of compounds and rescreening, it is possible to identify those molecules that bind at nearby sites as the ones that become competitive once a larger substituent is appended to one of the molecules. Once adjacent binding sites are realized, then the same methodology can be employed as in SAR by NMR, namely, attaching the two molecules to each other with linkers. This approach was applied to the discovery of lead compounds that bind to an RNA target for which no leads could be identified by conventional high-throughput screening methods.[108]

E.7 Peptidomimetics

In Section 2.2.E.5 on combinatorial chemistry, methods for the synthesis of libraries of peptides were described. Peptides are very important endogenous molecules that bind to a variety of receptors in their action as neurotransmitters, hormones, and neuromodulators,[109] and numerous enzymes are involved in the biosynthesis and catabolism of these peptides. Plants and animals,[110] including human skin,[111] contain a variety of antibiotic peptides. Endogenous peptides also function as analgesics,[112] antihypertensive agents,[113] and antitumor agents.[114] However, peptides do not make good drug candidates because they are rapidly proteolyzed in the GI tract and serum, and they are poorly bioavailable, rapidly excreted, and can bind to multiple receptors. What is needed is a compound that mimics or blocks the biological effect of a peptide by interacting with its receptor or enzyme, but does not have the undesirable characteristics of peptides; these are *peptidomimetics*.

Earlier in this chapter (Section 2.2.A, p. 17) morphine and morphine analogs (**2.21**) were discussed as potent binders to the μ opioid receptor. In the mid-1970s Hughes and coworkers[115] showed that the endogenous peptides, the enkephalins and β-endorphin, also bound to the same site on the opioid receptor as did morphine. A remarkable resemblance was demonstrated between the N-terminal tyrosine structure of these opioid peptides and the morphine phenol ring system, which suggested why they all interacted with these receptors in a similar way.[116] Farmer then proposed that this may be a general phenomenon and that other nonpeptide structures may mimic natural peptide effectors.[117] His postulate was that peptide mimetics (which later became "peptidomimetics") could be designed that would replace peptide backbones while retaining the appropriate topography for binding to a receptor; this initiated the field of peptidomimetics.

The design of peptidomimetics can be a lead optimization approach, which uses the desired peptide as the lead compound and modifies it to minimize (or preferably, eliminate) the undesirable pharmokinetic properties. The generation of peptidomimetics is based on the conformational, topochemical, and electronic properties of the lead peptide when bound to its target receptor or enzyme.[118] The goal is to replace as much of the peptide backbone as possible with nonpeptide fragments while still maintaining the pharmacophoric groups (usually the amino acid side chains) of the peptide. This makes the compound more lipophilic, which increases its bioavailability. Replacement of the amide bond with alternative groups prevents proteolysis and promotes metabolic stability. Initially, conformational flexibility has to be retained to allow the pharmacophoric groups a better opportunity to find their binding sites, but further lead refinement should favor the formation of more conformationally restricted analogs that hold appropriate pharmacophoric groups in the bioactive conformation for binding to the target receptor.[119]

Increased lipophilicity and conformational modification of amino acids can be designed into the peptidomimetic. These groups may not be recognized by peptidases. For example, conformationally restricted analogs of phenylalanine shown in Figure 2.9[120] can be incorporated into peptidomimetic receptor ligands. Likewise, conformational restriction and lipophilicity can be incorporated into peptides (Figure 2.10).[121]

Another approach involves the design of conformationally restricted analogs that mimic characteristics of the receptor-bound conformation of the endogenous peptide,[122] such as β-turns (**2.61**,[123] **2.62**,[124] Figure 2.11), α-helices (**2.63**),[125] Ω-loops (**2.64**),[126] and β-strands (**2.65**).[127] This idea can be extended to *scaffold peptidomimetics* in which important pharmacophoric residues are held in the appropriate orientation by a rigid template. Compounds that block the binding of fibrinogen to its receptor (glycoprotein IIb/IIIa) can prevent platelet aggregation and are of potential value in the treatment of strokes and heart attacks.[128]

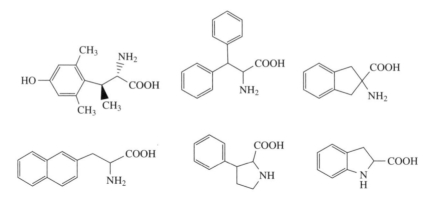

Figure 2.9 ▶ Conformationally restricted phenylalanine analogs

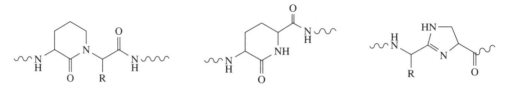

Figure 2.10 ▶ Conformationally restricted peptide analogs

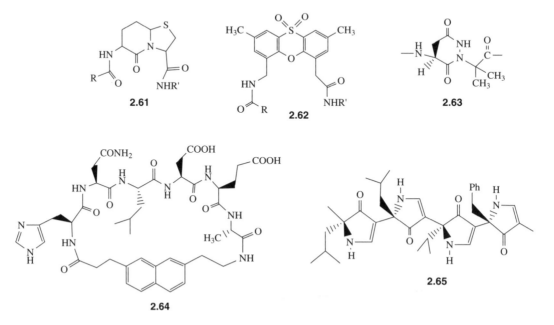

Figure 2.11 ▶ Conformationally restricted secondary structure peptidomimetics

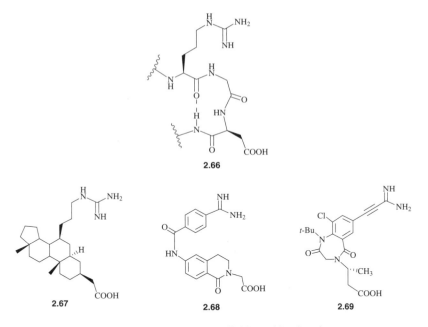

Figure 2.12 ▶ RGD scaffold peptidomimetics

A common β-turn motif that has been found to bind to GPIIbIIIa is arginine-glycine-aspartic acid (or in the one-letter amino acid code, RGD, **2.66**, Figure 2.12). Consequently, a variety of scaffold peptidomimetics for RGD have been designed based on the hypothesis that the glycine residue only represents a spacer between the two important recognition residues, arginine and aspartate. Several potent binders to this receptor have been found by replacement of the glycine with more rigid mimics, such as steroid (**2.67**),[129] isoquinolone (**2.68**),[130] and benzodiazepinedione (**2.69**)[131] spacers. A β-D-glucose-based nonpeptide scaffold (**2.71**) was designed as a mimic (note the darkened groups) of the potent somatostatin agonist (see Chapter 3, Section 3.2.C) **2.70**.[132] A target peptide for the treatment of cognitive disorders, such as Alzheimer's disease, is thyrotropin-releasing hormone (TRH, pyroGlu-His-ProNH$_2$, **2.72**);[133] a scaffold peptidomimetic for this hormone is **2.73**.[134]

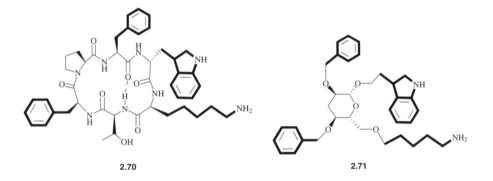

TABLE 2.5 ▶ Peptide Backbone Isosteres for Peptidomimetics

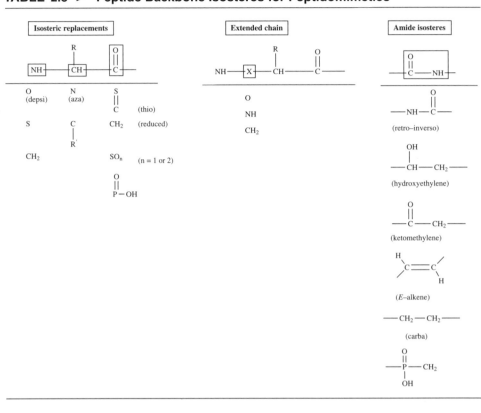

Isosteric replacements			Extended chain	Amide isosteres

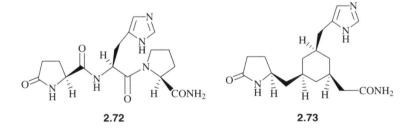

2.72   **2.73**

A common and important approach for the conversion of a peptide lead into a peptidomimetic is the use of peptide backbone isosteres (Table 2.5). Peptides in which the amide bonds are replaced with alternative groups are known as *pseudopeptides*.[135] These isosteric replacements remove the peptide linkage (thereby stabilizing the peptidomimetics to metabolism) and/or make them less polar and more lipophilic. The hydroxymethylene (also called statine)[136] isostere is one of the early mimetics used in the design of inhibitors of proteases, particularly of HIV protease.[137] Other variants of azapeptides (**2.74**, in which one or more of the α-carbons are replaced by N)[138] include azatides (**2.75**, azapeptides in which *all* of the α-carbons are replaced by N)[139] and peptoids (**2.76**, in which the α-CHR groups are replaced by NR units and the NH groups are replaced by CH_2 units, i.e., an amino acid sequence that is opposite that in peptides).[140]

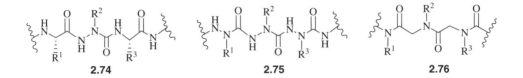

2.74   **2.75**   **2.76**

2.2.F Structure Modifications to Increase Oral Bioavailability

About three-quarters of drug candidates do not make it to clinical trials because of problems with pharmacokinetics in animals.[141] Less than 10% of drug candidates entering clinical trials become marketed products. About 40% of the molecules that fail in clinical trials do so because of pharmacokinetic problems, such as poor oral bioavailability or short plasma half-lives.[142] Because of the huge waste of time and resources by having a drug candidate fail late in the drug discovery process, a more recent trend is to examine pharmacokinetic aspects of molecules as early as possible in this process.[143] The use of mass spectrometry for this purpose is discussed in Chapter 7 (Section 7.3.C).

Low water solubility of a compound (high lipophilicity) can be a limiting factor in oral bioavailability,[144] and highly lipophilic compounds also are easily metabolized (see Chapter 7) or bind to plasma proteins. However, low lipophilicity is typically more of a problem, because that leads to poor permeability through membranes. Membrane permeability for a number of drugs is known.[145] In this section I try to assess how to incorporate better pharmacokinetic properties into lead modification design.

Several of the lead modification approaches discussed earlier were directed at improving both pharmacodynamics as well as pharmacokinetics, such as homologation, chain branching, ring-chain transformations, and bioisosterism. Increases in potency *in vivo* using these approaches could be explained either by pharmacodynamics (enhanced binding to a receptor) or by pharmacokinetics (increased lipophilicity, leading to improved absorption and distribution).

Because of the importance of lipophilicity in drug design,[146] it is essential to understand not only how to determine lipophilicities of compounds but also how to determine lipophilicities of substituents so that the correct substituent can be selected in lead modification approaches. The basis for the determination of the lipophilicities of substituents, as presented by Corwin Hansch and coworkers,[147] is derived from the earlier postulate by L. P. Hammett on how the electronic effects of substituents affect the reactivity of organic molecules, known as the Hammett equation. Those of you who know how to derive this equation can skip the next section.

F.1 Electronic Effects: The Hammett Equation

Hammett's postulate was that the electronic effects (both the inductive and resonance effects) of a set of substituents should be similar for different organic reactions. Therefore, if values could be assigned to substituents in a standard organic reaction, these same values could be used to estimate rates in a new organic reaction. This was the first approach that allowed the prediction of reaction rates. Hammett chose benzoic acids as the standard system.

Consider the reaction shown in Scheme 2.6. Intuitively, it seems reasonable that as X becomes electron withdrawing (relative to H), the equilibrium constant (K_a) should increase (the reaction should be favored to the right) because X is inductively pulling electron density

Scheme 2.6 ▶ Ionization of substituted benzoates

Scheme 2.7 ▶ Saponification of substituted ethyl benzoates

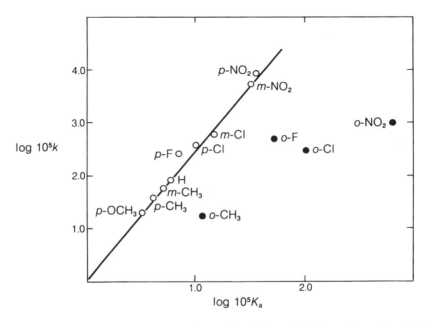

Figure 2.13 ▶ Linear free-energy relationship for the dissociation of substituted benzoic acids in water at 25°C (K_a) against the rates of alkaline hydrolysis of substituted ethyl benzoates in 85% ethanol-water at 30°C (k). [Reprinted with permission from Roberts, J. D. and Caserio, M. C. (1977). *Basic Principles of Organic Chemistry*, 2nd ed., p. 1331. W. A. Benjamin, Menlo Park, CA. Copyright ©1977 Benjamin/Cummings Publishing Company.]

from the carboxylic acid group, making it more acidic (reactant argument); it also is stabilizing the negative charge on the carboxylate group of the product (product argument). Conversely, when X is electron donating, the equilibrium constant should decrease. A similar relationship should exist for a rate constant (k); an electron-withdrawing substituent would stabilize a negative change in the transition state, thereby lowering the activation energy, and increasing the rate, and an electron-donating group would destabilize the transition state, decreasing the rate. Hammett chose the reaction shown in Scheme 2.7 as the standard system to determine electronic effects of substituents on the rate constant of a reaction.

If K_a is measured from Scheme 2.6 and k from Scheme 2.7 for a series of substituents X, and the data are expressed in a double logarithm plot (Figure 2.13), then a straight line can be drawn through most of the data points. This is known as a *linear free-energy relationship*. When X is a meta- or para-substituent, then virtually all of the points fall on the straight line;

the ortho-substituent points are badly scattered. The initial Hammett relationship does not hold for ortho-substituents because of steric interactions and polar effects. The linear correlation for the meta- and para-substituents is observed for rate and equilibrium constants for a wide variety of organic reactions. The straight line can be expressed by Equation 2.3,

$$\log k = \rho \log K + C \tag{2.3}$$

where the two variables are $\log k$ and $\log K$. The slope of the line is ρ, and the intercept is C. When there is no substituent, i.e., when X = H, then Equation 2.4 holds:

$$\log k_0 = \rho \log K_0 + C \tag{2.4}$$

Subtraction of Equation 2.4 from Equation 2.3 gives Equation 2.5,

$$\log k/k_0 = \rho \log K/K_0 \tag{2.5}$$

where k and K are the rate and equilibrium constants, respectively, for compounds with a substituent X, and k_0 and K_0 are the rate and equilibrium constants, respectively, for the parent compound (X = H). If $\log K/K_0$ is defined as σ, then Equation 2.5 reduces to Equation 2.6, the *Hammett equation*:

$$\log k/k_0 = \rho\sigma \tag{2.6}$$

The *electronic parameter*, σ, depends on the electronic properties and position of the substituent on the ring and, therefore, is also called the *substituent constant*. The more electron withdrawing a substituent, the more positive its σ value (relative to H, which is set at 0.0); conversely, the more electron donating, the more negative its σ value. The meta σ constants result from inductive effects, but the para σ constants correspond to the net inductive and resonance effects. Therefore, σ_{meta} and σ_{para} for the same substituent, generally, are not the same.

The ρ values (the slope) depend on the particular type of reaction and the reaction conditions (e.g., temperature and solvent) and, therefore, are called *reaction constants*. The importance of ρ is that it is a measure of the sensitivity of the reaction to the electronic effects of the meta- and para-substituents. A large ρ, either positive or negative, indicates great sensitivity to substituent effects. Reactions that are favored by electron donation in the transition state (such as reactions that proceed via carbocation intermediates) have negative ρ values (i.e., the linear free-energy relationship has a negative slope); reactions that are aided by electron withdrawal (such as reactions that proceed via carbanion intermediates) have positive ρ values.

F.2 Lipophilicity Effects

a. Importance of Lipophilicity

Hansch believed that, just as the Hammett equation relates the electronic effects of substituents to reaction rates, there should be a linear free-energy relationship between lipophilicity and biological activity. Hansch proposed that the first step in the overall drug process was a random walk, a diffusion process, in which the drug made its way from a dilute solution outside of the cell to a particular site in the cell. This was visualized as being a relatively slow process, the rate of which is highly dependent on the molecular structure of the drug. For the drug to reach the site of action, it must be able to interact with two different environments, lipophilic (e.g., membranes) and aqueous (the exobiophase, such as the cytoplasm). The cytoplasm of a cell is essentially a dilute solution of salts in water; all living cells are surrounded by a nonaqueous phase, the membrane. The functions of membranes are to protect the cell from water-soluble substances, to form a surface to which enzymes and other proteins can attach to produce a localization and structural organization, and to separate solutions of different electrochemical potentials (e.g., in nerve conduction). One of the most important membranes is known as the *blood–brain barrier*, a membrane that surrounds the capillaries of the circulatory system in the brain and protects it from passive diffusion of undesirable polar chemicals from the bloodstream. This is an important prophylactic boundary, but it also can block the delivery of central nervous system drugs to their site of action.

Although the structure of membranes has not been resolved, the most widely accepted model is the fluid mosaic model (Figure 2.14).[148] In this depiction integral proteins are embedded in a lipid bilayer; peripheral proteins are associated with only one membrane surface. The structure of the membrane is primarily determined by the structure of the lipids of which it is comprised. The principal classes of lipids found in membranes are neutral cholesterol (**2.77**) and the ionic phospholipids, e.g., phosphatidylcholine (**2.78**, $R = CH_2CH_2N(CH_3)_3^+$, phosphatidylethanolamine (**2.78**, $R = CH_2CH_2NH_3^+$), phosphatidylserine (**2.78**, $R = CH_2CH(NH_3^+)COO^-$), phosphatidylinositol (**2.78**, $R = $ inositol), and sphingomyelin (**2.79**, $R = {}^-OPO_2CH_2CH_2N(CH_3)_3^+$; $R'CO$ and $R''CO$ in **2.78** and **2.79**

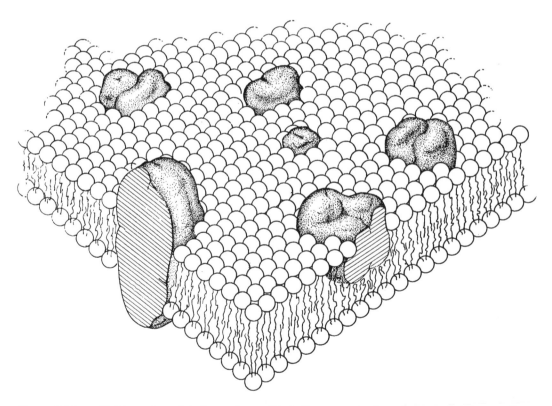

Figure 2.14 ▶ Fluid mosaic model of a membrane. The balls represent polar end groups, and the wavy lines are the hydrocarbon chains of the lipids. The masses embedded in the lipid bilayer are proteins. [Reprinted with permission from Singer, S. J. and Nicolson, G. L. (1972). *Science* 175, 720. Copyright ©1972 by American Association for the Advancement of Science.]

are derived from fatty acids. Glycolipids (**2.79**, R = sugar) also are important membrane constituents.

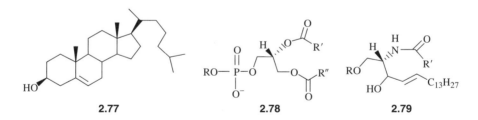

2.77	**2.78**	**2.79**

All of these lipids are amphipathic, which means that one end of the molecule is hydrophilic (water soluble) and the other is hydrophobic or, if you wish, lipophilic (water insoluble; soluble in organic solvents). Thus, the hydroxyl group in cholesterol, the ammonium groups in the phospholipids, and the sugar residue in the glycolipids are the polar, hydrophilic ends, and the steroid and hydrocarbon moieties are the lipophilic ends. The hydrocarbon part (R′ and R″) actually can be a mixture of chains from 14 to 24 carbon atoms long; approximately 50% of the chains contain a double bond. The polar groups of the lipid bilayer are in contact with the aqueous phase; the hydrocarbon chains project toward each other in the interior with a space between the layers. The stability of the membrane arises from the stabilization of the ionic charges by ion–dipole interactions (see Chapter 3, Section 3.2.B.3, p. 125) with the water and from association of the nonpolar groups. The hydrocarbon chains are relatively free to move; therefore, the core is similar to a liquid hydrocarbon.

b. Measurement of Lipophilicities

It occurred to Hansch that the fluidity of the hydrocarbon region of the membrane may explain the correlation noted by Richet,[149] Overton,[150] and Meyer[151] between lipid solubility and biological activity. He first set out to measure the lipophilicities of various compounds and then to determine the lipophilicities of substituents. But how should the lipophilicities be measured? The most relevant approach would be to determine their solubility in membranes or vesicles. However, as an organic chemist, Hansch probably realized that if he set a scale of lipophilicities based on membrane solubility, which required the researcher to prepare membranes or vesicles, there was no way organic chemists, especially in the 1960s, would ever bother to use this method, and it would become very limited. So he decided to propose a model for a membrane and determine lipophilicities by a simple methodology that organic chemists would not hesitate to employ. The model for the first step in drug action (transport to the site of action) would be the solubility of the compound in 1-octanol, which simulates a lipid membrane, relative to that in water (or aqueous buffer, the model for the cytoplasm). 1-Octanol has a long saturated alkyl chain, a hydroxyl group for hydrogen bonding, and it dissolves water to the extent of 1.7 M (saturation). This combination of lipophilic chains, hydrophilic (head) groups, and water molecules gives 1-octanol properties very close to those of natural membranes and macromolecules.

As a measure of lipophilicity, Hansch proposed the *partition coefficient*, P, a measure of the solubility in 1-octanol versus water,[152,153] and P was determined by Equation 2.7,

$$P = \frac{[\text{compound}]_{\text{oct}}}{[\text{compound}]_{\text{aq}}(1 - \alpha)} \qquad (2.7)$$

where α is the degree of dissociation of the compound in water calculated from ionization constants. (Ionization makes the compound more soluble in water than the structure appears, so that must be taken into account.) The partition coefficient is derived experimentally by placing a compound in a shaking device (like a separatory funnel) with varying volumes of 1-octanol and water, determining the concentration of the compound in each layer after mixing (by gas chromatography or HPLC), and employing Equation 2.7 to calculate P.[154] The value of P varies slightly with temperature and concentration of the solute, but with neutral molecules in dilute solutions (<0.01 M) and small temperature changes ($\pm5°$C), variations in P are minor.

Collander[155] had shown previously that the rate of movement of a variety of organic compounds through cellular material was approximately proportional to the logarithm of their partition coefficients between an organic solvent and water. Therefore, as a model for a drug traversing through the body to its site of action, the relative potency of the drug, expressed as log $1/C$, where C is the concentration of the drug that produces some standard biological effect, was related by Hansch *et al.*[156] to its lipophilicity by the parabolic expression shown in Equation 2.8:

$$\log 1/C = -k(\log P)^2 + k'(\log P) + k'' \tag{2.8}$$

On the basis of Equation 2.7, it is apparent that if a compound is more soluble in water than in 1-octanol, $P < 1$, and, therefore, log P is negative. Conversely, a molecule more soluble in 1-octanol has a $P > 1$, and the log P is positive. Therefore, the more positive the log P, the more lipophilic it is. The larger the value of P, the more there will be an interaction of the drug with the lipid phase (i.e., membranes). As P approaches infinity, micelles will form and/or the drug interaction will become so great that the drug will not be able to cross the aqueous phase, and it will localize in the first lipophilic phase with which it comes into contact. As P approaches zero, the drug will be so water soluble that it will not be capable of crossing the lipid phase and will localize in the aqueous phase. Somewhere between $P = 0$ and $P = \infty$, there will be a value of P such that drugs having this value will be least hindered in their journey through macromolecules to their site of action. This value is called log P_0, the logarithm of the optimum partition coefficient for biological activity. This random walk analysis supports the parabolic relationship (Equation 2.8) between potency (log $1/C$) and log P (Figure 2.15). Note the correlation of Figure 2.15 with the generalization regarding homologous series of compounds (Section 2.2.E.1, p. 26; Figure 2.3, p. 26). An increase in the alkyl chain length increases the lipophilicity of the molecule; the log P_0 generally occurs in the range of 5–9 carbon atoms. Hansch *et al.*[157] found that a number of series of

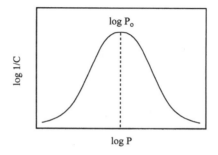

Figure 2.15 ▶ Effect of log P on biological response. P is the partition coefficient, and C is the concentration of the compound required to produce a standard biological effect

nonspecific hypnotics had similar log P_0 values, approximately 2, and they suggested that this is the value of log P_0 needed for penetration into the central nervous system (CNS), i.e., for crossing the blood–brain barrier. If a hypnotic agent has a log P considerably different from 2, then its activity probably is derived from mechanisms other than just lipid transport. If a lead compound has modest CNS activity and has a log P value of 0, it would be reasonable to synthesize an analog with a higher log P.

Because of the problem associated with ionization of compounds, which leads to greater water solubility than predicted from the neutral structure, often the term log D (the log of the distribution coefficient, generally between 1-octanol and aqueous buffer) is used to describe the lipophilicity of an ionizable compound.[158] Because ionization is a function of the pK_a of the compound and the pH of the solution in which the compound is dissolved, log D describes the log P of an ionizable compound at a particular pH. For example, log $D_{4.5}$ is the log P of an ionizable compound at pH 4.5. The log D value will change for ionizable compounds as a function of pH, whereas log P of nonionizable compounds will be independent of pH. For example, Table 2.6 shows how the log D for the antihypertensive drug metoprolol (**2.80**, Toprol-XL) changes as a function of pH. Note that at low pH values, the amine is protonated, lowering the log D value. As the pH is increased, the equilibrium starts to favor the neutral free base form, which is more lipophilic.

Although it is valuable to be able to determine the lipophilicity of a molecule, for lead modification purposes you need to be able to predict, prior to synthesis of the compound,

TABLE 2.6 ▶ Change in log D as a Function of pH for Metoprolol (2.80)[a]

2.80

pH	log D
2.0	−1.31
3.0	−1.31
4.0	−1.31
5.0	−1.28
5.5	−1.21
6.0	−1.05
6.5	−0.75
7.0	−0.34
7.5	0.12
8.0	0.59
8.5	1.03
9.0	1.39
10.0	1.73

[a] The author is grateful to Karolina Nilsson and Ola Fjellström (AstraZeneca) for providing the log D values as a function of pH using ACD software.

what the lipophilicity of an unknown molecule will be. To do that it is necessary to know the lipophilicities of substituents and atoms. In the same way that substituent constants were derived by Hammett for the electronic effects of atoms and groups (σ constants), Hansch and coworkers[159] derived substituent constants for the contribution of individual atoms and groups to the partition coefficient. The lipophilicity substituent constant, π, is defined by Equation 2.9,

$$\pi = \log P_X - \log P_H = \frac{\log P_X}{\log P_H} \tag{2.9}$$

which has the same derivation as the Hammett equation. The term P_X is the partition coefficient for the compound with substituent X, and P_H is the partition coefficient for the parent molecule (X = H). As in the case of the Hammett substituent constant σ, π is additive and constitutive. *Additive* means that multiple substituents exert an influence equal to the sum of the individual substituents. *Constitutive* indicates that the effect of a substituent may differ depending on the molecule to which it is attached or on its environment. Alkyl groups are some of the least constitutive groups. For example, methyl groups attached at the meta- or para-positions of 15 different benzene derivatives had π_{CH_3} values with a mean and standard derivation of 0.50 ± 0.04. Because of the additive nature of π values, π_{CH_2} can be determined as shown in Equation 2.10, where the log P values are obtained from standard tables:[160]

$$\pi_{CH_2} = \log P_{nitroethane} - \log P_{nitromethane}$$
$$= 0.18 - (-0.33) = 0.51 \tag{2.10}$$

Because, by definition, $\pi_H = 0$, then $\pi_{CH_2} = \pi_{CH_3}$. However, be aware that π_{CH_2OH} does not equal π_{CH_3O} because there is a difference in hydrogen bonding for substituents with and without hydroxyl groups, and, therefore, a difference in water solubility between these two substituents. Note that π represents the lipophilicity of a substituent and log P is the lipophilicity of a compound.

As was alluded to in Section 2.2.E.2 on molecular modification, branching in an alkyl chain lowers the log P or π as a result of larger molar volumes and shapes of branched compounds. As a rule of thumb, the log P or π is lowered by 0.2 unit per branch. For example, the π_{i-Pr} in 3-isopropylphenoxyacetic acid is 1.30; π_{n-Pr} is 3(0.5) = 1.50, or 0.2 greater than π_{i-Pr}.

Another case where π values are fairly constant is conjugated systems, as exemplified by $\pi_{CH=CHCH=CH}$ in Table 2.7.

Inductive effects are quite important to lipophilicity.[161] In general, electron-withdrawing groups increase π when a hydrogen-bonding group is involved. For example, π_{CH_2OH} varies as a function of the proximity of an electron-withdrawing phenyl group (Eq. 2.11), and π_{NO_2} varies as a function of the inductive effect of the nitro group on the hydroxyl group (Eq. 2.12). The electron-withdrawing inductive effects of the phenyl group (Eq. 2.11) and the nitro group (Eq. 2.12)

$$\pi_{CH_2OH} = \log P_{Ph(CH_2)_2OH} - \log P_{PhCH_3} = -1.33$$
$$\pi_{CH_2OH} = \log P_{PhCH_2OH} - \log P_{PhH} = -1.03 \tag{2.11}^{[162]}$$

$$\pi_{NO_2} = \log P_{PhNO_2} - \log P_{PhH} = -0.28$$
$$\pi_{NO_2} = \log P_{4-NO_2PhCH_2OH} - \log P_{PhCH_2OH} = 0.11 \tag{2.12}^{[163]}$$

make the nonbonded electrons on the OH group less available for hydrogen bonding, thereby reducing the affinity of this functional group for the aqueous phase. This, then, increases

TABLE 2.7 ▶ Constancy of π for $-CH=CH-CH=CH-$ [a]

$$\pi_{CH=CHCH=CH}$$

log P	$-$log P		$= 2.14 - 0.75 = 1.39$
log P	$-$log P		$= 2.03 - 0.65 = 1.38$
log P	$-$log P		$= 3.40 - 2.03 = 1.37$
log P	$-$log P		$= 4.12 - 2.67 = 1.45$
log P	$-$log P		$= 3.12 - 1.81 = 1.31$
log P	$-$log P		$= 3.45 - 2.13 = 1.32$
2/3 log P			$= 2/3(2.13) = 1.42$
log P	$-$log P		$= 2.84 - 1.46 = 1.38$
		ave.	1.38 ± 0.046

[a] Hansch, C.; Steward, A. R.; Anderson, S. M.; Bentley, D. *J. Med. Chem.* **1968**, *11*, 1.

the log P or π. Also note in Equations 2.11 and 2.12 that, because $\pi_H = 0$ by definition, log $P_{benzene} = \pi_{Ph}$. These examples enforce the notion of the constitutiveness of π values.

Resonance effects also are important to the lipophilicity, much the same way as are inductive effects. Delocalization of nonbonded electrons into aromatic systems decreases their availability for hydrogen bonding with the aqueous phase and, therefore, increases the π. This is supported by the general trend that aromatic π_X values are greater than aliphatic π_X values, again emphasizing the constitutive nature of π and log P.

Steric effects are variable. If a group sterically shields nonbonded electrons, then aqueous interactions will decrease, and the π value will increase. However, crowding of functional groups involved in hydrophobic interactions (Chapter 3, Section 3.2.B.6, p. 129) will have the opposite effect.

Conformational effects also can affect the π value. The π_X values for $Ph(CH_2)_3X$ are consistently lower (more water soluble) than π_X values for $CH_3(CH_2)_3X$ (Table 2.8). This phenomenon is believed to be the result of folding of the side chain onto the phenyl ring (**2.81**), which means a smaller apolar surface for organic solvation. The folding may be caused by the interaction of the CH_2–X dipole with the phenyl π-electrons and by intramolecular hydrophobic interactions.

TABLE 2.8 ▶ Effect of Folding of Alkyl Chains on π

X	π_X (aromatic)[a]	π_X (aliphatic)[b]	$\Delta\pi_X$
OH	−1.80	−1.16	0.64
F	−0.73	−0.17	0.56
Cl	−0.13	0.39	0.52
Br	0.04	0.60	0.56
I	0.22	1.00	0.78
COOH	−1.26	−0.67	0.59
CO$_2$CH$_3$	−0.91	−0.27	0.64
COCH$_3$	−1.26	−0.71	0.55
NH$_2$	−1.85	−1.19	0.66
CN	−1.47	−0.84	0.63
OCH$_3$	−0.98	−0.47	0.51
CONH$_2$	−2.28	−1.71	0.57
		Average	0.60 ± 0.05

[a] $\log P_{Ph(CH_2)_{3x}} - \log P_{Ph(CH_2)_3H}$
[b] $\log P_{CH_3(CH_2)_{3x}} - \log P_{CH_3(CH_2)3H}$

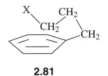

2.81

Two examples follow to show the additivity of π constants in predicting $\log P$ values. A calculation of the $\log P$ for the anticancer drug diethylstilbestrol (**2.82**, DES) is shown in Equation 2.13.

$$
\begin{aligned}
\text{Calc. } \log P &= 2\pi_{CH_3} + 2\pi_{CH_2} + \pi_{CH=CH} + 2\log P_{PhOH} - 0.40 \\
&= 2(0.50) + 2(0.50) + 0.69 + 2(1.46) - 0.40 \\
&= 5.21
\end{aligned}
\tag{2.13}
$$

In Equation 2.13, $\pi_{CH=CH} = 1/2(\pi_{CH=CHCH=CH})$, which was shown in Table 2.7 to be 1.38; −0.40 is added into the equation to account for two branching points (each end of the alkene). The calculated $\log P$ value of 5.21 is quite remarkable considering that the experimental $\log P$ value is 5.07.

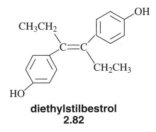

**diethylstilbestrol
2.82**

A calculation of the $\log P$ for the antihistamine diphenhydramine (**2.83**, Benedryl), is shown in Equation 2.14.

$$\text{Calc.} \log P = 2\pi_{\text{Ph}} + \pi_{\text{CH}} + \pi_{\text{OCH}_2} + \pi_{\text{CH}_2} + \pi_{\text{NMe}_2} - 0.2$$
$$= 2(2.13) + 0.50 - 0.73 + 0.50 - 0.95 - 0.2$$
$$= 3.38 \tag{2.14}$$

In this equation, 2.13 is $\log P$ for benzene, which is the same as π_{Ph}; 0.50 is π_{CH} (same as π_{CH_3}); -0.73 was obtained by subtracting 1.50 ($2\pi_{\text{CH}_3} + \pi_{\text{CH}_2}$) from $\log P_{\text{CH}_3\text{CH}_2\text{OCH}_2\text{CH}_3}$ ($= 0.77$); -0.95 is the value for π_{NMe_2} obtained by subtracting $\pi_{\text{Ph}(\text{CH}_2)_3}(2.13 + 3(0.5) = 3.63)$ from $\log P_{\text{Ph}(\text{CH}_2)_3\text{NMe}_2}(2.68)$; -0.2 is for branching at the CH. [Note that there is no branching at the $\text{N}(\text{CH}_3)_2$ because we used that whole substituent to obtain π.] The experimental $\log P$ value is 3.27.

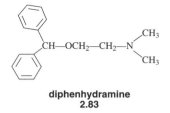

diphenhydramine
2.83

A more rapid approach than the standard shake-flask method[164] for determination of $\log P$ values that was described above has been reported[165] for neutral compounds. The reversed-phase HPLC method takes about 20 minutes per compound with a wide range of lipophilicities (6 $\log P$ units) with good accuracy and excellent reproducibility. The value obtained by this method is referred to as the $E \log P_{oct}$. A reversed-phase HPLC method for determination of $\log D$ values ($E \log D_{oct}$) also was devised by the same Pfizer group.[166]

c. Computerization of Log *P* Values

The determination of $\log P$ values has become less of a chore as a result of computerization of the method.[167] A nonlinear regression model for the estimation of partition coefficients was developed by Bodor *et al.*[168] using the following molecular descriptors: molecular surface, volume, weight, and charge densities. It was shown to have excellent predictive power for the estimation of the $\log P$ for complex molecules. A semiquantitative method for calculating $\log P$ values ($M \log P$) was developed by Moriguchi *et al.*[169] using a multiple regression analysis of 1230 organic molecules having a wide variety of structures; excellent correlation was observed between the observed $\log P$ and the calculated $\log P$.

Probably the simplest way to get $\log P$ values for unknown compounds is with the use of one of the numerous software packages that are now commercially available, such as those from Daylight ($C \log P$; the one developed by the medicinal chemistry group at Pomona College), Advanced Chemistry Development (ACD/log P DB), CTIS (AUTOLOG^{TM}), Scivision (Sci log P), and Bio-Rad (PredictIt^{TM} log P and log D). The problem with the software packages, however, is that the results can differ widely (2 or more $\log P$ units) and differ from the experimental value.[170] The reason is related to the fact that, as mentioned above, π values are constitutive; depending on the structure of the compound, the π value can differ. Also, ionization of groups varies with concentration and counter ions. So, no software package can account for π values for substituents on *every* scaffold. Most software packages will always give an answer, but $C \log P$ will not calculate a $\log P$ when it does not have sufficient data (e.g., if an atom is poorly parameterized). Of course, it is frustrating when a computer tells you it cannot compute (which may be detrimental to the software company sales), but it may be the most honest approach. One way to obtain the most accurate predictive results for your particular family of compounds is to determine experimentally the actual $\log P$ value for one of the members of the family, then ask a variety of software packages to predict the $\log P$ value, and use the program that comes closest to the experimental value for that family of compounds.

d. Membrane Lipophilicity

Although the $\log P$ values determined from 1-octanol/water partitioning are excellent models for *in vivo* lipophilicity, it has been found that for a variety of aromatic compounds whose $\log P$ values are greater than 5.5 (very lipophilic) or whose molar volumes are greater than $230 \, cm^3/mol$, there is a breakdown in the correlation of these values with those determined from partitioning between L-α-phosphatidylcholine dimyristoyl membrane vesicles and water.[171] Above $\log P$ of 5.5, the solvent solubility for these molecules is greater than their membrane solubility. As the compound increases in size, more energy per unit volume is required to form a cavity in the structured membrane phase. This is consistent with observations that branched molecules have lower $\log P$ values than their straight-chain counterparts, and that this effect is even greater in membranes than in organic solvents.

Note that although $\log P$ values are most commonly determined with 1-octanol/water mixtures, this is not universal because hydrophilicity resulting from acceptance of a hydrogen bond is not reflected well by partitioning in 1-octanol, which can accept hydrogen bonds almost as well as does water.[172] Consequently, this gives an apparently higher lipophilicity value than is reflected in membrane partitioning. Other nonhydroxylic solvents, such as cyclohexane, can provide insights into these processes. Because of this, Seiler[173] introduced a new additive constitutive substituent constant for solvents other than 1-octanol. Therefore, when using $\log P$ values, it is important to be aware of the solvent used to obtain the $\log P$ data.

F.3 Effects of Ionization on Lipophilicity and Oral Bioavailability

Receptors are typically proteins, comprised of amino acids with varying ionization states depending on the pH of the environment. For example, anionic groups in proteins include carboxylic acids (aspartic and glutamic acids, pK_a of 4–4.5), phenols (tyrosine, pK_a of 9.5–10), sulfhydryls (cysteine, pK_a of 8.5–9), and hydroxyls (serine and threonine, pK_a of 13.5–14). Cationic groups in proteins include imidazole (histidine, pK_a of 6–6.5), amino (lysine, pK_a of 10–10.5), and guanidino (arginine, pK_a of 12–13) groups. At physiological pH (pH 7.4), even the mildly acidic groups, such as carboxylic acid groups, will essentially be completely in the carboxylate anionic form; phenolic hydroxyl groups may be partially ionized. Likewise, basic groups, such as amines, will be partially or completely protonated to give the cationic form. The same is true for a drug; the ionization state of a drug will depend on the pH of the medium with which it has to interact and the pK_a values of the ionizable groups. Ionization will have a profound effect not only on its interaction with a receptor, but also on its lipophilicity. Consequently, it is important to appreciate the effects of ionization in lead modification approaches.

What if the drug you are attempting to discover binds at an ionized site in the receptor, so ionization of your drug favors binding to the receptor (see Chapter 3, Section 3.2.B.2),

$$RNH_2 + H^+ \rightleftharpoons RNH_3^+$$

$$RCOOH \rightleftharpoons RCOO^- + H^+$$

Scheme 2.8 ▶ Ionization equilibrium for an amine base and a carboxylic acid

but ionization of the drug also blocks its ability to cross various membranes prior to reaching the receptor? How is it possible to design a compound that is neutral when it needs to cross membranes, but ionized when it finally reaches the target receptor? This is possible because an equilibrium is established between the neutral and ionized form of a molecule or group that depends on the pH of the medium and the pK_a of the ionizable group (Scheme 2.8). When the pH of the medium equals the pK_a of the molecule, half of the molecules are in the neutral form and half in the ionized form. The ones that are neutral may be able to cross membranes, but once on the other side, the equilibrium with the ionized form is reestablished (the equilibrium mixture will again depend on the pH on the other side of the membrane), so there are now ionized molecules on the other side of the membrane that can interact with the target receptor. The ionized molecules that did not cross the membrane also reestablish an equilibrium and become a mixture of ionized and neutral molecules, so more neutral molecules can get across the membrane. If the equilibria could be reestablished indefinitely, eventually all of the molecules would cross the membranes and bind to the target receptor. However, drugs get metabolized and excreted (see Chapter 7), so they may never get across the membrane before they are excreted. To adjust the ionization equilibrium of the lead compound, you need to add electron-withdrawing or electron-donating groups to vary the pK_a of the molecule. Electron-withdrawing groups will lower the pK_a, making acids more ionizable and bases less ionizable; the opposite holds for electron-donating groups.

The importance of ionization was recognized in 1924 when Stearn and Stearn[174] suggested that the antibacterial activity of stabilized triphenylmethane cationic dyes was related to an interaction of the cation with some anionic group in the bacterium. Increasing the pH of the medium also increased the antibacterial effect, presumably by increasing the ionization of the receptors in the bacterium. Albert and coworkers[175] made the first rigorous proof that a correlation between ionization and biological activity existed. A series of 101 aminoacridines, including the antibacterial drug, 9-aminoacridine or aminacrine (**2.84**, Monacrin), all having a variety of pK_a values, was tested against 22 species of bacteria.

2.84

A direct correlation was observed between ionization (formation of the cation) of the aminoacridines and antibacterial activity. However, at lower pH values, protons can compete with these cations for the receptor, and antibacterial activity is diminished. When this was realized, Albert[176] quips, the Australian Army during World War II was advised to pretreat wounds with sodium bicarbonate to neutralize any acidity prior to treatment with aminacrine. This, apparently, was quite effective in increasing the potency of the drug. The mechanism of action of aminoacridines is discussed in Chapter 6 (Section 6.3.A.3.a, p. 349).

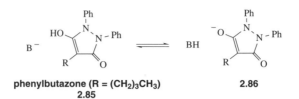

phenylbutazone (R = (CH$_2$)$_3$CH$_3$)
2.85

2.86

Scheme 2.9 ▶ Ionization equilibrium for phenylbutazone

Antihistamines and antidepressants tend to have pK_a values of about 9. The great majority of alkaloids that act as neuroleptics, local anesthetics, and barbiturates have pK_a values between 6 and 8; consequently, both neutral and cationic forms are present at physiological pH. This may allow them to penetrate membranes in the neutral form, and exert their biological action in the ionic form.

The uricosuric drug (increases urinary excretion of uric acid) phenylbutazone (**2.85**, Scheme 2.9; Butazolidine), R = (CH$_2$)$_3$CH$_3$ has a pK_a of 4.5 and is active as the anion (**2.86**). However, because the pH of urine is 4.8, suboptimal concentrations of the anion were found in the urinary system. Sulfinpyrazone (**2.85**, R = CH$_2$CH$_2$SOPh; Anturane) has a lower pK_a (2.8) and is about 20 times more potent than phenylbutazone; the anionic form blocks reabsorption of uric acid by renal tubule cells.[177]

The antimalarial drug pyrimethamine (**2.87**, Daraprim) has a pK_a of 7.2 and is best absorbed from solutions of sufficient alkalinity that it has a high proportion of molecules in the neutral form (to cross membranes). Its mode of action, the inhibition of the parasitic enzyme dihydrofolate reductase, however, requires that it be in the protonated cationic form.

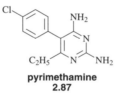

pyrimethamine
2.87

The effect of ionization can be rationalized either from a pharmacokinetic or pharmacodynamic perspective. For example, if changing the pK_a increases its potency, it could be because the neutral form becomes more prevalent and, therefore, crossing membranes becomes favored (pharmacokinetic argument), or it could be because there is a hydrophobic pocket in the receptor that the neutral form prefers to bind into (pharmacodynamic argument). How can the relative importance of these two properties be determined? If the drugs act on microbial systems, one way is to compare results of assaying the test compounds in a cell-free system (in which there are no membranes to cross) and in an intact cell system (in which it is necessary to cross a membrane to get to the receptor). For example, the pharmacokinetics of the antibacterial agent sulfamethoxazole (**2.88**, Scheme 2.10; Bactrim) depend on their nonionized form (**2.88**), but the pharmacodynamics depend on the anionic form (**2.89**). In a cell-free system the antibacterial activity of **2.88** and other sulfonamides is directly proportional to the degree of ionization, supporting the importance of ionization on pharmacodynamics, but in intact cells, where the drug must cross a membrane to get to the site of action, the antibacterial activity also is dependent on the neutral form,[178] supporting the notion that the neutral form is not important to pharmacodynamics, only to pharmacokinetics.

Scheme 2.10 ▶ Ionization equilibrium for sulfamethoxazole

The structure and function of a receptor and of a drug can be strongly dependent on the pH of the medium, especially if an *in vitro* assay is being used. However, you must be careful when trying to assess pK_a values of groups within a binding site of a receptor, because these values can be quite variable, and will depend on the microenvironment. On the basis of molecular dynamics simulations of several proteins in water, the interiors of these proteins were calculated to have dielectric constants of about 2–3,[179] which is comparable to the dielectric constant of nonpolar solvents such as benzene ($\varepsilon = 2.28$) or *p*-dioxane ($\varepsilon = 2.21$). This is quite different from the dielectric constant of water ($\varepsilon = 78.5$), which is a result of the strong dipole moment of the O–H bonds. If a carboxyl group is in a nonpolar region, its pK_a will rise because the anionic form will be destabilized. Glutamate-35 in the lysozyme-glycolchitin complex has a pK_a of 8.2[180] the pK_a of glutamate in water is 4–4.5. The pK_a of Asp-99 in a nonpolar region of 3-oxo-Δ^5-steroid isomerase is a remarkable 9.5![181] That is a change in equilibrium of a factor of 10^5 (remember, pK_a is a logarithm) in favor of the neutral form! If the carboxylate forms a salt bridge, it will be stabilized, and its pK_a will be lowered. If a carboxylic acid group is near an essential active site carboxylic acid, the anionic form will be destabilized, and its pK_a will be raised.[182a] Likewise, an amino group buried in a nonpolar microenvironment will have a lower pK_a because protonation will be disfavored (to avoid the polar cationic character). If the ammonium group of lysine forms a salt bridge, it will be stabilized, deprotonation will be inhibited, and the pK_a will rise. If basic residues are adjacent, the pK_a will drop to avoid two neighboring cations; the ε-amino group of the active site lysine residue in the enzyme acetoacetate decarboxylase has a pK_a of 5.9,[182b] whereas in water, it is about 10–10.5. Given this large change in pK_a values in different microenvironments, it is worthwhile to make large changes in pK_a values of compounds in a lead modification library to see how the potency changes in both *in vitro* and *in vivo* assays. Once it is established whether the potency of a molecule is favored in the neutral or ionized form, then pK_a considerations can be employed in further lead modification approaches.

F.4 Other Properties that Influence Oral Bioavailability and Ability to Cross the Blood–Brain Barrier

Pharmacokinetics is as important to drug discovery as pharmacodynamics, so Lipinski[183] proposed "the rule of five" as a guide to improve oral bioavailability during lead modification. Based on a large database of known drugs, the *rule of five* states that it is highly likely (>90% probability) that compounds with two or more of the following characteristics will have **poor** oral absorption and/or distribution properties:

▶ The molecular weight is >500.

▶ The log P is >5.

▶ There are more than 5 H-bond donors (expressed as the sum of OH and NH groups).

▶ There are more than 10 H-bond acceptors (expressed as the sum of N and O atoms).

Antibiotics, antifungals, vitamins, and cardiac glycosides are the exception because they often have active transporters to carry them across membranes, so lipophilicity is not relevant.

Therefore, when low potency of compounds in an *in vivo* assay is observed, the rule of five should be applied to determine if low potency is the result of a pharmacokinetic problem. To get a drug across the blood–brain barrier, the upper limits really should be 3 H-bond donors and 6 H-bond acceptors.[184]

In contrast to the rule of five, Veber and coworkers[185] measured the oral bioavailability of 1100 drug candidates and found that reduced molecular flexibility, as determined by the number of rotatable bonds (10 or fewer), and low polar surface area ($\leq 140\,\text{Å}^2$) or total hydrogen bond count (less than or equal to a total of 12 donors and acceptors) are important predictors of good oral bioavailability, *independent of molecular weight*. Both the number of rotatable bonds and hydrogen bond count tend to increase with molecular weight, which may explain Lipinski's first rule. Reduced polar surface area was found to correlate better with an increased membrane permeation rate than did lipophilicity. Nonetheless, molecular weight and lipophilicity were shown to have the greatest influence on getting a drug to the market.[186a]

Ajay and coworkers[186b] carried out computations to determine what drug properties were important for crossing the blood–brain barrier and for CNS activity. CNS-active and -inactive compounds were selected from the Comprehensive Medicinal Chemistry (CMC) and the MDDR (a database from MDL Inc.; MDL Drug Data Report) databases. Each molecule was described by seven 1-D descriptors (e.g., molecular weight, number of hydrogen bond donors, and number of hydrogen bond acceptors) and 166 2-D descriptors. Using all of these descriptors, 83% of the CNS-active compounds and 79% of the CNS-inactive compounds in these databases were correctly predicted. In general, they concluded that if the molecular weight, the degree of branching, the number of rotatable bonds, or the number of hydrogen bond acceptors is increased, the compound will be *less* likely to be CNS active. If the aromatic density, number of hydrogen bond donors, or log P is increased, the compound is *more* likely to be CNS active.

Absorption, distribution, metabolism, and excretion (ADME) characteristics of compounds are very important because a large percentage of drug candidates that reaches clinical trials are discontinued as a result of ADME and toxicity problems. If these properties could be predicted, much time and expense would be saved in designing, synthesizing, and testing compounds. Numerous computational methods have been devised that deal with these properties,[187] but reliable predictive capabilities are still lacking. A graphical model for estimating high, medium, or low oral bioavailability of drugs in humans, rats, dogs, and guinea pigs, based on both their permeability through human intestinal epithelial (Caco-2) cells and their *in vitro* liver enzyme metabolic stability rates, gave excellent results.[188]

Up to this point we have been discussing more or less random molecular modifications to make qualitative differences in a lead compound. In 1868 Crum-Brown and Fraser[189] predicted that some day a mathematical relationship between structure and activity would be expressed. It was not for almost 100 years that this prediction began to be realized, and a new era in drug design was born. In 1962 Corwin Hansch attempted to quantify the effects of particular substituent modifications, and from his studies the area of quantitative structure–activity relationships developed.[190]

2.2.G Quantitative Structure–Activity Relationships

G.1 Historical

The concept of quantitative drug design is based on the fact that the biological properties of a compound are a function of its physicochemical parameters, that is, physical properties, such as solubility, lipophilicity, electronic effects, ionization, stereochemistry, and so forth, that have a profound influence on the chemistry of the compounds. The first attempt to relate a physicochemical parameter to a pharmacological effect was reported in 1893 by Richet.[191] He observed that the narcotic action of a group of organic compounds was inversely related to their water solubility (*Richet's rule*). Overton[192] and Meyer[193] related tadpole narcosis induced by a series of nonionized compounds added to the water in which the tadpoles were swimming to the ability of the compounds to partition between oil and water. These early observations regarding the depressant action of structurally nonspecific drugs were rationalized by Ferguson.[194] He reasoned that when in a state of equilibrium, simple thermodynamic principles could be applied to drug activities, and that the important parameter for correlation of narcotic activities was the relative saturation (termed *thermodynamic activity* by Ferguson) of the drug in the external phase or extracellular fluids. This is known as *Ferguson's principle*, which is useful for the classification of the general mode of action of a drug and for predicting the degree of its biological effect. The numerical range of the thermodynamic activity for structurally nonspecific drugs is 0.01 to 1.0, indicating that they are active only at relatively high concentrations. Structurally specific drugs have thermodynamic activities considerably less than 0.01 and normally below 0.001.

In 1951 Hansch et al.[195] noted a correlation between the plant growth activity of phenoxyacetic acid derivatives and the electron density at the *ortho* position (lower electron density gave increased activity). They made an attempt to quantify this relationship by the application of the Hammett σ functions (see Section 2.2.F.1, p. 51), but this was unsuccessful.

The crucial breakthrough in QSAR came when Hansch and coworkers[196] conceptualized the action of a drug as depending on two processes. The first process is the journey of the drug from its point of entry into the body to the site of action (pharmacokinetics), and the second process is the interaction of the drug with the specific site (pharmacodynamics). Because of the importance of pharmacokinetics to the success of a drug, he developed the octanol-water scale for lipophilicity (see Section 2.2.F.2.b, p. 55) as a measurable physicochemical parameter to consider in addition to electronic effects that were developed by Hammett (see Section 2.2.F.1, p. 51). Another physicochemical parameter that Hansch thought should be important in lead discovery/lead optimization was steric effects, particularly for receptor binding.

G.2 Steric Effects: The Taft Equation and Other Equations

Because interaction of a drug with a receptor involves the mutual approach of two molecules, another important parameter for QSAR is the steric effect. In much the same way that Hammett derived quantitative electronic effects (see Section 2.2.F.1, p. 51), Taft[197] defined the steric parameter E_s as shown in Equation 2.15:

$$E_s = \log k_{XCO_2Me} - \log k_{CH_3CO_2Me} = \log k_X / k_0 \qquad (2.15)$$

Taft used for the reference reaction the relative rates of the acid-catalyzed hydrolysis of α-substituted acetates (XCH_2CO_2Me). This parameter is normally standardized to the methyl group ($XCH_2 = CH_3$) so that $E_s(CH_3) = 0.0$; it is possible to standardize it to hydrogen by adding 1.24 to every methyl-based E_s value.[198] Hancock et al.[199] claimed that this model reaction was under the influence of hyperconjugative effects and, therefore, developed corrected E_s values for the hyperconjugation of α-hydrogen atoms (Equation 2.16):

$$E_s^c = E_s + 0.306(n - 3) \qquad (2.16)$$

where E_s^c is the corrected E_s value and n is the number of α-hydrogen atoms.

Two other steric parameters worth mentioning are molar refractivity (MR) and the Verloop parameter. *Molar refractivity,*[200] the molar volume corrected by the refractive index which represents the size and polarizability of a fragment or molecule, is defined by the Lorentz-Lorenz equation:

$$MR = \frac{n^2 - 1}{n^2 + 2} \frac{MW}{d} \qquad (2.17)$$

where n is the index of refraction at the sodium D line, MW is the molecular weight, and d is the density of the compound. The greater the positive MR value of a substituent, the larger its steric or bulk effect. This parameter also measures the electronic effect and, therefore, may reflect dipole–dipole interactions at the receptor site.

The *Verloop steric parameters*[201] are used in a program called STERIMOL to calculate the steric substituent values from standard bond angles, van der Waals radii, bond lengths, and user-determined reasonable conformations. Five parameters are involved. One (L) is the length of the substituent along the axis of the bond between the substituent and the parent molecule. Four width parameters (B_1–B_4) are measured perpendicular to the bond axis. These five parameters describe the positions, relative to the point of attachment and the bond axis, of five planes that closely surround the group. In contrast to E_s values which, because of the reaction on which they are based, cannot be determined for many substituents, the Verloop parameters are available for any substituent.

G.3 Methods Used to Correlate Physicochemical Parameters with Biological Activity

Now that we can obtain numerous *physicochemical parameters* (also called *descriptors*) for any substituent, how do we use these parameters to gain information regarding what compound to synthesize next in an attempt to optimize the lead compound? First, several (usually, many) compounds related to the lead are synthesized, and the biological activities are determined in some screen. These data, then, can be manipulated by a number of QSAR methods. I present Hansch analysis first. If you are not interested in an overview of computational methods, you can skip sections 2.2.G.3 and 2.2.G.4, pp. 68–78.

a. Hansch Analysis: A Linear Multiple Regression Analysis

With the realization that there are (at least) two considerations for biological activity, namely, lipophilicity (required for the journey of the drug to the site of action) and electronic factors (required for drug interaction with the site of action), and that lipophilicity is a parabolic function, Hansch and Fujita[202] expanded Equation 2.8 to that shown in either Equation 2.18a or 2.18b, known as the *Hansch equation,*

$$\log 1/C = -k\pi^2 + k'\pi + \rho\sigma + k'' \qquad (2.18a)$$

$$\log 1/C = -k(\log P)^2 + k'(\log P) + \rho\sigma + k'' \qquad (2.18b)$$

where C is the molar concentration (or dose) that elicits a standard biological response (e.g., ED_{50}, the dose required for 50% of the maximal effect; IC_{50}, the concentration that gives 50% inhibition of an enzyme or antagonism of a receptor; LD_{50}, the lethal dose for 50% of the animal population); $k, k', \rho,$ and k'' are the regression coefficients derived from statistical curve fitting; and π and σ are the lipophilicity and electronic substituent constants, respectively. The reciprocal of the concentration ($1/C$) reflects the fact that greater potency is associated with a lower dose, and the negative sign for the π^2 [or $(\log P)^2$] term reflects the expectation of an optimum lipophilicity, i.e., the π_0 or $\log P_0$.

Because of the importance of steric effects and other shape factors of molecules for receptor interactions, an E_s term and a variety of other shape, size, or topography terms (S) have been added to the Hansch equation:

$$\log 1/C = -a\pi^2 + b\pi + \rho\sigma + cE_s + dS + e \qquad (2.19)$$

The way these parameters are used is by the application of the method of linear multiple regression analysis.[203] The best least-squares fit of the dependent variable (the biological activity) to a linear combination of the independent variables (the descriptors) is determined. Hansch analysis, also called the *extrathermodynamic method*, then, is a linear free-energy approach to drug design in congeneric series in which equations are set up involving different combinations of the physicochemical parameters; the statistical methodology allows the best equation to be selected and the statistical significance of the correlation to be assessed. Once this equation has been established, it can be used to predict the activities of untested compounds. Problems associated with the use of multiple regression analysis in QSAR studies have been discussed by Deardon.[204]

Several assumptions must be made when the extrathermodynamic method is utilized: Conformational changes in receptors can be ignored, metabolism does not interfere, linear free-energy terms relevant to receptor affinity are additive, the potency–lipophilicity relationship is parabolic or linear, and correlation implies a causal relationship. According to Martin[205] and Tute,[206] there is a balance of assets and liabilities to the extrathermodynamic method. The strengths are several-fold: (1) The use of descriptors (π, σ, E_s, MR, and so forth) permits data collected from simple organic chemical model systems to be utilized for the prediction of biological activity in complex systems, (2) the predictions are quantitative with statistical confidence limits, (3) the method is easy to use and is inexpensive, and (4) conclusions that are reached may have application beyond the substituents included in the particular analysis.

The weaknesses of this method are that (1) parameter values must be available for the substituents in the data set; (2) a large number of compounds must be included in the analysis to have confidence in the derived equations; (3) expertise in statistics and computer use is essential; (4) small molecule interactions are imperfect models for biological systems; (5) in contrast to chemical reactions in which you know the atoms that interact with the reagent, steric effects in biological systems may not be relevant, since it is often not certain which atoms in the drug interact with the receptor; (6) organic reactions used to determine the descriptors usually are studied under acidic or basic conditions when all analogs are fully protonated or deprotonated, but in biological systems, the drug may be partially protonated; (7) because QSAR is empirical, it is a retrospective technique that depends on the pharmacological activity of compounds belonging to the same structural type, and, therefore, new types of active compounds are not discovered (i.e., it is a lead optimization technique, not a lead discovery approach); and (8) like other empirical relationships, extrapolations frequently lead to false predictions.

Despite the weaknesses of this approach, it is used, and several successes in drug design attributable to Hansch analysis have been reported.[207] As pointed out in Chapter 3 (Section 3.2.E.2, p. 143), however, caution should be used when applying QSAR methods to racemic mixtures if only one enantiomer is active. Other important statistical approaches are mentioned briefly.

b. Free and Wilson or *de novo* Method

Not long after Hansch proposed the extrathermodynamic approach, Free and Wilson[208] reported a general mathematical method for assessing the occurrence of additive substituent effects and for quantitatively estimating their magnitude. It is a method for the optimization of substituents within a given molecular framework that is based on the (tenuous) assumption that the introduction of a particular substituent at any one position in a molecule always changes the relative potency by the same amount, regardless of what other substituents are present in the molecule. A series of linear equations of the form shown here is constructed:

$$BA = \sum a_i X_i + \mu \tag{2.20}$$

where BA is the magnitude of the biological activity, X_i is the ith substituent with a value of 1 if present and 0 if not, a_i is the contribution of the ith substituent to the BA, and μ is the overall average activity of the parent skeleton. These linear equations are solved by the method of least squares for the a_i and μ. All activity contributions at each position of substitution must sum to zero. The pros and cons of the Free-Wilson method have been discussed.[209] Fujita and Ban[210] suggested two modifications of the Free-Wilson approach on the assumption that the effect on the activity of a certain substituent at a certain position in a compound is constant and additive. First, that the biological activity should be expressed as log A/A_0, where A and A_0 represent the magnitude of the activity of the substituted and unsubstituted compounds, respectively, and that a_i is the log activity contribution of the ith substituent relative to H. This allows the derived substituent constants to be compared directly with other free-energy-related parameters that are additive. Second, that μ become analogous to the theoretically predicted (calculated) activity of the parent compound of the series. Both of these modifications have been widely accepted.

  As an example of the Free-Wilson approach, consider the hypothetical compound **2.90**.[211] If in one pair of analogs for which R^1, R^2, R^3, and R^4 are constant and R^5 is Cl or CH_3, the methyl compound is one-tenth as potent as the chloro analog, then the Free-Wilson method assumes that every R^5 methyl analog (where R^1–R^4 are varied) will be one-tenth as potent as the corresponding R^5 chloro analog. A requirement for this approach, then, is a series of compounds that have changes at more than one position. In addition, each type of substituent must occur more than once at each position in which it is found. The outcome is a table of the contribution to potency of each substituent at each position. If the free-energy relationships of the extrathermodynamic method are linear or position specific, then Free-Wilson calculations will be successful.

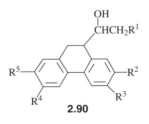

2.90

  The interaction model[212] is a mathematical model similar to that of the Free-Wilson additive model with an additional term ($e_X e_Y$) that is to account for possible interactions between substituents X and Y.

c. Enhancement Factor

One of the earliest QSAR observations resulted from a retrospective analysis of a large number of synthetic corticosteroids.[213] Examination of the biological properties of steroids prepared by the introduction of halogen, hydroxyl, alkyl, or double bond modifications revealed that each substituent affects the activity of the molecule in a quantitative sense and almost independently of other groups. The effect (whether positive or negative) of each substituent was assigned a numerical value termed the *enhancement factor*. Multiplication of the enhancement factor for each substituent by the biological activity of the unsubstituted compound gave the potency of the modified steroid.

d. Manual Stepwise Methods: Topliss Operational Schemes and Others

Because of the lack of easy access to computers by chemists in the early 1970s, Topliss[214] developed a nonmathematical, nonstatistical, and noncomputerized (hence, manual) guide to the use of the Hansch principles. This method is most useful when the synthesis of large numbers of compounds is difficult and when biological testing of compounds is readily available. It is an approach for the efficient optimization of the potency of a lead compound with the minimization of the number of compounds needed to be synthesized. The only prerequisite for the technique is that the lead compound must contain an unfused benzene ring. However, according to literature surveys at the time that this method was published, 40% of all reported compounds[215] contained an unfused benzene ring and 50% of drug-oriented patents[216] were concerned with substituted benzenes. This approach relies heavily on π and σ values and to a much lesser degree E_s values. The methodology will be outlined here; a more detailed discussion can be found in the Topliss papers.

Consider that your lead compound is benzenesulfonamide (**2.91**, R = H) and its potency has been measured in whatever screen is being used. Because many systems are $+\pi$ dependent, that is, the potency increases with increasing π values, then a good choice for your first analog would be one with a substituent having a $+\pi$ value. Because $\pi_{4\text{-Cl}} = 0.71$ and $\sigma_{4\text{-Cl}} = 0.23$ (remember, $\pi_H = \sigma_H = 0$), the 4-chloro analog (**2.91**, R = Cl) should be synthesized and tested. There are three possible outcomes of this effort, namely, the 4-chloro analog is more potent (M), equipotent (E), or less potent (L) than the parent compound. If it is more potent, then it can be attributed to a $+\pi$ effect, a $+\sigma$ effect, or to both. To determine which is important, one term could be held more or less constant and the other varied. For example, the 4-phenylthio analog ($\pi_{4\text{-PhS}} = 2.32$, $\sigma_{4\text{-PhS}} = 0.18$) would be a good test of the importance of lipophilicity, and the 4-trifluoromethyl analog ($\pi_{4\text{-CF}_3} = 0.88$, $\sigma_{4\text{-CF}_3} = 0.54$) would test the importance of electron withdrawal. If the 4-phenylthio analog is more potent than the 4-chloro analog, further increases in lipophilicity would be desirable. At this point a potency tree, termed a *Topliss decision tree*, could be constructed (Figure 2.16), and additional analogs could be made.

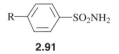

2.91

What if the 4-chloro analog was equipotent with the parent compound? This could result from a favorable $+\pi$ effect counterbalanced by an unfavorable $+\sigma$ effect or vice versa. If this is the case, then the 4-methyl analog ($\pi_{4\text{-Me}} = 0.56$, $\sigma_{4\text{-Me}} = -0.17$) should show enhanced potency. Enhancement of potency by the 4-methyl analog would suggest that the synthesis

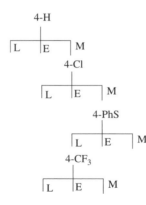

Figure 2.16 ▶ Topliss decision tree (M, more potent; E, equipotent; L, less potent)

of analogs with increasing π values and decreasing σ values would be propitious. If the 4-methyl analog is worse than the 4-chloro analog, perhaps the equipotency of the 4-chloro compound was the result of a favorable σ effect and an unfavorable π effect. The 4-nitro analog ($\pi_{4\text{-NO}_2} = -0.28$, $\sigma_{4\text{-NO}_2} = 0.78$) would, then, be a wise next choice.

If the 4-chloro analog was less potent than the lead, then there may be a steric problem at the 4 position or increased potency depends on $-\pi$ and $-\sigma$ values. The 3-chloro analog ($\pi_{3\text{-Cl}} = 0.71$, $\sigma_{3\text{-Cl}} = 0.37$) could be synthesized to determine if a steric effect is the problem. Note that the σ constant for the 3-Cl substituent is different from that for the 4-Cl one because these descriptors are constitutive. If there is no steric effect, then the 4-methoxy compound ($\pi_{4\text{-OMe}} = -0.04$, $\sigma_{4\text{-OMe}} = -0.27$) could be prepared to investigate the effect of adding a $-\pi$ and $-\sigma$ substituent. Increased potency of the 4-OMe substituent would suggest that other substituents with more negative π and/or σ constants be tried.

This analysis was based almost exclusively on π and σ values, and other factors such as steric effects have been neglected. Another way to increase both π and σ values would be by synthesizing the 3,4-dichloro analog ($\pi_{3,4\text{-Cl}_2} = 1.25$, $\sigma_{3,4\text{-Cl}_2} = 0.52$). Again, the 3,4-dichloro analog could be more potent, equipotent, or less potent than the 4-chloro compound. If it is more potent, then determination of whether $+\pi$ or $+\sigma$ is more important could be made by selection of appropriate substituents with higher π and/or σ values. If the 3,4-dichloro compound was less potent than the 4-chloro analog, it could be that the optimum values of π and σ were exceeded or that the 3-chloro group has an unfavorable steric effect. The latter hypothesis could be tested by the synthesis of the 4-trifluoromethyl analog ($\pi_{4\text{-CF}_3} = 0.88$, $\sigma_{4\text{-CF}_3} = 0.54$) which has no 3-substituent, but has a high σ and intermediate π value.

Topliss extended the operational scheme for side-chain problems when the group is adjacent to a carbonyl, amino, or amide functionality, i.e., $-COR$, $-NHR$, $-CONHR$, and $-NHCOR$, where R is the variable substituent. This approach is applicable to a variety of situations other than direct substitution on the aromatic nucleus. In this case, the parent molecule is the one where R = CH_3, and π, σ, and E_s parameters are used. Note that in the Topliss operational scheme, as in the other methods in this section, the procedure is stepwise; that is, the next compound is determined on the basis of the results obtained with the previous one.

Three other manual, stepwise methods are mentioned briefly: Craig plots,[217] the Fibonacci search method,[218] and sequential simplex strategy.[219] The Topliss decision tree approach evolved from the work of Craig, who pointed out the utility of a simple graphical plot of π versus σ (or any two parameters) to guide the choice of a substituent (Figure 2.17). Once the Hansch equation has been expressed for an initial set of compounds, the sign and

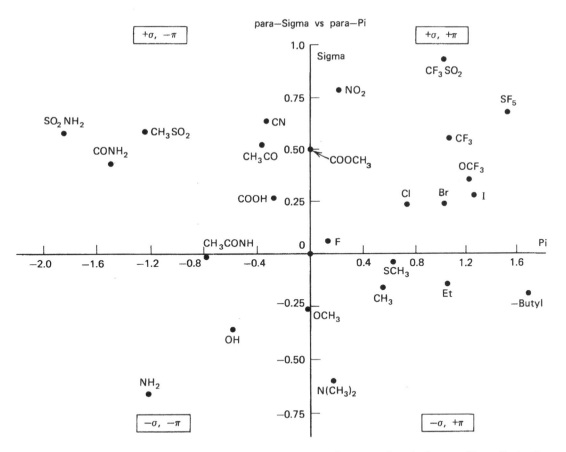

Figure 2.17 ► Craig plot of σ constants versus π values for aromatic substituents. [From Craig, P. N. (1980). In *Burger's Medicinal Chemistry*, (M. E. Wolff, ed.), 4th ed., Part I, p. 343. Wiley, New York. Copyright ©1980 John Wiley & Sons, Inc. This material is used by permission of John Wiley & Sons, Inc.]

magnitude of the π and σ regression coefficients determine the particular quadrant of the Craig plot that is to be used to direct further synthesis. Thus, if both the π and σ terms have positive coefficients, then substituents in the upper right-hand quadrant of the plot (Figure 2.17) should be selected for future analogs.

The Fibonacci search technique is a manual method to discover the optimum of some parabolic function, such as potency versus log P, in a minimum number of steps. Sequential simplex strategy is another stepwise technique suggested when potency depends on two physicochemical parameters such as π and σ.

e. Batch Selection Methods: Batchwise Topliss Operational Scheme, Cluster Analysis, and Others

The inherent problem with the Topliss operational scheme described above is its stepwise nature. Provided that pharmacological results can be obtained quickly, this is probably not much of a problem; however, sometimes biological evaluation is slow. Topliss[220] proposed an alternative scheme that uses batchwise analysis of small groups of compounds. Substituents were grouped by Topliss according to π, σ, π^2, and a variety of $x\pi$- and $y\sigma$-weighted combinations. The approach starts with the synthesis of five derivatives, the unsubstituted (4-H);

4-chloro; 3,4-dichloro; 4-methyl; and 4-methoxy compounds. After these five analogs have been screened, they are ranked in order of decreasing potency. The potency order determined for these analogs is, then, compared with the rankings in Table 2.9 to determine which parameter or combination of parameters is most dominant. If, for example, the potency order is 4-OCH$_3$ > 4-CH$_3$ > H > 4-Cl > 3,4-Cl$_2$, then $-\sigma$ is the dominant parameter. Once the parameter dependency is determined, Table 2.10 is consulted to discover what substituents should be investigated next. In the above example, 4-N(C$_2$H$_5$)$_2$, 4-N(CH$_3$)$_2$, 4-NH$_2$, 4-NHC$_4$H$_9$, 4-OH, 4-OCH(CH$_3$)$_2$, 3-CH$_3$, and 4-OCH$_3$ would be suitable choices. The major weakness of this approach is that it is difficult to extend the method to additional parameters unless computers are used.

A computer-based batch selection method, known as *cluster analysis*, was introduced by Hansch *et al.*[221] Substituents were grouped into clusters with similar properties according

TABLE 2.9 ▶ Potency Order for Various Parameter Dependencies [With permission from Topliss, J. G. (1977). Reprinted with permission from *J. Med. Chem.* 20, 463. Copyright © 1977 American Chemical Society.]

Parameters

Substituent	π	$2\pi - \pi^2$	σ	$-\sigma$	$\pi + \sigma$	$2\pi - \sigma$	$\pi - \sigma$	$\pi - 2\sigma$	$\pi - 3\sigma$	E_4^a
3,4-Cl$_2$	1	1-2	1	5	1	1	1-2	3-4	5	2-5
4-Cl	2	1-2	2	4	2	2-3	3	3-4	3-4	2-5
4-CH$_3$	3	3	4	2	3	2-3	1-2	1	1	2-5
4-OCH$_3$	4-5	4-5	5	1	5	4	4	2	2	2-5
H	4-5	4-5	3	3	4	5	5	5	3-4	1

^{a} Unfavorable steric effect from 4-substitution.

TABLE 2.10 ▶ New Substituent Selections [With permission from Topliss, J. G. (1977). Reprinted with permission from *J. Med. Chem.* 20, 463. Copyright 1977 American Chemical Society.]

Probable operative parameters	New substituent selection
$\pi, \pi + \sigma, \sigma$	3-CF$_3$, 4-Cl; 3-CF$_3$, 4-NO$_2$; 4-CF$_3$, 2,4-Cl$_2$; 4-c-C$_5$H$_9$; 4-c-C$_6$H$_{11}$
$\pi, 2\pi - \sigma, \pi - \sigma$	4-CH(CH$_3$)$_2$; 4-C(CH$_3$)$_3$; 3,4-(CH$_3$)$_2$; 4-O(CH$_2$)$_3$CH$_3$; 4-OCH$_2$Ph; 4-N(C$_2$H$_5$)$_2$
$\pi - 2\sigma, \pi - 3\sigma, -\sigma$	4-N(C$_2$H$_5$)$_2$; 4-N(CH$_3$)$_2$; 4-NH$_2$; 4-NHC$_4$H$_9$; 4-OH; 4-OCH(CH$_3$)$_2$; 3-CH$_3$, 4-OCH$_3$
$2\pi - \pi^2$	4-Br; 3-CF$_3$; 3,4-(CH$_3$)$_2$; 4-C$_2$H$_5$; 4-O(CH$_2$)$_2$CH$_3$; 3-CH$_3$, 4-Cl; 3-Cl; 3-CH$_3$; 3-OCH$_3$; 3-N(CH$_3$)$_2$; 3-CF$_3$; 3,5-Cl$_2$
Ortho effect	2-Cl; 2-CH$_3$; 2-OCH$_3$; 2-F
Other	4-F; 4-NHCOCH$_3$; 4-NHSO$_2$CH$_3$; 4-NO$_2$; 4-COCH$_3$; 4-SO$_2$CH$_3$; 4-CONH$_2$; 4-SO$_2$NH$_2$

TABLE 2.11 ▶ Typical Members of Clusters Based on $\sigma, \pi, F, R, MR,$ **and** MW **[With permission from Martin, Y. C. (1979). Reprinted from** *Drug Design*, **Vol. VIII, E. J. Ariëns, ed., "Advances in the Methodology of Quantitative Drug Design", pp 2–72, Copyright ©1979, with permission from Elsevier.]**

Cluster number[a]	Typical members
1	Me, H, 3,4-(OCH_2O), CH_2CH_2COOH, $CH=CH_2$, Et, CH_2OH
2	$CH=CHCOOH$
3a	CN, NO_2, CHO, COOH, COMe
3b	C + CH, CH_2Cl, Cl, NNN, SH, Sme, $CH=NOH$, CH_2CN, OCOMe, SCOMe, COOMe, SCN
4a	$CONH_2$, CONHMe, SO_2NH_2, SO_2Me, SOMe
4b	NHCHO, NHCOMe, $NHCONH_2$, $NHCSNH_2$, $NHSO_2Me$
5	F, OMe, NH_2, $NHNH_2$, OH, NHMe, NHEt, NMe_2
6	Br, OCF_3, CF_3, NCS, I, SF_5, SO_2F
7	CH_2Br, SeMe, $NHCO_2Et$, SO_2Ph, OSO_2Me
8	NHCOPh, $NHSO_2Ph$, OSO_2Ph, COPh, N=NPh, OCOPh, PO_2Ph
9	3,4-$(CH_2)_3$, 3,4-$(CH_2)_4$, Pr, i-Pr, 3,4-$(CH)_4$, NHBu, Ph, CH_2Ph, t-Bu, OPh
10	Ferrocenyl, adamantyl

[a] Clusters 3 and 4 contain many of the common substituents used in medicinal chemistry; hence, these clusters are further subdivided according to their cluster membership when 20 clusters have been made.

to their σ, π, π^2, E_s, F (field constant), R (resonance constant), MR (molar refractivity), and MW (molecular weight) values. Some of the clusters are shown in Table 2.11.[222] One member of each cluster would be selected for substitution into the lead compound, and the compounds would be synthesized and tested. If a substituent showed dominant potency, then other substituents from that cluster would be selected for further investigation. The important advantage of the batch selection methods is that the initial batch of analogs prepared is derived from the widest range of parameters possible so that the dominant physicochemical property can be revealed early in the lead modification process.

The initial promise of these computational methods has yet to be realized. They seem to be just additional examples of potentially exciting new approaches for which little success has been forthcoming. These early computational methods have largely been supplanted by what is known as 3D-QSAR and molecular modeling approaches.

G.4 Computer-Based Methods of QSAR Related to Receptor Binding: 3D-QSAR

Three-dimensional quantitative structure–activity relationships (3D-QSAR) permit correlations between a series of diverse molecular structures and their biological functions at a particular target. The general approach of 3D-QSAR is to select a group of molecules, each of which has been assayed for a particular activity; align the molecules according to some predetermined orientation rules; calculate a set of spatially dependent parameters for each molecule determined in the receptor space surrounding the aligned series; derive a function that relates each molecule's spatial parameters to their respective biological property; and establish self-consistency and predictability of the derived function. A variety of computer-based methods have been used to correlate molecular structure with receptor binding, and, therefore, activity. Some are mentioned here; many more are listed in the General References at the end of the chapter.

Crippen and coworkers[223,224] devised a linear free-energy model, termed the *distance geometry* approach, for calculating QSAR from receptor binding data. The distances between various atoms in the molecule, compiled into a table called the *distance matrix*, define the conformation of the molecule. Rotations about single bonds change the molecular conformation and, therefore, these distances; consequently, an upper and lower distance limit is set on each distance. Experimentally determined free energies of binding of a series of compounds to the receptor are used with the distance matrix of each molecule in a computerized method to deduce possible binding sites in terms of geometry and chemical character of the site, thereby defining a three-dimensional pharmacophore. Although this approach requires more computational effort and adjustable parameters than Hansch analysis, it is thought to give good results on more difficult data sets.

The distance geometry approach was extended by Sheridan *et al.*[225] to treat two or more molecules as a single ensemble. The ensemble approach to distance geometry can be used to find a common pharmacophore for a receptor with unknown structure from a small set of biologically active molecules. Once the pharmacophore has been, at least, partially identified, new molecular scaffolds can be revealed which contain that pharmacophore embedded in their structure by *three-dimensional database* (or *similarity*) *searching*.[226] In this method you start from a receptor ligand, enzyme substrate, or other molecule whose pharmacophoric groups are known for a particular target. Then a database of compounds (e.g., the company's library of compounds, the CMC database, or any database of compounds) is searched to determine which ones have a similar three-dimensional structure as the pharmacophore. The top virtual "hits" are visually inspected to determine which ones might be the best candidates, and then they are tested. This was the approach taken to identify inhibitors of human immunodeficiency virus type 1 integrase (HIV-1 IN) as potential anti-AIDS drugs.[227] HIV-1 IN mediates the integration of HIV-1 DNA into host chromosomal targets and is essential for effective viral replication. From a known inhibitor of HIV-1 IN, a pharmacophore hypothesis was proposed. Based on this hypothesis, a three-dimensional search of the National Cancer Institute (NCI) database of compounds was performed, which produced 267 structures that matched the pharmacophore; 60 of those were tested against HIV-1 IN, and 19 were found to be active. The relevance of the proposed pharmacophore was tested using a small three-dimensional validation database of known HIV-1 IN inhibitors, which had no overlap with the group of compounds found in the initial search. This new three-dimensional search supported the existence of the postulated pharmacophore and also suggested a possible second pharmacophore. Using the second pharmacophore in another three-dimensional search of the NCI database, 10 novel, structurally diverse HIV-1 IN inhibitors were found.

Hopfinger[228] has developed a set of computational procedures termed *molecular shape analysis* for the determination of the active conformations and, thereby, molecular shapes during receptor binding. Common pairwise overlap steric volumes calculated from low-energy conformations of molecules are used to obtain three-dimensional molecular shape descriptors that can be treated quantitatively and used with other physicochemical parameter descriptors.

Two other descriptors for substructure representation, the atom pair[229] and the topological torsion,[230] have been described by Venkataraghavan and coworkers. These descriptors characterize molecules in fundamental ways that are useful for the selection of potentially active compounds from hundreds of thousands of structures in a database. The atom pair method can select compounds from diverse structural classes that have atoms within the entire molecule similar to those of a particular active structure. The topological torsion descriptor is complementary to the atom pair descriptor, and focuses on a local environment of a molecule for comparison with active structures.

One of the most widely used computer-based 3D-QSAR methodologies, developed by Cramer and coworkers,[231] is termed *Comparative Molecular Field Analysis* (CoMFA).[232] In this method the molecule–receptor interaction is represented by the steric and electrostatic fields exerted by each molecule. A series of active compounds is identified, and three-dimensional structural models are constructed. These structures are superimposed on one another and placed within a regular three-dimensional grid. A probe atom, with its own energetic values, is placed at lattice points on the grid, where it is used to calculate the steric and electrostatic potentials between itself and each of the superimposed structures. At each lattice point one steric value, one electrostatic value, and one inhibition value are saved for each inhibitor in the series. The results are represented as a three-dimensional contour map in which contours of various colors represent locations on the structure where lower or higher steric or electrostatic interactions would increase binding. However, because simple steric and electrostatic fields are unlikely to represent a complete description of a drug–receptor interaction, alternative and modified forms have been proposed.[233] Because it is assumed that the molecules bind with similar orientations in the receptor, which may not necessarily be the case, correct alignments are almost impossible, particularly for compounds with a large number of rotatable bonds, which limits the applicability of CoMFA. Other approaches have been developed that do not depend on a common alignment of the molecules, such as *Comparative Molecular Moment Analysis* (CoMMA),[234] EVA,[235] and WHIM;[236] these approaches provide 3D descriptors that are independent of the orientation of the molecules in space, so they do not have to be aligned. However, it is not possible to give a 3D display of the resulting model. Goodford's program called *GRID* uses a grid force field that includes a very good description of hydrogen bonding.[237] Because the energetics, as well as the shape complementarity, of a drug–receptor complex are vital to its stability, this method simultaneously displays the energy contour surfaces and the macromolecular structure on the computer graphics system. This allows both the energy and shape to be considered together when considering the design of molecules that have an optimal fit to the receptor, and it determines probable interaction sites between various functional groups on the ligand and the enzyme surface. The program *HINT* (Hydrophobic INTeractions) maps potential hydrophobic and polar interactions between a molecule and a receptor.[238]

Another useful methodology is the *hypothetical active site lattice (HASL) technique*, which creates a QSAR model from a composite lattice generated from a series of regular orthogonal 3D grids established for each molecule.[239] These points are restricted to locations embedded in the van der Waals radii of a molecule and are kept in the analysis dependent on some feature of a proximal atom (for example, hydrophobicity). Each molecule's biological activity value is then averaged over all points on its respective lattice. A composite lattice is constructed from these partial activity values by averaging over all molecules that share common points.

Using an iterative optimization scheme, the partial activity values are gradually adjusted to yield a composite lattice best fitting the molecular series. Other pharmacophore-based design algorithms include Caveat,[240] Aladdin,[241] and Spacer Skeletons.[242]

The anti-Alzheimer's drug donepezil hydrochloride (**2.92**, Aricept) was discovered using a variety of 3D-QSAR methods.[243]

donepezil hydrochloride
2.92

2.2.H   Molecular Graphics-Based Drug Design

QSAR studies have relied heavily on the use of computers from the beginning for statistical calculations involving multiparameter equations. Researchers soon realized that drug design could be aided significantly if structures of receptors and drugs could be displayed on a computer terminal, and molecular processes could be observed. *Molecular graphics* is the visualization and manipulation of 3D representations of molecules on a graphics display device. The origins of molecular graphics have been traced by Hassall[244] to the project MAC (Multiple Access Computer),[245] which produced molecular graphics models of macromolecules for the first time. The potential to apply this technology to protein crystallography was quickly realized, and by the early 1970s electron density data from X-ray diffraction studies could be presented and manipulated in stick or space-filling multicolored representations on a computer terminal.[246] The number of X-ray crystal structures available in the protein data bank (PDB)[247] went from about 200 in 1990 to more than 20,000 by 2003.

Medicinal chemists saw the potential of this approach in drug design as well. These approaches are known as *structure-based drug design* (SBDD), *computer-assisted drug design* (CADD), or *computer-assisted molecular design* (CAMD). A variety of commercial software packages are available for structure-based drug design, for example, Sybyl (Tripos), Insight II (Molecular Simulations Inc.), and Gold (Cambridge Crystallographic Data Centre). It is now possible for a synthetic chemist to carry out his or her own molecular modeling without having to become a computer scientist.

Stick (Dreiding) and space-filling (CPK) molecular models have been used extensively by organic chemists for years for small molecules, but these handheld models have major disadvantages.[248] Space-filling models often obscure the structure of the molecule, and wire or plastic models can give false impressions of molecular flexibility and tend to change into unfavorable conformations at inopportune moments. Plastic models of proteins are much too cumbersome to work with. A three-dimensional computer graphics representation of a protein that can be manipulated in three dimensions allows the operator to visualize the interactions of small molecules with biologically important macromolecules. Superimposition of structures, which is cumbersome at best with manual models, can be performed easily by molecular graphics. Also, some systems have the capability to synthesize graphically new structures by the assemblage of appropriate molecular fragments from a fragment file.

Numerous molecular graphics systems are available,[249] but the typical system, which has not changed much over the years, utilized by every major pharmaceutical company in the United States, Western Europe, and Japan, consists of a mainframe or supermini computer linked to a high-resolution graphics terminal with local intelligence. The graphics terminal may be equipped with a variety of peripheral devices such as graphic tablets, light pens, function keys, and dials to effect the molecular display and three-dimensional manipulations. The mainframe or minicomputer executes all of the molecular calculations, such as calculations of bond lengths, bond angles, and quantum chemical or force field calculations.

A variety of approaches can be taken to utilize molecular modeling for drug design; *direct design* approaches are used when the structure of the target receptor is known, and *indirect design* approaches are used when the receptor structure is not known. The basic premise in the utilization of molecular graphics is that the better the complementary fit of the drug to the receptor, the more potent the drug will be. This is the lock-and-key hypothesis of Fischer[250] in which the receptor is the lock into which the key (i.e., the drug) fits. To apply this concept most effectively, the structure of the receptor (either X-ray crystal structure or NMR solution structure) should be known; then, different drug analogs can be docked into the receptor. *Docking* is a molecular graphics term for the computer-assisted movement of a terminal-displayed molecule into its receptor. It cannot be assumed that the lowest energy structure of the molecule binds to the receptor; the bioactive conformation can be a higher energy conformation of the molecule.[251]

The most effective use of molecular modeling is when a high-resolution crystal structure (or NMR solution structure) of a receptor with a ligand bound is available. Molecular graphics visualization of the electron density map of this complex may reveal empty pockets in the complex that could be filled by appropriate modification of a lead compound. An important example of structure-based drug design is the discovery of zanamivir (**2.93**, Relenza), an antiviral agent used against influenza A and B infections.[252] The hemagglutinin at the surface of the virus binds to sialic acid (**2.94**) residues on receptors at the host cell surface. The virus enters the cell and replicates in the nucleus. The progeny virus particles escape the cell and stick to the sialic acid residues on the cell surface as well as to each other. *Neuraminidase* (also known as *sialidase*) is a key viral surface enzyme that catalyzes the cleavage of terminal sialic acid residues from the cell surface, which releases the virus particles to spread into the respiratory tract and infect new cells. The important feature of this enzyme that made it an attractive target for drug design is that its active site is lined with amino acids that are invariant in neuraminidases of all known strains of influenza A and B. Therefore, inhibition of this enzyme should be effective against all strains of influenza A and B. Random screening did not produce any potent inhibitors of the enzyme, although a nonselective neuraminidase inhibitor (**2.95**, R = OH) was identified. The breakthrough came when the crystal structures of the influenza A neuraminidase[253] with inhibitors bound[254] were obtained. The active site of the enzyme with **2.95** (R = OH) bound was probed computationally using Goodford's GRID program (see Section 2.2.G.4). Predictions by GRID of energetically favorable substitutions suggested replacement of the 4-hydroxyl group of **2.95** (R = OH) by an amino group (**2.95**, R = NH_2), which when protonated would form a favorable electrostatic interaction with Glu-119 (Figure 2.18a). It was apparent from the crystal structure that extension of the 4-ammonium group with a 4-guanidinium group (**2.93**) would produce an even tighter affinity because of the increase in basicity of the guanidinium group and also because it could interact with both Glu-119 and Glu-227 (Figure 2.18b).

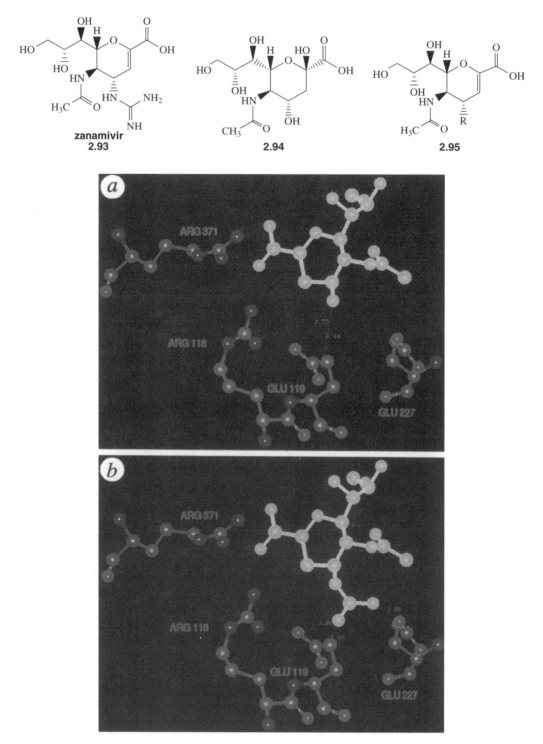

zanamivir
2.93

2.94

2.95

Figure 2.18 ▶ Crystal structure of neuraminidase active site with inhibitors bound. (a) Interaction of the protonated amino group of 2.95 ($R = NH_3^+$) with Glu-119. (b) Interaction of the protonated guanidinium group of **2.93** with Glu-119 and Glu-227 [Reprinted with permission from *Nature* 363, 418. Copyright ©1993 MacMillan Magazines Ltd.]. Reproduced in color between pages 172 and 173

Typically, the ideal compound is not realized so quickly. Rather, an idea for lead modification manifests from the crystal structure of the lead bound to the receptor. This modified lead is synthesized and tested; maybe only a minor improvement in potency is produced (or maybe lower potency). If there is some improvement, a new crystal structure is obtained, additional molecular modeling is carried out for further refinement of ideas, and new compounds are synthesized and tested. This process is reiterated with further rounds of design, synthesis, testing, and crystal structure until higher potency analogs are obtained. A beautiful example of how the iterative combination of molecular modeling, crystallography, and combinatorial and traditional medicinal chemistry synthesis was used to modify a lead neuraminidase inhibitor and enhance its potency 72,500-fold was described by the group at Abbott Laboratories.[255]

Sometimes even a crystal structure with the ligand bound is not sufficient. A high-resolution crystal structure of thymidylate synthase with a ligand bound did not properly account for a ligand-induced enzyme conformational change during structure-based drug design.[256] As a result the structure imparted an improper bias into the design of novel ligands.

Earlier in the chapter, the SAR of paclitaxel was described (**2.36**, Section 2.2.C, p. 23). By overlaying the molecular graphics depiction of the crystal structure of paclitaxel with those of four other natural products also found to promote stabilization of microtubules in competition with paclitaxel (Figure 2.19, Taxol), a common pharmacophore was proposed (Figure 2.20).[257] This gives a new perspective to lead modification, and permits the construction of new synthetic analogs having hybrid structures of each of the four unrelated scaffolds. Based on this pharmacophore model, **2.96** was synthesized and was shown to stabilize microtubules as well. Other 3D computer models of paclitaxel binding to microtubules have been promoted as well.[258] Without the molecular graphics capabilities, it would be very difficult to make this sort of comparison and design a new hybrid scaffold.

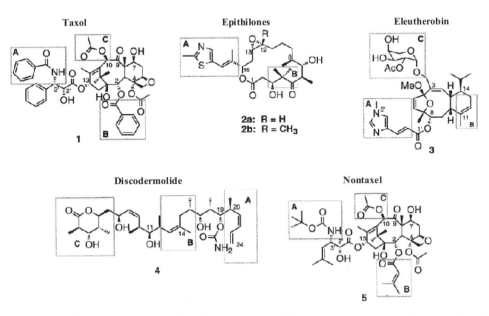

Figure 2.19 ► Five natural products found to promote stabilization of microtubules. The boxed sections were used to identify a common pharmacophore. [With permission from I. Ojima (1999). Reprinted with permission from *PNAS* 96, 4256. Copyright ©1999 National Academy of Science, U.S.A.]

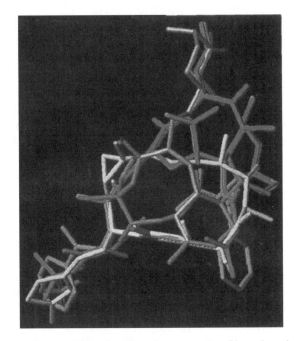

Figure 2.20 ▶ Common pharmacophore based on the composite of boxed sections in Figure 2.19. [With permission from I. Ojima (1999). Reprinted with permission from *PNAS* 96, 4256. Copyright ©1999 National Academy of Sciences, U.S.A.]. Reproduced in color between pages 172 and 173

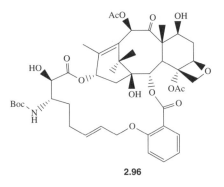

2.96

Kuntz *et al.*[259] reported on an algorithm called *DOCK* that was designed to fit small molecules into their macromolecular receptors for lead discovery.[260] This shape-matching method, which was originally restricted to rigid ligands (receptor-bound molecules) and receptors, was modified[261] for flexible ligands where a ligand is approximated as a small set of rigid fragments. Ideally, a high-resolution structure (X-ray crystal structure or NMR spectral structure) of the receptor *with a ligand bound* should be available. The ligand is removed from the binding site in the graphic display, then DOCK fills the binding site with sets of overlapping spheres, where a set of sphere centers serves as the negative image of the binding site. When a crystal structure of a receptor is available, but without a ligand bound, DOCK characterizes the entire surface of the receptor with regard to grooves that could potentially form target binding sites, which are filled with the overlapping spheres. Next DOCK matches X-ray or computer-derived structures of putative ligands to the image of the receptor on the basis of a comparison of internal distances. Then the program searches 3D databases of small molecules and ranks each candidate on the basis of the best orientations that can be found for a particular molecular conformation.[262] Various databases are available to search, such as the Cambridge Structural Database (CSD), a compendium of >200,000 small molecules whose crystal structures are known and the Fine Chemicals Directory (FCD) distributed by Molecular Design Limited, but the best ones to use are those containing commercially available compounds, such as the Available Chemicals Directory (ACD), so that any virtual hits can be purchased and assayed to determine quickly the effectiveness of the search. The drawbacks of this approach are the assumptions that binding is determined primarily by shape complementarity and that only small changes in the shape of the receptor occur on ligand binding. An important advantage, though, is that this method is not limited to docking of known ligands. A library of molecular shapes can be scanned to determine which shapes best fit a particular receptor binding site. In fact, DOCK was used to identify fullerenes as potential inhibitors of HIV-1 protease.[263] The high-resolution NMR structure of the aminoglycoside antibiotic paromomycin (**2.97**, Humatin) bound to the A site of the bacterial ribosomal RNA was used to perform a DOCK search of the CSD and the National Cancer Institute 3D database (a total of 273,000 compounds).[264] The compounds that emerged from this search formed the basis for the design of seven composite structures with additional features added to suppress resistance. As a result, several of these compounds were found to have enhanced activity *in vitro* and *in vivo* against a variety of pathogenic bacteria resistant to aminoglycosides.

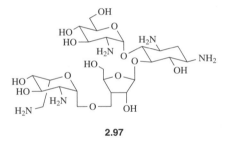

2.97

The program *LUDI* uses statistical analyses of nonbonded contacts in crystal packings of organic molecules to establish a set of rules that define the possible nonbonded contacts between proteins and ligands.[265] Using these rules it also can search databases to find structures that fit a particular binding site in a protein based not on shape, as in DOCK, but on physicochemical properties, such as hydrogen bonding, ionic interactions, and hydrophobic interactions.

Another program for *de novo* molecular design, GrowMol[266] (called AlleGrow[267] in the latest version), approaches the problem of receptor binding from a different direction. Instead of docking known molecules into the binding site, it generates molecules with steric and chemical complementarity to the three-dimensional structure of the receptor binding site by evaluating each new atom according to its chemical complementarity to the nearby receptor atoms. The program also connects a newly grown atom to a previously grown atom in the growing structure to make ring systems. The principal "liability" to this method is that it generates too many diverse structures, and it is necessary to evaluate each one visually and determine which ones are best to try first. Often the decision comes from a synthetic perspective or a knowledge of potential oral bioavailability. Having too many choices, of course, also can be an asset, and a variety of other criteria could be set up to search the database of newly generated compounds for specific beneficial properties.

In the modified method for docking flexible ligands into a receptor described above, an X-ray structure of the receptor is not necessary to characterize the shape of the receptor binding site. Rather, the receptor binding site can be deduced from the shapes of active ligands. This technique, which is useful for identification of the pharmacophore geometry, is called *receptor mapping*.[268] It too is founded on the premise that receptor topography is complementary to that of drugs, but in this case the structure of the lock is deduced from the shape of the keys that fit it. A variety of receptor mapping techniques have been described. An approach termed *steric mapping*[269] uses molecular graphics to combine the volumes of compounds known to bind to the desired receptor. This composite volume generates an enzyme-excluded volume map, which defines that region of the binding site available for binding by drug analogs and, therefore, not occupied by the receptor itself. The same procedure is, then, carried out for similar molecules that are inactive. The composite volume is inspected for regions of volume overlap common to all of the inactive analogs. These are the enzyme-essential regions, sites required by the receptor itself and unavailable for occupancy by ligands. Any other molecule that overlaps with these regions should be inactive. Drug design, then, would involve the synthesis of compounds with the appropriate pharmacophore that filled the enzyme-excluded regions and that avoided the enzyme-essential regions. Another approach that does not require the structure of the target receptor, known as *homology modeling*, deduces the topography of the unknown receptor site from that of a known related receptor structure.[270]

A major improvement in the use of molecular modeling came when high-throughput crystallography was coupled with combinatorial chemistry approaches.[271] Because structure-based drug design usually involves targets whose structures are already known, but which have different ligands bound, only the part of the structure where the ligand binds needs to be resolved. Software such as AutoSolve (Astex Technology) analyzes and interprets electron density data automatically without the need for an expert crystallographer, so hundreds of receptor complex crystals can be analyzed in just a few days.[272] With solid-phase synthetic methodology to make many analogs rapidly, the two processes produce large numbers of crystal structures with various ligands bound very rapidly.

The initial expectation for structure-based drug design—that potent receptor binders would be designed rapidly leading to the discovery of many new drugs—has not yet become a reality. Several problems with this approach may contribute to its less than optimal effectiveness. Table 2.12 lists various advantages to the use of molecular modeling approaches and its many limitations. Although the ease of visualization is appealing, the main problems are (1) that the structure of the molecular model may be completely different from the actual structure in the living organism; (2) even if the structure were correct, the resolution of the structure is insufficient to make an accurate assessment of ligand binding; and (3) the bioactive conformation of the ligands is not known, so the appropriate small molecules may not be used in docking experiments. Another important reason why there has not been a large increase in the number of new drugs being developed by molecular modeling techniques derives from the fact that pharmacokinetics are ignored by this method. Prior to the drug candidate interacting with a receptor, it must be properly absorbed, it must reach the receptor without metabolic or chemical degradation (unless it is a prodrug; see Chapter 8), excretion must be at an appropriate rate, and the drug candidate and metabolites must not be toxic or lead to undesirable side effects.

Because of the uncertainty involved with this method, the process of molecular modeling, synthesis, testing, and molecular modeling again needs to undergo many iterations. Structure-based drug design has to be taken as yet another tool available to the medicinal chemist; it is not the answer to drug discovery, but it can be an important part of the process.

TABLE 2.12 ▶ Advantages and Limitations to the Use of Molecular Modeling in Lead Modification

Advantages

▶ Proteins can be visualized in 3D and every amino acid can be located.

▶ The structure can be manipulated so that it can be observed from any direction in 3D.

▶ Particular regions, e.g., the binding site, can be enlarged for better viewing.

▶ The physicochemical properties, e.g., hydrophobic, polar, positive or negative charge, etc., of each part of the receptor can be viewed.

▶ Distances between groups can be determined.

▶ Small molecules can be docked into various regions to determine their fit and interactions.

▶ Residues that are most suitable to mutate for mechanism studies can be determined.

Limitations

▶ The coordinates from an X-ray crystal structure or NMR solution structure are required.

▶ Crystals are obtained by crystallization of proteins under nonphysiological conditions, such as at low or high pH, well below 37°C, and in the presence of additives, such as buffers or detergents. Are the proteins really in the same conformation as in the living cell?

▶ Crystal structures represent the thermodynamically most stable conformation of the protein under these nonphysiological conditions. Therefore, the crystal structure may depict the protein in a conformation very different from that in a living cell.

▶ Often crystal structures with ligands bound are obtained by soaking the ligand into the preformed crystal. If binding of the ligand in solution results in a conformational change, it is highly unlikely that it will occur in the crystalline state because the crystal packing forces will favor the preexisting conformation.

▶ The protein structure is considered to be rigid, but small conformational changes of side chains can induce large changes in the size, shape, and interaction pattern of binding pockets. Typically, when a small molecule binds to a protein, there is some movement of side chains.

▶ Resolutions of crystal structures are generally in the range of 2–2.5 Å; some at 1.5–2.0 Å; rarely more resolved (although <1.0 Å has been reported).[a] Therefore, there is *much* uncertainty as to the exact position of each atom. A rule of thumb is that the positional error of atoms is about one-sixth of the resolution,[b] so a structure at 2.4 Å resolution has an uncertainty of every atom of 0.4 Å.

▶ Small molecules in the ground state are generally energy minimized to give the lowest energy conformers prior to docking them into the structure, but a ligand does not have to bind in the lowest energy conformation, and it can be quite different from the ground state conformation. Also, solvent effects generally are not taken into account.

▶ For highly flexible molecules with several torsional angles, there may be many different geometries having the same conformational energy, but significantly different shapes.

▶ You tend to believe what you see in your molecular model and think it is accurate! This leads to many wrong assumptions.

---

[a] For example, Betzel, C.; Gourinath, S.; Kumar, P.; Kaur, P.; Perbrandt, M.; Eschenburg, S.; Singh, T. P. *Biochemistry* **2001**, *40*, 3080.
[b] Böhm, H.-J.; Klebe, G. *Angew. Chem. Int. Ed. Engl.* **1996**, *35*, 2588.

2.2.I Epilogue

On the basis of what was discussed in this chapter, it appears that even if you uncover a lead, it may be a fairly slow and random process to optimize its potency. The cost to get a drug on the market has increased from $4 million in 1962 to $350 million in 1996, $500 million in 2000,[273a] and about $800 million in 2003.[273b] Between 1960 and 1980, the time for development of a compound from synthesis to the market almost quadrupled, but the time has remained fairly constant since 1980 at about 12–15 years of research. The main cause for the increase in the length of time to bring a drug on the market occurred in 1962 as a result of the devastating effects of the drug thalidomide, a hypnotic drug shown to cause severe fetal limb abnormalities (phocomelia) when taken in the first trimester of pregnancy (see Chapter 3, Section 3.2.E.2). This tragedy led to the passing of the Harris-Kefauver Amendments to the Food, Drug, and Cosmetic Act in 1962, which required sufficient pharmacological and toxicological research in animals before a drug could be tested in humans; the data of the animal studies had to be submitted to the FDA in an application for approval of an investigational new drug (IND) before human testing could begin. After 1–5 years (average 2.6 years) of animal testing, three phases of clinical (human) trials were adopted (lasting from 4 to 10 years; see the first paragraph of this chapter for a description of the phases of clinical trials) before a new drug application (NDA) could be submitted for commercial approval of a new drug.[274]

It has been estimated that, in 1950, 7000 compounds had to be isolated and tested for each one that made it to the market; by 1979 that number rose to 10,000 compounds, and now it is greater than 20,000 compounds. There are only about 6000 known drugs in the Comprehensive Medicinal Chemistry (CMC) database of the estimated 10^{60} possible compounds that could be drug-like, and these 6000 drugs interact with only about 120 targets[275a] (40% receptors, 40% enzymes, and the rest ion channels and other) or <1% of the human *proteome* (the expressed proteins). Prior to the sequencing of the human genome, it was estimated that the number of potential drug targets may be between 5000 and 10,000,[275b] but it is now thought that it may only be 600–1500.[275a] Therefore, *genomics* (identifying and analyzing new gene targets from a genome) and *proteomics* (identifying and analyzing proteins expressed by the genes in the genome) have become very important aspects to drug discovery.[276] Once a new target from the proteome is identified, *bioinformatics*, in which databases of known proteins are scanned to find known proteins with similar structures to that of the new target, is employed. When the similarities are known, inhibitors of the known protein can be tested with the new target protein. In addition to these biological methodologies, which appear to be increasing the rate of lead discovery, other rational approaches to lead discovery and lead optimization, based on chemical and biochemical principles, must be used. Between 1995 and 2000 it was estimated at Bristol-Myers Squibb that there had been a threefold to fourfold increase in new drug candidates going into development, a 50% lower chemistry staff requirement per drug candidate, and a 40% reduction in lead optimization time, believed to be the result of combinatorial approaches.[277] Other companies have not enjoyed the predicted success of combinatorial chemistry, and some have even dropped their combinatorial chemistry groups and returned to only traditional medicinal chemistry efforts. The importance of combinatorial chemistry to drug discovery will not be known for at least 10 years when we find out if there is a direct link of this approach to new drugs entering the market or if it is just another false hope. However, in 2002, for the first time in the United States, the market share of nongeneric drugs was surpassed by that of generic drugs. Also in that year the number of new chemical entity approvals by the FDA, normally in the twenties or thirties per year, hit a 20-year low of only 16, although the R&D spending by the pharmaceutical industry had tripled in the previous decade.[278] Maybe Thomas Edison said it best: "I have not failed. I've just found 10,000 ways that won't work."

2.3 General References

Combinatorial Chemistry

Books and Reviews

Affleck, R. L. Solutions for library encoding to create collections of discrete compounds. *Curr. Opin. Chem. Biol.* **2001**, *5*, 257–263.

Barnes, C.; Balasubramanian, S. Recent developments in the encoding and deconvolution of combinatorial libraries. *Curr. Opin. Chem. Biol.* **2000**, *4*, 346–350.

Czarnik, A. W. Encoding methods for combinatorial chemistry. *Curr. Opin. Chem. Biol.* **1997**, *1*, 60–66.

Czarnik, A. W. (Ed.) *Solid-Phase Organic Synthesis*, Wiley, New York, Vol. 1, 2001 (annual series).

Fenniri, H. *Combinatorial Chemistry. A Practical Approach*, Oxford University Press, Oxford, UK, 2000.

Gordon, E. M.; Kerwin, J. F. (Eds.) *Combinatorial Chemistry and Molecular Diversity in Drug Discovery*, Wiley, New York, 1998.

Jung, G. (Ed.) *Combinatorial Chemistry*, Wiley-VCH, Weinheim, 1999.

Jung, G. (Ed.) *Combinatorial Peptide and Nonpeptide Libraries: A Handbook*, Wiley-VCH, Weinheim, 1996.

Wilson, S. R.; Czarnik, A. W. (Eds.) *Combinatorial Chemistry: Synthesis and Application*, Wiley, New York, 1997.

Yan, B., Czarnik, A. W. (Eds.) *Optimization of Solid-Phase Combinatorial Synthesis*, Marcel Dekker, New York, 2002.

Journals

Combinatorial Chemistry & High Throughput Screening
Journal of Combinatorial Chemistry
Journal of Medicinal Chemistry
Molecular Diversity

Webpages

http://www.5z.com
http://www.warr.com/ombichem.html (yes, "ombichem")
http://www.combi-web.com

Peptidomimetics

Bursavich, M. G.; Rich, D. H. *J. Med. Chem.* **2002**, *45*, 541.
Giannis, A.; Kolter, T. *Angew. Chem. Int. Ed. Engl.* **1993**, *32*, 1244.
Ripka, A. S.; Rich, D. H. *Current Opin. Chem. Biol.* **1998**, *2*, 441.

Pharmacokinetics

Hardman, J. G.; Limbird, L. E.; Gilman, A. G. *Goodman and Gilman's The Pharmacological Basis of Therapeutics*, 10th ed., McGraw-Hill, New York, 2001.

Smith, D. A.; van de Waterbeemd, H.; Walker, D. K.; Mannhold, R.; Kubinyi, H.; Timmerman, H. *Pharmacokinetics and Metabolism in Drug Design*, Wiley, New York, 2000.

QSAR

Books

Devillers, J. (Ed.) *Comparative QSAR*, Taylor and Francis, Washington, DC, 1998.
Hansch, C.; Leo, A. *Exploring QSAR*, Vol. 1, *Fundamentals and Applications in Chemistry and Biology*, ACS Publ., Washington, DC, 1995.
Kubinyi, H. *QSAR: Hansch Analysis and Related Approaches*, VCH, Weinheim, 1993.
Kubinyi, H. (Ed.) *3D-QSAR in Drug Design. Theory, Methods and Applications*, ESCOM, Leiden, Netherlands, 1993.
Leo, A.; Hansch, C.; Hoekman, D. *Exploring QSAR*, Vol. 2, *Hydrophobic, Electronic, and Steric Constants*, ACS Publ., Washington, DC, 1995.
Lipkowitz, K. B., Boyd, D. B. (Eds.) *Reviews in Computational Chemistry*, Wiley-VCH, New York (entire series of volumes).

Journals

Journal of Chemical Information and Computer Science
Journal of Medicinal Chemistry

Molecular Graphics

Books and Reviews

Antel, J. *Curr. Opin. Drug Discovery Dev.* **1999**, *2*, 224.
Bohacek, R. S.; McMartin, C.; Guida, W. C. *Med. Res. Rev.* **1996**, *16*, 3.
Böhm, H.-J. *Prog. Biophys. Mol. Biol.* **1996**, *66*, 197.
Böhm, H.-J.; Stahl, M. *Curr. Opin. Chem. Biol.* **2000**, *4*, 283.
Gubernator, K.; Böhm, H.-J. (Eds.) *Structure-Based Ligand Design*, Wiley-VCH, Weinheim, 1998.
Kirkpatrick, D. L.; Watson, S.; Ulhaq, S. *Combinatorial Chemistry. High Throughput Screening* **1999**, *2*, 211.
Kuntz, I. D.; Meng, E. C.; Shoichet, B. K. *Acc. Chem. Res.* **1994**, *27*, 117–123.
Ooms, F. *Curr. Med. Chem.* **2000**, *7*, 141.

Journals

Annual Reports in Combinatorial Chemistry and Molecular Design
Journal of Computer-Aided Molecular Design
Journal of Medicinal Chemistry
Journal of Molecular Graphics

Software

Sybyl™ (Tripos, Inc.)
Insight II™ (MDL, Inc.)

Molecular Conceptor^{TM} Courseware (Synergix, Ltd.)

CaChe^{TM} Software (Fujitsu, Inc.)

Webpages

http://www.netsci.org/Resources/Software/Modeling/CADD

Computer-Based Drug Design Methodologies[a]

Active Site Analysis

MCSS

Caflish, A.; Miranker, A.; Karplus, M. Multiple copy simultaneous search and construction of ligands in binding sites: application to inhibitors of HIV-1 aspartic proteinase. *J. Med. Chem.* **1993**, *36*, 2142–2167.

Stultz, C. M.; Karplus, M. MCSS functionality maps for a flexible protein. *Proteins* **1999**, *37*, 512–529.

GRID

Boobbyer, D. N. A.; Goodford, P. J.; McWhinnie, P. M.; Wade, R. C. New hydrogen-bond potentials for use in determining energetically favorable binding sites on molecules of known structure. *J. Med. Chem.* **1989**, *32*, 1083–1094.

Goodford, P. J. A computational procedure for determining energetically favorable binding sites on biologically important macromolecules. *J. Med. Chem.* **1985**, *28*, 849–857.

Wade, R. C.; Goodford, P. J. Further development of hydrogen bond functions for use in determining energetically favorable binding sites on molecules of known structure. 1. Ligand probe groups with the ability to form two hydrogen bonds. *J. Med. Chem.* **1993**, *36*, 140–147.

Wade, R. C.; Goodford, P. J. Further development of hydrogen bond functions for use in determining energetically favorable binding sites on molecules of known structure. 2. Ligand probe groups with the ability to form more than two hydrogen bonds. *J. Med. Chem.* **1993**, *36*, 148–156.

Applications of MCSS and GRID

Bitetti-Putzer, R.; Joseph-McCarthy, D.; Hogle, J. M.; Karplus, M. Functional group placement in protein binding sites: a comparison of GRID and MCSS. *J. Comput.-Aided Mol. Des.* **2001**, *15*, 935–960.

Powers, R. A.; Shoichet, B. K. Structure-based approach for binding site identification on AmpC β-lactamase. *J. Med. Chem.* **2002**, *45*, 3222–3234.

MCSS Applied to Structure-Based Ligand Design

Joseph-McCarthy, D.; Tsang, S. K.; Filman, D. J.; Hogle, J. M.; Karplus, M. Use of MCSS to design small targeted libraries: application to picornavirus ligands. *J. Am. Chem. Soc.* **2001**, *123*, 12758–12769.

Stultz, C. M.; Karplus, M. Dynamic ligand design and combinatorial optimization: designing inhibitors to endothiapepsin. *Proteins* **2000**, *40*, 258–289.

---

[a] Many thanks to Dr. Haitao Ji for compiling these methodologies.

GRID Applied to 3D-QSAR

Bohm, M.; Klebe, G. Development of new hydrogen-bond descriptors and their application to comparative molecular field analysis. *J. Med. Chem.* **2002**, *45*, 1585–1597.

Davis, A. M.; Gensmantel, N. P.; Johansson, E.; Marriott, D. P. The use of the GRID program in the 3D-QSAR analysis of a series of calcium-channel agonists. *J. Med. Chem.* **1994**, *37*, 963–972.

Cruciani, G.; Watson, K. A. Comparative molecular field analysis using GRID force-field and GOLPE variable selection methods in a study of inhibitors of glycogen phosphorylase b. *J. Med. Chem.* **1994**, *37*, 2589–2601.

Pastor, M.; Cruciani, G.; Mclay, I.; Pickett, S.; Clemente, S. Grid-INdependent Descriptors (GRIND): a novel class of alignment-independent 3-D molecular descriptors. *J. Med. Chem.* **2000**, *43*, 3233–3243.

GRID Applied to Selectivity Analysis

Kastenholz, M. A.; Pastor, M.; Gruciani, G.; Haaksma, E. E.; Fox, T. GRID/CPCA: A new computational tool to design selective ligand. *J. Med. Chem.* **2000**, *43*, 3033–3044.

X-SITE

Laskowski, R. A.; Thornton, J. M.; Humblet, C.; Singh, J. X-SITE: use of empirically derived atomic packing preferences to identify favourable interaction regions in the binding sites of proteins. *J. Mol. Biol.* **1996**, *259*, 175–201.

Molecular Docking

DOCK, GOLD, and FlexX are used as virtual screening tools when the 3-D structure of the binding site of the receptor is known.

Fast Shape Matching (DOCK)

Kuntz, I. D.; Blaney, J. M.; Oatley, S. J.; Langridge, R.; Ferrin, T. E. A geometric approach to macromolecule-ligand interactions. *J. Mol. Biol.* **1982**, *161*, 269–288.

Incremental Construction (FlexX, Hammerhead, Surflex)

Jain, A. N. Surflex: fully automatic flexible molecular docking using a molecular similarity-based search engine. *J. Med. Chem.* **2003**, 46, 499–511.

Rarey, M.; Kramer, B.; Lengauer, T.; Klebe, G. A fast flexible docking method using as incremental construction algorithm. *J. Mol. Biol.* **1996**, *261*, 470–489.

Welch, W.; Ruppert, J.; Jain, A. N. Hammerhead: fast fully automated docking of flexible ligands to protein binding sites. *Chem. Biol.* **1996**, *3*, 449–462.

Tabu Search (Pro_Leads)

Baxter, C. A.; Murray, C. W.; Clark, D. E.; Westhead, D. R.; Eldridge, M. D. Flexible docking using Tabu search and an empirical estimate of binding affinity. *Proteins* **1998**, *33*, 367–382.

Genetic Algorithm (GOLD, AutoDock 3.0)

Jones, G.; Wilett, P.; Glen, R. C.; Leach, A. R.; Taylor, R. Development and validation of a genetic algorithm for flexible docking. *J. Mol. Biol.* **1997**, *267*, 727–748.

Morris, G. M.; Goodsell, D. S.; Halliday, R.; Huey, R.; Hart, W. E.; Belew, R. K.; Olson, A. J. Automated docking using a Lamarckian genetic algorithm and an empirical binding free energy function. *J. Comput. Chem.* **1998**, *19*, 1639–1662.

Monte Carlo Simulations (ICM, MCDOCK, QXP)

Abagyan, R. A.; Totrov, M. M. Biased probability Monte Carlo conformational searches and electrostatic calculations for peptides and proteins. *J. Mol. Biol.* **1994**, *235*, 983–1002.

Liu, M.; Wang, S. MCDOCK: a Monte Carlo simulation approach to the molecular docking problem. *J. Comput-Aided Mol. Des.* **1999**, *13*, 435–451.

McMartin, C.; Bohacek, R. S. QXP: powerful, rapid computer algorithms for structure-based drug design. *J. Comput-Aided Mol. Des.* **1997**, *11*, 333–344.

Simulated Annealing (AutoDock 2.4)

Goodsell, D. S.; Olson, A. J. Automated docking of substrates to proteins by simulated annealing. *Proteins* **1990**, 8, 195–202.

Postdocking Treatments

Free-Energy Perturbation (FEP)

FEP is an accurate method, but it is very time consuming.

Kollman, P. A. Advances and continuing challenges in achieving realistic and predictive simulations of the properties of organic and biological molecules. *Acc. Chem. Res.* **1996**, *29*, 461–469.

Kollman, P. A. Free-energy calculations—applications to chemical and biochemical phenomena. *Chem. Rev.* **1993**, *93*, 2395–2417.

Toba, S.; Damodaran, K. V.; Merz, Jr., K. M. Binding preferences of hydroxamate inhibitors of the matrix metalloproteinase human fibroblast collagenase. *J. Med. Chem.* **1999**, *42*, 1225–1234.

OWFEG

OWFEG can be used to simplify the FEP method.

Pearlman, D. A. Free energy grids: a practical qualitative application of free energy perturbation to ligand design using the OWFEG method. *J. Med. Chem.* **1999**, *42*, 4313–4324.

Thermodynamic Integration (TI)

Guimaraes, C. R. W.; Bicca de Alencastro, R. Thermodynamic analysis of thrombin inhibition by benzamidine and *p*-methylbenzamidine via free-energy perturbations: inspection of intraperturbed-group contributions using the finite difference thermodynamic integration (FDTI) algorithm *J. Phys. Chem. B.* **2002**, *106*, 466–476.

McDonald, J. J.; Brooks, C. L. Theoretical approach to drug design. 2. Relative thermodynamics of inhibitor binding by chicken dihydrofolate reductase to ethyl derivatives of trimethoprim substituted at 3-, 4-, and 5-positions. *J. Am. Chem. Soc.* **1991**, *113*, 2295–2301.

Reddy, M. R.; Viswanadhan, V. N.; Weinstein, J. N. Relative differences in the binding free energies of human immunodeficiency virus 1 protease inhibitors: a thermodynamic cycle-perturbation approach. *Proc. Natl. Acad. Sci. USA* **1991**, *88*, 10287–10291.

Scoring Methods for Virtual Screening
Force Field Scoring Functions

DOCK
Kuntz, I. D.; Blaney, J. M.; Oatley, S. J.; Langridge, R.; Ferrin, T. E. A geometric approach to macromolecule-ligand interactions. *J. Mol. Biol.* **1982**, *161*, 269–288.

GOLD
Jones, G.; Wilett, P.; Glen, R. C.; Leach, A. R.; Taylor, R. Development and validation of a genetic algorithm for flexible docking. *J. Mol. Biol.* **1997**, *267*, 727–748.

VALIDATE
Head, R. D.; Smythe, M. L.; Opera, T. I.; Waller, C. L.; Green, S. M.; Marshall, G. R. VALIDATE: a new method for the receptor-based prediction of binding affinities of novel ligands. *J. Am. Chem. Soc.* **1996**, *118*, 3959–3969.

Empirical Free-Energy Scoring Functions

LUDI
Bohm, H. J. The development of a simple empirical scoring function to estimate the binding constant for a protein-ligand complex of known three-dimensional structure. *J. Comput-Aided Mol. Des.* **1994**, *8*, 243–256.

Chemscore
Eldrige, M.; Murray, C. W.; Auton, T. A.; Paolini, G. V.; Lee, R. P. Empirical scoring functions: I. the development of a fast empirical scoring function to estimate the binding affinity of ligands in receptor complexes. *J. Comput.-Aided Mol. Des.* **1997**, *11*, 425–445.

FlexX
Rarey, M.; Kramer, B.; Lengauer, T.; Klebe, G. A fast flexible docking method using as incremental construction algorithm. *J. Mol. Biol.* **1996**, *261*, 470–489.

Score
Wang, R.; Liu, L.; Lai, L.; Tang, Y. Score: a new empirical method for estimating the binding affinity of ligands in receptor complexes. *J. Comput.-Aided Mol. Des.* **1997**, *11*, 425–445.

Fresno
Rognan, D.; Lauemoller, S. L.; Holm, A.; Buus, S.; Tschinke, V. Predicting binding affinities of protein ligands from three-dimensional models: application to peptide binding to class I major histocompatibility proteins. *J. Med. Chem.* **1999**, *42*, 4650–4658.

Knowledge-Based Scoring Functions

PMF
Muegge, I.; Martin, Y. C. A general and fast scoring function for protein–ligand interactions: a simplified potential approach. *J. Med. Chem.* **1999**, *42*, 791–804.

DrugScore

Gohlke, H.; Hendlich, M.; Klebe, G. Knowledge-based scoring function to predict protein–ligand interaction. *J. Mol. Biol.* **2000**, *295*, 337–356.

AFMoC

Gohlke, H.; Klebe, G. DrugScore meets CoMFA: adaptation of fields for molecular comparison (AFMoC) or how to tailor knowledge-based pair-potentials to a particular protein. *J. Med. Chem.* **2002**, *45*, 4153–4170.

SmoG2001

Ishchenko, A. V.; Shakhnovich, E. I.; Small molecule growth 2001 (SmoG2001): An improved knowledge-based scoring function for protein–ligand interactions. *J. Med. Chem.* **2002**, *45*, 2770–2780.

Consensus Scoring

Charifson, P. S.; Corkey, J. J. Murcko, M. A.; Walters, W. P. Consensus scoring: a method for obtaining improved hit rates from docking databases of three-dimensional structure into proteins. *J. Med. Chem.* **1999**, *42*, 5100–5109.

Terp, G. E.; Johansen, B. N.; Christensen, I. T. ; Jorgensen, F. S. A new concept for multidimensional selection of ligand conformations (MultiSelect) and multidimensional scoring (MultiScore) of protein-ligand binding affinities. *J. Med. Chem.* **2001**, *44*, 2333–2343.

De Novo Lead Design

Site-Point Connection

LUDI

Bohm, H. J. LUDI: rule-based automatic design of new substituents for enzyme inhibitor leads. *J. Comput.-Aided Mol. Des.* **1992**, *6*, 593–606.

CLIX

Lawrence, M. C.; Davis, P. C. CLIX: a search algorithm for finding novel ligands capable of binding proteins of known three-dimensional structure. *Proteins* **1992**, *12*, 31–41.

Fragment-Joining Method

SAR by NMR

Hajduk, P. J.; Meadows, R. P.; Fesik, S. W. Discovering high-affinity ligands for proteins. *Science* **1997**, *278*, 497–499.

Shuker, S. B.; Hajduk, P. J.; Meadows, R. P.; Fesik, S. W. Discovering high-affinity ligands for proteins: SAR by NMR. *Science* **1996**, *274*, 1531–1534.

MCSS/HOOK

Eisen, M. B.; Wiley, D. C.; Karplus, M.; Hubbard, R. E. HOOK: a program for finding novel molecular architectures that satisfy the chemical and steric requirements of a macromolecule binding site. *Proteins* **1994**, *19*, 199–221.

CAVEAT

Lauri, G.; Bartlett, P. A. CAVEAT: a program to facilitate the design of organic molecules. *J. Comput-Aided Mol. Des.* **1994**, *8*, 51–66.

PRO-LIGAND

Clark, D. E.; Frenkel, D.; Levy, S. A.; Li, J.; Murray, C. W.; Robson, B.; Waszkowycz, B.; Westhead, D. R. PRO-LIGAND: an approach to *de novo* molecular design. 1. Application to the design of organic molecules. *J. Comput.-Aided Mol. Des.* **1995**, *9*, 13–32.

Waszkowycz, B.; Clark, D. E.; Frenkel, D.; Li, J.; Murray, C.W.; Robson, B.; Westhead, D. R. PRO-LIGAND: an approach to *de novo* molecular design. 2. design of novel molecules from molecular field analysis (MFA) models and pharmacophores. *J. Med. Chem.* **1994**, *37*, 3994–4002.

NEWLEAD

Tschinke, V.; Cohen, N. C. The NEWLEAD program: a new method for the design of candidate structures from pharmacophoric hypotheses. *J. Med. Chem.* **1993**, *36*, 3863–3870.

Sequential Buildup (Atom Grow)

LEGEND

Honma, T.; Hayashi, K.; Aoyama, T.; Hashimoto, N.; Machida, T.; Fukasawa, K.; Iwama, T.; Ikeura, C.; Ikuta, M.; Suzuki-Takahashi, I.; Iwasawa, Y.; Hayama, T.; Nishimura, S.; Morishima, H. Structure-based generation of a new class of potent CDK4 inhibitors: new *de novo* design strategy and library design. *J. Med. Chem.* **2001**, *44*, 4615–4627.

Nishibata, Y.; Itai, A. Automatic creation of drug candidate structures based on receptor structure. Starting point for artificial lead generation. *Tetrahedron* **1991**, *47*, 8985–90.

Nishibata, Y.; Itai, A. Confirmation of usefulness of a structure construction program based on three-dimensional receptor structure for rational lead generation. *J. Med. Chem.* **1993**, *36*, 2921–8.

LeapFrog from SYBYL

GROW

Moon, J. B.; Howe, W. J. Computer design of bioactive molecules: a method for receptor-based *de novo* ligand design. *Proteins* **1991**, *11*, 314–328.

GROWMOL (more recent version is called AlleGrow; http://www.bostondenovo.com)

Bohacek, R. S.; McMartin, C. Multiple highly diverse structures complementary to enzyme binding sites: results of extensive application of a *de novo* design method incorporating combinatorial growth. *J. Am. Chem. Soc.* **1994**, *116*, 5560–5571.

SPROUT

Gillet, V. J.; Newell, W.; Mata, P.; Myatt, G.; Sike, S.; Zsoldos, Z.; Johnson, A. P. SPROUT: Recent developments in the *de novo* design of molecules. *J. Chem. Inf. Comput. Sci.* **1994**, *34*, 207–217.

Random Connections

MCSS/DLD

Stultz, C. M.; Karplus, M. Dynamic ligand design and combinatorial optimization: designing inhibitors to endothiapepsin. *Proteins* **2000**, *40*, 258–289.

CONCERTS

Pearlman, D. A.; Murcko, M. A. CONCERTS: dynamic connection of fragments as an approach to *de novo* ligand design. *J. Med. Chem.* **1996**, *39*, 1651–1663.

Virtual Combinatorial Screening

Methods for Virtual Combinatorial Screening

Legion from SYBYL

PRO_SELECT

Liebeschuetz, J. W.; Jones, S. D.; Morgan, P. J.; Murray, C. W.; Rimmer, A. D.; Roscoe, J. M.; Waszkowycz, B.; Welsh, P. M.; Wylie, W. A.; Young, S. C.; Martin, H.; Mahler, J.; Brady, L.; Wilkinson, K. PRO_SELECT: combining structure-based drug design and array-based chemistry for rapid lead discovery. 2. The development of a series of highly potent and selective factor Xa inhibitors. *J. Med. Chem.* **2002**, *45*, 1221–1232.

Murray, C. W.; Clark, D. E.; Auton, T. R.; Firth, M. A.; Li, J.; Sykes, R. S.; Waszkowycz, B.; Westhead, D. R.; Young. S. C. PRO_SELECT: Combining structure-based drug design and combinatorial chemistry for rapid lead discovery. 1. Technology. *J. Comput.-Aided Mol. Des.* **1997**, *11*, 193–207.

Combinatorial Library Design

Drug-Likeness

MoSELECT

Gillet, V. J.; Khatib, W.; Willett, P.; Fleming, P. J.; Green, D. V. Combinatorial library design using a multiobjective genetic algorithm. *J. Chem. Inf. Comput. Sci.* **2002**, *42*, 375–385.

Gillet, V. J.; Willett, P.; Fleming, P. J.; Green, D. V. Designing focused libraries using MoSELECT. *J. Mol. Graph Model* **2002**, *20*, 491–498.

REOS

Walters, W. P.; Murcko, M. A. Prediction of "drug-likeness." *Adv. Drug Deliv. Rev.* **2002**, *54*, 255–271.

Molecular Diversity

Martin, Y. C. Challenges and prospects for computational aids to molecular diversity. *Perspectives in Drug Discovery and Design* **1997**, 7/8, 159–172.

DiverseSolution from SYBYL

Cramer, R. D.; Clark, R. D.; Petterson, D. E.; Ferguson, A. M. Bioisosterism as a molecular diversity descriptor: steric fields of single "topomeric" conformers. *J. Med. Chem.* **1996**, *39*, 3060.

Patterson, D. E.; Cramer, R. D.; Ferguson, A. M.; Clark, R. D.; Weinberger, L. E. Neighborhood behavior: a useful concept for validation of molecular diversity descriptors. *J. Med. Chem.* **1996**, *39*, 3049.

FlexSim-S

Briem, H.; Lessel, U. F. *In vitro* and *in silico* affinity fingerprints: finding similarities beyond structural classes. *Perspective in Drug Discovery and Design* **2000**, *20*, 231–244.

LASSOO

Koehler, R. T.; Dixon, S. L.; Villar, H. O. LASSOO: A generalized directed diversity approach to the design and enrichment of chemical libraries. *J. Med. Chem.* **1999**, *42*, 4695–4704.

Others

Andrews, K. M.; Cramer, R. D. Toward general methods of targeted library design: topomer shape similarity searching with diverse structures as queries. *J. Med. Chem.* **2000**, *43*, 1723–1740.

Makara, G. M. Measuring molecular similarity and diversity: total pharmacophore diversity. *J. Med. Chem.* **2001**, *44*, 3563–3571.

Mason, J. S.; Morize, I.; Menard, P. R.; Cheney, D. L.; Hulme, C.; Labaudiniere, R. F. New 4-point pharmacophore method for molecular similarity and diversity applications: overview of the method and applications, including a novel approach to the design of combinatorial libraries containing privileged substructures. *J. Med. Chem.* **1999**, *42*, 3251–3264.

Mount, J.; Ruppert, J.; Welch, W.; Jian, A. N. Ice Pick: A flexible surface-based system for molecular diversity. *J. Med. Chem.* **1999**, *42*, 60–66.

Srinivasan, J.; Castellino, A.; Bradley E. K.; Eksterowicz, J. E.; Grootenhuis P. D. J.; Putta, S.; Stanton R. Evaluation of a novel shape-based computational filter for lead evolution: application to thrombin inhibitors. *J. Med. Chem.* **2002**, *45*, 2494–2500.

Synthetic Accessibility

Gasteiger, J.; Pfortner, M.; Sitzmann, M.; Hollering, R.; Sacher, O.; Kostka, T.; Karg, N. Computer-assisted synthesis and reaction planning in combinatorial chemistry. *Perspective in Drug Discovery and Design* **2000**, *20*, 245–264.

Gillet, V. J.; Nicolotti, O. Evaluation of reactant-based and product-based approaches to the design of combinatorial libraries. *Perspective in Drug Discovery and Design* **2000**, *20*, 265–287.

Lewell, X. Q.; Judd, D. B.; Watson, S. P.; Hann, M. M. RECAP—retrosynthetic combinatorial analysis procedure: a powerful new technique for identifying privileged molecular fragments with useful applications in combinatorial chemistry. *J. Chem. Inf. Comput. Sci.* **1998**, *38*, 511–522.

Ligand Optimization

ADME (Pharmacokinetics)

Property-Based Ligand Design

van De Waterbeemd, H; Smith, D. A.; Beaumont, K.; Walker, D. K. Property-based design: optimization of drug absorption and pharmacokinetics. *J. Med. Chem.* **2001**, *44*, 1313–1333.

VolSurf

Cruciani, G.; Pastor, M.; Guba, W. VolSurf: a new tool for the pharmacokinetic optimization of lead compounds. *Eur. J. Pharm. Sci.* **2000**, *Suppl 2*, S29–39.

ChemGPS

Alifrangis, L. H.; Christensen, I. T.; Berglund, A.; Sandberg, M.; Hovgaard, L.; Frokjaer, S. Structure-property model for membrane partitioning of oligopeptides. *J. Med. Chem.* **2000**, *43*, 103–113.

Crivori, P.; Cruciani, G.; Carrupt, P. A.; Testa, B. Predicting blood–brain barrier permeation from three-dimensional molecular structure. *J. Med. Chem.* **2000**, *43*, 2204–2216.

Cruciani, G.; Pastor, M.; Mannhold, R. Suitability of molecular descriptors for database mining. A comparative analysis. *J. Med. Chem.* **2002**, *45*, 2685–2694.

Egan, W. J.; Merz Jr., K. M.; Baldwin, J. J. Prediction of drug absorption using multivariate statistics. *J. Med. Chem.* **2000**, *43*, 3867–3877.

Ooms, F.; Weber, P.; Carrupt, P. A.; Testa, B. A simple model to predict blood–brain barrier permeation from 3D molecular fields. *Biochim. Biophys. Acta* **2002**, *1587*, 118–125.

Oprea, T. I.; Zamora, I.; Ungell, A. L. Pharmacokinetically based mapping device for chemical space navigation. *J. Comb. Chem.* **2002**, *4*, 258–266.

QSAR

2D-QSAR

Hansch method

Domine, D.; Devillers J.; Chastrette, M. A nonlinear map of substituent constants for selecting test series and deriving structure–activity relationships. 1. aromatic series. *J. Med. Chem.* **1994**, *37*, 973–980.

Domine, D.; Devillers, J.; Chastrette, M. A nonlinear map of substituent constants for selecting test series and deriving structure–activity relationships. 2. aliphatic series. *J. Med. Chem.* **1994**, *37*, 981–987.

Manallack, D. T.; Ellis, D. D.; Livingstone D. J. Analysis of linear and nonlinear QSAR data using neural networks. *J. Med. Chem.* **1994**, *37*, 3639–3654.

3D-QSAR

CoMFA (Comparative Molecular Field Analysis)

Cramer, R. D. Topomer CoMFA: a design methodology for rapid lead optimization. *J. Med. Chem.* **2003**, *46*, 374–388.

Cramer, R. D.; Paterson, D. E.; Bunce, J. D. Comparative molecular field analysis (CoMFA). I. Effect of shape on binding of steroids to carried proteins. *J. Am. Chem. Soc.* **1988**, *110*, 5959–5967.

Robinson, D. D.; Winn P. J.; Lyne P. D.; Richards, W. G. Self-organizing molecular field analysis: a tool for structure–activity studies. *J. Med. Chem.* **1999**, *42*, 573–583.

So, S.-S.; Karplus, M. Three-dimensional QSAR from molecular similarity matrices and genetic neural networks. 1. method and validations. *J. Med. Chem.* **1997**, *40*, 4347–4359.

So, S.-S.; Karplus, M. Three-dimensional QSAR from molecular similarity matrices and genetic neural networks. 2. applications. *J. Med. Chem.* **1997**, *40*, 4360–4371.

CoMSIA (Comparative Molecular Similarity Analysis)

Klebe, G.; Abraham, U.; Mietzner, T. Molecular similarity indices in a comparative analysis (CoMSIA) of drug molecules to correlate and predict their biological activity. *J. Med. Chem.* **1994**, *37*, 4130–4146.

CoMMA (Comparative Molecular Moment Analysis)

Silverman, B. D.; Platt, D. E. Comparative molecular moment analysis (CoMMA): 3D-QSAR without molecular superposition. *J. Med. Chem.* **1996**, *39*, 2129–2140.

COMBINE (Comparative Binding Energy)

Ortiz, A. R.; Pisabarro, M. T.; Gago, F.; Wade R. C. Prediction of drug binding affinities by comparative binding energy analysis. *J. Med. Chem.* **1995**, *38*, 2681–2691.

Perez, C.; Pastor, M.; Ortiz, A. R. Gago F. Comparative binding energy analysis of HIV-1 protease inhibitors: incorporation of solvent effects and validation as a powerful tool in receptor-based drug design. *J. Med. Chem.* **1998**, *41*, 836–852.

GRID/GOLPE

Cho, S. J.; Tropsha A. Cross-validated R^2-guided region selection for comparative molecular field analysis: a simple method to achieve consistent results. *J. Med. Chem.* **1995**, *38*, 1060–1066.

Cruciani, G.; Watson, K. A. Comparative molecular field analysis using GRID force-field and GOLPE variable selection methods in a study of inhibitors of glycogen phosphorylase b. *J. Med. Chem.* **1994**, *37*, 2589–2601.

4D-QSAR

Klein, C. D.; Hopfinger, A. J. Pharmacological activity and membrane interactions of antiarrhythmics: 4D-QSAR/QSPR analysis. *Pharm. Res.* **1998**, *15*, 303–311.

5D-QSAR

Vedani, A.; Dobler, M. 5D-QSAR: the key for simulating induced fit? *J. Med. Chem.* **2002**, *45*, 2139–49.

Molecular Superposition

FlexS

Lemmen, C.; Lengauer, T.; Klebe, G. FLEXS: a method for fast flexible ligand superposition. *J. Med. Chem.* **1998**, *41*, 4502–20.

SQ

Miller, M. D.; Sheridan, R. P.; Kearsely, S. K. SQ: A program for rapidly producing pharmacophorically relevant molecular superpositions. *J. Med. Chem.* **1999**, *42*, 1505–1514.

QXP

McMartin, C.; Bohacek, R. S. QXP: powerful, rapid computer algorithms for structure-based drug design. *J. Comput.-Aided Mol. Des.* **1997**, *11*, 333–344.

Pharmacophore Elucidation and Pharmacophore-Based Database Screening

Martin, Y. C. 3D database searching in drug design. *J. Med. Chem.* **1992**, 35, 2145–2154.

DISCO from SYBYL

Marriott, D. P.; Dougall, I. G.; Meghani, P. Liu Y.-J.; Flower, D. R. Lead generation using pharmacophore mapping and three-dimensional database searching: application to muscarinic M3 receptor antagonists. *J. Med. Chem.* **1999**, *42*, 3210–3216.

Martin, Y. C. DISCO: what we did right and what we missed. *IUL Biotechnology Series* **2000**, *2*, 49–68.

Martin, Y. C.; Bures, M. G.; Danaher, E. A.; DeLazzer, J.; Lico, I.; Pavlik, P. A. A fast new approach to pharmacophore mapping and its application to dopaminergic and benzodiazepine agonists. *J. Comput.-Aided Mol. Des.* **1993**, *7*, 83–102.

GASP from SYBYL

Holliday, J. D.; Willett, P. Using a genetic algorithm to identify common structural features in sets of ligands. *J. Mol. Graph. Model.* **1997**, *15*, 221–232.

Jones, G.; Willett, P.; Glen, R. C. GASP: genetic algorithm superimposition program. *IUL Biotechnology Series* **2000**, *2*, 85–106.

RECEPTOR from SYBYL

Josien, H.; Convert, O.; Berlose, J.-P.; Sagan, S.; Brunissen, A.; Lavielle, S.; Chassaing, G. Topographic analysis of the S7 binding subsite of the tachykinin neurokinin-1 receptor. *Biopolymers* **1996**, *39*, 133–147.

UNITY from SYBYL
Catalyst

Kurogi, Y.; Guner, O. F. Pharmacophore modeling and three-dimensional database searching for drug design using catalyst. *Curr. Med. Chem.* **2001**, *8*, 1035–1055.

2.4 PROBLEMS

(Answers can be found in Appendix at the end of the book.)

1. What are some of the advantages and disadvantages of random screening?

2. Nitric oxide synthase catalyzes the conversion of *L*-arginine to *L*-citrulline and nitric oxide. If you wanted to interfere with the production of NO, design some potential leads.

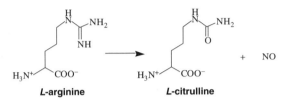

3. A. Compound **1** was found to be a lead compound at Pfizer for enhancement of cytotoxic effects of cancer drugs. (see Canan Koch, S. S. *et al., J. Med. Chem.* **2002**, *45*, 4961–74). Suggest an approach you would take if you wanted to determine the pharmacophoric groups, the groups interfering with receptor (in this case an enzyme) binding, and those not involved at all.

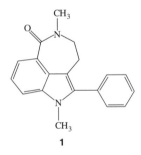

1

B. How would you interpret the following results:

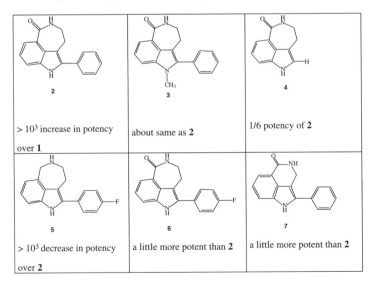

4. A new protein was identified that appears to be abundant in individuals prone to a certain type of cancer, and you are trying to identify a molecule that will bind to this protein as a lead compound, but you do not know the structure or function of the protein. How would you proceed?

5. The LD_{50} for a potential antiobesity compound was found to be 10 mg/kg and the ED_{50} was 2 mg/kg. Is this an important drug candidate? Why?

6. Candesartan cilexetil is an antihypertensive drug that antagonizes the AT_1 angiotensin receptor. Within its structure are four lead modification approaches with which you should be familiar. Working backwards, draw a lead molecule from which this drug may have been derived and point out where the lead modifications occurred.

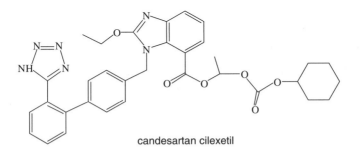

candesartan cilexetil

7. What potential problems arise when using bioisosteric replacements for lead modification?

8. What problems arise when using combinatorial chemistry approaches for lead (or drug) discovery?

9.　A. Use the split synthesis method to make a combinatorial library of all dipeptides of Phe, Gly, and Glu.

　　B. Show the steps in a Merrifield synthesis of *one* of the dipeptides.

10. A combinatorial library of tetrapeptides containing Lys, Ser, Glu, and His was constructed using Merrifield's resin and Still's tag method for encoding. The active bead was isolated and photolyzed, and electron capture gas chromatography gave the following result:

Tags 2, 4, 9, 11, 12 were detected

Use Still's encoding method to determine the structure of the active peptide on the polymer bead.

Assume:　001 = Lys　010 = Ser　011 = Glu　100 = His

11. Compounds **1** and **2** below were the optimized leads determined from an SAR by NMR study of a new receptor. Based on this analysis, **3** was synthesized, and *n* was varied, but all of the compounds made had much lower potency than either **1** or **2**.

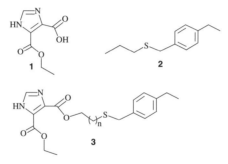

A. What conclusions can you draw from this result? (No, the experiment *was* done correctly.)

B. What structure would you try next? Why?

12. Design three peptidomimetics for Glu-Tyr-Val, one using a ring-chain transformation, one a scaffold peptidomimetic, and one having at least one bioisosteric replacement.

13. Based on your knowledge of how the Hammett equation was developed (and basic organic mechanisms), show a mechanism and explain how a change in X will affect the rate of the following reaction.

14. Use the partial list of log *P* values at the end of the problem set to calculate the log *P* for the given molecules.

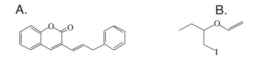

C. Why can there be many correct, but different answers to these calculations?

15. A. In general, why does log *P* change with pH for some compounds, but not others?

B. Rationalize the change in log *D* versus pH for omeprazole shown below. Refer to the structure when discussing the log *D* changes with pH.

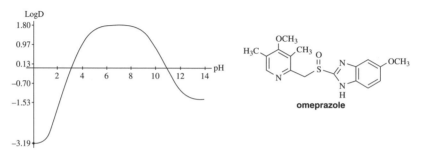

Reproduced in color between pages 172 and 173

16. Do you predict that the compound below will have good oral bioavailability? Why or why not?

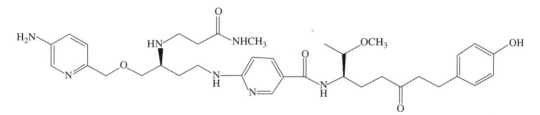

17. The following compound has potent antifungal activity in a cell-free system, but has poor activity in mice.

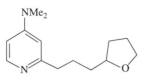

A. Why is it not effective in mice?

B. Suggest a structural modification that might increase antifungal activity in mice.

18. You just discovered a lead compound for the treatment of chemistryphobia with structure **4**.

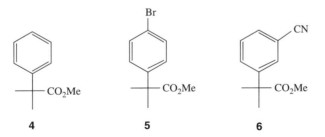

A. Compound **5** was synthesized and found to be less potent than **4**. Offer explanations and suggest other compounds to synthesize to improve the potency.

B. Compound **6** was prepared and also was less potent than **4**, but more potent than **5**. Explain and suggest other compounds (with rationalizations) to synthesize.

19. *Briefly* describe the basis for Kuntz's DOCK program and Cramer's CoMFA.

20. For computer modeling approaches in drug design, what could be the problems associated with using a crystal structure of the target receptor without a small molecule bound to it?

21. Steric, electronic, lipophilic, and H-bonding effects are important parameters of molecules employed in computer-aided drug design. Why are each of these effects important in drug design?

Partial List of Log P Values for Use with Problem 14
From Leo, A.; Hansch, C.; Elkins, D. *Chem Rev.* **1971**, *71*, 525.

Compound	log P_{oct}	Compound	log P_{oct}	Compound	log P_{oct}
CH_3OH	−0.66	$CH_2{=}CHCOOH$	0.43	(cyclopropyl)–O–CH_3	1.20
CH_3NH_2	−0.57	CH_3CH_2CN	0.16	$CH_2{=}CH{-}OCH_2CH_3$	1.04
CCl_3COOH	1.49	$H_3C{-}\overset{O}{\overset{\|}{C}}{-}CH_3$	−0.24	$CH_3CH_2CH_2COOH$	0.79
$BrCH_2COOH$	0.41	$CH_2{=}CHCH_2OH$	0.17	$CH_3CH_2CH_2CH_2OH$	0.83
$ClCH_2COOH$	0.47	CH_3CH_2CHO	0.38	$CH_3CH_2OCH_2CH_3$	0.77
FCH_2COOH	−0.12	CH_3CO_2Me	0.18	$CH_3CH_2OCH_2CH_2OH$	−0.54
ICH_2COOH	0.87	CH_3CH_2COOH	0.33	$CH_3CH_2NHCH_2CH_3$	0.57
CH_3CN	−0.34	CH_3OCH_2COOH	−0.55	(piperidine)	0.85
CH_3CHO	0.43	$CH_3CH_2CH_2Br$	2.10	$CH_3CH_2CH_2CH_2CH_2F$	2.33
CH_3COOH	−0.17	$CH_3CH_2CH_2NO_2$	0.65	$PhCH_2OH$	1.10
$HOCH_2COOH$	−1.11	$CH_3OCH_2OCH_3$	0.00	$PhCH_2NH$	1.09
CH_3CH_2Br	1.74	$CH_3OCH_2CH_2OH$	−0.60	(phthalimide)	1.15
CH_3CH_2Cl	1.54	Me_3N	0.27	$PhCH_2COOH$	1.41
CH_3CH_2I	2.00	CH_3I	1.69	$PhOCH_2COOH$	1.26
CH_3CONH_2	−1.46	CH_3NO_2	−0.33	(benzene)	2.13
$CH_3CH_2NO_2$	0.18	(uracil)	−1.07	(1-ethynylcyclohexanol)	1.73
CH_3CH_2OH	−0.32	$HOOCCH{=}CHCOOH$	0.28	(cyclohexanone)	0.81
Me_2NH	−0.23	(succinimide)	−1.21	(coumarin)	1.39
$CH_3CH_2NH_2$	−0.19	$CH_2{=}CH{-}O{-}CH{=}CH_2$	1.81	(1,3-indandione)	0.61
$HOCH_2CH_2NH_2$	−1.31	$CH_3CH{=}CHCOOH$	0.72	(quinoline)	2.03
$HC{\equiv}CCO_2H$	0.46	$HOOCCH_2CH_2COOH$	−0.59	(naphthalene)	3.37
$CH_2{=}CHCN$	−0.92	$CH_2{=}CHCH_2OCH_3$	0.94	(indole)	2.00

◼ 2.5 References

1. Fleming, A. *Brit. J. Exp. Pathol.* **1929**, *10*, 226.

2. Stone, T.; Darlington, G. *Pills, Potions and Poisons. How Drugs Work*, Oxford University Press, Oxford, 2000, p. 255.

3. Hare, R. *The Birth of Penicillin*, Allen & Unwin, London, 1970.

4. Beveridge, W. I. B. *Seeds of Discovery*, W. W. Norton, New York, 1981.

5. Abraham, E. P.; Chain, E.; Fletcher, C. M.; Gardner, A. D.; Heatley, N. G.; Jennings, M. A.; Florey, H. W. *Lancet* **1941**, *2*, 177.

6. Florey, H. W.; Chain, E.; Heatley, N. G.; Jennings, M. A.; Sanders, A. G.; Abraham, E. P.; Florey, M. E. *Antibiotics*, Oxford University Press, London, 1949, Vol. II.

7. Moyer, A. J.; Coghill, R. D. *J. Bacteriol.* **1946**, *51*, 79.

8. (a) Sheehan, J. C. *The Enchanted Ring: The Untold Story of Penicillin*, MIT Press, Cambridge, MA, 1982. (b) Williams, T. I. *Robert Robinson: Chemist Extraordinary*, Clarendon Press, Oxford, 1990. (b) Todd, A. R.; Cornforth, J. S. *Biograph. Memoirs Roy. Soc.* **1976**, *22*, 490.

9. Hodgkin, D. C.; Bunn, C.; Rogers-Low, B.; Turner-Jones, A. In *Chemistry of Penicillin*, Clarke, H. T.; Johnson, J. R.; Robinson, R. (Eds.) Princeton University Press, Princeton, NJ, 1949.

10. Sternbach, L. H. *J. Med. Chem.* **1979**, *22*, 1.

11. (a) Dunn, D.; Orlowski, M.; McCoy, P.; Gastgeb, F.; Appell, K.; Ozgur, L.; Webb, M.; Burbaum, J. *J. Biomol. Screen.* **2000**, *5*, 177. (b) Kenny, B. A.; Bushfield, M.; Parry-Smith, D. J.; Fogarty, S.; Treherne, J. M. *Prog. Drug Res.* **1998**, *51*, 245.

12. Drews, J. *Science* **2000**, *287*, 1960.

13. Lahana, R. *Drug Discovery Today* **1999**, *4*, 447.

14. Wong, K. K.; Kuo, D. W.; Chabin, R. M.; Founier, C.; Gegnas, L. D.; Waddell, S. T.; Marsilio, F.; Leiting, B.; Pompliano, D. L. *J. Am. Chem. Soc.* **1998**, *120*, 13527.

15. (a) Gao, J.; Cheng, X.; Chen, R.; Sigal, G. B.; Bruce, J. E.; Schwartz, B. L.; Hofstadler, S. A.; Anderson, G. A.; Smith, R. D.; Whitesides, G. M. *J. Med. Chem.* **1996**, *39*, 1949. (b) Rossi, D. T.; Sinz, M. W. (Eds.) *Mass Spectrometry in Drug Discovery*, Marcel Dekker, New York, 2002. (c) Ganem, B.; Henion, J. D. *Bioorg. Med. Chem.* **2003**, *11*, 311.

16. (a) Hajduk, P. J.; Olejniczak, E. T.; Fesik, S. W. *J. Am. Chem. Soc.* **1997**, *119*, 12257. (b) Hajduk, P. J.; Gerfin, T.; Boehlen, J.-M.; Häberli, M.; Marek, D.; Fesik, S. W. *J. Med. Chem.* **1999**, *42*, 2315.

17. Oprea, T. I.; Davis, A. M.; Teague, S. J.; Leeson, P. D. *J. Chem. Inf. Comput. Sci.* **2001**, *41*, 1308.

18. Shen, T. Y. In *Clinoril in the Treatment of Rheumatic Disorders*, Huskisson, E. C.; Franchimont, P. (Eds.), Raven Press, New York, 1976.

19. Gay, L. N.; Carliner, P. E. *Science* **1949**, *109*, 359.

20. Strickland, B. A., Jr.; Hahn, G. L. *Science*, **1949**, *109*, 359.

21. (a) Corbin, J. D.; Francis, S. H. *J. Biol. Chem.* **1999**, *274*, 13729. (b) Palmer, E. *Chem. Brit.* **1999**, *35*, 24.

22. Shen, T. Y.; Winter, C. A. *Adv. Drug Res.* **1977**, *12*, 89.

23. Grewe, R. *Naturwissenschaften* **1946**, *33*, 333.

24. Schnider, O.; Grüssner, A. *Helv. Chem. Acta*, **1949**, *32*, 821.

25. May, E. L.; Murphy, J. G. *J. Org. Chem.* **1955**, *20*, 257.

26. Schaumann, O. *Naunyn-Schmiedebergs Arch. Pharmacol. Exp. Pathol.* **1940**, *196*, 109.

27. (a) Bentley, K. W.; Hardy, D. G. *J. Am. Chem. Soc.* **1967**, *89*, 3281. (b) Bentley, K. W.; Hardy, D. G. *J. Am. Chem. Soc.* **1967**, *89*, 3273. (c) Bentley, K. W.; Hardy, D. G. *J. Am. Chem. Soc.* **1967**, *89*, 3267.

28. Andrews, P. R.; Craik, D. J.; Martin, J. L. *J. Med. Chem.* **1984**, *27*, 1648.

29. Plobeck, N.; Delorme, D.; Wei, Z.-Y.; Yang, H.; Zhou, F.; Schwarz, P.; Gawell, L.; Gagnon, H.; Pelcman, B.; Schmidt, R.; Yue, S. Y.; Walpole, C.; Brown, W.; Zhou, E.; Labarre, M.; Payza, K.; St-Onge, S.; Kamassah, A.; Morin, P.-E.; Projean, D.; Ducharme, J.; Roberts, E. *J. Med. Chem.* **2000**, *43*, 3878.

30. (a) Crum-Brown, A.; Fraser, T. R. *Trans. Roy. Soc. Edinburgh* **1868–1869**, *25*, 151, 693. (b) Crum-Brown, A.; Fraser, T. R. *Proc. Roy. Soc. Edinburgh* **1872**, *7*, 663.

31. Richardson, B. W. *Med. Times Gaz.* **1869**, *18*, 703.

32. Northey, E. H. *The Sulfonamides and Allied Compounds*, American Chemical Society Monograph Series, Van Nostrand Reinhold, New York, 1948.

33. Loubatieres, A. In *Oral Hypoglycemic Agents*, Campbell, G. D. (Ed.), Academic Press, New York, 1969.

34. High-ceiling diuretics are highly efficacious inhibitors of Na^+-K^+-$2Cl^-$ symport. Jackson, E. D. In *Goodman & Gilman's The Pharmacological Basis of Therapeutics*, 9th ed., Hardman, J. G.; Limbird, L. E.; Molinoff, P. B.; Ruddon, R. W.; Gilman, A. G. (Eds.), McGraw-Hill, New York, 1996, p. 685.

35. Sprague, J. M. In *Topics in Medicinal Chemistry*; Robinowitz, J. L.; Myerson, R. M. (Eds.), Wiley, New York, 1968, Vol. 2.

36. (a) He, L.; Jagtap, P. G.; Kingston, D. G. I.; Shen, H.-J.; Orr, G. A.; Horwitz, S. *Biochemistry* **2000**, *39*, 3972. (b) Díaz, J. F.; Strobe, R.; Engelborghs, Y.; Souto, A. A.; Andreu, J. M. *J. Biol. Chem.* **2000**, *275*, 26265. (c) Yvon, A. C.; Wadsworth, P.; Jordan, M. *Mol. Biol. Cell* **2000**, *10*, 947.

37. (a) Kingston, D. *J. Nat. Prod.* **2000**, *63*, 726. (b) Dubois, J.; Thoret, S.; Guéritte, F.; Guénard, D. *Tetrahedron Lett.* **2000**, *41*, 3331. (c) Yuan, H.; Kingston, D. *Tetrahedron* **1999**, *55*, 9089.

38. Evans, B. E.; Rittle, K. E.; Bock, M. G.; DiPardo, R. M.; Freidinger, R. M.; Whitter, W. L.; Lundell, G. F.; Veber, D. F.; Anderson, P. S.; Chang, R. S. L.; Lotti, V. J.; Cerino, D. J.; Chen, T. B.; Kling, P. J.; Kunkel, K. A.; Springer, J. P.; Hirshfield, J. *J. Med. Chem.* **1988**, *31*, 2235.

39. (a) Ariëns, E. J.; Beld, A. J.; Rodrigues de Miranda, J. F.; Simonis, A. M. In *The Receptors: A Comprehensive Treatise*, O'Brien, R. D. (Ed.), Plenum Press, New York, 1979, p. 33. (b) Ariëns, E. J. *Med. Res. Rev.* **1987**, *7*, 367.

40. (a) Patchett, A. A.; Nargund, R. P. *Annu. Rep. Med. Chem.* **2000**, *35*, 289. (b) Hajduk, P. J.; Bures, M.; Praestgaard, J.; Fesik, S. W. *J. Med. Chem.* **2000**, *43*, 3443. (c) Fecik, R. A.; Frank, K. E.; Gentry, E. J.; Menon, S. R.; Mitscher, L. A.; Telikepalli, *Med. Res. Rev.* **1998**, *18*, 149. (d) Horton, D. A.; Bourne, G. T.; Smythe, M. L. *J. Computer-Aided Mol. Des.* **2002**, *16*, 415. (e) Klabunde, T.; Hessler, G. *Chembiochem.* **2002**, *3*, 928. (f) Matter, H.; Baringhaus, K. H.; Naumann T.; Klabunde T.; Pirard B. *Comb. Chem. High Throughput Screen.* **2001**, *4*, 453. (g) Horton, D. A.; Bourne, G. T.; Smythe, M. L. *Chem. Rev.* **2003**, *103*, 893.

41. Bemis, G. W.; Murcko, M. A. *J. Med. Chem.* **1996**, *39*, 2887.

42. Bemis, G. W.; Murcko, M. A. *J. Med. Chem.* **1999**, *42*, 5095.

43. Ajay; Walters, W. P.; Murcko, M. A. *J. Med. Chem.* **1998**, *41*, 3314.

44. This is an electronic database of Volume 6 of *Comprehensive Medicinal Chemistry* (Pergamon Press) available from MDL Information Systems, Inc., San Leandro, CA 94577.

45. Sadowski, J.; Kubinyi, H. *J. Med. Chem.* **1998**, *41*, 3325.

46. The ACD is available from MDL Information Systems, Inc., San Leandro, CA, and contains specialty and bulk commercially available chemicals.

47. The WDI is from Derwent Information.

48. (a) Walters, W. P.; Stahl, M. T.; Murcko, M. A. *Drug Discovery Today* **1998**, *3*, 160. (b) Walters, W. P.; Ajay; Murcko, M. A. *Curr. Opin. Chem. Biol.* **1999**, *3*, 384. (c) Teague, S. J.; Davis, A. M.; Leeson, P. D.; Oprea, T. *Angew. Chem. Int. Ed. Engl.* **1999**, *38*, 3743. (d) Oprea, T. I. *J. Comput.-Aided Mol. Des.* **2000**, *14*, 251. (e) Gillet, V. J.; Willett, P. L.; Bradshaw, J. *J. Chem. Inf. Comput. Sci.* **1998**, *38*, 165. (f) Wagener, M.; vanGeerestein, V. J. *J. Chem. Inf. Comput. Sci.* **2000**, *40*, 280. (g) Ghose, A. K.; Viswanadhan, V. N.; Wendoloski, J. J. *J. Comb. Chem.* **1999**, *1*, 55. (h) Xu, J.; Stevenson, J. *J. Chem. Inf. Comput. Sci.* **2000**, *40*, 1177. (i) Muegge, I.; Heald, S. L.; Brittelli, D. *J. Med. Chem.* **2001**, *44*, 1841. (j) Anzali, S.; Barnickel, G.; Cezanne, B.; Krug, M.; Filimonov, D.; Poroikiv, V. *J. Med. Chem.* **2001**, *44*, 2432. (k) Brüstle, M.; Beck, B.; Schindler, T.; King, W.; Mitchell, T.; Clark, T. *J. Med. Chem.* **2002**, *45*, 3345.

49. (a) McGovern, S. L.; Caselli, E.; Grigorieff, N.; Shoichet, B. K. *J. Med. Chem.* **2002**, *45*, 1712. (b) McGovern, S. L.; Helfand, B. J.; Feng, B.; Shoichet, B. K. *J. Med. Chem.* **2003**, *46*, 4265.

50. Huang, Z.; Yang, G.; Lin, Z.; Huang, J. *Bioorg. Med. Chem. Lett.* **2001**, *11*, 1099.

51. Richardson, B. W. *Med. Times Gaz.* **1869**, *18*, 703.

52. Dohme, A. R. L.; Cox, E. H.; Miller, E. *J. Am. Chem. Soc.* **1926**, *48*, 1688.

53. Funcke, A. B. H.; Ernsting, M. J. E.; Rekker, R. F.; Nauta, W. T. *Arzneimittel. Forsch.* **1953**, *3*, 503.

54. Thornber, C. W. *Chem. Soc. Rev.* **1979**, *8*, 563.

55. Burger, A. In *Medicinal Chemistry*, 3rd ed., Burger, A. (Ed.), Wiley, New York, 1970.

56. Korolkovas, A. *Essentials of Molecular Pharmacology: Background for Drug Design*, Wiley, New York, 1970, pp. 54–57.

57. Lipinski, C. A. *Annu. Rep. Med. Chem.* **1986**, *21*, 283.

58. (a) Grimm, H. G. *Z. Elektrochem.* **1925**, *31*, 474. (b) Grimm, H. G. *Z. Elektrochem.* **1928**, *34*, 430.

59. Erlenmeyer, H. *Bull. Soc. Chim. Biol.* **1948**, *30*, 792.

60. (a) Patani, G. A.; LaVoie, E. J. *Chem. Rev.* **1996**, *96*, 3147. (b) Burger, A. *Prog. Drug Res.* **1991**, *37*, 287. (c) Lipinski, C. A. *Annu. Rep. Med. Chem.* **1986**, *21*, 283.

61. Song, Y.; Connor, D. T.; Doubleday, R.; Sorenson, R. J.; Sercel, A. D.; Unangst, P. C.; Roth, B. D.; Gilbertsen, R. B.; Chan, K.; Schrier, D. J.; Guglietta, A.; Bornemeier, D. A.; Dyer, R. D. *J. Med. Chem.* **1999**, *42*, 1151.

62. (a) Gallop, M. A.; Barrett, R. W.; Dower, W. J.; Fodor, S. P. A.; Gordon, E. M. *J. Med. Chem.* **1994**, *37*, 1233. (b) Gordon, E. M.; Barrett, R. W.; Dower, W. J.; Fodor, S. P. A.; Gallop, M. A. *J. Med. Chem.* **1994**, *37*, 1385. (c) Balkenhohl, F.; von dem Bussche-Hünnefeld, C.; Lansky, A.; Zechel, C. *Angew. Chem. Int. Ed. Engl.* **1996**, *35*, 2289. (d) Thompson, L. A.; Ellman, J. A. *Chem. Rev.* **1996**, *96*, 555. (e) Antel, J. *Curr. Opin. Drug Discovery Develop.* **1999**, *2*, 224. (f) Martin, E. J.; Blaney, J. M.; Siani, M. A.; Spellmeyer, D. C.; Wong, A. K.; Moos, W. H. *J. Med. Chem.* **1995**, *38*, 1431 (g) Smith, G. P.; Scott, J. K. *Methods Enzymol.* **1993**, *217*, 228. (h) Fecik, R. A.; Frank, K. E.; Gentry, E. J.; Menon, S. R.; Mitscher, L. A.; Telikepalli, H. *Med. Res. Rev.* **1998**, *18*, 149.

63. Kaldor, S. W. *Tetrahedron Lett.* **1996**, *37*, 7193.

64. Han, H.; Wolfe, M. M.; Brenner, S.; Janda, K. D. *Proc. Natl. Acad. Sci. USA* **1995**, *92*, 6419.

65. Brown, D. *Mol. Diversity* **1996**, *2*, 217.

66. Martin, Y. C. *Persp. Drug Discovery Design* **1997**, *7/8*, 159.

67. Furka, A. *Notariell Beglaubigtes Dokument Nr 36237/1982*, Budapest, Hungary, 1982.

68. Geysen, H. M.; Meloen, R. H.; Barteling, S. J. *Proc. Natl. Acad. Sci. USA* **1984**, *81*, 3998.

69. Houghten, R. A. *Proc. Natl. Acad. Sci. USA* **1985**, *82*, 5131.

70. Zuckermann, R. N.; Martin, E. J.; Spellmeyer, D. C.; Stauber, G. B.; Shoemaker, K. R.; Karr, J. M.; Figliozzi, G. M.; Goff, D. A.; Siani, M. A.; Simon, R. J.; Banville, S. C.; Brown, E. G.; Wang, L.; Richter, L. S.; Moos, W. H. *J. Med. Chem.* **1994**, *37*, 2678.

71. (a) Thompson, L. A.; Ellman, J. A. *Chem. Rev.* **1996**, *96*, 555. (b) Ellman, J. A. *Acc. Chem. Res.* **1996**, *29*, 132.

72. Terrett, N. K.; Gardner, M.; Gordon, D. W.; Kobylecki, R. J.; Steele, J. *Tetrahedron* **1995**, *51*, 8135.

73. (a) Furka, A.; Sebestyen, F.; Asgedom, M.; Dibo, G. *Int. J. Pept. Protein Res.* **1991**, *37*, 487. (b) Lam, K. S.; Salmon, S. E.; Hersh, E. M.; Hruby, V. J.; Kazmierski, W. M.; Knapp, R. J. *Nature* **1991**, *354*, 82. (c) Zuckermann, R. N.; Kerr, J. M.; Siani, M. A.; Banville, S. C. *Int. J. Pept. Protein Res.* **1992**, *40*, 498.

74. (a) Brummel, C. L.; Lee, I. N. W.; Zhou, Y.; Benkovic, S. J.; Winograd, N. *Science* **1994**, *264*, 399. (b) Zambias, R. A.; Boulton, D. A.; Griffin, P. R. *Tetrahedron Lett.* **1994**, *35*, 4283. (c) Youngquist, R. S.; Fuentes, G. R.; Lacey, M. P.; Keough, T. *J. Am. Chem. Soc.* **1995**, *117*, 3900.

75. Eichler, J.; Appel, J. R.; Blondelle, S. E.; Dooley, C. T.; Dörner, B.; Ostresh, J. M.; Pérez-Payá, E.; Pinilla, C.; Houghten, R. A. *Med. Res. Rev.* **1995**, *15*, 481.

76. Dooley, C. T.; Chung, N. N.; Schiller, P. W.; Houghten, R. S. *Proc. Natl. Acad. Sci. USA* **1993**, *90*, 10811.

77. Dooley, C. T.; Chung, N. N.; Wilkes, B. C.; Schiller, P. W.; Bidlack, J. M.; Pasternak, G. W.; Houghten, R. A. *Science* **1994**, *266*, 2019.

78. (a) Affleck, R. L. *Curr. Opin. Chem. Biol.* **2001**, *5*, 257. (b) Xiao, X. Y. *Front. Biotechnol. Pharm.* **2000**, *1*, 114. (c) Czarnik, A. W. *Curr. Opin. Chem. Biol.* **1997**, *1*, 60. (d) Lam, K. S.; Krchnak, Y.; Lebl, M. *Chem. Rev.* **1997**, *97*, 411.

79. Oldenburg, K. R.; Loganathan, D.; Goldstein I. J.; Schultz, P. G.; Gallop, M. A. *Proc. Natl. Acad. Sci. USA* **1992**, *89*, 5393.

80. Brenner, S.; Lerner, R. A. *Proc. Natl. Acad. Sci. USA* **1992**, *89*, 5381.

81. (a) Ohlmeyer, M. H.; Swanson, R. N.; Dillard, L. W.; Reader, J. C.; Asouline, G.; Kobayashi, R.; Wigler, M.; Still, W. C. *Proc. Natl. Acad. Sci. USA* **1993**, *90*, 10922. (b) Nestler, H. P.; Bartlett, P. A.; Still, W. C. *J. Org. Chem.* **1994**, *59*, 4723.

82. Ni, Z. J.; Maclean, D.; Holmes, C. P.; Ruhland, B.; Murphy, M. M.; Jacobs, J. W.; Gordon, E. M.; Gallop, M. A. *J. Med. Chem.* **1996**, *39*, 1601.

83. (a) Geysen, H. M.; Wagner, C. D.; Bodnar, W. M.; Markworth, C. J.; Parke, G. J.; Schoenen, F. J.; Wagner, D. S.; Kinder, D. S. *Chem. Biol.* **1996**, *3*, 679. (b) Edwards, P. N.; Main, B. G.; Shute, R. E. *U. K. Patent Appl. GB 2,297,551 A* (1996).

84. Still, W. C. *Acc. Chem. Res.* **1996**, *29*, 155.

85. Liu, R.; Marik, J.; Lam, K. S. *J. Am. Chem. Soc.* **2002**, *124*, 7678.

86. Liu, G.; Lam, K. S. In *Combinatorial Chemistry. A Practical Approach Series*, Fenniri, H. (Ed.), Oxford University Press, New York, 2000, pp. 33–49.

87. Song, A.; Zhang, J.; Lebrilla, C. B.; Lam, K. S. *J. Am. Chem. Soc.* **2003**, *125*, 6180.

88. (a) Moran, E. J.; Sarshan, S.; Cargill, J. F.; Shahbaz, M. M.; Lio, A.; Mjalli, A. M. M.; Armstrong, R. W. *J. Am. Chem. Soc.* **1995**, *117*, 10787. (b) Nicolaou, K. C.; Xiao, X. Y.; Paradoosh, Z.; Senyei, A.; Nova, M. P. *Angew. Chem. Int. Ed. Engl.* **1995**, *34*, 2289.

89. Vaino, A. R.; Janda, K. D. *Proc. Natl. Acad. Sci. USA* **2000**, *97*, 7692.

90. Winograd, N.; Braun, R. M. *PharmaGenomics* **2002**, *2*, 34.

91. Dixon, St. L.; Villar, H. O. *J. Chem. Inf. Comp. Sci.* **1998**, *38*, 1192.

92. Borman, S. *Chem. Eng. News* **1998** (April 6) 47.

93. Plunkett, M.J.; Ellman, J. A. *J. Am. Chem. Soc.* **1995**, *117*, 3306.

94. Henkel, T.; Brunne, R. M.; Müller, H.; Reichel, F. *Angew. Chem. Int. Ed. Engl.* **1999**, *38*, 643.

95. (a) Harvey, A. *Drug Discovery Today* **2000**, *5*, 294. (b) Feher, M.; Schmidt, J. M. *J. Chem. Inf. Comput. Sci.* **2003**, *43*, 218.

96. Martin, Y. C.; Kofron, J. L.; Traphagen, L. M. *J. Med. Chem.* **2002**, *45*, 4350.

97. Ng, H. P.; May, K.; Bauman, J. G.; Ghannam, A.; Islam, I.; Liang, M.; Horuk, R.; Hesselgesser, J.; Snider, R. M.; Perez, H. D.; Morrissey, M. M. *J. Med. Chem.* **1999**, *42*, 4680.

98. (a) Nicolaou, K. C.; Pfefferkorn, J. A.; Roecker, A. J.; Cao, G.-Q.; Barluenga, S.; Mitchell, H. J. *J. Am. Chem. Soc.* **2000**, *122*, 9939. (b) Nicolaou, K. C.; Pfefferkorn, J. A.; Mitchell, H. J.; Roecker, A. J.; Barluenga, S.; Cao, G.-Q.; Affleck, R. L.; Lillig, J. E. *J. Am. Chem. Soc.* **2000**, *122*, 9954. (c) Nicolaou, K. C.; Pfefferkorn, J. A.; Barluenga, S.; Mitchell, H. J.; Roecker, A. J.; Cao, G.-Q. *J. Am. Chem. Soc.* **2000**, *122*, 9968.

99. Shuker, S. B.; Hajduk, P. J.; Meadows, R. P.; Fesik, S. W. *Science* **1996**, *274*, 1531.

100. (a) Whittaker, M.; Floyd, C. D.; Brown, P.; Gearing, J. H. *Chem. Rev.* **1999**, *99*, 2735. (b) Woessner, J. F., Jr. *FASEB J.* **1991**, *5*, 2145.

101. (a) Hajduk, P. J.; Sheppard, G.; Nettesheim, D. G.; Olejniczak, E. T.; Shuker, S. B.; Meadows, R. P.; Steinman, D. H.; Carrera, G. M.; Marcotte, P. A.; Severin, J.; Walter, K.; Smith, H.; Gubbins, E.; Simmer, R.; Holtzman, T. F.; Morgan, D. W.; Davidsen, S. K.; Summers, J. B.; Fesik, S. W. *J. Am. Chem. Soc.* **1997**, *119*, 5818. (b) Olejniczak, E. T.; Hajduk, P. J.; Marcotte, P. A.; Nettesheim, D. G.; Meadows, R. P.; Edalji, R.; Holtzman, T. F.; Fesik, S. W. *J. Am. Chem. Soc.* **1997**, *119*, 5828.

102. Pervushin, K.; Rick, R.; Wider, G.; Wüthrich, K. *Proc. Natl. Acad. Sci. USA* **1997**, *94*, 12366.

103. (a) Petsko, G. A. *Nature* **1996**, *384 (Supp. 7)*, 7. (b) Blundell, T. L. *Nature* **1996**, *384 (Supp. 7)*, 23. (c) Martin, J. L. *Curr. Med. Chem.* **1996**, *3*, 419.

104. Hajduk, P. J.; Gerfin, T.; Böhlen, J.-M.; Häberli, M.; Marek, D.; Fesik, S. W. *J. Med. Chem.* **1999**, *42*, 2315.

105. Dalvit, C.; Flocco, M.; Knapp, S.; Mostardini, M.; Perego, R.; Stockman, B. J.; Veronesi, M.; Varasi, M. *J. Am. Chem. Soc.* **2002**, *124*, 7702.

106. Maly, D. J.; Choong, I. C.; Ellman, J. A. *Proc. Natl. Acad. Sci. USA* **2000**, *97*, 2419.

107. (a) Griffey, R. H.; Hofstadler, S. A.; Sannes-Lowery, K. A.; Ecker, D. J.; Crooke, S. T. *Proc. Natl. Acad. Sci. USA* **1999**, *96*, 10129. (b) Hofstadler, S. A.; Sannes-Lowery, K. A.; Crooke, S. T.; Ecker, D. J.; Sasmor, H.; Manalili, S.; Griffey, R. H. *Anal. Chem.* **1999**, *71*, 3436. (c) Griffey, R. H.; Sannes-Lowery, K. A.; Drader, J. J.; Mohan, V.; Swayze, E. E.; Hofstadler, S. A. *J. Am. Chem. Soc.* **2000**, *122*, 9933.

108. Swayze, E. E.; Jefferson, E. A.; Sannes-Lowery, K. A.; Blyn, L. B.; Risen, L. M.; Arakawa, S.; Osgood, S. A.; Hofstadler, S. A.; Griffey, R. H. *J. Med. Chem.* **2002**, *45*, 3816.

109. (a) Swanson, H. H. In *Brain Mechanism Psychotropic Drugs*, Baskys, A.; Remington, G. (Eds.), CRC Press, Boca Raton, FL, 1996, pp. 131–150. (b) Calza, L.; Giardino, L.; Ceccatelli, S.; Hokfelt, T.; Zanni, M.; Velardo, A. In *Clinical Perspective on Endogenous Opioid Peptides*, Wiley, Chichester, 1992, pp. 233–254.

110. Boman, H. G. *Annu. Rev. Immunol.* **1995**, *13*, 61.

111. Harder, J.; Bartels, J.; Christophers, E.; Schröder, J.-M. *Nature* **1997**, *387*, 861.

112. (a) Horvat, S. *Curr. Med. Chem.: Cent. Nerv. Syst. Agents* **2001**, *1*, 133. (b) Vaccarino, A. L.; Kastin, A. J. *Peptides* **2000**, *21*, 1975. (c) Roques, B. P.; Noble, F.; Fournie-Zaluski, M.-C. In *Opioids Pain Control*, Stein, C. (Ed.), Cambridge University Press, Cambridge, UK, 1999, pp. 21–45. (d) Stein, C.; Cabot, P. J.; Schafer, M. In *Opioids Pain Control*, Stein, C. (Ed.), Cambridge University Press, Cambridge, UK, 1999, pp. 96–108.

113. Sagnella, G. A. *Cardivasc. Res.* **2001**, *51*, 416.

114. Boccardo, F.; Amoroso, D. *Chemotherapy* **2001**, *47 (Supp. 2)*, 67.

115. Hughes, J.; Smith, T. W.; Kosterlitz, H. W.; Fothergill, L. A.; Morgan, B. A.; Morris, H. R. *Nature* **1975**, *258*, 577.

116. (a) Smith, G. S.; Griffin, J. F. *Science* **1978**, *199*, 1214. (b) Bradbury, A. F.; Smyth, D. G.; Snell, C. R. *Nature* **1976**, *260*, 165.

117. Farmer, P. S. In *Drug Design*, Ariëns, E. J. (Ed.), Academic Press, New York, 1980.

118. (a) Ripka, A. S.; Rich, D. H. *Curr. Opin. Chem. Biol.* **1998**, *2*, 441. (b) Estiarte, M. A.; Rich, D. H. In *Burger's Medicinal Chemistry and Drug Discovery*, 6th ed., Abraham, D. (Ed.), John Wiley, New York, 2002, Vol. I.

119. Giannis, A.; Kolter, T. *Angew. Chem. Int. Ed. Engl.* **1993**, *32*, 1244.

120. (a) Hsieh, K.-H.; LaHann, T. R.; Speth, R. C. *J. Med. Chem.* **1989**, *32*, 898. (b) Corey, E. J.; Link, J. O. *J. Am. Chem. Soc.* **1992**, *114*, 1906. (c) Schiller, P. W.; Weltrowka, G.; Nguyen, T. M. D.; Lemieux. C.; Chung, N. N.; Marken, B. J.; Wilke, B. C. *J. Med. Chem.* **1991**, *34*, 3125. (d) Holladay, M. W.; Lin, C. W.; May, C.; Garvey, D.; Witte, D. G.; Miller, T. R.; Wolfram, C. A. W.; Nadzan, A. M. *J. Med. Chem.* **1991**, *34*, 455.

121. Yanagisawa, H.; Ishihara, S.; Ando, A.; Kanazaki, T.; Miyamoto, S.; Koike, H.; Iijima, Y.; Oizumi, K.; Matsushita, Y.; Hata, T. *J. Med. Chem.* **1987**, *30*, 1984. (b) Giannis, A.; Kolter, T. *Angew. Chem. Int. Ed. Engl.* **1993**, *32*, 1244.

122. Olson, G. L.; Bolin, D. R.; Bonner, M. P.; Bös, M.; Cook, C. M.; Fry, D. C.; Graves, B. J.; Hatada, M.; Hill, D. E.; Kahn, M.; Madison, V. S.; Rusiecki, V. K.; Sarabu, R.; Sepinwall, J.; Vincent, G. P.; Voss, M. E. *J. Med. Chem.* **1993**, *36*, 3039.

123. (a) Nagai, U.; Sato, K.; Nakamura, R.; Kato, R. *Tetrahedron* **1993**, *49*, 3577. (b) Sato, K.; Nagai, U. *J. Chem. Soc. Perkin Trans 1*, **1986**, 1231.

124. (a) Brandmeier, V.; Sauer, W. H. B.; Feigel, M. *Helv. Chim. Acta* **1994**, *77*, 70. (b) Wagner, G.; Feigel, M. *Tetrahedron* **1993**, *49*, 10831.

125. (a) Kemp, D. S.; Curran, T. P.; Davis, W. M.; Boyd, J. G.; Muendel, C. *J. Org. Chem.* **1991**, *56*, 6672. (b) Kemp, D. S.; Curran, T. P.; Boyd, J. G.; Allen, T. J. *J. Org. Chem.* **1991**, *56*, 6683.

126. Sarabu, R.; Lovey, K.; Madison, V. S.; Fry, D. C.; Greeley, D. W.; Cook, C. M.; Olson, G. L. *Tetrahedron* **1993**, *49*, 3629.

127. Smith, A. B. III; Keenan, T. P.; Holcomb, R. C.; Sprengeler, P. A.; Guzman, M. C.; Wood, J. L.; Carroll, P. J.; Hirschmann, R. *J. Am. Chem. Soc.* **1992**, *114*, 10672.

128. Lincoff, A. M.; Califf, R. M.; Topol, E. J. *J. Am. Coll. Cardiol.* **2000**, *35*, 1103.

129. Hirschmann, R.; Sprengeler, P. A.; Kawasaki, T.; Leahy, J. W.; Shakespeare, W. C.; Smith, A. B. III *J. Am. Chem. Soc.* **1992**, *114*, 9699.

130. Fisher, M. J.; Gunn, B.; Harms, C. S.; Kline, A. D.; Mullaney, J. T.; Nunes, A.; Scarborough, R. M.; Arfsten, A. E.; Skelton, M. A.; Um, S. L.; Utterback, B. G.; Jakubowski, J. A. *J. Med. Chem.* **1997**, *40*, 2085.

131. Blackburn, B. K.; Lee, A.; Baier, M.; Kohl, B.; Olivero, A. G.; Matamoros, R.; Robarge, K. D.; McDowell, R. S. *J. Med. Chem.* **1997**, *40*, 717.

132. Hirschmann, R.; Nicolaou, K. C.; Pietranico, S.; Salvino, J.; Leahy, E. M.; Sprengeler, P. A.; Furst, G.; Smith, A. B. III.; Strader, C. D.; Cascieri, M. A.; Candelore, M. R.; Donaldson, C.; Vale, W.; Maechler, L. *J. Am. Chem. Soc.* **1992**, *114*, 9217.

133. Miyamoto, M.; Yamazaki, N.; Nagaoka, A.; Nigawa, Y. *Ann. N.Y. Acad. Sci.* **1989**, 508.

134. Olson, G. L.; Cheung, H.-C.; Chiang, E.; Madison, V. S.; Sepinwall, J.; Vincent, G. P.; Winokur A.; Gary, K. A. *J. Med. Chem.* **1995**, *38*, 2866.

135. Spatola, A. F. *Chem. Biochem. Amino Acids Pept. Prot.* **1983**, *7*, 267.

136. Wiley, R. A.; Rich, D. H. *Med. Res. Rev.* **1993**, *13*, 327.

137. (a) Ren, S.; Lien, E. J. *Prog. Drug Res.* **1998**, *51*, 1. (b) Tomasselli, A. G.; Heinrikson, R. L. *Biochim. Biophys. Acta* **2000**, *1477*, 189.

138. Gante, J. *Synthesis* **1989**, 405.

139. Han, H.; Janda, K. D. *J. Am. Chem. Soc.* **1996**, *118*, 2539.

140. (a) Simon, R. J.; Kania, R. S.; Zuckermann, R. N.; Huebner, V. D.; Jewell, D. A.; Banville, S.; Ng, S.; Wang, L.; Rosenberg, S.; Marlowe, C. K.; Spellmeyer, D. C.; Tan, R. Y.; Frankel, A. D.; Santi, D. V.; Cohen, F. E.; Bartlett, P. A. *Proc. Natl. Acad. Sci. USA* **1992**, *89*, 9367. (b) Zuckermann, R. N.; Kerr, J. M.; Kent, S. B. H.; Moos, W. H. *J. Am. Chem. Soc.* **1992**, *114*, 10646.

141. Edwards, R. A.; Zhang, K.; Ferth, L. *Drug Discovery World* **2002**, *3*, 67.

142. (a) Prentis, R. A.; Lis, Y.; Walker, S. R. *Brit. J. Clin. Pharmacol.* **1988**, *25*, 387. (b) Kennedy, T. *Drug Discovery Today* **1997**, *2*, 436.

143. Cole, M. J.; Janiszewski, J. S.; Fouda, H. G. In *Practical Spectroscopy*, Pramanik, B. N.; Ganguly, A. K.; Gross, M. L. (Eds.), Marcel Dekker, New York 2002, Vol. 32, pp. 211–249.

144. Gorswant, C. V.; Thoren, P.; Engstrom, S. *J. Pharm. Sci.* **1998**, *87*, 200.

145. Yee, S. *Pharm. Res.* **1997**, *6*, 763.

146. Smith, D. A.; Jones, B. C.; Walker, D. K. *Med. Res. Rev.* **1996**, *16*, 243.

147. (a) Hansch, C.; Maloney, P. P.; Fujita, T.; Muir, R. M. *Nature* **1962**, *194*, 178. (b) Fujita, T.; Iwasa, J.; Hansch, C. *J. Am. Chem. Soc.* **1964**, *86*, 5175.

148. Singer, S. J.; Nicolson, G. L. *Science* **1972**, *175*, 720.

149. Richet, M. C. *Compt. Rend. Soc. Biol.* **1893**, *45*, 775.

150. Overton, E. Z. *Phys. Chem.* **1897**, *22*, 189.

151. Meyer, H. *Arch. Exp. Pathol. Pharmacol.* **1899**, *42*, 109.

152. Hansch, C.; Maloney, P. P.; Fujita, T.; Muir, R. M. *Nature* **1962**, *194*, 178.

153. Fujita, T.; Iwasa, J.; Hansch, C. *J. Am. Chem. Soc.* **1964**, *86*, 5175.

154. Sangster, J. *Octanol-Water Partition Coefficients: Fundamentals and Physical Chemistry*, Wiley, New York, 1997, pp. 79–112.

155. (a) Collander, R. *Physiol. Plant* **1954**, *7*, 420. (b) Collander, R. *Acta Chem. Scand.* **1951**, *5*, 774.

156. Hansch, C.; Steward, A. R.; Anderson, S. M.; Bentley, D. *J. Med. Chem.* **1968**, *11*, 1.

157. Hansch, C.; Steward, A. R.; Anderson, S. M.; Bentley, D. *J. Med. Chem.* **1968**, *11*, 1.

158. (a) Scherrer, R. A.; Howard, S. M. *J. Med. Chem.* **1977**, *20*, 53. (b) Stopher D.; McClean, S. *J. Pharm. Pharmacol.* **1990**, *42*, 144.

159. (a) Hansch, C.; Maloney, P. P.; Fujita, T.; Muir, R. M. *Nature* **1962**, *194*, 178. (b) Hansch, C.; Fujita, T. *J. Am. Chem. Soc.* **1964**, *86*, 1616. (c) Fujita, T.; Iwasa, J.; Hansch, C. *J. Am. Chem. Soc.* **1964**, *86*, 5175.

160. Hansch, C.; Leo, A. *Substituent Constants for Correlation Analysis in Chemistry and Biology*, Wiley, New York, 1979.

161. Leo, A.; Hansch, C.; Elkins, D. *Chem. Rev.* **1971**, *71*, 525.

162. Iwasa, J.; Fujita, T.; Hansch, C. *J. Med. Chem.* **1965**, *8*, 150.

163. Leo, A.; Hansch, C.; Elkins, D. *Chem. Rev.* **1971**, *71*, 525.

164. Sangster, J. *Octanol-Water Partition Coefficients: Fundamentals and Physical Chemistry*, Wiley, New York, 1997, pp. 79–112.

165. Lombardo, F.; Shalaeva, M. Y.; Tupper, K. A.; Gao, F.; Abraham, M. H. *J. Med. Chem.* **2000**, *43*, 2922.

166. Lombardo, F.; Shalaeva, M. Y.; Tupper, K. A.; Gao, F.; Abraham, M. H. *J. Med. Chem.* **2001**, *44*, 2490.

167. (a) Chou, J. T.; Jurs, P. C. *J. Chem. Inf. Comput. Sci.* **1979**, *19*, 171. (b) Pomona College Medicinal Chemistry Project; see Hansch, C.; Björkroth, J. P.; Leo, A. *J. Pharm. Sci.* **1987**, *76*, 663.

168. (a) Bodor, N.; Gabanyi, Z.; Wong, C.-K. *J. Am. Chem. Soc.* **1989**, *111*, 3783. (b) Bodor, N.; Gabanyi, Z.; Wong, C.-K. *J. Am. Chem. Soc.* **1989**, *111*, 8062.

169. Moriguchi, I.; Hirono, S.; Liu, Q.; Nakagome, I.; Matsushita, Y. *Chem. Pharm. Bull.* **1992**, *40*, 127.

170. Moriguchi, I.; Hirono, S.; Liu, Q.; Nakagome, Y.; Matsushita, Y. *Chem. Pharm. Bull.* **1994**, *42*, 976. (b) Leo, A. J. *Chem. Pharm. Bull.* **1995**, *43*, 512.

171. Gobas, F. A. P. C.; Lahittete, J. M.; Garofalo, G.; Shiu, W. Y.; Mackay, D. *J. Pharm. Sci.* **1988**, *77*, 265.

172. Abraham, M. H.; Lieb, W. R.; Franks, N. P. *J. Pharm. Sci.* **1991**, *80*, 719.

173. Seiler, P. *Eur. J. Med. Chem.* **1974**, *9*, 473.

174. Stearn, A.; Stearn, E. *J. Bacteriol.* **1924**, *9*, 491.

175. (a) Albert, A.; Rubbo, S.; Goldacre, R. *Nature (London)* **1941**, *147*, 332. (b) Albert, A.; Rubbo, S.; Goldacre, R.; Davey, M.; Stone, J. *Brit. J. Exp. Pathol.* **1945**, *26*, 160. (c) Albert, A.; Goldacre, R. *Nature (London)* **1948**, *161*, 95. (d) Albert, A. *The Acridines, Their Preparation, Properties, and Uses*, 2nd ed., Edward Arnold, London, 1966.

176. Albert, A. *Selective Toxicity*, 7th ed., Chapman and Hall, London, 1985, p. 398.

177. Burns, J.; Yuü, T.; Dayton, P.; Gutman, A.; Brodie, B. *Ann. N.Y. Acad. Sci.* **1960**, *86*, 253.

178. Miller, G.; Doukos, P.; Seydel, J. *J. Med. Chem.* **1972**, *15*, 700.

179. Simonson, T.; Brooks, C. L. III *J. Am Chem. Soc.* **1996**, *118*, 8452.

180. (a) Parsons, S. M.; Raftery, M. A. *Biochemistry* **1972**, *11*, 1623. (b) Parsons, S. M.; Raftery, M. A. *Biochemistry* **1972**, *11*, 1630. (c) Parsons, S. M.; Raftery, M. A. *Biochemistry* **1972**, *11*, 1633.

181. (a) Wu, Z. R.; Ebrahimian, S.; Zawrotny, M. E.; Thornburg, L. D.; Perez-Alvarado, G. C.; Brothers, P.; Pollack, R. M.; Summers, M. F. *Science* **1997**, *276*, 415. (b) Cho, H.-S.; Choi, G.; Choi, K. Y.; Oh, B.-H. *Biochemistry* **1998**, *37*, 8325.

182. (a) Harris, J. M.; McIntosh, E. M.; Muscat, G. E. O. *J. Mol. Biol.* **1999**, *288*, 275. (b) Schmidt, D. E.; Westheimer, F. H. *Biochemistry* **1971**, *10*, 1249.

183. (a) Lipinski, C. A.; Lombardo, F.; Dominy, B.-W.; Feeney, P. J. *Adv. Drug Delivery Rev.* **1997**, *23*, 3. (b) Lipinski, C. A. *J. Pharmacol. Toxicol. Meth.* **2001**, *44*, 235. (c) Lipinski, C. A.; Lombardo, F.; Dominy, B. W.; Feeney, P. J. *Adv. Drug Delivery Rev.* **2001**, *46*, 3.

184. Lipinski, C. A., personal communication.

185. Veber, D. F.; Johnson, S. R.; Cheng, H.-Y.; Smith, B. R.; Ward, K. W.; Kopple, K. D. *J. Med. Chem.* **2002**, *45*, 2615.

186. (a) Wenlock, M. C.; Austin, R. P.; Barton, P.; Davis, A. M.; Leeson, P. D. *J. Med. Chem.* **2003**, *46*, 1250. (b) Ajay; Bemis, G. W.; Murcko, M. A. *J. Med. Chem.* **1999**, *42*, 4942.

187. (a) Wessel, M. D.; Mente, S. *Annu. Rept. Med. Chem.* **2001**, *36*, 257. (b) Blake, J. F. *Curr. Opin. Biotechnol.* **2000**, *11*, 104. (c) Clark, D. E.; Pickett, S. D. *Drug Discovery Today* **2000**, *5*, 49. (d) Kramer, S. D. *Pharm. Sci. Technol. Today* **1999**, *2*, 373.

188. Mandagere, A.; Thompson, T. N.; Hwang, K.-K. *J. Med. Chem.* **2002**, *45*, 304.

189. (a) Crum-Brown, A.; Fraser, T. R. *Trans. Roy. Soc. Edinburgh* **1868–1869**, *25*, 151 and 693. (b) Crum-Brown, A.; Fraser, T. R. *Proc. Roy. Soc. Edinburgh* **1869**, *6*, 556. (c) Crum-Brown, A.; Fraser, T. R. *Proc. Roy. Soc. Edinburgh* **1872**, *7*, 663.

190. Hansch, C.; Maloney, P. P.; Fujita, T.; Muir, R. M. *Nature* **1962**, *194*, 178.

191. Richet, M. C. *Compt. Rend. Soc. Biol.* **1893**, *45*, 775.

192. Overton, E. *Z. Phys. Chem.* **1897**, *22*, 189.

193. Meyer, H. *Arch. Exp. Pathol. Pharmacol.* **1899**, *42*, 109.

194. Ferguson, J. *Proc. Roy. Soc. London, Ser. B* **1939**, *127*, 387.

195. Hansch, C.; Muir, R. M.; Metzenberg, R. L., Jr. *Plant Physiol.* **1951**, *26*, 812.

196. (a) Hansch, C.; Maloney, P. P.; Fujita, T.; Muir, R. M. *Nature* **1962**, *194*, 178. (b) Hansch, C.; Fujita, T. *J. Am. Chem. Soc.* **1964**, *86*, 1616.

197. Taft, R. W., In *Steric Effects in Organic Chemistry*; Neuman, M. S. (Ed.), Wiley, New York, 1956, pp. 556–675.

198. Unger, S. H.; Hansch, C. *Prog. Phys. Org. Chem.* **1976**, *12*, 91.

199. Hancock, C. K.; Meyers, E. A.; Yager, B. J. *J. Am. Chem. Soc.* **1961**, *83*, 4211.

200. Hansch, C.; Leo, A.; Unger, S. H.; Kim, K. H.; Nikaitani, D.; Lien, E. J. *J. Med. Chem.* **1973**, *16*, 1207.

201. (a) Verloop, A.; Hoogenstraaten, W.; Tipker, J. In *Drug Design*, Ariens, E. J. (Ed.), Academic Press, New York, 1976, Vol. VII, pp. 165–207. (b) Draber, W. Z. *Naturforsch., C: Biosci.* **1996**, *51*, 1.

202. Hansch, C.; Fujita, T. *J. Am. Chem. Soc.* **1964**, *86*, 1616.

203. (a) Daniel, C.; Wood, F. S. *Fitting Equations to Data*, Wiley, New York, 1971. (b) Draper, N. R.; Smith, H. *Applied Regression Analysis*, Wiley, New York, 1966. (c) Snedecor, G. W.; Cochran, W. G. *Statistical Methods*, Iowa State University Press, Ames, 1967.

204. Deardon, J. C. In *Trends in Medicinal Chemistry*, Mutschler, E.; Winterfeldt, E. (Eds.), VCH, Weinheim, 1987, pp. 109–123.

205. Martin, Y. C. *Quantitative Drug Design: A Critical Introduction*, Marcel Dekker, New York, 1978, Chap. 2.

206. Tute, M. S. In *Physical Chemical Properties of Drugs*, Yalkowsky, S. H.; Sinkula, A. A.; Valvani, S. C. (Eds.), Marcel Dekker, New York, 1980, p. 141.

207. (a) Unger, S. H. In QSAR in *Design of Bioactive Compounds*; Kuchar, M. (Ed.), Prous, Barcelona, 1984, pp. 1–9. (b) Hopfinger, A. J. *J. Med. Chem.* **1985**, *28*, 1133. (c) Fujita, T. In *Drug Design: Fact or Fantasy?*, Jolles, G.; Wooldridge, K. R. H. (Eds.), Academic Press, London, 1984, Chap. 2. (c) Kubinyi, H. *QSAR: Hansch Analysis and Related Approaches*; Vol. 1 of *Methods and Principles in Medicinal Chemistry*, Mannhold R. (Ed.), VCH, Weinheim, 1993.

208. (a) Free, S. M., Jr.; Wilson, J. W. *J. Med. Chem.* **1964**, *7*, 395. (b) Tomic, S.; Nilsson, L.; Wade, R. C. *J. Med. Chem.* **2000**, *43*, 1780.

209. (a) Blankley, C. J. In *Quantitative Structure–Activity Relationships of Drugs*, Topliss, J. G. (Ed.), Academic Press, New York, 1983, Chap. 1. (b) Schaad, L. J.; Hess, B. A., Jr.; Purcell, W. P.; Cammarata, A.; Franke, R.; Kubinyi, H. *J. Med. Chem.* **1981**, *24*, 900.

210. Fujita, T.; Ban, T. *J. Med. Chem.* **1971**, *14*, 148.

211. (a) Martin, Y. C. *Quantitative Drug Design: A Critical Introduction*, Marcel Dekker, New York, 1978, Chap. 2. (b) Tute, M. S. In *Physical Chemical Properties of Drugs*, Yalkowsky, S. H.; Sinkula, A. A.; Valvani, S. C. (Eds.), Marcel Dekker, New York, 1980, p. 141.

212. (a) Bocek, K.; Kopecký, J.; Krivucová, M.; Vlachová, D. *Experientia* **1964**, *20*, 667. (b) Kopecký, J.; Bocek, K.; Vlachová, D. *Nature (London)* **1965**, *207*, 981.

213. Fried, J.; Borman, A. *Vitam. Horm.* **1958**, *16*, 303.

214. Topliss, J. G. *J. Med. Chem.* **1972**, *15*, 1006.

215. Granito, C. E.; Becker, G. T.; Roberts, S.; Wiswesser, W. J.; Windlinz, K. J. *J. Chem. Doc.* **1971**, *11*, 106.

216. Goodford, P. J. *Adv. Pharmacol. Chemother.* **1973**, *11*, 51.

217. (a) Craig, P. N. *J. Med. Chem.* **1971**, *14*, 680. (b) Craig, P. N. In *Burger's Medicinal Chemistry*, 4th ed., Wolff, M. E. (Ed.), Wiley, New York, 1980, Part I, Chap. 8.

218. (a) Bustard, T. M. *J. Med. Chem.* **1974**, *17*, 777. (b) Santora, N. J.; Auyang, K. *J. Med. Chem.* **1975**, *18*, 959. (c) Deming, S. N. *J. Med. Chem.* **1976**, *19*, 977.

219. Darvas, F. *J. Med. Chem.* **1974**, *17*, 799.

220. Topliss, J. G. *J. Med. Chem.* **1977**, *20*, 463.

221. Hansch, C.; Unger, S. H.; Forsythe, A. B. *J. Med. Chem.* **1973**, *16*, 1217.

222. Martin, Y. C. In *Drug Design*, Ariens, E. J. (Ed.), Academic Press, New York, 1979, Vol. VIII, p. 5.

223. Srivastava, S.; Richardson, W. W.; Bradley, M. P.; Crippen, G. M. In *3D-QSAR in Drug Design: Theory, Methods, and Applications*, Kubinyi, H. (Ed.), ESCOM, 1993, pp. 409–430.

224. (a) Crippen, G. M. *J. Med. Chem.* **1979**, *22*, 988. (b) Ghose, A. K.; Crippen, G. M. *J. Med. Chem.* **1982**, *25*, 892. (c) Crippen, G. M. *Distance Geometry and Conformational Calculations*, Research Studies Press, New York, 1981.

225. (a) Sheridan, R. P.; Nilakantan, R.; Dixon, J. S.; Venkataraghavan, R. *J. Med. Chem.* **1986**, *29*, 899. (b) Sheridan, R. P.; Venkataraghaven, R. *Acc. Chem. Res.* **1987**, *20*, 322.

226. (a) Martin, Y. C.; Bures, M. G.; Willett, P. In *Reviews in Computational Chemistry*, Lipkowitz, K. B.; Boyd, D. B. (Eds.), Wiley-VCH, New York, 1990, Vol. 1, p. 213. (b) Kearsley, S. K.; Underwood, D. J.; Sheridan, R. P.; Miller, M. D. *J. Comput. Aided Mol. Design* **1994**, *8*, 565.

227. (a) Nicklaus, M. C.; Neamati, N.; Hong, H.; Mazumder, A.; Sunder, S.; Chen, J.; Milne, G. W.; Pommier, Y. *J. Med. Chem.* **1997**, *40*, 920. (b) Hong, H.; Neamati, N.; Wang, S.; Nicklaus, M. C.; Mazumder, A. Zhao, H.; Burke, T. R. J.; Pommier, Y.; Milne, G. W. *J. Med. Chem.* **1997**, *40*, 930.

228. (a) Hopfinger, A. J. *J. Am. Chem. Soc.* **1980**, *102*, 7196. (b) Hopfinger, A. J. *J. Med. Chem.* **1981**, *24*, 818. (c) Hopfinger, A. J. *J. Med. Chem.* **1983**, *26*, 990.

229. Carhart, R. E.; Smith, D. H.; Venkataraghavan, R. *J. Chem. Inf. Comput. Sci.* **1985**, *25*, 64.

230. Nilakantan, R.; Bauman, N.; Dixon, J. S.; Venkataraghavan, R. *J. Chem. Inf. Comput. Sci.* **1987**, *27*, 82.

231. Cramer, R. D. III; Patterson, D. E.; Bunce, J. D. *J. Am. Chem. Soc.* **1988**, *110*, 5959.

232. Kim, K. H.; Greco, G.; Novellino, E. In *3D QSAR in Drug Design*, Kybinyi, H.; Folkers, G.; Martin, Y. C. (Eds.), Kluwer Academic, Dordrecht, 1998, Vol. 3, p. 257.

233. Green, S. M.; Marshall, G. R. *TIPS* **1995**, *16*, 285.

234. Silverman, B. D.; Platt, D. E.; Pitman, M.; Rigoutsos, I. In *3D QSAR in Drug Design*, Kybinyi, H.; Folkers, G.; Martin, Y. C. (Eds.), Kluwer Academic, Dordrecht, 1998, Vol. 3, p. 183.

235. Heritage, T. W.; Ferguson, A. M.; Turner, D. B.; Willett, P. In *3D QSAR in Drug Design*, Kybinyi, H.; Folkers, G.; Martin, Y. C. (Eds.), Kluwer Academic, Dordrecht, 1998, Vol. 2, p. 381.

236. Todeschini, R.; Gramatica, P. In *3D QSAR in Drug Design*, Kybinyi, H.; Folkers, G.; Martin, Y. C. (Eds.), Kluwer Academic, Dordrecht, 1998, Vol. 2, p. 355.

237. Goodford, P. J. *J. Med. Chem.* **1985**, *28*, 849.

238. (a) Kellogg, G. E.; Semus, S. F.; Abraham, D. J. *J. Comput. Aided Mol. Des.* **1991**, *5*, 545. (b) Wireko, F. C.; Kellogg, G. E.; Abraham, D. J. *J. Med. Chem.* **1991**, *34*, 758.

239. Doweyko, A. M. *J. Med. Chem.* **1988**, *31*, 1396.

240. Bartlett, P. A.; Shea, G. T.; Telfer, S. J.; Waterman, S. In *Chemical and Biological Problems in Molecular Recognition*, Roberts, S. M.; Ley, S. V.; Campbell, M. M. (Eds.), Royal Society of Chemistry, London, 1989, pp. 182–196.

241. VanDrie, J. H.; Weininger, D.; Martin, Y. C. *J. Comput. Aided Mol. Des.* **1989**, *3*, 225.

242. Lewis, R. A.; Dean, P. M. *Proc. Roy. Soc. London* **1989**, *B236*, 125 and 141.

243. Kawakami, Y.; Inoue, A.; Kawai, T.; Wakita, M.; Sugimoto, H.; Hopfinger, A. J. *Bioorg. Med. Chem.* **1996**, *4*, 1429.

244. Hassall, C. H. *Chem. Brit.* **1985**, *21*, 39.

245. Levinthal, C. *Sci. Am.* **1966**, *214*, 42.

246. (a) Barry, C. D.; North, A. C. T. *Cold Spring Harbor Quant. Biol.* **1971**, *36*, 577. (b) Barry, C. D. *Nature (London)* **1971**, *232*, 236.

247. http://www.rcsb.org/pdb

248. Tollenaere, J. P.; Janssen, P. A. *J. Med. Res. Rev.* **1988**, *8*, 1.

249. (a) Gund, P.; Halgren, T. A.; Smith, G. M. *Annu. Rep. Med. Chem.* **1987**, *22*, 269. (b) Gund, T.; Gund, P. In *Molecular Structure and Energetics*, Liebman, J. F.; Greenberg, A. (Eds.), VCH, Weinheim, 1987, Vol. 4, Chap. 10.

250. Fischer, E. *Ber. Deut. Chem. Ges.* **1894**, *27*, 2985.

251. (a) Klebe, G. *Perspectives in Drug Discovery and Design* **1995**, *3*, 85. (b) Beusen, D. D.; Shands, E. F. B. *Drug Discovery Today* **1996**, *1*, 429.

252. von Itzstein, M.; Wu, W.-Y.; Kok, G. B.; Pegg, M. S.; Dyason, J. C.; Jin, B.; van Phan, T.; Smythe, M. L.; White, H. F.; Oliver, S. W.; Colman, P. M.; Varghese, J. N.; Ryan, D. M.; Woods, J. M.; Bethell, R. C.; Hotham, V. J.; Cameron, J. M.; Penn, C. R. *Nature* **1993**, *363*, 418.

253. (a) Varghese, J. N.; Laver, W. G.; Colman, P. M. *Nature* **1983**, *303*, 35. (b) Colman, P. M.; Varghese, J. N.; Laver, W. G. *Nature* **1983**, *303*, 41.

254. (a) Varghese, J. N.; Colman, P. M. *J. Mol. Biol.* **1991**, *221*, 473. (b) Varghese, J. N.; McKimm-Breschkin, J. Caldwell, J. B.; Kortt, A. A.; Colman, P. M. *Proteins* **1992**, *14*, 327.

255. Stoll, V.; Stewart, K. D.; Maring, C. J.; Muchmore, S.; Giranda, V.; Gu, Y.-g. Y.; Wang, G.; Chen, Y.; Sun, M.; Zhao, C.; Kennedy, A. L.; Madigan, D. L.; Xu, Y.; Saldivar, A.; Kati, W.; Laver, G.; Sowin, T.; Sham, H. L.; Greer, J.; Kempf, D. *Biochemistry* **2003**, *42*, 718.

256. Anderson, A.; O'Neill, R.; Surti, T.; Stroud, R. *Chem. Biol.* **2001**, *8*, 445.

257. Ojima, I.; Chakravarty, S.; Inoue, T.; Lin, S.; He, L.; Horwitz, S. B.; Kuduk, S. D.; Danishefsky, S. J. *Proc. Natl. Acad. Sci. USA* **1999**, *96*, 4256.

258. (a) Wang, M.; Xia, X.; Kim, Y.; Hwang, D.; Jansen, J. M.; Botta, M.; Liotta, D. C.; Snyder, J. P. *Org. Lett.* **1999**, *1*, 43. (b) Metaferia, B. B.; Hoch, J.; Glass, T. E.; Bane, S. L.; Chatterjee, S. K.; Snyder, J. P.; Lakdawala, A.; Cornett, B.; Kingston, D. G. I. *Org. Lett.* **2001**, *3*, 2461.

259. Kuntz, I. D.; Blaney, J. M.; Oatley, S. J.; Langridge, R.; Ferrin, T. E. *J. Mol. Biol.* **1982**, *161*, 269.

260. (a) Ewing, T. J.; Makino, S.; Skillman, A. G.; Kuntz, I. D. *J. Comput. Aided Mol. Des.* **2001**, *15*, 411. (b) Lorber, D. A.; Shoichet, B. K. *Protein Sci.* **1998**, *7*, 938.

261. (a) DesJarlais, R. L.; Sheridan, R. P.; Dixon, J. S.; Kuntz, I. D. Venkatarghavan, R. *J. Med. Chem.* **1986**, *29*, 2149. (b) DesJarlais, R. L.; Sheridan, R. P.; Seibel, G. L.; Dixon, J. S.; Kuntz, I. D.; Venkataraghavan, R. *J. Med. Chem.* **1988**, *31*, 722.

262. Kuntz, I. D. *Science* **1992**, *257*, 1078.

263. (a) Friedman, S. H.; DeCamp, D. L.; Sijbesma, R. P.; Srdanov, G.; Wudl, F.; Kenyon, G. L. *J. Am. Chem. Soc.* **1993**, *115*, 6506. (b) Friedman, S. H.; Ganapathi, P. S.; Rubin, Y.; Kenyon, G. L. *J. Med. Chem.* **1998**, *41*, 2424.

264. Haddad, J.; Kotra, L. P.; Llano-Sotelo, B.; Kim, C.; Azucena, E. F., Jr.; Liu, M.; Vakulenko, S. B.; Chow, C. S.; Mobashery, S. *J. Am. Chem. Soc.* **2002**, *124*, 3229.

265. (a) Böhm, H.-J. *J. Comput. Aided Mol. Des.* **1992**, 6, 593. (b) Böhm, H.-J. *J. Comput. Aided Mol. Des.* **1992**, 6, 61. (c) Böhm, H.-J. *J. Comput. Aided Mol. Des.* **1994**, 8, 623.

266. (a) Bohacek, R. S.; McMartin, C. *J. Am. Chem. Soc.* **1994**, *116*, 5560. (b) Bohacek, R. S.; McMartin, C. *Curr. Opin. Chem. Biol.* **1997**, *1*, 157.

267. regine@bostondenovo.com

268. Guner, O. F. (Ed.) *Pharmacophore, Perception, Development, and Use in Drug Design*, IUL Biotechnology Series, International University Line, La Jolla, CA, 2000.

269. (a) Sufrin, J. R.; Dunn, D. A.; Marshall, G. R. *Mol. Pharmacol.* **1981**, *19*, 307. (b) Humblet, C.; Marshall, G. R. *Annu. Rep. Med. Chem.* **1980**, *15*, 267. (c) Marshall, G. R. *Ann. N.Y. Acad. Sci.* **1985**, *439*, 162.

270. (a) Brinkworth, R. I.; Fairlie, D. P.; Leung, D.; Young, P. R. *J. Gen. Virol.* **1999**, *80*, 1167. (b) Naus, J. L.; Reid, R. H.; Sadegh-Nasseri, S. *J. Biomol. Struct. Dyn.* **1995**, *12*, 1213. (c) Carlson, G. M.; MacDonald, R. J.; Meyer, E. F., Jr. *J. Theor. Biol.* **1986**, *119*, 107.

271. (a) Kirkpatrick, D. L.; Watson, S.; Ulhaq, S. *Combinatorial Chem. High Throughput Screen.* **1999**, *2*, 211. (b) Antel, J. *Curr. Opin. Drug Discovery Develop.* **1999**, *2*, 224.

272. Henry, C. M. *Chem. Eng. News* **2001** (June 4), 69.

273. (a) Ooms, F. *Curr. Med. Chem.* **2000**, *7*, 141. (b) DiMasi, J. A.; Hansen, R. W.; Grabowski, H. G. *J. Health Econ.* **2003**, *22*, 151.

274. Nies, A. S.; Spielberg, S. P. In *Goodman & Gilman's The Pharmacological Basis of Therapeutics*, 9th ed., Hardman, J. G.; Limbird, L. E.; Molinoff, P. B.; Ruddon, R. W.; Gilman, A. G. (Eds.), McGraw Hill, New York, 1996, p. 43.

275. (a) Hopkins, A. L.; Groom, C. R. *Nat. Rev. Drug Discov.* **2002**, *1*, 727. (b) Drews, J. *Science* **2000**, *287*, 1960.

276. (a) Wang, J. H.; Hewick, R. M. *Drug Discov. Today* **1999**, *4*, 129. (b) Borman, S. *Chem. Eng. News* **2000** (July 31), 31. (c) Debouck, C.; Metcalf, B. *Annu. Rev. Pharmacol. Toxicol.* **2000**, *40*, 193. (d) Brown, T. A. *Genomes*, John Wiley & Sons, NY, 2002.

277. Borman, S. *Chem. Eng. News* **2000** (May 15), 53.

278. Wolff, M. E. *J. Med. Chem.* **2003**, *46*, 3178.

Receptors

3.1 Introduction

Up to this point in our discussion it appears that a drug is taken, it travels through the body to a target site and elicits a pharmacological effect. The site of drug action, which is ultimately responsible for the pharmaceutical effect, is a *receptor*. Allusions were made in Chapter 2 to the binding of a drug to a receptor, which constitutes *pharmacodynamics*. In this chapter the emphasis will be placed on pharmacodynamics of general receptors, that is, how a drug interacts with a receptor; in Chapter 4, a special class of proteins that have catalytic properties, called *enzymes*, will be discussed; and in Chapter 6 a nonprotein receptor, *DNA*, will be the topic of discussion. The drug–receptor properties discussed in this chapter will also apply to drug–enzyme and drug–DNA complexes.

In 1878 John N. Langley, a physiology student at Cambridge University, while studying the mutually antagonistic action of the alkaloids atropine (**3.1**; now used as a smooth muscle relaxant in a variety of drugs, such as Prosed) and pilocarpine (**3.2**; Salagen causes sweating and salivation) on cat salivary flow, suggested that both of these chemicals interacted with some yet unknown substance (no mention of "receptors" was made) in the nerve endings of the gland cells.[1] Langley, however, did not follow up this notion for more than 25 years.

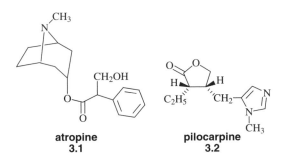

atropine
3.1

pilocarpine
3.2

Paul Ehrlich worked for a dye manufacturing company and was fascinated by the observation that dyes could attach so tightly to fabrics that they could not be removed by washing. He also was intrigued by why different bacteria caused different diseases and thought that the toxins generated by bacteria might produce their effects by attaching tightly to specific sites in the cells of the body, just as dyes attach to fabrics. In 1897 Ehrlich suggested his *side-chain theory*.[2] According to this hypothesis cells have side chains attached to them that contain specific groups capable of combining with a particular group of a toxin. Ehrlich termed these side chains *receptors*. Another ground-breaking facet of this hypothesis was that when toxins combined with the side chains, excess side chains are produced and released into the bloodstream. In today's biochemical vernacular these excess side chains would be called *antibodies*, and they combine with macromolecular toxins stoichiometrically.

In 1905 Langley[3] studied the antagonistic effects of curare (a generic name for a variety of South American quaternary alkaloid poisons that cause muscular paralysis) on nicotine stimulation of skeletal muscle. He concluded that there was a *receptive substance* that received the stimulus and, by transmitting it, caused muscle contraction. This was really the first time that attention was drawn to the two fundamental characteristics of a receptor, namely, a *recognition capacity* for specific molecules and an *amplification component*, the ability of the complex between the molecule and the receptor to initiate a biological response.

Receptors are mostly membrane-bound proteins that selectively bind small molecules, referred to as *ligands*, that elicit some physiological response. Receptors are generally integral proteins that are embedded in the phospholipid bilayer of cell membranes (see Figure 2.14). They, typically, function in the membrane environment; consequently, their properties and mechanisms of action depend on the phospholipid milieu. Vigorous treatment of cells with detergents is required to dissociate these proteins from the membrane. Once they become dissociated, however, they generally lose their integrity. Because they usually exist in minute quantities and can be unstable, few membrane-bound receptors have been purified and little structural information is known about them. Advances in molecular biology have permitted the isolation, cloning, and sequencing of receptors,[4] and this is leading to further advances in molecular characterization of these proteins. However, these receptors, unlike many enzymes, are still typically characterized in terms of their function rather than by their structural properties. The two functional components of receptors, the recognition component and the amplification component, may represent the same or different sites on the same protein. Various hypotheses regarding the mechanisms by which drugs may initiate a biological response are discussed in Section 3.2.D.

3.2 Drug–Receptor Interactions

3.2.A General Considerations

To appreciate mechanisms of drug action, it is important to understand the forces of interaction that bind drugs to their receptors. Because of the low concentration of drugs and receptors in the bloodstream and other biological fluids, the law of mass action alone cannot account for the ability of small doses of structurally specific drugs to elicit a total response by combination with all, or practically all, of the appropriate receptors. One of my favorite calculations, shown below, supports the notion that something more than mass action is required to get the desired drug–receptor interaction.[5] One mole of a drug contains 6.02×10^{23} molecules (Avogadro's number). If the molecular weight of an average drug is 200 g/mol, then 10 mg (often an effective dose) will contain $6.02 \times 10^{23}(10 \times 10^{-3})/200 = 3 \times 10^{19}$ molecules of drug. The human organism is composed of about 3×10^{13} cells. Therefore, each cell will be acted on by $3 \times 10^{19}/3 \times 10^{13} = 10^{6}$ drug molecules. One erythrocyte cell contains about 10^{10} molecules. On the assumption that the same number of molecules is found in all cells, then for each drug molecule, there are $10^{10}/10^{6} = 10^{4}$ molecules of the human body! With this ratio of human molecules to drug molecules, Le Chatelier would have a difficult time explaining how the drug could interact and form a stable complex with the desired receptor.

The driving force for the drug–receptor interaction can be considered as a low energy state of the drug–receptor complex (Scheme 3.1), where k_{on} is the rate constant for formation of the drug–receptor complex, which depends on the concentrations of the drug and the receptor, and k_{off} is the rate constant for breakdown of the complex, which depends on the concentration of the drug–receptor complex as well as other forces. The biological activity of a drug is related to its affinity for the receptor, i.e., the stability of the drug–receptor complex. This stability is commonly measured by how difficult it is for the complex to dissociate, which is measured by its K_d, the dissociation constant for the drug–receptor complex at equilibrium:

$$K_d = \frac{[\text{drug}]\,[\text{receptor}]}{[\text{drug–receptor complex}]} \qquad (3.1)$$

$$\text{drug} + \text{receptor} \underset{k_{\text{off}}}{\overset{k_{\text{on}}}{\rightleftharpoons}} \text{drug–receptor complex}$$

Scheme 3.1 ▶ Equilibrium between a drug, a receptor, and a drug–receptor complex

Note that because K_d is a dissociation constant, the smaller the K_d, the larger the concentration of the drug–receptor complex, the more stable is that complex, and the greater is the affinity of the drug for the receptor. Formation of the drug–receptor complex involves an elaborate equilibrium. Solvated ligands (such as drugs) and solvated proteins (such as receptors) generally exist as an equilibrium mixture of several conformers each. To form a complex, the solvent molecules that occupy the binding site of the receptor must be displaced by the drug to produce a solvated complex in which the interactions between the drug and the receptor are stronger than the interactions between the drug and receptor with the solvent molecules.[6] Drug–receptor complex formation also is entropically unfavorable; it causes a loss in conformational degrees of freedom for both the protein and the ligand, as well as the loss of three rotational and three translational degrees of freedom.[7] Therefore, highly favorable enthalpic contacts (interactions) between the receptor and the drug must compensate for the entropic loss.

3.2.B   Interactions (Forces) Involved in the Drug–Receptor Complex

The interactions involved in the drug–receptor complex are the same forces experienced by all interacting organic molecules and include covalent bonding, ionic (electrostatic) interactions, ion–dipole and dipole–dipole interactions, hydrogen bonding, charge-transfer interactions, hydrophobic interactions, and van der Waals interactions. Weak interactions usually are possible only when molecular surfaces are close and complementary, that is, when bond strength is distance dependent. The spontaneous formation of a bond between atoms occurs with a decrease in free energy, that is, ΔG is negative. The change in free energy is related to the binding equilibrium constant (K_{eq}) as follows:

$$\Delta G^\circ = -RT \ln K_{\text{eq}} \tag{3.2}$$

Therefore, at physiological temperature (37°C) changes in free energy of a few kcal/mol can have a major effect on the establishment of good secondary interactions. In fact, if the K_{eq} were only 0.01 (i.e., 1% of the equilibrium mixture in the form of the drug–receptor complex), then a ΔG° of interaction of -5.45 kcal/mol would shift the binding equilibrium constant to 100 (i.e., 99% in the form of the drug–receptor complex).

   In general, the bonds formed between a drug and a receptor are weak noncovalent interactions; consequently, the effects produced are reversible. Because of this, a drug becomes inactive as soon as its concentration in the extracellular fluids decreases. Often it is desirable for the drug effect to last only a limited time so that the pharmacological action can be terminated. In the case of CNS stimulants and depressants, for example, prolonged action could be harmful. Sometimes, however, the effect produced by a drug should persist, and even be irreversible. For example, it is most desirable for a *chemotherapeutic agent*, a drug that acts selectively on a foreign organism or tumor cell, to form an irreversible complex with its receptor so that the drug can exert its toxic action for a prolonged period.[8] In this case, a covalent bond would be desirable. In the following subsections the various types of drug–receptor interactions are discussed briefly. These interactions are applicable to all types of receptors, including enzymes and DNA, that are described in this book.

B.1 Covalent Bonds

The *covalent bond* is the strongest bond, generally worth anywhere from -40 to -110 kcal/mol in stability. It is seldom formed by a drug–receptor interaction, except with enzymes and DNA. These bonds will be discussed further in Chapters 5 and 6.

B.2 Ionic (or Electrostatic) Interactions

For protein receptors at *physiological pH* (generally taken to mean pH 7.4) basic groups such as the amino side chains of arginine, lysine, and, to a much lesser extent, histidine are protonated and, therefore, provide a cationic environment. Acidic groups, such as the carboxylic acid side chains of aspartic acid and glutamic acid, are deprotonated to give anionic groups.

Drug and receptor groups will be mutually attracted provided they have opposite charges. This *ionic interaction* can be effective at distances farther than those required for other types of interactions, and they can persist longer. A simple ionic interaction can provide a $\Delta G^\circ = -5$ kcal/mol, which declines by the square of the distance between the charges. If this interaction is reinforced by other simultaneous interactions, the ionic interaction becomes stronger ($\Delta G^\circ = -10$ kcal/mol) and persists longer. The antidepressant drug pivagabine (Tonerg) is used as an example of a molecule that can hypothetically participate in an ionic interaction with an arginine residue (Figure 3.1).

B.3 Ion–Dipole and Dipole–Dipole Interactions

As a result of the greater electronegativity of atoms such as oxygen, nitrogen, sulfur, and halogens relative to that of carbon, C–X bonds in drugs and receptors, where X is an electronegative atom, will have an asymmetric distribution of electrons; this produces electronic dipoles. These dipoles in a drug molecule can be attracted by ions (*ion–dipole interaction*) or by other dipoles (*dipole–dipole interaction*) in the receptor, provided charges of opposite sign are properly aligned. Because the charge of a dipole is less than that of an ion, a dipole–dipole interaction is weaker than an ion–dipole interaction. In Figure 3.2 the insomnia drug zaleplon (Sonata) is used to demonstrate these interactions, which can provide a $\Delta G^\circ = -1$ to -7 kcal/mol.

B.4 Hydrogen Bonds

Hydrogen bonds are a type of dipole–dipole interaction formed between the proton of a group X–H, where X is an electronegative atom, and other electronegative atoms (Y) containing a pair of nonbonded electrons. The only significant hydrogen bonds occur in molecules where

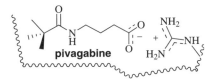

Figure 3.1 ▶ Example of an electrostatic (ionic) interaction. Wavy line represents the receptor surface

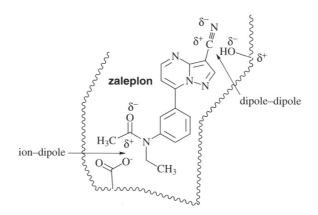

Figure 3.2 ▶ Examples of ion–dipole and dipole–dipole interactions. Wavy line represents the receptor surface

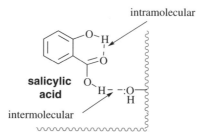

Figure 3.3 ▶ Examples of hydrogen bonds. Wavy line represents the receptor surface

X and Y are N, O, or F. X removes electron density from the hydrogen so it has a partial positive charge, which is strongly attracted to the nonbonded electrons of Y. The interaction is denoted as a dotted line, $-$X$-$H $\cdots$ Y$-$, to indicate that a covalent bond between X and H still exists, but that an interaction between H and Y also occurs. When X and Y are equivalent in electronegativity and degree of ionization, the proton can be shared equally between the two groups, i.e., $-$X $\cdots$ H $\cdots$ Y$-$, referred to as a *low-barrier hydrogen bond*.[9]

The hydrogen bond is unique to hydrogen because it is the only atom that can carry a positive charge at physiological pH while remaining covalently bonded in a molecule, and which also is small enough to allow close approach of a second electronegative atom. The strength of the hydrogen bond is related to the Hammett σ constants.[10]

There are *intramolecular* and *intermolecular* hydrogen bonds; the former are stronger (see salicylic acid, used in wart removal remedies, in Figure 3.3). Intramolecular hydrogen bonding is an important property of molecules that may have a significant effect on lead modification approaches. As discussed in Chapter 2, Section 2.2.A, p. 17, the bioactive conformation of a molecule, the conformation that binds optimally to the receptor, is the ideal conformation. When there are oxygen and/or nitrogen atoms in a compound that have the possibility of forming 5- or 6-membered conformers containing intramolecular hydrogen bonds, those bonds will produce a stable conformation that may or may not approximate the bioactive conformation. This becomes increasingly important if a bioisosteric replacement of an oxygen atom in an ether (capable of forming strong hydrogen bonds) is replaced by a sulfur atom in a thioether (which forms very weak or no hydrogen bonds); this could have a major impact on the potency and even activity of the compound if an intramolecular hydrogen bond changes the conformation of the molecule. This same difference between oxygen or nitrogen and sulfur also becomes important in intermolecular bonding between the drug molecule and the receptor. Intramolecular hydrogen bonding also may mask the binding of a pharmacophoric group. For example, methyl salicylate (**3.3**, wintergreen oil), an active ingredient in many muscle pain remedies and at least one antiseptic, is a weak antibacterial agent. The corresponding *para*-isomer, methyl *p*-hydroxybenzoate (**3.4**), however, is considerably more potent as an antibacterial agent and is used as a food preservative. It is believed that the antibacterial activity of **3.4** is derived from the phenolic hydroxyl group. In **3.3** this group is masked by intramolecular hydrogen bonding.[11]

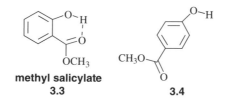

methyl salicylate
3.3   **3.4**

Hydrogen bonds are essential to maintain the structural integrity of α-helix (**3.5**) and β-sheet (**3.6**) conformations of peptides and proteins and the double helix of DNA (**3.7**) (Figure 3.4). As is discussed in Chapter 6, many antitumor agents act by alkylation of the DNA bases, thereby preventing hydrogen bonding. This disrupts the double helix and destroys the DNA.

The $\Delta G°$ for hydrogen bonding can be between -1 and -7 kcal/mol, but usually is in the range -3 to -5 kcal/mol. Binding affinities increase by about one order of magnitude per hydrogen bond.

B.5 Charge-Transfer Complexes

When a molecule (or group) that is a good electron donor comes into contact with a molecule (or group) that is a good electron acceptor, the donor may transfer some of its charge to the acceptor. This forms a *charge-transfer complex*, which, in effect, is a molecular dipole–dipole interaction. The potential energy of this interaction is proportional to the difference between the ionization potential of the donor and the electron affinity of the acceptor.

Donor groups contain π-electrons, such as alkenes, alkynes, and aromatic moieties with electron-donating substituents, or groups that contain a pair of nonbonded electrons, such as oxygen, nitrogen, and sulfur moieties. Acceptor groups contain electron-deficient π-orbitals, such as alkenes, alkynes, and aromatic moieties having electron-withdrawing substituents, and weakly acidic protons. There are groups on receptors that can act as electron donors, such as the aromatic ring of tyrosine or the carboxylate group of aspartate, as electron acceptors, such as cysteine, and as electron donors and acceptors, such as histidine, tryptophan, and asparagine.

Charge-transfer interactions are believed to provide the energy for intercalation of certain planar aromatic antimalarial drugs, such as chloroquine (**3.8**, Aralen), into parasitic DNA (see Chapter 6). The fungicide, chlorothalonil, is used in Figure 3.5 as a hypothetical example for a charge-transfer interaction with a tyrosine.

3.5

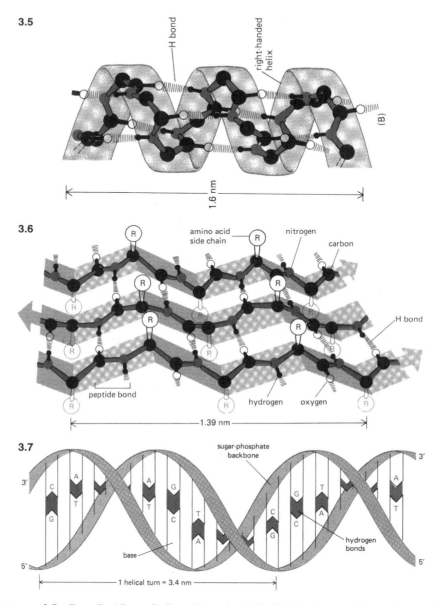

3.6

3.7

Figure 3.4 ▶ **3.5**—From B. Alberts, D. Bray, J. Lewis, M. Raff, K. Roberts, and J. D. Watson, *Molecular Biology of the Cell*, 2nd ed., p. 110, Garland Publishing, New York, 1989, with permission. Copyright ©️ 1989 Garland Publishing.
3.6—From B. Alberts, D. Bray, J. Lewis, M. Raff, K. Roberts, and J. D. Watson, *Molecular Biology of the Cell*, 2nd ed., p. 109, Garland Publishing, New York, 1989, with permission. Copyright ©️ 1989 Garland Publishing.
3.7—From B. Alberts, D. Bray, J. Lewis, M. Raff, K. Roberts, and J. D. Watson, *Molecular Biology of the Cell*, 2nd ed., p. 99, Garland Publishing, New York, 1989, with permission. Copyright ©️ 1989 Garland Publishing. Reproduced in color between pages 172 and 173

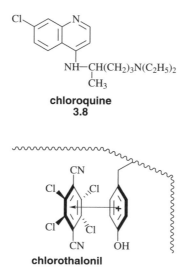

chloroquine
3.8

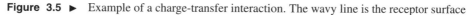

chlorothalonil

Figure 3.5 ► Example of a charge-transfer interaction. The wavy line is the receptor surface

The $\Delta G°$ for charge-transfer interactions also can range from -1 to $-7\,$kcal/mol.

B.6 Hydrophobic Interactions

In the presence of a nonpolar molecule or region of a molecule, the surrounding water molecules orient themselves and, therefore, are in a higher energy state than when only other water molecules are around. When two nonpolar groups, such as a lipophilic group on a drug and a nonpolar receptor group, each surrounded by ordered water molecules, approach each other, these water molecules become disordered in an attempt to associate with each other. This increase in entropy, therefore, results in a decrease in the free energy ($\Delta G = \Delta H - T\Delta S$) that stabilizes the drug–receptor complex. This stabilization is known as a *hydrophobic interaction* (see Figure 3.6). Consequently, this is not an attractive force of two nonpolar groups "dissolving" in one another, but, rather, is the decreased free energy of the nonpolar group because of the increased entropy of the surrounding water molecules. Jencks[12] has suggested that hydrophobic forces may be the most important single factor responsible for noncovalent intermolecular interactions in aqueous solution. Hildebrand,[13] on the other hand, is convinced that there is no hydrophobic effect between water and alkanes; instead, he believes that there just is not enough hydrophilicity to break the hydrogen bonds of water and allow alkanes to go into solution without the assistance of other polar groups. Addition of a single methyl group that can occupy a receptor binding pocket improves binding by 0.7 kcal/mol.[14] In Figure 3.7 the topical anesthetic butamben is depicted in a hypothetical hydrophobic interaction with an isoleucine group.

B.7 Van der Waals or London Dispersion Forces

Atoms in nonpolar molecules may have a temporary nonsymmetrical distribution of electron density, which results in the generation of a temporary dipole. As atoms from different molecules (such as a drug and a receptor) approach each other, the temporary dipoles of one molecule induce opposite dipoles in the approaching molecule. Consequently, intermolecular attractions, known as *van der Waals forces*, result. These weak universal forces only become

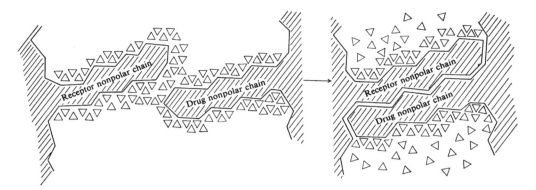

Figure 3.6 ▶ Formation of hydrophobic interactions. [From Korolkovas, A. (1970). *Essentials of Molecular Pharmacology*, p. 172. Wiley, New York. Copyright ©1970 John Wiley & Sons, Inc. This material is used by permission of John Wiley & Sons, Inc.; and by permission of Kopple, K. D. 1966, *Peptides and Amino Acids*. Addison-Wesley, Reading, MA.]

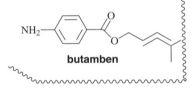

butamben

Figure 3.7 ▶ Example of hydrophobic interactions. The wavy line represents the receptor surface

significant when there is a close surface contact of the atoms; however, when there is molecular complementarity, numerous atomic interactions result (each interaction contributing about -0.5 kcal/mol to the $\Delta G°$), which can add up to a significant overall drug–receptor binding component.

B.8 Conclusion

Because noncovalent interactions are generally weak, cooperativity by several types of interactions is critical. To a first approximation, enthalpy terms will be additive. Once the first interaction has taken place, translational entropy is lost. This results in a much lower entropy loss in the formation of the second interaction. The effect of this cooperativity is that several rather weak interactions may combine to produce a strong interaction. This phenomenon is the basis for why the SAR by NMR approach to lead modification (see Chapter 2, Section 2.2.E.6) can produce such high-affinity ligands from two moderate or poor affinity ligands. Because several different types of interactions are involved, selectivity in drug–receptor interactions can result. In Figure 3.8 the local anesthetic dibucaine is used as an example to show the variety of interactions that are possible.

    The binding constants for 200 drugs and potent enzyme inhibitors were used to calculate the average strength of noncovalent bonds (i.e., the binding energy) associated with 10 common functional groups in an average drug–receptor environment.[15] As suggested above, charged groups bind more strongly than polar groups, which bind more tightly than nonpolar groups; ammonium ions form the best electrostatic interactions (11.5 kcal/mol), then

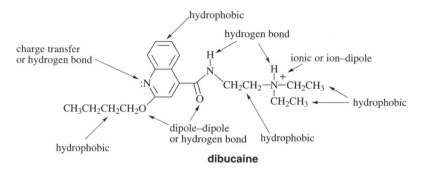

Figure 3.8 ▶ Examples of potential multiple drug–receptor interactions. The van der Waals interactions are excluded

phosphate (10.0 kcal/mol), then carboxylate (8.2 kcal/mol). For loss of rotational and translational entropy, 14 kcal/mol of binding energy has to be subtracted and 0.7 kcal/mol of energy is subtracted for each degree of conformational freedom restricted.[16] Compounds that bind to a receptor exceptionally well have measured binding energies that exceed the calculated average binding energy, and those whose binding energy is less than the average calculated value fit poorly into the receptor.

3.2.C Determination of Drug–Receptor Interactions

Hormones and neurotransmitters are important endogenous molecules that are responsible for the regulation of a myriad of physiological functions. These molecules interact with a specific receptor in a tissue and elicit a specific characteristic response. For example, the activation of a muscle by the central nervous system is mediated by release of the excitatory neurotransmitter acetylcholine (ACh; **3.9**). If a plot is made of the logarithm of the concentration of acetylcholine added to a muscle tissue preparation versus the percentage of total muscle contraction, the graph shown in Figure 3.9 may result. This is known as a *dose–response* or *concentration–response curve*. The low concentration part of the curve results from too few neurotransmitter molecules available for collision with the receptor. As the concentration increases, it reaches a point where a linear relationship is observed between the logarithm of the neurotransmitter concentration and the biological response. As most of the receptors become occupied, the probability of a drug and receptor molecule interacting diminishes, and the curve deviates from linearity (the high concentration end). Dose–response curves are a means of measuring drug–receptor interactions, and the standard method for comparing the potencies of various compounds that interact with a particular receptor. The K_d is the concentration of the test compound that produces 50% of the maximal activity. Any measure of a response can be plotted on the ordinate, such as LD_{50}, ED_{50}, or percentage of a physiological effect.

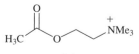

3.9

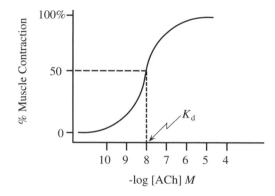

Figure 3.9 ▶ Effect of increasing the concentration of a neurotransmitter (acetylcholine [ACh]) on muscle contraction

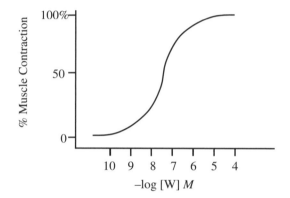

Figure 3.10 ▶ Dose–response curve for a full agonist (W)

If another compound (W) is added in increasing amounts to the same tissue preparation and the curve shown in Figure 3.10 results, the compound, which produces the same maximal response as the neurotransmitter, is called a *full agonist*.

A second compound (X) added to the tissue preparation shows no response at all (Figure 3.11A); however, if it is added to the neurotransmitter, the effect of the neurotransmitter is blocked until a higher concentration of the neurotransmitter is added (Figure 3.11B). Compound X is called a *competitive antagonist*. There are two general types of antagonists, competitive antagonists and noncompetitive antagonists. The former, which is the larger category, is one in which the degree of antagonism is dependent on the relative concentrations of the agonist and the antagonist; both bind to the same site on the receptor, or, at least, the antagonist directly interferes with the binding of the agonist. The degree of blocking of a *noncompetitive antagonist* (X') is independent of the amount of agonist present, so the K_d does not change with increasing neurotransmitter (Figure 3.11C). Two different binding sites may be involved; when the noncompetitive antagonist binds to its *allosteric* binding site, a site to which the endogenous ligand normally does not bind, it may cause a conformational change in the protein, which affects binding of the endogenous molecule. Only competitive antagonists will be discussed further in this text.

If a compound Y is added to the tissue preparation and some response is elicited, but not a full response, regardless of how high a concentration of Y is used, then Y is called a

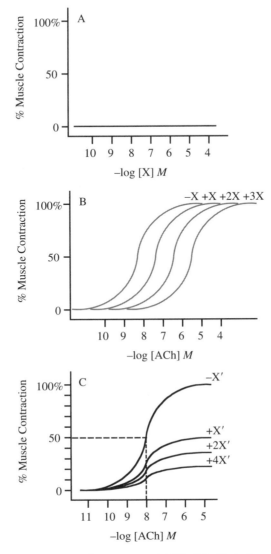

Figure 3.11 ▶ (A) Dose–response curve for an antagonist (X); (B) effect of a competitive antagonist (X) on the response of a neurotransmitter (acetylcholine [ACh]); and (C) effect of a noncompetitive antagonist (X′) on the response of the neurotransmitter

partial agonist (see Figure 3.12A). A partial agonist has properties of both an agonist and an antagonist. When Y is added to low concentrations of a neurotransmitter sufficient to give a response less than the maximal response of the partial agonist (for example, 20% as shown in Figure 3.12B), additive effects are observed as Y is increased, but the maximum response does not exceed that produced by Y alone. Under these conditions, the partial agonist is having an agonistic effect. However, if Y is added to high concentrations of a neurotransmitter sufficient to give full response of the neurotransmitter, then antagonistic effects are observed; as Y increases, the response decreases to the point of maximum response of the partial agonist (Figure 3.12C). If this same experiment is done starting with higher concentrations of the neurotransmitter, the same results are obtained except that the dose–response curves shift to the right, resembling the situation of adding an antagonist to the neurotransmitter (Figure 3.12C).

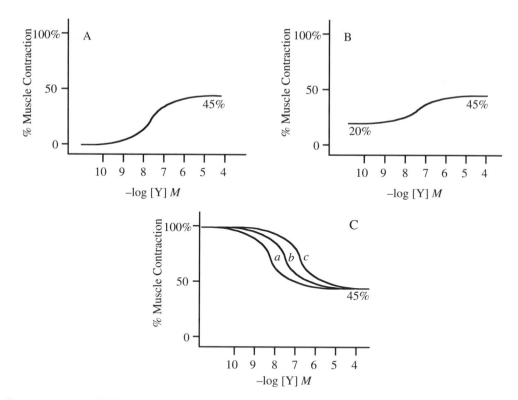

Figure 3.12 ▶ (A) Dose–response curve for a partial agonist (Y); (B) effect of a low concentration of neurotransmitter on the response of a partial agonist (Y); and (C) effect of a high concentration of neurotransmitter on the response of a partial agonist (Y). In (C) the concentration of the neurotransmitter (a,b,c) is $c > b > a$

In a hypothetical situation, compound Z is added to the tissue preparation and muscle relaxation occurs (the opposite effect of the agonist). This would be a *full inverse agonist*, a compound that binds to the receptor, but displays an effect opposite to that of the natural ligand (Figure 3.13A). Valium (see **2.7**), for example, binds to the benzodiazepine receptor and acts as an anticonvulsant; β-carbolines (**3.10**) bind to the same receptor, are inverse agonists, and act as convulsants.[17] Just as an antagonist can displace an agonist or natural ligand (Figure 3.11B), it also can displace an inverse agonist (Figure 3.13B). A *partial inverse agonist* (Z′) is one that, at any concentration, does not give 100% of the effect of a full inverse agonist (Figure 3.13C).

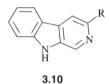

3.10

On the basis of the above discussion, if you wish to design a drug to effect a certain response of a receptor, an agonist would be desired; if you wish to design a drug to prevent a

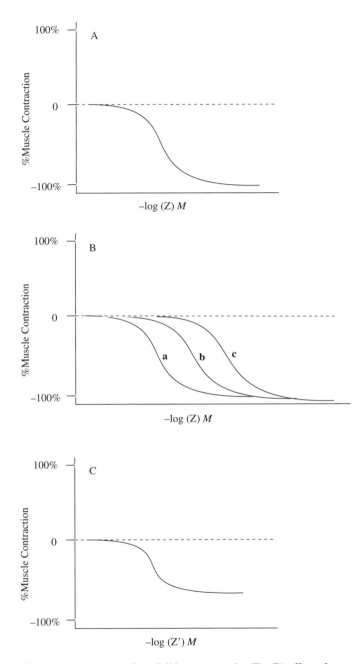

Figure 3.13 ▶ (A) Dose–response curve for a full inverse agonist (Z); (B) effect of a competitive antagonist or natural ligand on the response of a full inverse agonist (*a*, *b*, and *c* represent increasing concentrations of the added antagonist or natural ligand to Z); and (C) dose–response curve for a partial inverse agonist (Z′)

particular response of a natural ligand, an antagonist would be required; if you wish to design a drug that causes the opposite effect of the natural ligand, then an inverse agonist is what you want. Often, there are great structural similarities among a series of agonists, but little structural similarity exists in a series of competitive antagonists. For example, Table 3.1 shows some agonists and antagonists for histamine and epinephrine; a more detailed list of agonists and antagonists for specific receptors has been reported.[18] The differences in the structures of the antagonists is not surprising because a receptor can be blocked by an antagonist simply by its binding to a site near enough to the binding site for the agonist that it physically blocks the agonist from reaching its binding site. This may explain why antagonists are frequently much more bulky than the corresponding agonists. It is easier to design a molecule that blocks a receptor site than one that interacts with it in the specific way required to elicit a response. An agonist can be transformed into an antagonist by appropriate structural modifications.

How is it possible for an antagonist to bind to the same site as an agonist and not elicit a biological response? There are several ways that this may occur. Figure 3.14A shows an

TABLE 3.1 ▶ Agonists and Antagonists

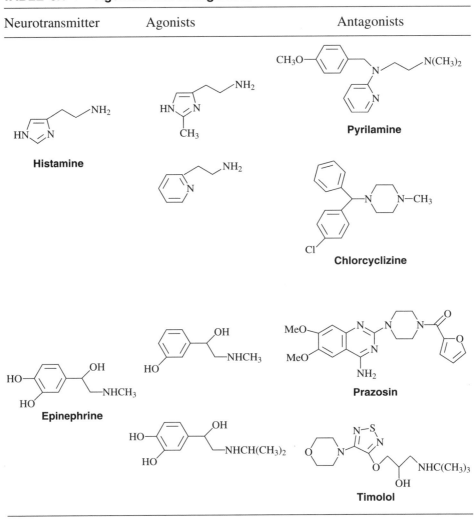

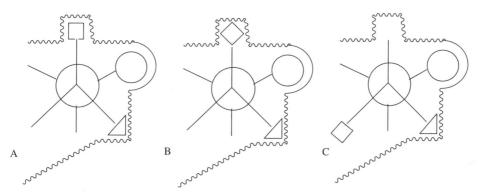

Figure 3.14 ▶ Inability of an antagonist to elicit a biological response. The wavy line is the receptor surface [Adapted with permission from W. O. Foye, Ed., 1989, *Principles of Medicinal Chemistry*, 3rd ed., p. 63. Copyright © 1989 Lea & Febiger, Philadelphia, PA.]

agonist (or natural ligand) with appropriate groups interacting with three receptor binding sites and eliciting a response. In Figure 3.14B the compound has two groups that can interact with the receptor, but one essential group is missing. In the case of enantiomers (Figure 3.14C shows the enantiomer of the compound in Figure 3.14A), only two groups are able to interact with the proper receptor sites. If appropriate groups must interact with all three binding sites in order for a response to be elicited, then the compounds depicted by Figures 3.14B and C would be antagonists.

There are two general categories of compounds that interact with receptors: (1) compounds that occur naturally within the body, such as hormones, neurotransmitters, and other agents that modify cellular activity (*autocoids*); and (2) *xenobiotics*, compounds that are foreign to the body. All naturally occurring chemicals are known to act as agonists, but most xenobiotics that interact with receptors are antagonists.

Receptor selectivity is very important, but often difficult to attain because receptor structures are often unknown. Many current drugs are pharmacologically active at multiple receptors, some of which are not associated with the illness that is being treated. This can lead to side effects. For example, the clinical effect of neuroleptics is believed to result from their antagonism of dopamine receptors.[19] In general, this class of drugs also blocks cholinergic and α-adrenergic receptors, and this results in side effects such as sedation and hypotension.

3.2.D Theories for Drug–Receptor Interactions

Over the years a number of hypotheses have been proposed to account for the ability of a drug to interact with a receptor and elicit a biological response. Several of the more important proposals are discussed here, starting from the earliest hypothesis (the occupancy theory) to the current one (the multistate model).

D.1 Occupancy Theory

The *occupancy theory* of Gaddum[20] and Clark[21] states that the intensity of the pharmacological effect is directly proportional to the number of receptors occupied by the drug. The response ceases when the drug–receptor complex dissociates. However, as discussed in Section 3.2.C, not all agonists produce a maximal response. Therefore, this theory does not rationalize partial agonists, and it does not explain inverse agonists.

Ariëns[22] and Stephenson[23] modified the occupancy theory to account for *partial agonists*, a term coined by Stephenson. These authors utilized the original Langley concept of a receptor, namely, that drug–receptor interactions involve two stages: First, there is a complexation of the drug with the receptor, which they both termed the *affinity*; second, there is an initiation of a biological effect which Ariëns termed the *intrinsic activity* and Stephenson called the *efficacy*. Affinity, then, is a measure of the capacity of a drug to bind to the receptor, and is dependent on the molecular complementarity of the drug and the receptor. Intrinsic activity (α) now refers to the maximum response induced by a compound relative to that of a given reference compound, and efficacy is the property of a compound that produces the maximum response or the ability of the drug–receptor complex to initiate a response.[24] Because of the slight change in definitions, we will use the term *efficacy* to refer to the ability of a compound to initiate a biological response. In the original theory this latter property was considered to be constant. Examples of affinity and efficacy are given in Figure 3.15. Figure 3.15A shows the theoretical dose–response curves for five drugs with the same affinity for the receptor ($pK_d = 8$), but having efficacies varying from 100% of the maximum to 20% of the maximum. The drug with 100% efficacy is a full agonist; the others are partial agonists. Figure 3.15B shows dose–response curves for four drugs with the same efficacy (all full agonists), but having different affinities varying from a pK_d of 9 to 6.

Antagonists can bind tightly to a receptor (great affinity), but be devoid of activity (no efficacy). Potent agonists may have less affinity for their receptors than partial agonists or antagonists. Therefore, these two properties, affinity and efficacy, are uncoupled. Also, the terms *agonist, partial agonist, antagonist*, and *inverse agonist* are biological system dependent and not necessarily properties of drugs. A compound that is an agonist for one receptor may be an antagonist or inverse agonist for another receptor. A particular receptor is considered to have an intrinsic *maximum response*; this is the largest magnitude of response that the receptor is capable of producing by any ligand. A compound that elicits the maximum response is a full agonist; a particular compound may be capable of exceeding the maximum response of a tissue, but the observed response can only be the maximum response of that particular tissue. A drug that is not capable of eliciting the maximum response of the tissue, which is dependent on the structure of the drug, is a partial agonist. A full agonist or partial agonist is said to display *positive efficacy*, an antagonist displays zero efficacy, and a full or partial inverse

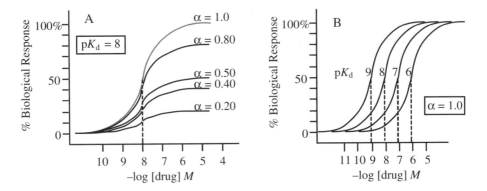

Figure 3.15 ▶ Theoretical dose–response curves illustrate (A) drugs with equal affinities and different efficacies and (B) drugs with equal efficacies but different affinities

agonist displays *negative efficacy* (depresses basal tissue response). The modified occupancy theory accounts for the existence of partial agonists and antagonists, but it does not account for why two drugs that can occupy the same receptor can act differently, i.e., one as an agonist, the other as an antagonist.

D.2 Rate Theory

As an alternative to the occupancy theory, Paton[25] proposed that the activation of receptors is proportional to the total number of encounters of the drug with its receptor per unit time. Therefore, the *rate theory* suggests that the pharmacological activity is a function of the rate of association and dissociation of the drug with the receptor, and not the number of occupied receptors. Each association would produce a quantum of stimulus. In the case of agonists, the rates of both association and dissociation would be fast (the latter faster than the former). The rate of association of an antagonist with a receptor would be fast, but the dissociation would be slow. Partial agonists would have intermediate drug–receptor complex dissociation rates. At equilibrium, the occupancy and rate theories are mathematically equivalent. As in the case of the occupancy theory, the rate theory does not rationalize why the different types of compounds exhibit the characteristics that they do.

D.3 Induced-Fit Theory

The *induced-fit theory* of Koshland[26] was originally proposed for the action of substrates and enzymes, but it could apply to drug–receptor interactions as well. According to this theory the receptor (enzyme) need not necessarily exist in the appropriate conformation required to bind the drug (substrate). As the drug (substrate) approaches the receptor (enzyme), a *conformational change* is induced that orients the essential binding (catalytic) sites (Figure 3.16). The conformational change in the receptor could be responsible for the initiation of the biological response. The receptor (enzyme) was suggested to be elastic, and it could return to its original conformation after the drug (product) was released. The conformational change need not occur only in the receptor (enzyme); the drug (substrate) also could undergo deformation, even if this resulted in strain in the drug (substrate). According to this theory, an agonist would induce a conformational change and elicit a response, an antagonist would bind without a conformational change, and a partial agonist would cause a partial conformational change. The induced-fit theory can be adapted to the rate theory. An agonist would induce a conformational change in the receptor, resulting in a conformation to which the agonist binds less tightly and from which it can dissociate more easily. If drug–receptor complexation does not cause a conformational change in the receptor, then the drug–receptor complex will be stable, and an antagonist will result.

Other theories evolved from the induced-fit theory, such as the macromolecular perturbation theory, the activation-aggregation theory, and multistate models.

D.4 Macromolecular Perturbation Theory

Having considered the conformational flexibility of receptors, Belleau[27] suggested that in the interaction of a drug with a receptor two general types of *macromolecular perturbations* could result: *specific conformational perturbation* makes possible the binding of certain molecules that produce a biological response (an agonist); *nonspecific conformational perturbation* accommodates other types of molecules that do not elicit a response (e.g., an antagonist). If

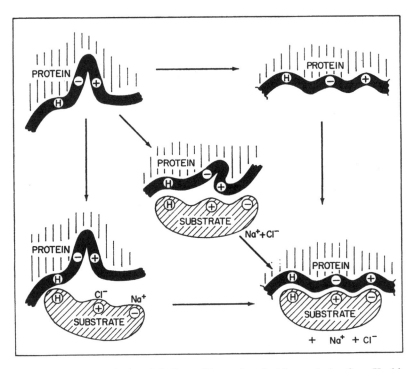

Figure 3.16 ▶ Schematic of the induced-fit theory [Reproduced with permission from Koshland, Jr., D. E., and Neet, K. E., *Annual Review of Biochemistry*, Vol. 37, © 1968 by Annual Reviews, Inc.]

the drug contributes to both macromolecular perturbations, a mixture of two complexes will result (a partial agonist). This theory offers a physicochemical basis for the rationalization of molecular phenomena that involve receptors, but does not address the concept of inverse agonism.

D.5 Activation-Aggregation Theory

An extension of the macromolecular perturbation theory (which also is based on the induced-fit theory) is the *activation-aggregation theory* of Monad, Wyman, and Changeux[28] and Karlin.[29] According to this theory, even in the absence of drugs, a receptor is in a state of dynamic equilibrium between an activated form (R_0), which is responsible for the biological response, and an inactive form (T_0). Using this theory, agonists bind to the R_0 form and shift the equilibrium to the activated form, antagonists bind to the inactive form (T_0), and partial agonists bind to both conformations. In this model the agonist binding site in the R_0 conformation can be different from the antagonist binding site in the T_0 conformation. If there are two different binding sites and conformations, then this could account for the structural differences in these classes of compounds and could rationalize why an agonist elicits a biological response but an antagonist does not. This theory can explain the ability of partial agonists to possess both the agonistic and antagonistic properties as depicted in Figure 3.12. In Figure 3.12B as the partial agonist interacts with the remaining unoccupied receptors, there is an increase in the response up to the maximal response for the partial agonist interaction. In Figure 3.12C the partial agonist competes with the neurotransmitter for the receptor sites. As the partial agonist displaces the neurotransmitter, it changes the amount of R_0 and T_0 receptor forms (T_0 increases and, therefore, the response decreases) until all of the receptors have the partial agonist bound. This theory, however, does not address inverse agonists.[30]

D.6 The Two-State (Multistate) Model of Receptor Activation

The concept of a conformational change in a receptor inducing a change in its activity has been viable for many years.[31] The Monod-Wyman-Changeux idea described above involves a two-state model of receptor activation, but it does not go far enough. This model was revised based mostly on observations with guanine nucleotide-binding regulatory protein (G-protein)-coupled receptors,[32] the largest class of receptors known, which are activated by a variety of ligands such as peptides, hormones, neurotransmitters, chemokines, lipids, glycoproteins, divalent cations, and light.[33] Binding of these ligands causes a conformational change in the structure of these cell surface receptors to facilitate interaction of the receptor with a member of the G-protein family. G-protein activation by the receptor results in the activation of intracellular signal transduction cascades, which leads to a change in the activity of ion channels and enzymes, thereby causing an alteration in the rate of production of intracellular second messengers.[34] Therefore, the G-protein-coupled receptors are involved in the control of every aspect of our behavior and physiology and are linked to numerous diseases, including cardiovascular problems, mental disorders, retinal degeneration, cancer, and AIDS. More than half of all drugs target G-protein-coupled receptors by either activating or inactivating them.

The revised *two-state model of receptor activation* proposes that, in the absence of the natural ligand or agonist, receptors exist in equilibrium (defined by equilibrium constant L; Figure 3.17) between an active state (R^*) which is able to initiate a biological response, and a resting state (R), which cannot. In the absence of a natural ligand or agonist, the equilibrium between R^* and R defines the basal activity of the receptor. A drug can bind to one or both of these conformational states, according to equilibrium constants K_d and K_d^* for formation of the drug–receptor complex with the resting (D·R) and active state (D·R*), respectively. Full agonists alter the equilibrium fully to the active state by binding to the active state and causing maximum response; partial agonists preferentially bind to the active state, but not to the extent that a full agonist does, so maximum response is not attained; full inverse agonists alter the equilibrium fully to the resting state by binding to the resting state, causing a negative

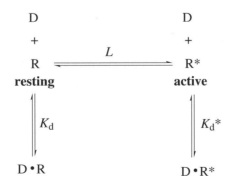

Figure 3.17 ▶ Two-state model of receptor activation. D is the drug, R is the receptor, L is the equilibrium between the resting (R) and the active (R^*) state of the receptor

efficacy (a decrease in the basal activity); partial inverse agonists preferentially bind to the resting state, but not to the extent that a full inverse agonist does; and antagonists have equal affinities for both states (i.e., have no effect on the equilibrium or basal activity, and, therefore, exhibit neither positive nor negative efficacy).[35] A competitive antagonist is able to displace either an agonist or inverse agonist from the receptor.

Leff and coworkers further extended the two-state receptor model to a *three-state receptor model*.[36] In this model there are two active conformations (this could become a multistate model by extension to more than two active conformations) and an inactive conformation. This accommodates experimental findings regarding variable agonist and inverse agonist behavior (both affinities and efficacies) in different systems containing the same receptor type (called *receptor promiscuity*). According to this hypothesis, the basis for differential agonist efficacies among different agonists is their different affinities for the different active states.

3.2.E Topographical and Stereochemical Considerations

Up to this point in our discussion of drug–receptor interactions, we have been concerned with what stabilizes a drug–receptor complex, how drug–receptor interactions are measured, and possible ways in which the drug–receptor complex may form. In this section we turn our attention to molecular aspects and examine the topography and stereochemistry of drug–receptor complexes.

E.1 Spatial Arrangement of Atoms

It was indicated in the discussion of bioisosterism (Chapter 2, Section 2.2.E.4, p. 29) and from SAR studies that many antihistamines have a common pharmacophore (Figure 3.18).[37] In Figure 3.18 Ar^1 is aryl, such as phenyl, substituted phenyl, or heteroaryl (2-pyridyl or thienyl); A^2 is aryl or arylmethyl. The two aryl groups also can be connected through a bridge (as in phenothiazines, **2.40**), and the CH_2CH_2N moiety can be part of another ring (as in chlorcyclizine, Table 3.1). X is CH–O–, N–, or CH–; C– C is a short carbon chain (2 or 3 atoms) which may be saturated, branched, contain a double bond, or be part of a ring system. These compounds are called *antihistamines* because they are antagonists of a histamine receptor known as the H_1 *histamine receptor*. When a sensitized person is exposed to an allergen, an antibody is produced, an antigen–antibody reaction occurs, and histamine is released. Histamine binding to the H_1 receptor can cause stimulation of smooth muscle and produce allergic and hypersensitivity reactions such as hay fever, pruritus (itching), contact and atopic dermatitis, drug rashes, urticaria (edematous patches of skin), and anaphylactic shock. Antihistamines are used widely to treat these symptoms. Unlike histamine (see Table 3.1 for structure), most H_1 antagonists contain tertiary amino groups, usually a dimethylamino or pyrrolidino group. At physiological pH, then, this group will be protonated, and it is believed that an ionic interaction with the receptor is a key binding contributor. The commonality of structures

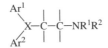

Figure 3.18 ▶   General structure of antihistamines

of antihistamines suggests that there are specific binding sites on the H_1 histamine receptor that have an appropriate topography for interaction with certain groups on the antihistamine, which are arranged in a similar configuration (see Section 3.2.B). It must be reiterated, however, that although the antihistamines are competitive antagonists of histamine for the H_1 receptor, the same set of atoms on the receptor need not interact with both histamine and the antagonists.[38] Consequently, it is difficult to make conclusions regarding the receptor structure on the basis of antihistamine structure–activity relationships. Because of the essentiality of various parts of antihistamine molecules, it is likely the minimum binding requirements include a negative charge on the receptor to interact with the ammonium cation and hydrophobic (van der Waals) interactions with the aryl groups. Obviously, many other interactions are involved.

From this very simplistic view of drug–receptor interactions it is not possible to rationalize the fact that enantiomers, i.e., compounds that are identical in all physical and chemical properties except for their effect on plane polarized light, can have quite different binding properties to receptors. This phenomenon is discussed in more detail in the next section.

E.2 Drug and Receptor Chirality

Histamine is an achiral molecule, but many of the H_1 receptor antagonists are chiral molecules. Proteins are polyamino acid macromolecules, and amino acids are chiral molecules (in the case of mammalian proteins, they are all *L*-isomers); consequently, proteins (receptors) also are chiral substances. Complexes formed between a receptor and two enantiomers are diastereomers, not enantiomers, and, as a result, they have different energies and chemical properties. This suggests that dissociation constants for drug–receptor complexes of enantiomeric drugs may differ, and may even involve different binding sites. The chiral antihistamine dexchlorpheniramine (**3.11**) is highly stereoselective (one stereoisomer is more potent than the other); the *S*-(+)-isomer is about 200 times more potent than the *R*-(−)-isomer.[39] According to the nomenclature of Ariëns,[40] when there is isomeric stereoselectivity, the more potent isomer is termed the *eutomer*; the less potent isomer is the *distomer*. The ratio of the more potent (higher affinity) enantiomer to the less potent enantiomer is termed the *eudismic ratio*. The *in vivo* eudismic ratio (−/+) for the analgesic agent etorphine (see **2.28**, R = CH_3, R′ = C_3H_7) is greater than 6666.[41]

***S*-(+)-dexchlorpheniramine**
3.11

High-potency antagonists are those having a high degree of complementarity with the receptor. When the antagonist contains a stereogenic center in the pharmacophore, a high eudismic ratio is generally observed for the stereoisomers because the receptor complementarity would not be retained for the distomer. This increase in eudismic ratio, with an increase in potency of the eutomer, is *Pfeiffer's rule*.[42] Small eudismic ratios are observed when the eutomer has low affinity for the receptor (poor molecular complementarity) or, in the case of chiral compounds, when the stereogenic center lies outside of the region critically involved in receptor binding, i.e., the pharmacophore.

The distomer actually should be considered as an impurity in the mixture, or, in the terminology of Ariëns, the *isomeric ballast*. It, however, may contribute to undesirable side effects and toxicity; in that case, the distomer for the biological activity may be the eutomer for the side effects. For example, *d*-ketamine (**3.12**; the asterisk marks the chiral carbon) is a hypnotic and analgesic agent; the *l*-isomer is responsible for the unde- sired side effects[43] (note that *d* is synonymous with (+) and *l* is synonymous with (−)). Probably the most horrendous example of toxicity by a distomer is that of thalido- mide (**3.13**, Contergan), a drug used in the late 1950s and early 1960s as a sedative and to prevent nausea during pregnancy, which was shown to cause severe fetal limb abnormalities (phocomelia) when taken in the first trimester of pregnancy. This tragedy led to the development of three phases of clinical trials and the requirement for FDA approval of drugs (see Chapter 2, Section 2.2.I, p. 86). Later, it was thought that the ter- atogenicity (birth defect) of thalidomide was caused by the (*S*)-isomer only,[44] but then it was found that the (*R*)-isomer was converted into the (*S*)-isomer *in vivo*.[45] Despite the potential danger of this drug, it is back on the market (as the racemate, Thalomid) for treatment of moderate or severe erythema nodosum leprosum in leprosy patients, but it is not administered to pregnant women and preferably only to those women beyond child-bearing age.

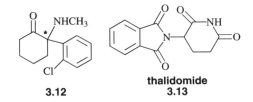

3.12   thalidomide
                              3.13

It also is possible that both isomers are biologically active, but only one contributes to the toxicity, such as the local anesthetic prilocaine (**3.14**, EMLA).[46]

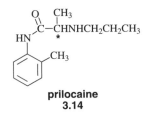

prilocaine
3.14

In some cases it is desirable to have both isomers present. Both isomers of bupiva- caine (**3.15**, Sensorcaine) are local anesthetics, but only the *l*-isomer shows vasoconstrictive activity.[47] The experimental diuretic (increases water excretion) drug indacrinone (**3.16**) has a uric acid retention side effect. The *d*-isomer of **3.16** is responsible (i.e., the eutomer) for both the diuretic activity and the uric acid retention side effect. Interestingly, however, the *l*-isomer acts as a uricosuric agent (reduces uric acid levels). Unfortunately, the ratio that gives the optimal therapeutic index (see Chapter 2, Section 2.2.E, p. 25) is 1:8 (*d:l*), not 1:1 as is present in the racemic mixture.[48]

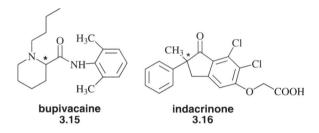

bupivacaine
3.15

indacrinone
3.16

Enantiomers may have different therapeutic activities as well.[49] Darvon (**3.17**), $2R, 3S$-(+)- dextropropoxyphene, is an analgesic drug, and its enantiomer, Novrad (**3.18**), $2S, 3R$-(−)- levopropoxyphene, is an antitussive (anticough) agent, an activity that is not compatible with analgesic action. Consequently, these enantiomers are marketed separately. You may have noticed that the trade names are enantiomeric as well! The (S)-(+)-enantiomer of the anti-inflammatory/analgesic drug ketoprofen (**3.19**, Orudis) is the eutomer; the (R)-(−)-isomer shows activity against bone loss in periodontal disease.

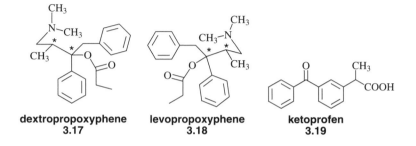

dextropropoxyphene
3.17

levopropoxyphene
3.18

ketoprofen
3.19

It, also, is possible for enantiomers to have opposite effects.[50] The (R)-(−)-enantiomer of 1-methyl-5-phenyl-5-propylbarbituric acid (**3.20**) is a narcotic, and the (S)-(+)-enantiomer is a convulsant![51] These are examples of inverse agonists (see Section 3.2.C, p. 131). The (+)-isomer of the experimental narcotic analgesic picenadol (**3.21**) is an opiate agonist, the (−)-isomer is a narcotic antagonist, and the racemate is a partial agonist.[52] This suggests a potential danger in studying racemic mixtures; one enantiomer may antagonize the other, and no effect will be observed. For example, the racemate of UH-301 (**3.22**) exhibits no serotonergic activity; (R)-UH-301 is an agonist of the 5-HT$_{1A}$ receptor, but (S)-UH-301 is an antagonist of the same receptor.[53] Consequently, no activity is observed with the racemate.

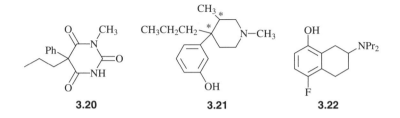

3.20

3.21

3.22

It is quite common for chiral compounds to show stereoselectivity with receptor action, and the stereoselectivity of one compound can vary for different receptors. For example, (+)-butaclamol (**3.23**) is a potent antipsychotic, but the (−)-isomer is essentially inactive; the eudismic ratio (+/−) is 1250 for the D$_2$-dopaminergic, 160 for the D$_1$-dopaminergic, and

73 for the α-adrenergic receptors. $(-)$-Baclofen (**3.24**) is a muscle relaxant that binds to the GABA$_B$ receptor; the eudismic ratio $(-/+)$ is 800.[54]

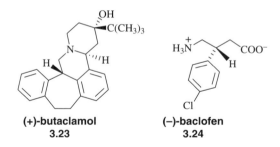

(+)-butaclamol   (–)-baclofen
3.23   **3.24**

Remember that the $(+)$- and $(-)$-nomenclature refers to the effect of the compound on the direction of rotation of plane polarized light and has nothing to do with the stereochemical configuration of the molecule. The stereochemistry about a stereogenic carbon atom is noted by the R, S convention of Cahn *et al.*[55] Because the R, S convention is determined by the atomic numbers of the substituents about the stereogenic center, two compounds having the same stereochemistry, but a different substituent can have opposite chiral nomenclatures. For example, the eutomer of the antihypertensive agent propranolol is the S-$(-)$-isomer (**3.25**, X = NHCH(CH$_3$)$_2$).[56] If X is varied so that the attached atom has an atomic number greater than that of oxygen, such as F, Cl, Br, or S, then the nomenclature rules dictate that the molecule is designated as an R isomer, even though there is no change in the stereochemistry. Note, however, that even though the absolute configuration about the stereogenic carbon remains unchanged after variation of the X group in **3.25**, the effect on plane polarized light cannot necessarily be predicted; the compound with a different substituent X can be either $+$ or $-$. The most common examples of this phenomenon in nature are some of the amino acids. (S)-Alanine, for example, is the $(+)$-isomer and (S)-serine (same absolute stereochemistry) is the $(-)$-isomer; the only difference is a CH$_3$ group for alanine and a CH$_2$OH group for serine.

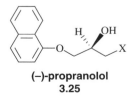

(–)-propranolol
3.25

Propranolol (**3.25**, X = NHCH(CH$_3$)$_2$, Inderal), the first member of a family of drugs known as β-*blockers* (Sir James W. Black shared the Nobel Prize in Medicine in 1988 for this discovery), is a competitive antagonist (blocker) of the β-adrenergic receptor, which triggers a decrease in blood pressure and regulates cardiac rhythm and oxygen consumption for those with cardiovascular disease. The β_1- and β_2-adrenergic receptors are important to cardiac and bronchial vasodilation, respectively; propranolol is nonselective in its antagonism for these two receptors. The eudismic ratio $(-/+)$ for propranolol is about 100; however, propranolol also exhibits local anesthetic activity for which the eudismic ratio is 1. The latter activity apparently is derived from some other mechanism than β-adrenergic receptor blockage. A compound of this type that has two separate mechanisms of action and, therefore, different therapeutic activities, has been called a *hybrid drug* by Ariëns.[57] (+)-Butaclamol (**3.23**), which interacts with a variety of receptors, is another hybrid drug. However, butaclamol has three chiral centers and, therefore, has eight possible isomeric forms. When multiple isomeric forms are involved in the biological activity, the drug is called a *pseudo hybrid drug*. Another important example of this type of drug is the antihypertensive agent, labetalol (Figure 3.19, Normodyne), which, as a result of having two stereogenic centers, exists in four stereoisomeric forms (two diastereomeric pairs of enantiomers), having the stereochemistries (RR), (SS), (RS), and (SR). This drug has α- and β-adrenergic blocking properties. The (RR)-isomer is predominantly the β-blocker (the eutomer for β-adrenergic blocking action), and the (SR)-isomer is mostly the α-blocker (the eutomer for α-adrenergic blocking); the other 50% of the isomers, the (SS)- and (RS)-isomers, are almost inactive (the isomeric ballast). Labetalol, then, is a pseudo hybrid drug, a mixture of isomers having different receptor-binding properties.

Labetalol also is an example of how relatively minor structural modifications of an agonist can lead to transformation into an antagonist. (l)-Epinephrine (**3.26**) is a natural hybrid molecule that induces both α- and β-adrenergic effects. Introduction of the phenylalkyl substituent on the nitrogen transforms the α-adrenergic activity of the agonist (l)-epinephrine into the α-adrenergic antagonist labetalol. The modification of one of the catechol hydroxyl groups of (l)-epinephrine to a carbamyl group of labetalol changes the β-adrenergic action (agonist) to a β-adrenergic blocking action (antagonist).

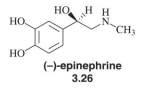

(–)-epinephrine
3.26

As pointed out by Ariëns[58] and by Simonyi,[59] it is quite common for mixtures of isomers, particularly racemates, to be marketed as a single drug, even though at least half of the mixture not only may be inactive for the desired biological activity, but may, in fact, be responsible for various side effects. In the case of β-adrenergic blockers, antiepileptics, and oral anticoagulants, about 90% of the drugs on the market are racemic mixtures, and for antihistamines, anticholinergics, and local anesthetics about 50% are racemic. In general, about a third of drugs are sold as racemic mixtures. The isomeric ballast, typically, is not removed for economic reasons; it can be quite expensive to separate the enantiomeric impurity

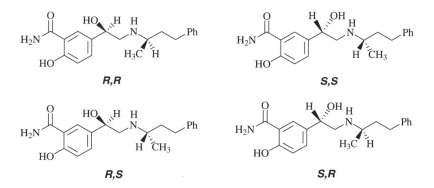

Figure 3.19 ▶ Four stereoisomers of labetalol

and may become economically prohibitive for a company to prepare sufficient quantities at a profit. The FDA (Food and Drug Administration) has been satisfied if the racemate has low toxicity in animal and human studies.

In 1992 the FDA finally announced official guidelines for the marketing of racemic drugs. Its position was to allow drug companies to choose whether to develop chiral drugs as racemates or as single enantiomers, but that they would have to furnish rigorous justification for FDA approval of racemates. To further encourage companies to prepare and market single-entity drugs, the concept of a *racemic switch* was introduced. This is the redevelopment in single enantiomer form of a drug that is being marketed as a racemate (the racemate is switched for the eutomer). Even if the racemate is currently covered by an active patent, the patent office would allow a new patent to a second company for the eutomer of the racemate. Of course, the same company also can be awarded a patent for a racemic switch as well, which is an interesting strategy to extend the life of exclusivity for a drug. For example, AstraZeneca markets the antiulcer drug omeprazole (**3.27**, Prilosec) as a racemate, but shortly before the patent expired, a new patent was issued to the same company for the active (*S*)-isomer, which was approved for marketing as esomeprazole (Nexium). Because the racemate has already been approved by the FDA, less testing is needed for the active enantiomer. Interestingly, the (*R*)-isomer is more potent than either the (*S*)-isomer or racemate in rats, the two enantiomers are equipotent in dogs, but the (*S*)-isomer is most potent in humans (apparently because of the higher bioavailability and consistent pharmacokinetics compared with the other enantiomer).[60]

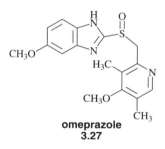

omeprazole
3.27

The use of a single enantiomer is generally expected to lower side effects and toxicity. For example, the antiasthma drug albuterol (**3.28**, Ventolin/Proventil) is an agonist for β_2-adrenergic receptors on airway smooth muscle, leading to bronchodilation. The racemic switch, levalbuterol (the *R*-(−)-isomer, Xopenex), appears to be solely responsible for the therapeutic effect. The (*S*)-isomer seems to produce side effects such as pulse rate increases, tremors, and decreases in blood glucose and potassium levels. Because of this advantage, single isomer drug sales have been steadily increasing worldwide; in 1999, they accounted for almost a third of the market, and in 2000, 40% of the drugs on the market were single enantiomers.[61]

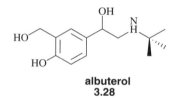

albuterol
3.28

However, it is not always best to use the single enantiomer of the drug. The antidepressant drug fluoxetine (**3.29**, Prozac) is marketed as the racemate (in this case both isomers are active as serotonin reuptake inhibitors). Clinical trials with just the (R)-isomer at a higher dosage, however, produced a cardiac side effect. Another unusual problem associated with the use of a single enantiomer may occur if the two enantiomers have synergistic pharmacological activities. For example, the (+)-isomer of the antihypertensive drug nebivolol (**3.30**, Nebilet) is a β-blocker (see above); the (−)-isomer is not a β-blocker, but it is still a vasodilating agent (via the nitric oxide pathway), so the drug is sold as a racemate to take advantage of two different antihypertensive mechanisms. Sometimes an unexpected side benefit is associated with the use of a racemic mixture. The racemic calcium ion channel blocker (see Section 3.2.F, p.158) verapamil (**3.31**, Calan) has long been used as an antihypertensive drug. The (S)-isomer is the eutomer, but the (R)-isomer has been found to inhibit the resistance of cancer cells to anticancer drugs.[62]

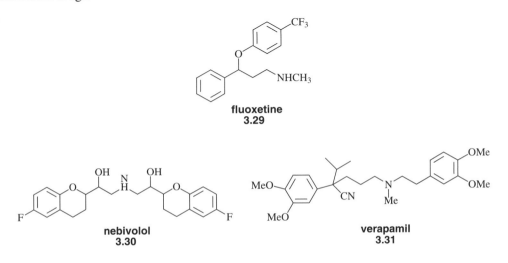

fluoxetine
3.29

nebivolol
3.30

verapamil
3.31

For cases in which the enantiomers are readily interconvertible *in vivo*, there is no reason to go to the expense of marketing a single enantiomer. Enantiomers of the antidiabetes drug rosiglitazone (**3.32**, Avandia) spontaneously racemize in solution, so it is sold as a racemate.

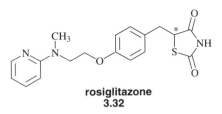

rosiglitazone
3.32

Because of potential vast differences in activities of two enantiomers, caution should be used when applying QSAR methods such as Hansch analyses (see Chapter 2, Section 2.2.G.3.a, p. 68) to racemic mixtures. These methods really should be applied to the separate isomers.[63]

It is quite apparent from the above discussion that receptors are capable of recognizing and selectively binding optical isomers. Cushny[64] was the first to suggest that enantiomers could have different biological activities because one isomer could fit into a receptor much better than the other. How are they able to accomplish this?

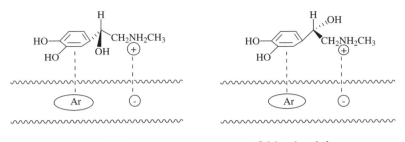

R-(–)-epinephrine   **S-(+)-epinephrine**

Figure 3.20 ▶ Binding of epinephrine enantiomers to a two-site receptor. The wavy lines are the receptor surfaces

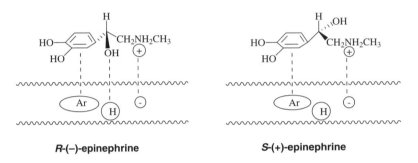

R-(–)-epinephrine   **S-(+)-epinephrine**

Figure 3.21 ▶ Binding of epinephrine enantiomers to a three-site receptor. The wavy lines are the receptor surfaces

If you consider two enantiomers, such as epinephrine, interacting with a receptor that has only two binding sites (Figure 3.20), it becomes apparent that the receptor cannot distinguish between them. However, if there are at least three binding sites (Figure 3.21), the receptor easily can differentiate them. The R-(–)-isomer has three points of interaction and is held in the conformation shown to maximize molecular complementarity. The S-(+)-isomer can have only two sites of interaction (the hydroxyl group cannot interact with the hydroxyl binding site, and may even have an adverse steric interaction); consequently it has a lower binding energy. Easson and Stedman[65] were the first to recognize this *three-point attachment* concept: A receptor can differentiate enantiomers if there are as few as three binding sites. As in the case of the β-adrenergic receptors discussed above, the structure of α-adrenergic receptors to which epinephrine binds is unknown. α-Adrenergic receptors appear to mediate vasoconstrictive effects of catecholamines in bronchial, intestinal, and uterine smooth muscle. The eudismic ratio (R/S) for vasoconstrictor activity of epinephrine is only 12–20,[66] indicating that there is relatively little difference in binding energy for the two isomers to the α-adrenergic receptor. Although the above discussion was directed at the enantioselectivity of receptor interactions, it should be noted that there also is enantioselectivity with respect to pharmacokinetics, i.e., absorption, distribution, metabolism, and excretion, which will be discussed in Chapter 7.[67]

As noted in Chapter 1 (Section 1.2), a relatively large percentage of anti-infectives and antitumor compounds are natural products or are analogs of natural products. The above discussion would suggest that the chirality of natural products should be very important to their biological activities. This is true; however, the unnatural enantiomer of the natural product could be even more potent than the natural product. For example, *ent*-(–)-roseophilin (**3.33**), the unnatural enantiomer of the natural antitumor antibiotic, is 2–10 times more potent than the natural (+)-isomer in cytotoxicity assays,[68] and *ent*-fredericamycin A (**3.34**) is as cytotoxic as its natural enantiomer.[69] Why should that be so? The organisms that produce these natural products may not be producing them for the purpose of protecting themselves from the disease state we have in mind for these compounds. After all, are these organisms really concerned with developing cancer? There are many possible mechanisms of action for antitumor agents, and the *ent*-natural product may bind to the relevant receptor better than the natural product does.

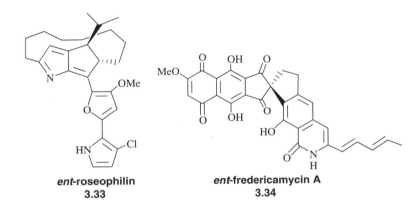

ent-roseophilin
3.33

ent-fredericamycin A
3.34

E.3 Geometric Isomers (Diastereomers)

Geometric isomers (*E*- and *Z*-isomers[70] and epimers) are diastereomers, stereoisomers having different spatial arrangements of atoms that are not mirror images; consequently, they are different compounds, having different energies and stabilities. As a result of their different configurations, receptor interactions will be different. Unlike enantiomers, which are relatively difficult to separate, diastereomers often can be easily separated by chromatography or recrystallization, so they should be tested separately. The antihistamine activity of *E*-triprolidine (**3.35a**, found in cold remedies, such as Actifed) was found to be 1000-fold greater than the corresponding *Z*-isomer (**3.35b**).[71] Likewise, the neuroleptic potency of the *Z*-isomer of the antipsychotic drug chlorprothixene (**3.36a**, Taractan) is more than 12 times greater than that of the corresponding *E*-isomer (**3.36b**).[72] On the other hand, the *E*-isomer of the anticancer drug diethylstilbestrol (**3.37a**) has 14 times greater estrogenic activity than the *Z*-isomer (**3.37b**), possibly because its overall structure and the interatomic distance between the two hydroxyls in the *E*-isomer are similar to that of estradiol (**3.38**).

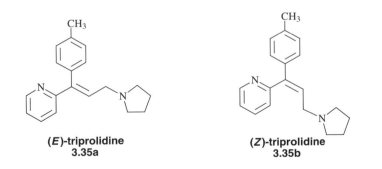

(*E*)-triprolidine
3.35a

(*Z*)-triprolidine
3.35b

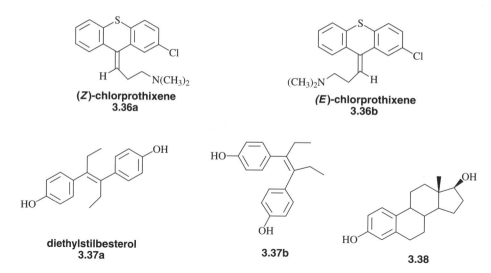

(Z)-chlorprothixene
3.36a

(E)-chlorprothixene
3.36b

diethylstilbesterol
3.37a

3.37b

3.38

Although in some cases the *cis* and *trans* nomenclature does correspond with Z and E, respectively, it should be kept in mind that these terminologies are based on different conventions, so there may be confusion. The Z, E nomenclature is unambiguous, and should be used.

E.4   Conformational Isomers

Geometric isomers and enantiomers can be separated, isolated, and screened individually. Some isomers, however, typically cannot be separated, namely, *conformational isomers* or *conformers* (isomers generated by a change in conformation). As a result of free rotation about single bonds in acyclic molecules and conformational flexibility in many cyclic compounds, a drug molecule can assume a variety of *conformations*, i.e., the location of the atoms in space without breakage of bonds. The pharmacophore of a molecule is defined not only by the configuration of a set of atoms, but also by the bioactive conformation in relation to the receptor binding site. A receptor may bind only one of the conformers. As was pointed out in Chapter 2, Section 2.2.H, p. 78, the conformer that binds to a receptor need *not* be the lowest energy conformer observed in the crystalline state, as determined by X-ray crystallography, or found in solution, as determined by NMR spectrometry, or determined theoretically by molecular mechanics calculations. The binding energy to the receptor may overcome the barrier to the formation of a higher energy conformation. In order for drug design to be efficient, it is essential to know the *bioactive conformation* (the conformation when bound to the receptor) in the drug–receptor complex. Figure 3.22 is a crystal structure of the antidiabetic drug rosiglitazone (see **3.32**, Avandia) bound to the peroxisome proliferator activated receptor gamma (PPARγ), a transcription factor.[73] Note that the bioactive conformation of the drug bound to the receptor is an inverted U shape, rather than an extended conformation. Compounds that cannot attain this inverted U structure will not be able to bind to that site.

If a lead compound has low potency, it only may be because the population of the active conformer in solution is low (higher in energy); for example, with PPARγ shown in Figure 3.22, an inverted U conformation of the compound is essential for high potency. The energy of a conformer will determine the relative population of that conformer in the equilibrium mixture of conformers. A higher energy conformer will be in lower concentration in the equilibrium mixture of conformers. Therefore, if the bioactive conformation is a high energy conformer, the

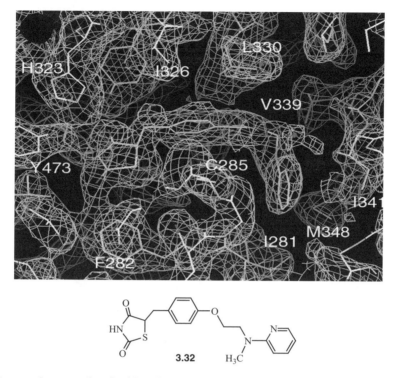

Figure 3.22 ▶ An example of a bioactive conformation. Rosiglitazone (**3.32**), an antidiabetic drug (green structure in the middle), bound to peroxisome proliferator-activated receptor gamma (PPARγ). Note the sickle-shaped conformation in the binding site to accommodate the shape formed by the active-site residues. [With permission from Xu, H.E, (2000). Reprinted from *Molecular Cell*, Vol. 5, "Asymmetry in the PPARγ/RXRα Crystal Structure Reveals the Molecular Basis of Hetrodimerization among Nuclear Receptors", pp. 545–555, Copyright © 2000, with permission of Elsevier.]. Reproduced in color between pages 172 and 173

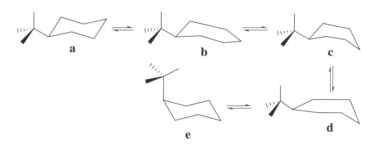

Scheme 3.2 ▶ Cyclohexane conformations: **a**, chair (substituent equatorial); **b**, half chair; **c**, boat; **d**, half chair; **e**, chair (substituent axial)

K_d for the molecule will appear high (poor affinity), not because the structure of the compound is incorrect, but because the population of the ideal conformation is so low. If the conformation of the ideal conformer were in higher concentration, the K_d would be much lower. To give you a simple example of conformational populations from organic chemistry, consider 1-*tert*-butylcyclohexane (Scheme 3.2). Cyclohexanes can exist in numerous conformations, including a chair form with the substituent in the equatorial position (**a**), a half-chair (**b**), a boat (**c**) (including twist boat), a different half chair (**d**), and a chair conformer with the substituent axial (**e**). The difference in free energy for the two chair conformers with the *tert*-butyl group either equatorial (**a**) or axial (**e**), which in a receptor binding site would make an enormous difference on binding effectiveness, is −5.4 kcal/mol, which translates into an equilibrium mixture ([equatorial]/[axial]) at 37°C of 6619 ($\Delta G° = -RT \ln K$). If the axial conformer were the bioactive conformation, and the mixture were 1 μM in 1-*tert*-butylcyclohexane, it would only be 0.00015 μM (i.e., 150 pM) in the axial conformer of 1-*tert*- butylcyclohexane. This would lead to the conclusion that 1-*tert*-butylcyclohexane was inactive, whereas, if only the axial conformer existed in solution, it could be the most potent binder ever observed for that receptor.

A unique approach has been taken to determine the bioactive conformation of a drug molecule in the drug–receptor complex. This approach involves the synthesis of *conformationally rigid analogs* of flexible drug molecules. The potential pharmacophore becomes locked into various configurations by judicious incorporation of cyclic or unsaturated moieties into the drug molecule. These conformationally rigid analogs are then tested, and the analog with the optimal activity (or potency) can be used as the prototype for further structural modification. Conformationally rigid analogs are propitious because key functional groups, presumably part of the pharmacophore, are constrained in one position, thereby permitting the determination of the *pharmacophoric conformation*. The major drawback to this approach is that in order to construct a rigid analog of a flexible molecule, usually additional atoms and/or bonds must be attached to the original compound, and these can affect the chemical and physical properties. Consequently, it is imperative that the conformationally rigid analog and the drug molecule be as similar as possible in size, shape, and mass.

First we will look at the conformationally rigid analog approach to determine the bioactive conformation of a natural ligand (a neurotransmitter); then we will apply this methodology to lead modification. An example of the use of conformationally rigid analogs for the elucidation of receptor binding site topography is the study of the interaction of the neurotransmitter, acetylcholine (ACh), with its receptors. There are at least two important receptors for ACh, one activated by the alkaloid muscarine (**3.39**) and the other by the alkaloid nicotine (**3.40**; presumably in the protonated pyrrolidine form).

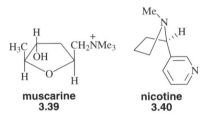

muscarine
3.39

nicotine
3.40

Acetylcholine has a myriad of conformations; four of the more stable possible conformers (groups staggered) are **3.41a–d**. There are also conformers with groups eclipsed that are higher in energy. Four different *trans*-decalin stereoisomers were synthesized[74] (**3.42a–d**) corresponding to the four ACh conformers shown in **3.41**. All four isomers exhibited low muscarinic receptor activity; however, **3.42a** (which corresponds to the most stable conformer, the *anti*-conformer) was the most potent (0.06 times the potency of ACh). The low potency of **3.42a** is believed to be the result of interference by the *trans*-decalin moiety. A comparison of *erythro*- (**3.43**) and *threo*-2,3-dimethylacetylcholine (**3.44**) gave the startling result that **3.43** was 14 times more potent than ACh, and **3.44** was 0.036 times as potent as ACh. Compound **3.42a** corresponds to the *threo*-isomer **3.44**, and, therefore, is expected to have low potency.

The corresponding *erythro* analog does not have a *trans*-decalin analogy, so it could not be tested. To minimize the number of extra atoms added to ACh, *trans*- (**3.45**) and *cis*-1-acetoxy-2-trimethylammoniocyclopropanes (**3.46**) were synthesized and tested[75] for *cholinomimetic properties*, i.e., production of a response resembling that of ACh. The (+)-*trans*-isomer (shown in **3.45**)[76] has about the same muscarinic activity as does ACh, thus indicating the importance of minimizing additional atoms; the (−)-trans-isomer is about 1/500th the potency of ACh. This strongly supports the *anti*-conformer (**3.41a**) as the bioactive conformer. Unfortunately, the other conformers cannot be modeled by substituted cyclopropane analogs; the *cis*-isomer (**3.46**) models an eclipsed conformer of ACh. Nonetheless, the racemic *cis*-isomer has negligible activity. The (+)-*trans*-isomer has the same absolute configuration as the active enantiomers of the two muscarinic receptor agonists muscarine and acetyl β-methylcholine. These results suggest that ACh binds in an extended form (**3.41a**). However, both the *cis*- and the *trans*-cyclopropyl isomers, as well as all of the *trans*-decalin stereoisomers (**3.42a–d**) were only weakly active with the nicotinic cholinergic receptor. Because **3.45**, the lowest energy conformer, is not active with the nicotinic ACh receptor, it can be concluded that a higher energy conformer is. This supports the concept that the lowest energy conformer does not have to be the one that binds to a receptor.

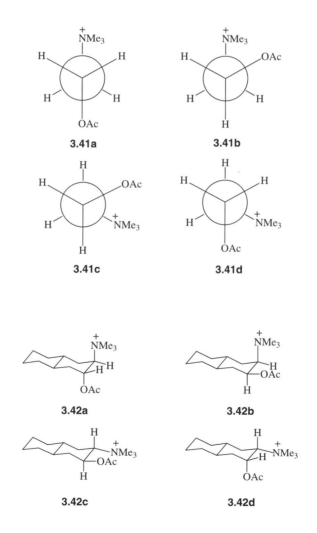

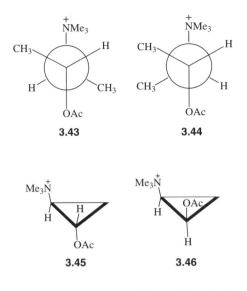

3.43   **3.44**

3.45   **3.46**

An example of the use of conformationally rigid analogs in drug design was reported by Li and Biel.[77] 4-(4-Hydroxypiperidino)-4′-fluorobutyrophenone (**3.47**) was found to have moderate tranquilizing activity in lower animals and man; however, unlike the majority of antipsychotic butyrophenone-type compounds, it only had minimal antiemetic (prevents vomiting) activity. The piperidino ring can exist in various conformations (**3.48a–d**, R = F–C$_6$H$_4$CO(CH$_2$)$_3$), including two chair forms (**3.48a** and **3.48d**) and two twist-boat forms (**3.48b** and **3.48c**).

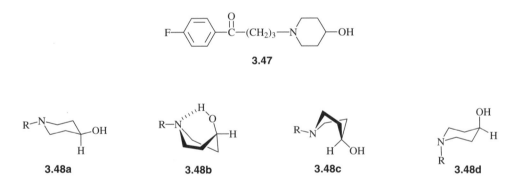

3.47

3.48a   **3.48b**   **3.48c**   **3.48d**

The difference in free energy between the axial and equatorial hydroxyl conformers of the related compound, *N*-methyl-4-piperidinol (**3.48**, R = Me) is 0.94 ± 0.05 kcal/mol at 40°C. (The equatorial conformer is favored by a factor of 4.56 over the axial conformer.)[78] Energies for the twist-boat conformers are about 6 kcal/mol higher, but because of hydrogen bonding, **3.48b** should be more stable than **3.48c**. On the assumption that the chair conformers are more likely, which may not be the case, three conformationally rigid chair analogs, **3.49–3.51** (R = F–C$_6$H$_4$CO(CH$_2$)$_3$), were synthesized to determine the effect on receptor binding of the hydroxyl group in the equatorial (**3.49**), axial (**3.50**), and both (**3.51**) positions. Of course, with **3.51**, it must be assumed that, if the hydroxyl group is involved in hydrogen bonding, it is as an acceptor, not as a donor. Also, for synthetic reasons, the conformationally rigid analog of **3.48d** could not be made; instead, the diastereomer (with the R group still equatorial, but the hydroxyl group axial) was synthesized. This study, then, provides data for the preference of the position of the hydroxyl group, not strictly for the conformer preference. When subjected to muscle relaxation tests, the order of potency was **3.50** > **3.51** > **3.49**, indicating again that the conformationally less stable compound with the axial hydroxyl group has better molecular complementarity with the receptor than does the more stable compound with the equatorial hydroxyl group. This suggests that further analogs should be prepared where the axial hydroxyl is the more stable conformer or where it can be held in that configuration.

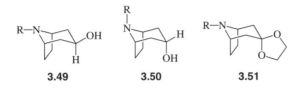

| **3.49**   **3.50**   **3.51** |

Another use of conformationally rigid analogs is to determine the appropriate orientation of pharmacophoric groups for binding to related receptors of unknown structure. The NMDA (*N*-methyl-D-aspartate) subclass of glutamate receptors are composed of an ion channel with multiple binding sites, including one for phencyclidine (PCP, **3.52**, Figure 3.23). Phencyclidine analogs can bind to the PCP site of the NMDA receptor, the σ receptor, and the dopamine-D_2 receptor.[79] Neither the physical nature nor endogenous ligands for the σ receptor has been identified, but several structurally unrelated ligands are known.[80] PCP is a flexible molecule that can undergo conformational ring inversion of both the cyclohexyl and piperidinyl rings as well as rotation of the phenyl group. The various conformations place the ammonium ion and the phenyl ring in different spacial orientations, which may be responsible for binding to the various receptor sites. Conformationally rigid analogs of PCP were synthesized that fixed the orientation of the ammonium center of the PCP with respect to the centrum of the phenyl ring to determine the importance of conformation on selectivity between the PCP and σ sites (Figure 3.23; ϕ is the angle defined by the darkened atoms in **3.52**).[81] In **3.53** ϕ is 0°, in **3.54** (actually the one lower homolog was made) ϕ is 30°, and in **3.55** ϕ is 60°. As the rigidity increases (**3.55** → **3.54** → **3.53**), the affinity for the PCP site is diminished, and none binds well to the PCP site (the best, **3.55**, only has 2% of the affinity of PCP). However, all three bind well to the σ site, almost twice as well as does PCP itself and fit a pharmacophore model for the σ receptor.[82]

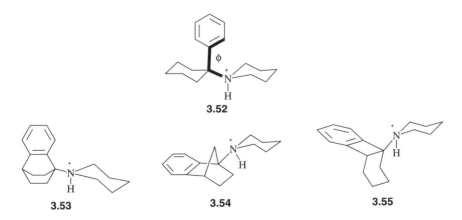

Figure 3.23 ▶ Phencyclidine (PCP, **3.52**) and three conformationally rigid analogs of PCP

E.5 Ring Topology

Tricyclic psychomimetic drugs show an almost continuous transition of activity in going from structures such as the tranquilizer chlorpromazine (**3.56**, Thorazine) through the antidepressant amitriptyline (**3.57**, Elavil), which has a tranquilizing side effect, to the pure antidepressant agent imipramine (**3.58**, Tofranil).[83] Stereoelectronic effects seem to be the key factor, even though tranquilizers and antidepressants have different molecular mechanisms. Three angles can be drawn to define the positions of the two aromatic rings in these compounds (Figure 3.24). The angle α (**3.59**) describes the bending of the ring planes; β (**3.60**) is the annellation angle of the ring axes that passes through carbon 1 and 4 of each aromatic ring; γ (**3.61**) is the torsional angle of the aromatic rings as viewed from the side of the molecule. In general, the tranquilizers have only a bending angle α; no β and γ angles. The mixed tranquilizer-antidepressants have both a bending (α) and annellation angle (β), but no γ angle. The pure antidepressants exhibit all three angles. The activities arise from the binding of the compounds to different receptors; these angles determine the overall three-dimensional structure of the pharmacophore of the compound, which dictates the binding affinities for various receptors.

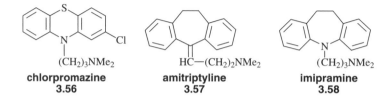

chlorpromazine	amitriptyline	imipramine
3.56	**3.57**	**3.58**

3.2.F Ion Channel Blockers

A receptor has two basic characteristics, recognition of a substance and ability to initiate a biological response. Ion channels, then, fulfill the definition of receptors: They selectively bind ions, and they mediate a response. An *ion channel* is a transmembrane pore that is composed of three elements, a *pore* responsible for the transit of the ion, and one or more *gates* that open and close in response to specific stimuli that are received by the *sensors*. Conformational mobility

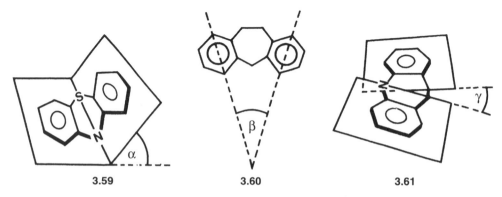

| 3.59 | 3.60 | 3.61 |

Figure 3.24 ▶   Ring topology of tricyclic psychomimetic drugs [Reproduced with permission from Nogrady, T. (1985). In *Medicinal Chemistry: A Biochemical Approach*, p. 29. Oxford University Press, New York. Copyright © 1985 Oxford University Press.]

is an integral component of the function of ion channels; the three states of a channel, closed, open, and activated, are all believed to be regulated by conformational changes. Ligands may gain access to the channel either by membrane permeation or through an open channel state.

The movement of calcium ions into cells is vital to the excitation and contraction of the heart muscle. When a cardiac cell potential reaches a threshold, a sodium ion channel allows rapid influx of sodium ions through the cell membrane. This is followed by a slower movement of calcium ions through a calcium ion channel; the calcium ions maintain the plateau phase of the cardiac action potential. *Calcium ion channel blockers* prevent the influx of calcium ions, which then alters the plateau phase and, therefore, the coronary blood flow. Consequently, calcium channel blockers such as verapamil (see **3.31**, Calan), nifedipine (**3.62**, Adalat), and diltiazem (**3.63**, Cardizem) are valuable drugs in the treatment of angina (resulting from reduced oxygen), cardiac arrhythmias, and hypertension. Given the large dissimilarity among these structures, it was originally thought that these compounds may bind to three different calcium ion channels. However, it was found that all three drugs bind to the L-type calcium ion channel, but at three different sites.[84] Often receptors have what is known as *allosteric sites*, alternative binding sites that may become available by conformational changes in the receptor. Multiple binding sites is another good reason to include random screening as a lead discovery approach.

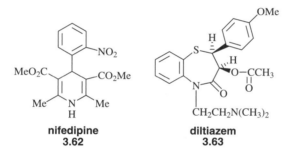

nifedipine
3.62

diltiazem
3.63

3.2.G Case History of Rational Drug Design of a Receptor Antagonist: Cimetidine

The antiulcer drug cimetidine (**3.64**, Tagamet) is a truly elegant example of drug discovery and the use of physical organic chemical principles, coupled with the various lead modification approaches discussed in Chapter 2, to uncover the first H_2 histamine receptor antagonist and an entirely new class of drugs. Cimetidine is one of the first drugs discovered by a rational approach, thanks to the valiant efforts of medicinal chemists C. Robin Ganellin and Graham Durant and pharmacologist James Black at Smith, Kline, & French Laboratories (now Glaxo SmithKline; Sir James W. Black shared the 1988 Nobel Prize in Physiology or Medicine for the discovery of propranolol and also is credited for the discovery of this drug; actually, the medicinal chemists would have made the discovery). This is a case, however, where neither QSAR nor molecular graphics approaches were utilized. As described in Section 3.2.E.1, p. 142, histamine binds to the H_1 receptor and causes allergic and hypersensitivity reactions, which antihistamines antagonize. Black found that another action of histamine is the stimulation of gastric acid secretion.[85] However, antihistamines have no effect on this activity; consequently, it was suggested that there was a second histamine receptor, which was termed the H$_2$ *receptor*. The H$_1$ and H$_2$ receptors can be differentiated by agonists and antagonists. 2-Methylhistamine (**3.65**) preferentially elicits H$_1$ receptor responses, and 4-methylhistamine (**3.66**) has the corresponding preferential effect on H$_2$ receptors. An antagonist of the histamine H$_2$ receptor would be beneficial to the treatment of hypersecretory conditions such as duodenal and gastric ulcers (peptic ulcers). Consequently, in 1964 Smith, Kline & French Laboratories in England initiated a search for a lead compound that would antagonize the H$_2$ receptor.[86] Actually, now there are at least four different histamine receptors known, each one responsible for a different physiological function.[87] The critically important challenge in drug design is to get selectivity of action of molecules.

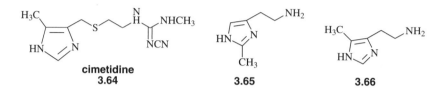

cimetidine
3.64   **3.65**   **3.66**

The first requirement for initiation of a lead discovery program for the H$_2$ receptor is an efficient bioassay (screen). Unfortunately, there were no high-throughput screens at that time, so a tedious *in vivo* screen was developed. Histamine was infused into anesthetized rats to stimulate gastric acid secretion, then the pH of the perfusate from the lumen of the stomach was measured before and after administration of the test compound.

The lead discovery approach that was taken involved a biochemical rationale. Because a histamine receptor antagonist was sought, histamine analogs were synthesized on the assumption that the receptor would recognize that general backbone structure. However, the structure had to be sufficiently different so as not to stimulate a response and defeat the purpose. After 4 years none of the 200 or so compounds made showed any H$_2$-receptor antagonistic activity. Then a new, more sensitive assay, was developed, and some of the same compounds were retested, which identified the first lead compound, N$^\alpha$-guanylhistamine (**3.67**). This compound was only very weakly active as an inhibitor of histamine stimulation; later it was determined to be a partial agonist, not an antagonist. An isostere, the isothiourea **3.68** was made, which was found to be more potent. The corresponding conformationally-rigid analog **3.69** (a ring-chain transformation), however, was less potent than **3.68**; consequently, it was thought that flexibility in the side chain was important. Many additional compounds were synthesized, but they acted as partial agonists. They could block histamine binding, but they could not inhibit acid secretion.

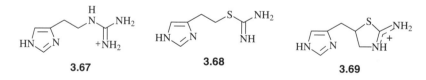

3.67   **3.68**   **3.69**

It, therefore, became necessary to separate the agonist and antagonist activities. The reason for their agonistic activity, apparently, was their structural similarity to histamine. Not only were these compounds imidazoles, but at physiological pH, the side chains were protonated and positively charged, just like histamine. Consequently, it was reasoned that the imidazole ring should be retained for receptor recognition, but the side chain could be modified to eliminate the positive charge. After numerous substitutions, the neutral homologous thiourea analog (**3.70**) was prepared having weak antagonistic activity without stimulatory activity. Further homologation of the side chain gave a purely competitive antagonist (**3.71**, R = H); no agonist effects were observed. Methylation and further homologation on the the thiourea nitrogen was carried out; the *N*-methyl analog (**3.71**, R = CH_3) called burimamide, was found to be highly specific as a competitive antagonist of histamine at the H_2 receptor. It was shown to be effective in the inhibition of histamine-stimulated gastric acid secretion in rat, cat, dog, and man. Burimamide was the first H_2-receptor antagonist tested in humans,[88] but it lacked adequate oral activity, so the search for more potent analogs was continued.

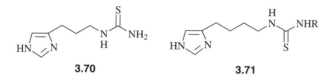

3.70   **3.71**

The poor oral potency of burimamide could be a pharmacokinetic problem or a pharmacodynamic problem. The Smith, Kline, & French group decided to consider the latter. In aqueous solution at physiological pH the imidazole ring can exist in three main forms (**3.72a–3.72c**, Figure 3.25; R is the rest of burimamide). The thioureido group can exist as four conformers (**3.73a–d**, Figure 3.26; R is the remainder of burimamide). The side chain can exist in a myriad of conformers. Therefore, it is possible that only a very small fraction of the molecules in equilibrium would have the bioactive conformation, and this could account for the low potency.

One approach taken to increase the potency of burimamide was to compare the population of the imidazole form in burimamide at physiological pH to that in histamine.[89] The

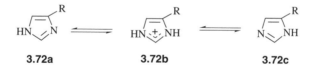

3.72a   **3.72b**   **3.72c**

Figure 3.25 ▶ Three principal forms of 5-substituted imidazoles at physiological pH

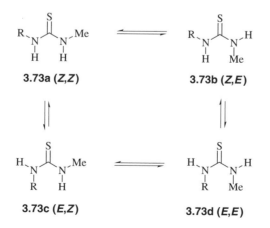

3.73a (Z,Z)   **3.73b (Z,E)**

3.73c (E,Z)   **3.73d (E,E)**

Figure 3.26 ▶ Four conformers of the thioureido group

population can be estimated from the electronic influence of the side chain, which alters the electron densities at the ring nitrogen atoms and, therefore, affects the proton acidity. This effect is more important at the nearer nitrogen atom so if R is electron donating, it would make the adjacent nitrogen more basic, and **3.72c** (Figure 3.25) should predominate; if R is electron withdrawing, it would make the adjacent nitrogen less basic, and **3.72a** should be favored. The fraction present as **3.72b** can be determined from the ring pK_a and the pH of the solution. The electronic effect of R can be calculated from the measured ring pK_a with the use of the Hammett equation applied as follows:

$$pK_a^R = pK_a^H + \rho\sigma_m \tag{3.3}$$

where pK_a^R is the pK_a of the substituted imidazole, pK_a^H is that of imidazole (R = H), σ_m is the meta electronic substituent constant, and ρ is the reaction constant (see Section 2.2.F.1, p. 51). Imidazole has a pK_a of 6.80 and at physiological temperature and pH 20% of the molecules are in the protonated form. The imidazole in histamine under these conditions has a pK_a of 5.90. This indicates that the side chain in histamine is electron withdrawing, thus favoring tautomer **3.72a** (to the extent of 80%), and only 3% of the molecules are in the cationic form (**3.72b**). The pK_a of the imidazole in burimamide, however, is 7.25, indicating it has an electron donating side chain that favors tautomer **3.72c**. The cation is one of the principal species, about 40% of the molecules. Therefore, even though the side chains in histamine and burimamide appear to be similar, they have opposite electronic effects on the imidazole ring. On the assumption that the desired form of the imidazole should resemble that in histamine, the Smith, Kline & French group decided to increase the electron-withdrawing effect of the side chain of burimamide; however, they did not want to make a major structural modification. Incorporation of an electron-withdrawing atom into the side chain near the imidazole ring was contemplated, and the isosteric replacement of a methylene by a sulfur atom to give thiaburimamide (**3.74**, R = H) was carried out. A comparison of the physical properties of the two compounds (**3.71**, R = CH₃, and **3.74**, R = H) shows that they have similar van der Waals radii and bond angles, although the C–S bond is slightly longer than the C–C bond and is more flexible. A sulfur atom also is more hydrophilic than a methylene group; the log P for thiaburimamide is 0.16 and for burimamide is 0.39. The pK_a of the imidazole in thiaburimamide was determined to be 6.25, indicating that the electron-withdrawing effect of the side chain increased, and more of the favored tautomeric form as in histamine was present (**3.72a**). Thiaburimamide is about three times more potent as an H₂ histamine receptor antagonist *in vitro* than is burimamide.

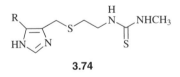

3.74

A second way to increase the population of tautomer **3.72a** would be to introduce an electron-donating substituent at the 4-position of the ring, because electron-donating groups increase the basicity of the adjacent nitrogen, which is chemically equivalent to putting an electron-withdrawing group at the side chain position in thiaburimamide. Because 4-methylhistamine (**3.66**) is a known H₂-receptor agonist, there should be no steric problem with a 4-methyl group. However, the addition of an electron-donating group should increase the pK_a of the ring, thereby increasing the population of the cation (**3.72b**). Although the increase in tautomer **3.72a** is somewhat offset by the decrease in the total uncharged population, the overall effect was favorable. Metiamide (**3.74**, R = CH$_3$) has a pK_a identical with that of imidazole, indicating that the effect of the electron-withdrawing side chain exactly balanced the effect of the electron-donating 4-methyl group; the percentage of molecules in the charged form was 20%. The important result, however, is that metiamide is 8 to 9 times more potent than burimamide.

Wouldn't you think that the tautomeric form would be shifted even more favorably toward **3.72a** by substitution of the side chain with a more electronegative oxygen atom instead of a sulfur atom? Theoretically, it should. This compound, oxaburimamide, was synthesized, but it was *less* potent than burimamide! The explanation for this unexpected result is that intramolecular hydrogen-bonding between the oxygen atom and the thiourea NH produces an unfavorable *conformationally restricted analog* (**3.75**). This is one of the problems associated with isosteric replacements; although CH$_2$, NH, O, and S can have similar biological activity, NH and O can participate in intramolecular (and intermolecular) hydrogen bonding, which changes the shape of the compound and may disfavor (although it may also favor) the bioactive conformation.

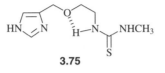

3.75

Metiamide was tested on 700 patients with duodenal ulcers and was found to produce significant increase in the healing rate with marked symptomatic relief. However, a few cases of granulocytopenia (reduction of the number of white blood cells in blood) developed. Even though this was a reversible side effect, it was undesirable (compromises the immune system), and it halted further clinical work with this compound.

The Smith, Kline & French group conjectured that the granulocytopenia that was associated with metiamide was caused by the thiourea group; consequently, alternative substituents were sought. An isosteric replacement approach was taken. The corresponding urea (**3.76**, X = 0) and guanidino (**3.76**, X = NH) analogs were synthesized and found to be 20 times less potent than metiamide. Of course, the guanidino analog would be positively charged at physiological pH, and that could be the cause for the lower potency. Charton[90] found a Hammett relationship between the σ and pK_a values for *N*-substituted guanidines; consequently, if guanidino basicity were the problem, then substitution of the guanidino nitrogen with electron-withdrawing groups could lower the pK_a. In fact, cyanoguanidine and nitroguanidine have pK_a values of −0.4 and −0.9, respectively (compared with −1.2 for thiourea), a drop of about 14 pK_a units from that of guanidine. The corresponding cyanoguanidine (**3.76**, X = N−CN; cimetidine, Tagamet) and nitroguanidine (**3.76**, X = N−NO$_2$) were synthesized in 1972, and both were potent H$_2$ antagonists, comparable in potency to that of metiamide, but without the granulocytopenia (cimetidine was slightly more potent than **3.76**, X = NNO$_2$).

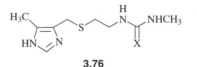

3.76

Because strong electron-withdrawing substituents on the guanidino group favor the imino tautomer, the cyanoguanidino and nitroguanidino groups correspond to the thiourea structure (**3.76**, X = NCN, NNO$_2$, and S, respectively). These three groups are actually bioisosteres; they are all planar structures of similar geometries, are weakly amphoteric (weakly basic and acidic), being un-ionized in the pH range 4–11, are very polar, and are hydrophilic. The crystal structures of metiamide (**3.74**, R = CH$_3$) and cimetidine (**3.76**, X = NCN) are almost identical. The major difference in the two groups is that whereas N, N'-disubstituted thioureas assume three stable conformers (Figure 3.26; Z, Z; Z, E; and E, Z), N, N'-disubstituted cyanoguanidines appear to assume only two stable conformers (Z, E and E, Z). This suggests that the most stable conformer, the Z, Z conformer, is not the bioactive conformation. An isocytosine analog (**3.77**) also was prepared (pK_a 4.0), which can exist only in the Z, Z and E, Z conformations. It was only about one-sixth as potent as cimetidine. However, the isocytosino group has a lower log P (more hydrophilic) than that of the N-methylcyanoguanidino group, and it was thought that lipophilicity may be an important physicochemical parameter. There was, indeed, a correlation found between the H$_2$-receptor antagonist activity *in vitro* and the octanol-water partition coefficient of the corresponding acid of the substituent Y (Figure 3.27). Although increased potency correlates with increased lipophilicity, all of these compounds are fairly hydrophilic. Because the correlation was determined in an *in vitro* assay, membrane transport is not a concern; consequently, these results probably reflect a property involved with receptor interaction, not with transport. Therefore,

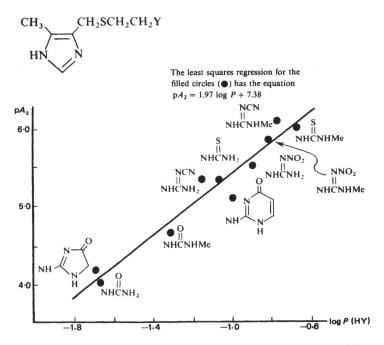

Figure 3.27 ▶ Linear free-energy relationship between H$_2$-receptor antagonist activity and the partition coefficient. The least-squares regression for the filled circles (●) has the equation pA_2 = 1.97 log P + 7.38 [Reproduced with permission from Ganellin, C. R., and Parsons, M. E. (1982). In *Pharmacology of Histamine Receptors*, p. 83. Wright-PSG, Bristol. Reprinted by permission of Elsevier Ltd.]

it is not clear if the lower potency of the isocytosine analog is structure or hydrophilicity dependent.

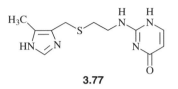

3.77

Cimetidine was first marketed in the United Kingdom in 1976; therefore, it took only 12 years from initiation of the H_2-receptor antagonist program to commercialization. Subsequent to the introduction of cimetidine onto the U.S. drug market, two other H_2-receptor antagonists were approved, ranitidine (**3.78**, Zantac, Glaxo Laboratories), which rapidly became the largest selling drug worldwide, and famotidine (**3.79**, Pepcid, Yamanouchi, Ltd.). It is apparent that an imidazole ring is not essential for H_2-receptor recognition and that a positive charge near the heterocyclic ring (the Me_2N- and guanidino groups of **3.78** and **3.79**, respectively, will be protonated at physiological pH) is not unfavorable.

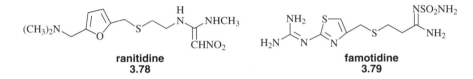

ranitidine   **famotidine**
3.78   **3.79**

Cimetidine became the first drug ever to achieve more than $1 billion a year in sales, thereby having the distinction of being the first *blockbuster drug*. The discovery of cimetidine is one of many examples now of how the judicious use of physical organic chemistry can result in lead discovery, if not in drug discovery. Next, we turn our attention to a special class of receptors called enzymes, which also are very important targets for drug design.

3.3 General References

Membranes and Receptors

Emmett, J. C. (Ed.) *Comprehensive Medicinal Chemistry*, Pergamon Press, Oxford, 1990, Vol. 3.

Drug-Receptor Interactions

Albert, A. *Selective Toxicity*, 7th ed., Chapman and Hall, London, 1985.
Cannon, J. G. *Pharmacology for Chemists*, Oxford University Press, London, 1999.
Kenakin, T. *Molecular Pharmacology: A Short Course*, Blackwell Science, Cambridge, MA, 1997.
Korolkovas, A. *Essentials of Molecular Pharmacology*, Wiley, New York, 1970.

Drug-Receptor Theories

O'Brien, R. D. (Ed.) *The Receptors*, Plenum, New York, 1979.
Smithies, J. R. Bradley, R. J. (Eds.) *Receptors in Pharmacology*, Marcel Dekker, New York, 1978.

Stereochemical Considerations

Smith, D. F. (Ed.) *CRC Handbook of Stereoisomers: Therapeutic Drugs*, CRC Press, Boca
   Raton, FL, 1989.

Ion Channels

Cahalan, M. D.; Chandy, K. G. *Curr. Opin. Biotechnol.* **1997**, *8*, 749.

Conn, P. M. (Ed.) *Ion Channels*, Academic Press, San Diego, 1998 and 1999; parts B and C
   (*Methods in Enzymology* **1999**, *293* and *294*).

Doggrell, S. A.; Brown, L. *Expert Opin. Invest. Drugs* **1996**, *5*, 495.

Hille, B. *Ionic Channels of Excitable Membranes*, Sinauer Assoc., Sunderland, MA, 1984.

Jones, S. W. *J. Bioenerg. Biomembr.* **1998**, *30*, 299.

Triggle, D. J. In *Trends in Medicinal Chemistry*, Mutschler, E.; Winterfeldt, E. (Eds.) VCH,
   Weinheim, 1987.

Triggle, D. J.; Janis, R. A. *Annu. Rev. Pharmacol. Toxicol.* **1987**, *27*, 347.

Histamine Receptors and Antagonists

Cooper, D. G.; Young, R. C.; Durant, G. J.; Ganellin, C. R. In *Comprehensive Medicinal
   Chemistry*, Emmett, J. C. (Ed.) Pergamon Press, Oxford, 1990, Vol. 3, p. 323.

Ganellin, C. R.; Parsons, M. E. (Eds.) *Pharmacology of Histamine Receptors*, Wright-PSG,
   Bristol, England, 1982.

Hill, S. J.; Ganellin, C. R.; Timmerman, H.; Schwartz, J. C.; Shankley, N.P.; Young, J. M.;
   Schunack, W.; Levi, R.; Haas, H. L. *Pharmacol. Rev.* **1997**, *49*, 253.

3.4 PROBLEMS

(Answers can be found in Appendix at the end of the book.)

   1. Indicate what drug–receptor interactions are involved at every arrow shown. More than
      one kind of interaction is possible for each letter.

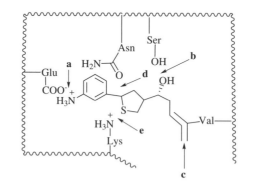

   a.

   b.

   c.

   d.

   e.

2. A receptor has lysine and histidine residues important to binding that do not interact with each other. The pK_a of the lysine residue is 6.4 (pK_a in solution is 10.5), and the pK_a of the histidine residue is 9.4 (pK_a in solution is 6.5). Based on the discussion in Chapter 2 about pK_a variabilities as a result of the environment, what can you say about possible other residues in the binding site to rationalize these observations?

3. Draw a dose–response curve for:

    A. a full agonist with a K_d of 10^{-9} M

    B. a mixture of a full agonist and a competitive antagonist

4. Draw dose–response curves (place on same plot) for a series of three compounds with the following properties:

	K_d(M)	α
1	10^{-6}	1.0
2	10^{-9}	0.8
3	10^{-9}	0.4

5. A series of dopamine analogs was synthesized and assayed for their effect on the dopaminergic D_2 receptor. The results are shown in Table 1.

Table 1 Compound	K_d (nM)	% Change in basal activity	% Change when 100 μM dopamine added
1	180	−12.1	−6.2
2	37	0.1	0.8
3	0.46	19	19
4	14.9	12.2	18.4
Dopamine	1.9	19	

    A. Compare the affinities of **1–4** to that of dopamine.

    B. Compare the efficacies of **1–4** to dopamine.

    C. What type of effect is produced by **1–4**?

6.   **A.** What problems are associated with administration of racemates?

    B. How can you increase the eudismic ratio?

7. Design confomationally rigid analogs for:

    A. 4-aminobutyric acid (GABA)

    B. Epinephrine

    C. Nicotine

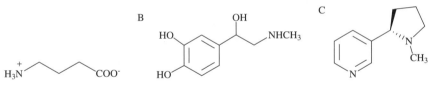

8. Based on generalizations about ring topology discussed in the chapter, would you expect the compound below to be a tranquilizer, have both antidepressant and tranquilizing properties, or be an antidepressant agent? Why?

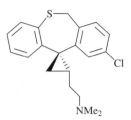

9. An isosteric series of compounds shown below, where $X = CH_2$, NH, O, S, was synthesized. The order of potency was $X = NH > O > S > CH_2$. How can you rationalize these results (you need to consider the three-dimensional structure)?

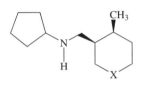

10. Tyramine binds to a receptor that triggers the release of norepinephrine, which can raise the blood pressure. If the tyramine receptor were isolated, and you wanted to design a new antihypertensive agent, discuss what you would do in terms of lead discovery and modification.

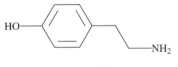

tyramine

3.5   References

1. Langley, J. N. *J. Physiol.* (*London*) **1878**, *1*, 339.

2. Ehrlich, P. *Klin. Jahr.* **1897**, *6*, 299.

3. Langley, J. N. *J. Physiol.* (*London*) **1905**, *33*, 374.

4. (a) Kenakin, T. *Molecular Pharmacology: A Short Course*, Blackwell Science, Cambridge, MA, 1997. (b) Lindstrom, J. In *Neurotransmitter Receptor Binding*, Yamamura, H. I.; Enna, S. J.; Kuhar, M. J. (Eds.), Raven Press, New York, 1985, p. 123. (c) Douglass, J.; Civelli, O.; Herbert, E. *Annu. Rev. Biochem.* **1984**, *53*, 665.

5. Litter, M. *Farmacologia*, 2nd ed., El Ateneo, Buenos Aires, 1961.

6. (a) Ringe, D. *Curr. Opin. Struct. Biol.* **1995**, *5*, 825. (b) Karplus, P. A.; Faerman, C. *Curr. Biol.* **1994**, *4*, 770.

7. (a) Searle, M. S.; Williams, D. H. *J. Am. Chem. Soc.* **1992**, *114*, 10690. (b) Babine, R. E.; Bender, S. L. *Chem. Rev.* **1997**, *97*, 1359.

8. Albert, A. *Selective Toxicity*, 7th ed., Chapman and Hall, London, 1985, p. 206.

9. (a) Cleland, W. W.; Kreevoy, M. M. *Science* **1994**, *264*, 1887. (b) Cleland, W. W.; Frey, P. A.; Gerlt, J. A. *J. Biol. Chem.* **1998**, *273*, 25529.

10. Jencks, W. P. *Catalysis in Chemistry and Enzymology*, McGraw-Hill, New York, 1969, p. 340.

11. Korolkovas, A. *Essentials of Molecular Pharmacology*, Wiley, New York, 1970, p. 159.

12. Jencks, W. P. *Catalysis in Chemistry and Enzymology*, McGraw-Hill, New York, 1969, p. 393.

13. Hildebrand, J. H. *Proc. Natl. Acad. Sci. USA* **1979**, *76*, 194.

14. Andrew, P. R.; Craik, D. J.; Martin, J. L. *J. Med. Chem.* **1984**, *27*, 1648.

15. Andrews, P. R.; Craik, D. J.; Martin, J. L. *J. Med. Chem.* **1984**, *27*, 1648.

16. (a) Page, M. I. *Angew. Chem. Int. Ed. Engl.* **1977**, *16*, 449. (b) Page, M. I. *In Quantitative Approaches to Drug Design*, Dearden, J. C. (Ed.), Elsevier, Amsterdam, 1983, p. 109.

17. Allen, M. S.; Tan, Y.-C.; Trudell, M. L.; Narayanan, K.; Schindler, L. R.; Martin, M. J.; Schultz, C.; Hagen, T. J.; Koehler, K. F.; Codding, P. W.; Skolnick, P.; Cook, J. M. *J. Med. Chem.* **1990**, *33*, 2343.

18. Williams, M.; Enna, S. J. *Annu. Rep. Med. Chem.* **1986**, *21*, 211.

19. Costall, B.; Naylor, R. J. *Life Sci.* **1981**, *28*, 215.

20. Gaddum, J. H. *J. Physiol.* (*London*) **1926**, *61*, 141.

21. Clark, A. J. *J. Physiol.* (*London*) **1926**, *61*, 530.

22. (a) Ariëns, E. J. *Arch. Intern. Pharmacodyn. Ther.* **1954**, *99*, 32. (b) van Rossum, J. M.; Ariëns, E. J. *Arch. Intern. Pharmacodyn. Ther.* **1962**, *136*, 385. (c) van Rossum, J. M. *J. Pharm. Pharmacol.* **1963**, *15*, 285.

23. Stephenson, R. P. *Brit. J. Pharmacol. Chemother.* **1956**, *11*, 379.

24. Wermuth, C.-G.; Ganellin, C. R.; Lindberg, P.; Mitscher, L. A. *Annu. Rep. Med. Chem.* **1998**, *33*, 385.

25. Paton, W. D. M. *Proc. Roy. Soc. London, Ser. B* **1961**, *154*, 21.

26. (a) Koshland, D. E., Jr. *Proc. Natl. Acad. Sci. USA* **1958**, *44*, 98. (b) Koshland, D. E., Jr. *Biochem. Pharmacol.* **1961**, *8*, 57. (c) Koshland, D. E., Jr.; Neet, K. E. *Annu. Rev. Biochem.* **1968**, *37*, 359.

27. (a) Belleau, B. *J. Med. Chem.* **1964**, *7*, 776. (b) Belleau, B. *Adv. Drug Res.* **1965**, *2*, 89.

28. Monad, J.; Wyman, J.; Changeux, J.-P. *J. Mol. Biol.* **1965**, *12*, 88.

29. Karlin, A. *J. Theor. Biol.* **1967**, *16*, 306.

30. Milligan, G.; Bond, R. A.; Lee, M. *Trends Pharmacol. Sci.* **1995**, *16*, 10.

31. del Castillo, J.; Katz, B. *Proc. Roy. Soc. London Ser. B* **1957**, *146*, 369.

32. (a) Leff, P. *Trends Pharmacol. Sci.* **1995**, *16*, 89. (b) Kenakin, T. P. *Trends Pharmacol. Sci.* 1995, *16*, 232. (c) Bond, R.; Milligan, G.; Bouvier, M. *Handb. Exp. Pharmacol.* **2000**, *148*, 167. (d) De Ligt, R. A. F.; Kourounakis, A. P.; Ijzerman, A. P. *Br. J. Pharmacol.* **2000**, *130*, 1.

33. (a) Milligan, G.; Rees, S. *Annu. Rept. Med. Chem.* **2000**, *35*, 271. (b) Hamm, H. E. *Proc. Natl. Acad. Sci. USA* **2001**, *98*, 4819.

34. (a) Schoneberg, T.; Schulz, A.; Gudermann, T. *Rev. Physiol. Biochem. Pharmacol.* **2002**, *144*, 143. (b) Lombardi, M. S.; Kavelaars, A.; Heijnen, C. J. *Crit. Rev. Immunol.* **2002**, *22*, 141.

35. Samama, P.; Cotecchia, S.; Costa, T.; Lefkowitz, R. J. *J. Biol. Chem.* **199**3, *268*, 4625.

36. Leff, P.; Scaramellini, C.; Law. C.; McKechnie, K. *Trends Pharmacol. Sci.* **1997**, *18*, 355.

37. Ganellin, C. R. In *Pharmacology of Histamine Receptors*, Ganellin, C. R.; Parsons, M. E. (Eds.), Wright-PSG, Bristol, 1982, Chap. 2.

38. Ariëns, E. J.; Simonis, A. M.; van Rossum, J. M. In *Molecular Pharmacology*, Ariëns, E. J. (Ed.) Academic, New York, 1964, Vol. I, pp. 212 and 225.

39. Roth, F. E.; Govier, W. M. *J. Pharmacol. Exp. Ther.* **1958**, *124*, 347.

40. (a) Ariëns, E. J. *Med. Res. Rev.* **1986**, *6*, 451. (b) Ariëns, E. J. *Med. Res. Rev.* **1987**, *7*, 367.

41. Jacobson, A. E. In *Problems of Drug Dependence 1989*; Harris, L. S. (Ed.) U.S. Government Printing Office, Washington, DC, 1989, p. 556.

42. Pfeiffer, C. *Science* **1956**, *124*, 29.

43. White, P.; Ham, J.; Way, W.; Trevor, A. *Anesthesiology* **1980**, *52*, 231.

44. Mason, S. *New Scientist* **1984**, *101*, 10.

45. Winter, W.; Frankus, E. *Lancet* **1992**, *339*, 365.

46. (a) Takada, T.; Tada, M.; Kiyomoto, A. *Nippon Yakurigaku Zasshi* **1966**, *62*, 64, (b) Takada, T.; Tada, M.; Kiyomoto, A. *Chem. Abstr.* **1967**, *67*, 72326s.

47. Aps, C.; Reynolds, F. *Br. J. Clin. Pharmacol.* **1978**, *6*, 63.

48. Tobert, J.; Cirillo, V.; Hitzenberger, G.; James, I.; Pryor, J.; Cook, T.; Buntinx, A.; Holmes, I.; Lutterbeck, P. *Clin. Pharmacol. Ther.* **1981**, *29*, 344.

49. Drayer, D. E. *Clin. Pharmacol. Ther.* **1986**, *40*, 125.

50. Knabe, J. In *Chirality and Biological Activity*, Alan R. Liss, New York, 1990, pp. 237–246.

51. Knabe, J.; Rummel, W.; Bch, H. P.; Franz, N. *Arzneim. Forsch.* **1978**, *28*, 1048.

52. Zimmerman, D.; Gesellchen, P. *Annu. Rep. Med. Chem.* **1982**, *17*, 21.

53. Hillver, S. E.; Bjrk, I.; Li, Y.-L.; Svensson, B.; Ross, S.; Andn, N.-E.; Hacksell, U. *J. Med. Chem.* **1990**, *33*, 1541.

54. Hill, D. R.; Bowery, N. G. *Nature* (*London*) **1981**, *290*, 149.

55. Cahn, R. S.; Ingold, C. K.; Prelog, V. *Angew. Chem. Int. Ed. Engl.* **1966**, *5*, 385.

56. Dukes, M.; Smith, L. H. *J. Med. Chem.* **1971**, *14*, 326.

57. Ariëns, E. J. *Med. Res. Rev.* **1988**, *8*, 309.

58. Ariëns, E. J. *Med. Res. Rev.* **1987**, *7*, 367.

59. Simonyi, M. *Med. Res. Rev.* **1984**, *4*, 359.

60. Lindberg, P.; Keeling, D.; Fryklund, J.; Andersson, T.; Lundborg, P.; Carlsson, E. *Aliment. Pharmacol. Ther.* **2003**, *17*, 481.

61. Stinson, S. C. *Chem. Eng. News* **2001**, Oct. 1, p. 79.

62. Stinson, S. C. *Chem. Eng. News* **1994**, Sept. 19, pp. 38, 40.

63. Lien, E. J.; Rodrigues de Miranda, J. F.; Ariëns, E. J. *Mol. Pharmacol.* **1976**, *12*, 598.

64. Cushny, A. *Biological Relations of Optically Isomeric Substances*, Williams and Wilkins, Baltimore, 1926.

65. Easson, L. H.; Stedman, E. *Biochem. J.* **1933**, *27*, 1257.

66. Blaschko, H. *Proc. Roy. Soc. B* **1950**, *137*, 307.

67. Jamali, F.; Mehvar, R.; Pasutto, F. M. *J. Pharm. Sci.* **1989**, *78*, 695.

68. Boger, D. L.; Hong, J. *J. Am. Chem. Soc.* **2001**, *123*, 8515.

69. Boger, D. L.; Hüter, O.; Mbiya, K.; Zhang, M. *J. Am. Chem. Soc.* **1995**, *117*, 11839.

70. Cross, L. C.; Klyne, W. *Pure Appl. Chem.* **1976**, *45*, 11.

71. Casy, A. F.; Ganellin, C. R.; Mercer, A. D.; Upton, C. *J. Pharm. Pharmacol.* **1992**, *44*, 791.

72. Kaiser, C.; Setler, P. E. In *Burger's Medicinal Chemistry*, 4th ed., Wolff, M. E. (Ed.) Wiley, New York, 1981, Part III, Chap. 56.

73. Gampe, R. T.; Montana, V. G.; Lambert, M. H.; Miller, A. B.; Bledsoe, R. K.; Milburn, M. V.; Kliewer, S. A.; Willson, T. M.; Xu, H. E. *Mol. Cell* **2000**, *5*, 545.

74. Smissman, E. E.; Nelson, W. L.; LaPidus, J. B.; Day, J. L. *J. Med. Chem.* **1966**, *9*, 458.

75. (a) Armstrong, P. D.; Cannon, J. G.; Long, J. P. *Nature (London)* **1968**, *220*, 65. (b) Chiou, C. Y.; Long, J. P.; Cannon, J. G.; Armstrong, P. D. *J. Pharmacol. Exp. Ther.* **1969**, *166*, 243.

76. Armstrong, P. D.; Cannon, J. G. *J. Med. Chem.* **1970**, *13*, 1037.

77. Li, J. P.; Biel, J. H. *J. Med. Chem.* **1969**, *12*, 917.

78. Chen, C.-Y.; LeFèvre, R. J. W. *Tetrahedron Lett.* **1965**, 4057.

79. (a) Carey, R. E.; Heath, R. G. *Life Sci.* **1976**, *18*, 1105. (b) Smith, R. C.; Meltzer, H. Y.; Arora, R. C.; Davis, J. M. *Biochem. Pharmacol.* **1977**, *26*, 1435.

80. (a) Maurice, T.; Lockhart, B. P. *Prog. Neuro-Psychopharmacol. Biol. Psychiat.* **1997**, *21*, 69. (b) Su, T. P.; London, E. D.; Jaffe, J. H. *Science* **1988**, *240*, 219. (c) Ross, S. B.; Gawaell, L.; Hall, H. *Pharmacol. Toxicol.* **1987**, *61*, 288.

81. Moriarity, R. M.; Enache, L. A.; Zhao, L.; Gilardi, R.; Mattson, M. V.; Prakash, O. *J. Med. Chem.* **1998**, *41*, 468.

82. Hudkins, R. L.; Mailman, R. B.; DeHaven-Hudkins, D. L. *J. Med. Chem.* **1994**, *37*, 1964.

83. Nogrady, T. *Medicinal Chemistry*, Oxford, New York, 1985, p. *28*.

84. Rampe, D.; Triggle, D. J. *Prog. Drug Res.* **1993**, *40*, 191.

85. Black, J. W.; Duncan, W. A. M.; Durant, C. J.; Ganellin, C. R.; Parsons, E. M. *Nature (London)* **1972**, *236*, 385.

86. (a) Ganellin, C. R. *J. Med. Chem.* **1981**, *24*, 913. (b) Ganellin, C. R.; Durant, G. J. In *Burger's Medicinal Chemistry*, 4th ed., Wolff, M. E. (Ed.), Wiley, New York, 1981, Part III, Chap. 48.

87. Hill, S. J.; Ganellin, C. R.; Timmerman, H.; Schwartz, J. C.; Shankley, N. P.; Young, J. M.; Schunack, W.; Levi, R.; Haas, H. L. *Pharmacol. Rev.* **1997**, *49*, 253.

88. Wyllie, J. H.; Hesselbo, T.; Black, J. W. *Lancet* **1972**, *2(7787)*, 1117.

89. Black, J. W.; Durant, G. J.; Emmett, J. C.; Ganellin, C. R. *Nature (London)* **1974**, *248*, 65.

90. Charton, M. *J. Org. Chem.* **1965**, *30*, 969.

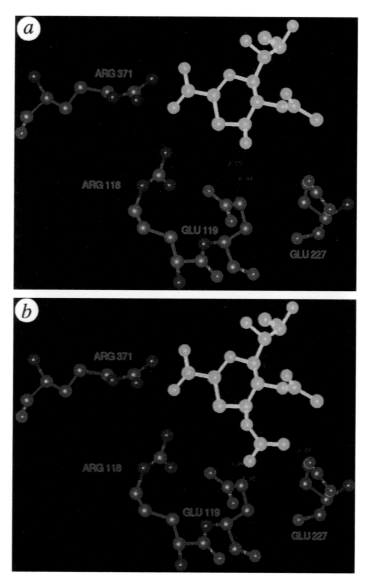

Figure 2.18 ▶ Crystal structure of neuraminidase active site with inhibitors bound. (a) Interaction of the protonated amino group of 2.95 (R = NH_3^+) with Glu-119. (b) Interaction of the protonated guanidinium group of **2.93** with Glu-119 and Glu-227 [Reprinted with permission from *Nature* 363, 418. Copyright ©1993 MacMillan Magazines Ltd.]

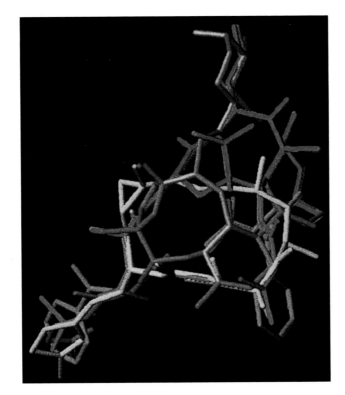

Figure 2.20 ► Common pharmacophore based on the composite of boxed sections in Figure 2.19. [With permission from I. Ojima (1999). Reprinted with permission from *PNAS* 96, 4256. Copyright © 1999 National Academy of Sciences, U.S.A.]

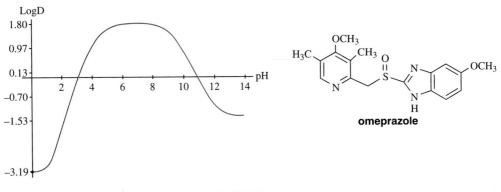

Problem 2.15

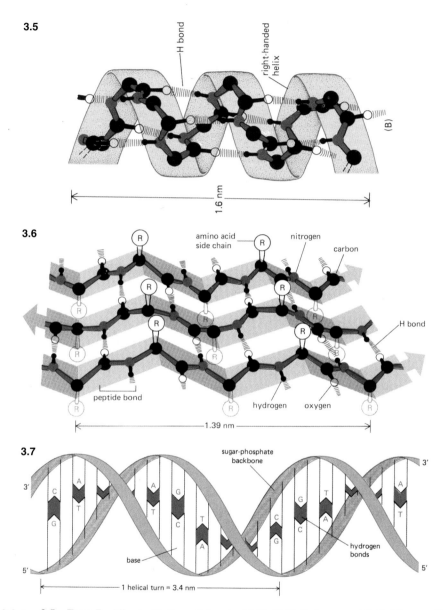

3.5

H bond

right-handed helix

(B)

1.6 nm

3.6

amino acid side chain

nitrogen

carbon

R

R

R

R

R

R

R

R

R

H bond

peptide bond

hydrogen

oxygen

R

R

R

1.39 nm

3.7

sugar-phosphate backbone

3'

3'

C

A

A

G

T

G

T

A

T

C

C

T

G

base

A

C

G

hydrogen bonds

5'

5'

1 helical turn = 3.4 nm

Figure 3.4 ▶ **3.5**—From B. Alberts, D. Bray, J. Lewis, M. Raff, K. Roberts, and J. D. Watson, *Molecular Biology of the Cell*, 2nd ed., p. 110, Garland Publishing, New York, 1989, with permission. Copyright © 1989 Garland Publishing.

3.6—From B. Alberts, D. Bray, J. Lewis, M. Raff, K. Roberts, and J. D. Watson, *Molecular Biology of the Cell*, 2nd ed., p. 109, Garland Publishing, New York, 1989, with permission. Copyright © 1989 Garland Publishing.

3.7—From B. Alberts, D. Bray, J. Lewis, M. Raff, K. Roberts, and J. D. Watson, *Molecular Biology of the Cell*, 2nd ed., p. 99, Garland Publishing, New York, 1989, with permission. Copyright © 1989 Garland Publishing

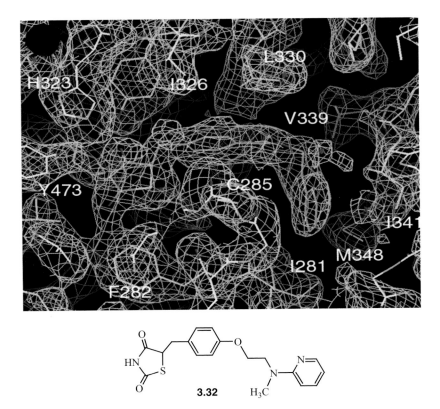

Figure 3.22 ▶ An example of a bioactive conformation. Rosiglitazone (**3.32**), an antidiabetic drug (green structure in the middle), bound to peroxisome proliferator-activated receptor gamma (PPARγ). Note the sickle-shaped conformation in the binding site to accommodate the shape formed by the active-site residues. [With permission from Xu, H.E, (2000). Reprinted from *Molecular Cell*, Vol. 5, "Asymmetry in the PPAR$_\gamma$/RXR$_\alpha$ Crystal Structure Reveals the Molecular Basis of Hetrodimerization among Nuclear Receptors", pp. 545–555, Copyright © 2000, with permission of Elsevier.]

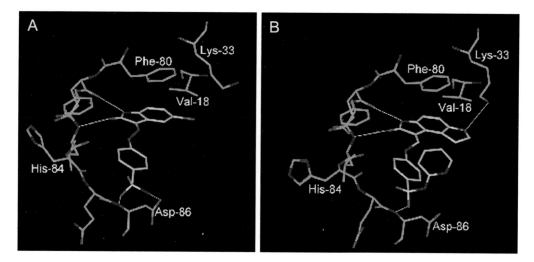

Figure 6.1 ▶ Crystal structure of (A) **6.2** (beige) and (B) **6.3** (purple) bound to cyclin-dependent kinase 2 [Reprinted with permission from *Science* 291, 134. Copyright © 2001 American Association for the Advancement of Science.]

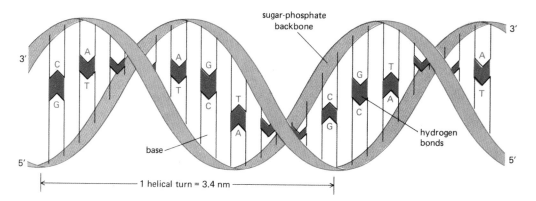

Figure 6.2 ▶ DNA structure. [Reproduced with permission from Alberts, B., Bray, D., Lewis, J., Raff, M., Roberts, K., and Watson, J. D. (1989). *Molecular Biology of the Cell*, 2nd ed., p. 99. Garland Publishing, New York. Copyright 1989 Garland Publishing.]

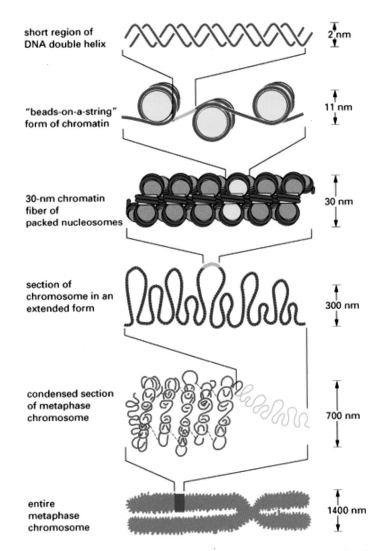

short region of
DNA double helix — 2 nm

"beads-on-a-string"
form of chromatin — 11 nm

30-nm chromatin
fiber of
packed nucleosomes — 30 nm

section of
chromosome in an
extended form — 300 nm

condensed section
of metaphase
chromosome — 700 nm

entire
metaphase
chromosome — 1400 nm

Figure 6.7 ▶ Stages in the formation of the entire metaphase chromosome starting from duplex DNA [With permission from Alberts, B. (1994). Copyright 1994 From *Molecular Biology of the Cell, 3rd Ed.* by Bruce Alberts, Dennis Bray, Julian Lewis, Martin Raff, Keith Roberts, and James D. Watson. Reproduced by permission of Routledge, Inc., part of The Taylor & Francis Group.]

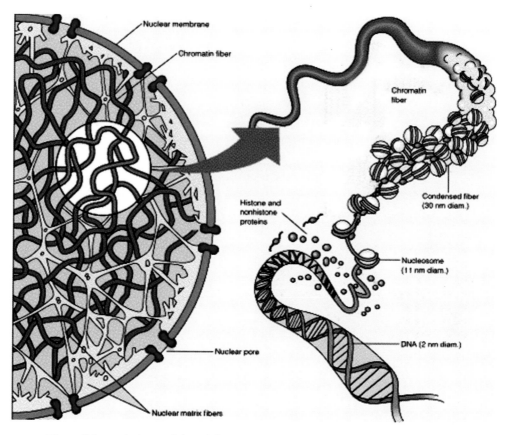

Figure 6.8 ▶ Artist rendition of the conversion of duplex DNA into chromatin fiber

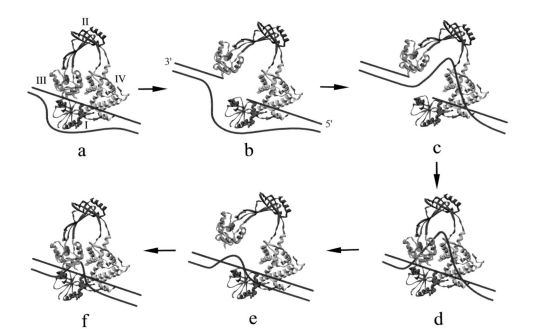

Figure 6.12 ▶ Artist rendition of a possible mechanism for a topoisomerase I reaction [With permission from Champoux, J. J. (2001). With permission from the *Annual Review of Biochemistry*, Volume 70 © 2001 by Annual Reviews www.annualreviews.org.]

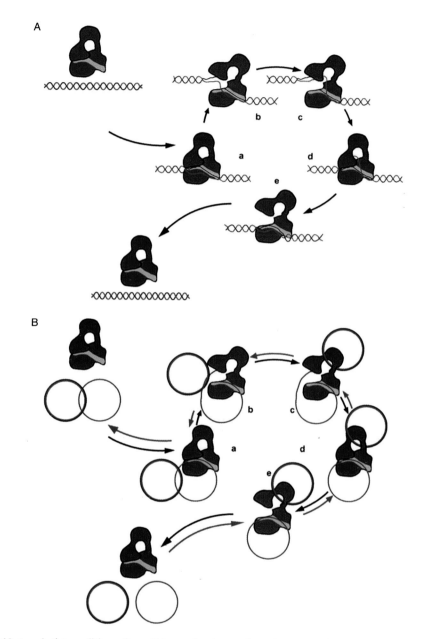

Figure 6.13 ▶ Artist rendition of possible mechanisms of topoisomerase IA-catalyzed relaxation of (A) supercoiled DNA and (B) decatenation of a DNA catenane [Reproduced with permission from Li, Z.; Mondragón, A.; DiGate, R. J., *Mol. Cell* **2001**, *7*, 301.]

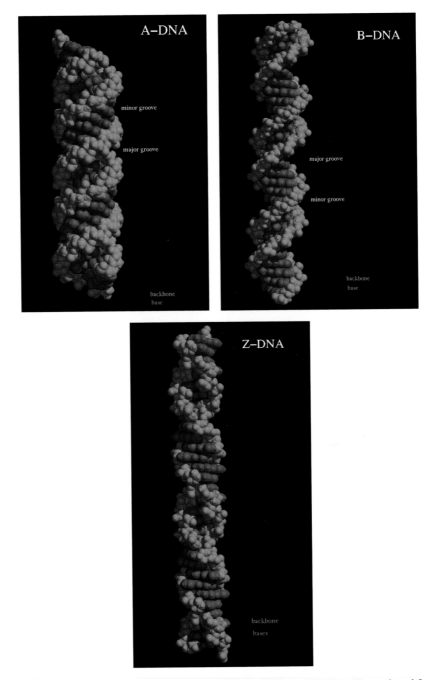

Figure 6.14 ▶ Computer graphics depictions of A-DNA, B-DNA, and Z-DNA [Reproduced from the IMB Jena Image Library of Biological Macromolecules; http://www.imb-jena.de/IMAGE.html.]

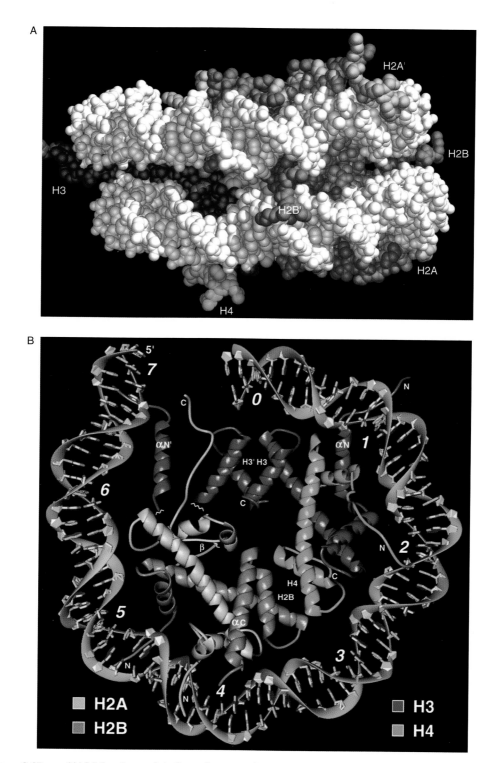

Figure 6.15 ▶ (A) Molecular model of a nucleosome. (B) Cutaway view of the nucleosome with the histones in the center and duplex DNA wrapped around them [With permission from Luger, K (1997). Reprinted with permission from *Nature* 398, 251–260. Copyright 1993 Macmillan Publishers Limited.]

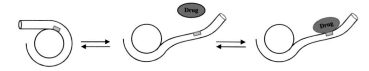

Figure 6.16 ▶ Schematic of how a drug could bind to DNA wrapped around histones in the nucleosome [Reproduced with permission from Polach K. J., Widom J. *J. Mol. Biol.* **1995**, *254*, 130.]

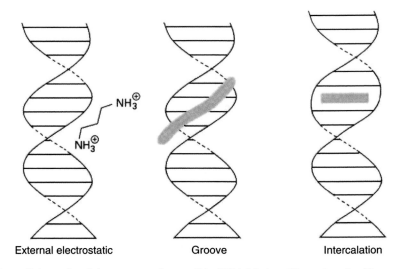

Figure 6.17 ▶ Schematic of three types of reversible DNA binders [Reproduced with permission from Blackburn G. M., Gait M. J., Eds. *Nucleic Acids in Chemistry and Biology*, 2nd ed., p. 332. Oxford University Press, Oxford; Copyright 1996 Oxford University Press.]

Chapter 4

Enzymes

4.1 Enzymes as Catalysts

Enzymes are special types of receptors. The receptors discussed in Chapter 3 are mostly membrane-bound proteins that interact with natural ligands and agonists to form complexes which then elicit a biological response. Subsequent to the response, the ligand is released intact. Enzymes, most of which are soluble and found in the cytosol of cells, interact with substrates to form complexes, but, unlike receptors, it is from these *enzyme–substrate complexes* that enzymes catalyze reactions, thereby transforming the substrates into products that are released. Therefore, the two characteristics of enzymes are their ability to recognize a substrate and to catalyze a reaction of it.

4.1.A   What Are Enzymes?

Enzymes are natural proteins that catalyze chemical reactions; RNA also can catalyze chemical reactions.[1] The first enzyme to be recognized as a protein was jack bean urease, which was crystallized in 1926 by Sumner[2] and was shown to catalyze the hydrolysis of urea to CO_2 and NH_3. It took almost 70 years more, however, before its crystal structure would be obtained by Andrew Karplus (for the enzyme from *Klebsiella aerogenes*).[3] By the 1950s hundreds of enzymes had been discovered and many were purified to homogeneity and crystallized. In 1960 Hirs, Moore, and Stein[4] were the first to sequence an enzyme, namely, ribonuclease A, having only 124 amino acids (molecular mass 13,680 Da). This was an elegant piece of work, and William H. Stein and Stanford Moore shared the Nobel Prize in Chemistry in 1972 for the methodology of protein sequencing which was developed to determine the ribonuclease A sequence.

Enzymes can have molecular masses of several thousand to several million daltons, yet catalyze transformations on molecules as small as carbon dioxide or nitrogen. Carbonic anhydrase from human erythrocytes, for example, has a molecular mass of about 31,000 Da and each enzyme molecule can catalyze the hydration of 1,400,000 molecules of CO_2 to H_2CO_3 per second! This is almost 10^8 times faster than the uncatalyzed reaction, which is actually on the low side of rate enhancements for an enzyme. Orotidine 5'-phosphate decarboxylase, for example, catalyzes the decarboxylation of orotidine 5'-phosphate to uridine 5'-phosphate 10^{17} times faster than the nonenzymatic rate.[5]

4.1.B   How Do Enzymes Work?

In general, enzymes function by lowering transition state energies and energetic intermediates and by raising the ground state energy. The transition state for an enzyme-catalyzed reaction, just as in the case of a chemical reaction, is a high-energy state having a lifetime of about 10^{-13} sec, the time for one bond vibration.[6] There is no spectroscopic method that can detect the transition state structure in an enzyme.

At least 21 different hypotheses for how enzymes catalyze reactions have been proposed.[7] The one common link among all of these proposals, however, is that an enzyme-catalyzed reaction always is initiated by the formation of an *enzyme–substrate* (or $E \cdot S$) *complex*, from which the catalysis takes place. The concept of an enzyme–substrate complex was originally proposed independently in 1902 by Brown[8] and Henri;[9] this idea is an extension of the 1894 *lock and key hypothesis* of Fischer[10] in which it was proposed that an enzyme is the lock into which the substrate (the key) fits. This interaction of the enzyme and substrate would account for the high degree of specificity of enzymes, but the lock and key hypothesis does not rationalize certain observed phenomena. For example, compounds whose structures are related to that of the substrate, but have *less* bulky substituents, often fail to be substrates, even though they should have fit into the enzyme. Some compounds with *more* bulky substituents are observed to bind *more* tightly to the enzyme than does the substrate. If the lock and key hypothesis were correct, it would be thought that a more bulky compound would not fit into the lock. Some enzymes that catalyze reactions between two substrates do not bind one substrate until the other one is already bound to the enzyme. These curiosities led Koshland[11] in 1958 to propose the *induced fit hypothesis* (discussed in relationship to receptors in Chapter 3, Section 3.2.D.3, p. 139), namely, that when a substrate begins to bind to an enzyme, interactions of various groups on the substrate with particular enzyme functional groups are initiated, and these mutual interactions induce a *conformational change* in the enzyme. This results in a change of the enzyme from a low catalytic form to a high catalytic form by destabilization of the enzyme and/or by inducing proper alignment of the groups involved in catalysis. The conformational change could serve as a basis for substrate specificity. Compounds resembling the substrate except with smaller or larger substituents may bind to the enzyme but may not induce the conformational change necessary for catalysis. Unlike the lock and key hypothesis, which implies a rigid active site, the induced fit hypothesis requires a flexible active site to accommodate different binding modes and conformational changes in the enzyme. Actually, the concept of a flexible active site was stated earlier by Pauling[12] who hypothesized that an enzyme is a flexible template that is most complementary to substrates at the transition state rather than at the ground state. This flexible model is consistent with many observations regarding enzyme action.

In 1930 Haldane[13] suggested that an E · S complex requires additional activation energy prior to enzyme catalysis, and this energy may be derived from substrate strain energy on the enzyme. As early as 1921 Polanyi proposed that transition state binding is essential for catalysis,[14] and he and his student Eyring developed transition state theory,[15] which is the basis for the above-mentioned hypothesis of Pauling. According to this hypothesis, the substrate does not bind most effectively in the E · S complex; as the reaction proceeds, the enzyme conforms to the transition state structure, leading to the tightest interactions (increased binding energy) with the transition state structure.[16] This increased binding energy, known as *transition state stabilization*, results in rate enhancement. Schowen has suggested[17] that all of the above-mentioned 21 hypotheses of enzyme catalysis (as well as other correct hypotheses) are just alternative expressions of transition state stabilization.

Similar to the case of receptors discussed in Chapter 3, in which the pharmacophore of the drug interacts with a relatively small part of the total receptor, the substrate likewise binds to only a small part of the enzyme known as the *active site* of the enzyme. There may be only a dozen or so amino acid residues that comprise the active site, and of these only three may be involved directly in substrate binding and/or catalysis. Because all of the catalysis takes place in the active site of the enzyme, you may wonder why it is necessary for enzymes to be so large. There are several hypotheses regarding the function of the remainder of the enzyme. One suggestion[18] is that the most effective binding of the substrate to the enzyme (the largest binding energy) results from close packing of the atoms within the protein; possibly, the remainder of the enzyme outside of the active site is required to maintain the integrity of the active site for effective catalysis. The protein may also serve the function of channeling the substrate into the active site.

Enzyme catalysis is characterized by two features: *specificity* and *rate acceleration*. The active site contains moieties that are responsible for both of these properties of an enzyme, namely, amino acid residues and, in the case of some enzymes, cofactors. A *cofactor*, also called a *coenzyme*, is an organic molecule or metal ion that binds to the active site, in some cases covalently and in others noncovalently, and is essential for the catalytic action of those enzymes that require cofactors.

B.1 Specificity of Enzyme-Catalyzed Reactions

Specificity refers to both specificity of binding and specificity of reaction. Certain active site constituents are involved in these binding interactions which are responsible for the binding specificity. These interactions are the same as those discussed in Chapter 3, Section 3.2.B, p. 124, for the interaction of an agonist with a noncatalytic receptor and include covalent, electrostatic, ion–dipole, dipole–dipole, hydrogen bonding, charge-transfer, hydrophobic, hydrophilic, and van der Waals interactions.

a. Binding Specificity

As indicated above, maximum binding interactions at the active site occur at the transition state of the reaction. An enzyme binds the transition state structure about 10^{12} times more tightly than it binds the substrate or products (the enzyme orotidine 5′-monophosphate decarboxylase binds the transition state 10^{17} times greater than substrate!).[19] Therefore, it is important that an enzyme does not bind to intermediate states excessively or this will increase the free-energy difference between the intermediate and transition state. The binding interactions set up the substrate for the reaction that the enzyme catalyzes (Scheme 4.1).

At one end of the spectrum, binding specificity can be absolute, that is, essentially only one substrate forms an E · S complex with a particular enzyme, which then leads to product formation. At the other end of the spectrum, binding specificity can be very broad, in which case many molecules of related structure can bind and be converted to product, such as the family of enzymes known as cytochrome P450, which protects us from all of the toxins we eat and breathe.[20] Specificity may involve E · S complex formation with only one enantiomer of a racemate or E · S complex formation with both enantiomers, but only one is converted to product. The reason enzymes can accomplish this enantiomeric specificity, just as receptors do with enantiomers (see Chapter 3, Section 3.2.E.2) is because they are chiral molecules (mammalian enzymes are comprised of only *L*-amino acids); therefore, interactions of an enzyme with a racemic mixture results in the formation of two diastereomic complexes. This is analogous to the principle behind the resolution of racemic mixtures with chiral reagents. If, for example (Scheme 4.2), a pure *R*-isomer of a chiral amine such as (*R*)-2-methylbenzylamine (**4.1**) is mixed with a racemic mixture of a carboxylic acid such as the nonsteroidal anti-inflammatory drug ibuprofen (**4.2**; Advil), then two diasteromeric salts, (*R* − *R*) and (*R* − *S*), will be formed. Because these salts are no longer enantiomers, they will have different properties, and can be separated by physical means. When an enzyme is

$$\text{E} + \text{S} \; \underset{}{\overset{K_{\text{s}}}{\rightleftharpoons}} \; \text{E} \cdot \text{S} \; \underset{}{\overset{k_{\text{cat}}}{\rightleftharpoons}} \; \text{E} \cdot \text{P} \; \rightleftharpoons \; \text{E} + \text{P}$$

Scheme 4.1 ▶ Generalized enzyme-catalyzed reaction

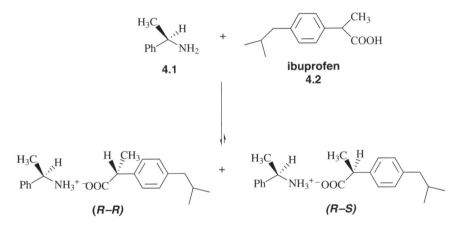

Scheme 4.2 ► Resolution of a racemic mixture

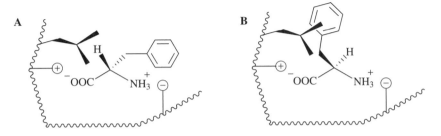

Figure 4.1 ► Differential binding interactions by enantiomers

exposed to a racemic mixture of a substrate, the binding energy for E · S complex formation with one enantiomer may be much higher than that with the other enantiomer either because of differential binding interactions as noted above or for steric reasons. For example, after the ammonium and carboxylate substituents of phenylalanine have interacted with active site groups (Figure 4.1), a third substituent at the stereogenic center (the benzyl group) has two possible orientations; in the case of the *S*-isomer, there is a binding pocket (Figure 4.1A), but the benzyl group of the *R*-isomer (Figure 4.1B) points in the other direction toward the leucine side-chain and causes steric hindrance (i.e., the *R*-isomer does not bind to the active site). If the binding energies for the two complexes are significantly different, then only one E · S complex may form (as would be the case in Figure 4.1). Alternatively, both E · S complexes may form, but for steric or electronic reasons only one E · S complex may lead to product formation. The enantiomer that forms the E · S complex that is not *turned over* (i.e., converted by the enzyme to product) is said to undergo *nonproductive binding* to the enzyme. Enzymes also can demonstrate complete stereospecificity with geometric isomers since these are diastereomers already.

b. Reaction Specificity

Reaction specificity also arises from constituents of the active site, namely, specific acid, base, and nucleophilic functional groups of amino acids (Section 4.2) and from cofactors

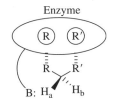

Figure 4.2 ► Enzyme specificity for chemically identical protons. R and R′ on the enzyme are groups that interact specifically with R and R′, respectively, on the substrate

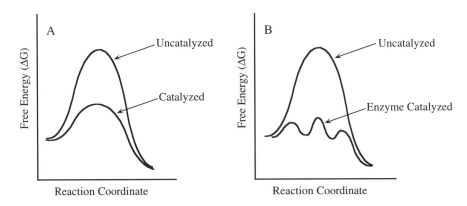

Figure 4.3 ► Effect of (A) a chemical catalyst and (B) an enzyme on the activation energy

(Section 4.3). Unlike reactions in solution, enzymes can show specificity for chemically identical protons (Figure 4.2). If there are specific binding sites for R and R′ at the active site of the enzyme, and a base (B⁻) of an amino acid side-chain is juxtaposed so that it can only reach proton H_a, then abstraction of H_a will occur stereospecifically, even though in a nonenzymatic reaction H_a and H_b would be chemically equivalent and, therefore, would have equal probability to be abstracted. The approach taken by synthetic chemists in designing chiral reagents for stereospecific reactions is modeled after this.

B.2 Rate Acceleration

In general, catalysts stabilize the transition state relative to the ground state, and this decrease in activation energy is responsible for the rate acceleration that results (Figure 4.3A). Jencks proposed that the fundamental feature that distinguishes enzymes from simple chemical catalysts is the ability of enzymes to utilize binding interactions away from the site of catalysis.[21] These binding interactions facilitate reactions by positioning substrates with respect to one another and with respect to the catalytic groups at the active site. Because an enzyme has numerous opportunities to invoke catalysis, for example, by stabilization of the transition states (thereby lowering the transition state energy), by destabilization of the E·S complex (thereby raising the ground state energy), by destabilization of intermediates, and during product release, multiple steps, each having small activation energies, may be involved (Figure 4.3B). As a result of these multiple catalytic steps, rate accelerations of 10^{10}–10^{14} over the corresponding nonenzymatic reactions are common. Wolfenden hypothesized that the rate acceleration produced by an enzyme is proportional to the affinity of the enzyme for the transition state structure of the bound substrate;[22] the reaction rate is proportional to the amount of substrate that is in the transition state complex. The enzyme has to be able to bind tightly only to the unstable transition state structure (with a lifetime of one bond vibration) and *not* to the substrate or the products. A conformational change in the protein structure plays an important role in this operation.

Enzyme catalysis does not alter the equilibrium of a reversible reaction. If an enzyme accelerates the rate of the forward reaction, it must accelerate the rate of the corresponding back reaction by the same amount; its effect is to accelerate the attainment of the equilibrium, but not the relative concentrations of substrates and products at equilibrium.

Typically, enzymes have *turnover numbers* (also termed k_{cat}), that is, the number of molecules of substrate converted to product per unit of time per molecule of enzyme active site, on the order of 10^3 s^{-1} (about 1000 molecules of substrate are converted to product every second!). The enzyme catalase is one of the most efficient enzymes,[23] having a turnover number of 10^7 s^{-1}. Because there are two other important steps to enzyme catalysis, namely, substrate binding and product release, high turnover numbers are only useful if these two physical steps occur at faster rates. This is not always the case.

4.2 Mechanisms of Enzyme Catalysis

Once the substrate binds to the active site of the enzyme via the interactions noted in Section 3.2.B, p. 124, there are a variety of mechanisms that the enzyme can utilize to catalyze the conversion of the substrate to product. The most common mechanisms[24–26] are approximation, covalent catalysis, general acid–base catalysis, electrostatic catalysis, desolvation, and strain or distortion. All of these act by stabilizing the transition state energy or destabilizing the ground state (which is generally not as important as transition state stabilization).

4.2.A Approximation

Approximation is rate enhancement by proximity, that is, the enzyme serves as a template to bind the substrates so that they are close to each other in the reaction center. This results in a loss of rotational and translational entropies of the substrate on binding to the enzyme; however, this entropic loss is offset by a favorable binding energy of the substrate, which provides the driving force for catalysis. Furthermore, since the catalytic groups are now an integral part of the same molecule, the reaction of enzyme-bound substrates becomes first order rather than second order when these compounds are free in solution. Holding the reaction centers in proximity and in the correct geometry for reaction is equivalent to increasing the concentration of the reacting groups. This phenomenon can be exemplified with nonenzymatic model studies. For example, consider the second-order reaction of acetate with an aryl acetate (Scheme 4.3). If the rate constant k for this reaction is set equal to 1.0 M^{-1} s^{-1}, and then the effect of decreasing rotational and translational entropy is determined by measuring the corresponding first-order rate constants for related molecules that can undergo the corresponding intramolecular reactions, it is apparent from Table 4.1 that forcing the reacting groups to be closer to

Scheme 4.3 ▶ Second-order reaction of acetate with aryl acetate

TABLE 4.1 ▶ Effect of Approximation on Reaction Rate

	k_{obs} (30°C)	relative rate, k	EM
	3.36×10^{-7} M^{-1}s^{-1}	1.0 M^{-1}s^{-1}	
	7.39×10^{-5} s^{-1}	220 s^{-1}	220 M
	1.71×10^{-2} s^{-1}	5.1×10^{4} s^{-1}	5.1×10^{4} M
	7.61×10^{-1} s^{-1}	2.3×10^{6} s^{-1}	2.3×10^{6} M
	3.93 s^{-1}	1.2×10^{7} s^{-1}	1.2×10^{7} M

each other increases the reaction rate.[27,28] Thirty-six years after the original experimental study of the effect of restricted rotation on rate acceleration, a theoretical investigation using MM3 calculations showed that when the nucleophile and electrophile are closely arranged, and the van der Waals surfaces are properly juxtaposed, the activation energy is lowered as a result of a decrease in the enthalpy of the reaction ($\Delta H°$), and the rate of the reaction really should increase, thereby supporting the earlier experimental observations.[29]

Although first- and second-order rate constants cannot be compared directly, the efficiency of an intramolecular reaction can be defined in terms of its *effective molarity* (EM),[30] the concentration of the reactant (or catalytic group) required to cause the intermolecular reaction to proceed at the observed rate of the intramolecular reaction. The EM is calculated by dividing the first-order rate constant for the intramolecular reaction by the second-order rate constant for the corresponding intermolecular reaction (see Table 4.1). What this indicates is that the acetate ion would have to be at a concentration of, for example, 220 M ($220 \, sec^{-1}/1 \, M^{-1} \, sec^{-1}$) for the intermolecular reaction of acetate and aryl acetate to proceed at a rate comparable to that of the glutarate monoester reaction. Of course, 220 M acetate ion is an imaginary number (pure water is only 55 M), so the effect of decreasing the enthalpy is quite significant. Effective molarities for a wide range of intramolecular reactions have been measured, and the conclusion is that the efficiency of intramolecular catalysis varies with structure, and can be as high as 10^{16} M for reactive systems. Therefore, holding groups proximal to each other, particularly when the reacting moieties in an enzyme–substrate complex are aligned correctly for reaction, can be an important contributor to catalysis.

4.2.B Covalent Catalysis

Some enzymes can use nucleophilic amino acid side-chains or cofactors in the active site to form covalent bonds to the substrate; in some cases, a second substrate then can react with this enzyme–substrate intermediate to generate the product. This is known as *nucleophilic catalysis* (Scheme 4.4), a subclass of *covalent catalysis* that involves covalent bond formation as a result of attack by an enzyme nucleophile at an electrophilic site on the substrate. For example, if Y in Scheme 4.4 is an amino acid or peptide and Z^- is a hydroxide ion, then the enzyme would be a peptidase (or protease). For nucleophilic catalysis to be most effective, Y should be converted into a better leaving group than X (for example, by protonation), and the covalent intermediate (**4.3**, Scheme 4.4) should be more reactive than the substrate. The most common active site nucleophiles are the thiol group of cysteine, the hydroxyl group of serine, the imidazole of histidine, the amino group of lysine, and the carboxylate group of aspartate or glutamate. These active site nucleophiles are generally activated by deprotonation, often by a neighboring histidine imidazole or by a water molecule that is deprotonated in a general base reaction (see Section 4.2.C, p. 182). The principal catalytic advantage of using an active site residue instead of water directly is that the former leads to a unimolecular reaction (because the substrate is bound to the enzyme, attack by the serine residue is equivalent to an intramolecular reaction), which is entropically favored over the bimolecular reaction with water. Also, alkoxides (ionized serine) and thiolates (ionized cysteine) are better nucleophiles than hydroxide ion. Amide bonds of peptides have low reactivity, but the nucleophile in the

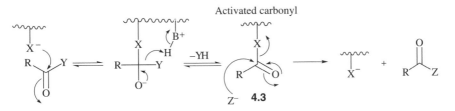

Scheme 4.4 ▶ Nucleophilic catalysis

active site could be made more nucleophilic in a nonpolar environment. Once **4.3** is generated, the carbonyl becomes much more reactive (if X is the hydroxyl of serine, then **4.3** is an ester; if X is the thiol of cysteine, then it is a thioester).

Nucleophilic catalysis is the enzymatic analogy to anchimeric assistance by neighboring groups in organic reaction mechanisms. *Anchimeric assistance* is the process by which a neighboring functional group assists in the expulsion of a leaving group by intermediate covalent bond formation.[31] This results in accelerated reaction rates. Scheme 4.5 shows how a neighboring sulfur atom makes the displacement of a β-chlorine a much more facile reaction than it would be without the sulfur atom. If the sulfur atom were part of an active site nucleophile, such as a methionine, the C–Cl bond were part of a substrate, and HO⁻ were generated by enzyme-catalyzed deprotonation of water, this would represent covalent catalysis in an enzyme-catalyzed reaction, where the covalent adduct is the episulfonium intermediate.

Typical enzymatic reactions where nucleophilic catalysis is important include many of the proteolytic enzymes, for example, the serine proteases (i.e., proteases that utilize a serine residue at the active site as the nucleophile) such as elastase (degrades elastin, a connective tissue prevalent in the lung) or plasmin (lyses blood clots), and the cysteine proteases (utilize an active site cysteine residue as the nucleophile) such as papain (found in papaya fruit and used in digestion).

4.2.C   General Acid–Base Catalysis

In any reaction where proton transfer occurs, *general acid catalysis* and/or *general base catalysis* can be an important mechanism for specificity and rate enhancement. There are two kinds of acid–base catalysis: specific catalysis and general catalysis. If catalysis occurs by a hydronium (H_3O^+) or hydroxide (HO⁻) ion and is determined only by the pH, not the buffer concentration, it is referred to as *specific acid* or *specific base catalysis*, respectively. As an example of how specific acid–base catalysis works, consider the hydrolysis of ethyl acetate (Scheme 4.6). This is an exceedingly slow reaction at neutral pH because both the nucleophile (H_2O) and the electrophile (the carbonyl of ethyl acetate) have low reactivity. The reaction rate could be accelerated, however, if the reactivity of either the nucleophile or the electrophile could be enhanced. An increase in the pH increases the concentration of hydroxide ion, which is a much better nucleophile than is water, and, in fact, the rate of hydrolysis at higher pH increases (Scheme 4.7). Likewise, a decrease in the pH increases the

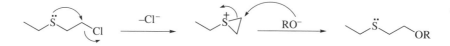

Scheme 4.5 ▶ Anchimeric assistance by a neighboring group

$$H_3C\overset{\displaystyle O}{\overset{\|}{-}C}-OC_2H_5 \quad + \quad H_2O \quad \rightleftharpoons \quad H_3C\overset{\displaystyle O}{\overset{\|}{-}C}-OH \quad + \quad C_2H_5OH$$

Scheme 4.6 ▶ Hydrolysis of ethyl acetate

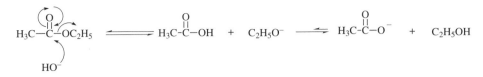

Scheme 4.7 ▶ Alkaline hydrolysis of ethyl acetate

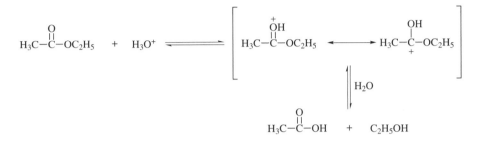

Scheme 4.8 ▶ Acid hydrolysis of ethyl acetate

concentration of the hydronium ion, which can protonate the ester carbonyl, thereby increasing its electrophilicity, and this also increases the hydrolysis rate (Scheme 4.8). That being the case, then the hydrolysis rate should be doubly increased if base *and* acid are added together, right? Of course not. Addition of an acid to a base would only lead to neutralization and loss of any catalytic effect.

General acid–base catalysis, on the other hand, occurs when the reaction rate increases with increasing buffer concentration at a constant pH and ionic strength, and shows a larger increase with a buffer that contains a more concentrated acid or base component. Because the hydronium or hydroxide ion concentration is not increasing (the pH is constant), it must be the buffer that is catalyzing the reaction. This is *general acid catalysis* (if acids other than hydronium ion, such as active site acid groups, accelerate the reaction rate) or *general base catalysis* (if bases other than hydroxide ion, such as active site basic groups, accelerate the rate).

Unlike reactions in solution, however, an enzyme can utilize acid and base catalysis simultaneously (Scheme 4.9) for even greater catalysis. The protonated base in Scheme 4.9 is an acidic amino acid side-chain and the free base is a basic residue. As was discussed in Chapter 2, Section 2.2.F.3, p. 62), it is important to appreciate the fact that the pK_a values of amino acid side-chain groups within the active site of enzymes are not necessarily the same as those measured in solution. Also, pK_a values can change drastically in hydrophobic environments. Therefore, removal of seemingly higher pK_a protons from substrates by active site bases may not be as unreasonable as would appear if only solution chemistry were taken into consideration.

As an example of general acid–base (and covalent) catalysis, consider the enzyme *α-chymotrypsin*, a serine protease, which means that it utilizes an active site serine residue in a covalent catalytic cleavage of peptide bonds (Scheme 4.10). Because the nucleophilic group of serine is hydroxyl, it should be a poor nucleophile. However, aspartic acid and histidine residues nearby have been implicated in the conversion of the serine to an alkoxide by a mechanism called the *charge relay system* by Blow and coworkers,[32] the discoverers of

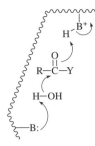

Scheme 4.9 ▶ Simultaneous acid and base enzyme catalysis

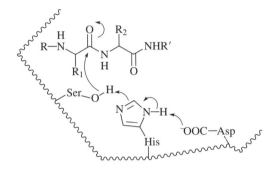

Scheme 4.10 ▶ Charge relay system for activation of an active site serine residue

the existence of the hydrogen bonding network involving Asp-102, His-57, and Ser-195. This *catalytic triad* involves the aspartate carboxylate (pK_a of the acid is 3.9 in solution) removing a proton from the histidine imidazole (pK_a 6.1 in solution), which, in turn, removes a proton from the serine hydroxyl group (pK_a 14 in solution). Any respectable organic chemist would find that suggestion absurd; how can a base such as aspartate, whose conjugate acid is two pK_a units lower than that of the histidine imidazole, remove the imidazole proton efficiently, and, then, how can this imidazole remove the proton from the hydroxyl group of serine, which is eight pK_a units higher than the (protonated) imidazole of histidine? The equilibrium is only 1% in favor of the first proton transfer, and the equilibrium for the second proton transfer favors the back direction by a factor of 10^8! One explanation (see Chapter 2, Section 2.2.F.3, p. 62) could be that the pK_a values of some of these acids and bases at the active site are different from those in solution. Furthermore, because these groups are held close together at the active site, as the proton is beginning to be removed from the serine hydroxyl group, the charge density proceeds to the next step (attack of the alkoxide at the peptide carbonyl), thereby driving the equilibrium in the forward direction. This is the beauty of enzyme-catalyzed reactions; the approximation of the groups and the fluidity of the active site residues working in concert permit reactions to occur that would be nearly impossible in solution.

4.2.D Electrostatic Catalysis

An enzyme catalyzes a reaction by stabilization of the transition state and by destabilization of the ground state. Stabilization of the transition state may involve the presence of an ionic charge or partial ionic charge at the active site to interact with an opposite charge developing on the substrate at the transition state of the reaction (Scheme 4.11). In the case of the tetrahedral

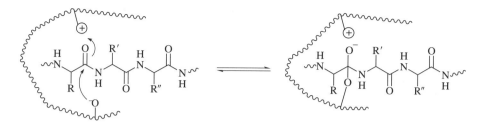

Scheme 4.11 ▶ Electrostatic stabilization of the transition state

intermediate shown in Scheme 4.11, the site in the enzyme that leads to this stabilization is referred to as the *oxyanion hole*.[33] Electrostatic interactions may not be as pronounced as is shown in Scheme 4.11; instead of a full positive charge on the enzyme, there may be one or more local dipoles having partial positive charges directed at the incipient transition state anion or a protonated group available for hydrogen bonding. In the case of the serine protease subtilisin it has been suggested that the lowering of the free energy of the activated complex is the result of hydrogen bonding of the developing oxyanion with protein residues.[34] When the suspected active site proton donor was replaced by a leucine residue using site-directed mutagenesis, the k_{cat} greatly diminished, but the K_m remained the same, indicating the importance of hydrogen bonding to catalysis. Furthermore, a *mutant* (i.e., a form of the enzyme that has one or more of its amino acid residues changed, generally by site-directed mutagenesis) of the protease subtilisin, in which all three of the catalytic triad residues (serine-221, histidine-64, and aspartate-32) were replaced by alanine residues by site-directed mutagenesis, still was able to hydrolyze amides 10^3 times faster than the uncatalyzed hydrolysis rate, albeit 2×10^6 times slower than the *wild-type enzyme* (i.e., the nonmutated form).[35] This suggests that factors other than nucleophilic and general base catalysis must also be important.

4.2.E Desolvation

Ground state destabilization could occur by desolvation, that is, removal of water molecules from charged groups at the active site on substrate binding. This exposes the substrate to a lower dielectric constant (possibly hydrophobic) environment, which would destabilize a charged group on the substrate. Desolvation also could expose a water-bonded charged group at the active site so it can more effectively participate by electrostatic catalysis in stabilization of a charge generated at the transition state. Because electrostatic interactions are much stronger in low dielectric media than in water, the presence of positive and negative charges (or partial charges) within the low dielectric medium of the enzyme active site more strongly stabilizes developing charges at the transition state of a reaction (Scheme 4.11).

4.2.F Strain or Distortion

In organic chemistry *strain* and *distortion* play an important role in the reactivity of molecules. The much higher reactivity of epoxides relative to other ethers demonstrates this phenomenon. Cyclic phosphate ester hydrolysis is another example. Considerable ring strain in **4.4** (Scheme 4.12) is released on alkaline hydrolysis; the rate of hydrolysis of **4.4** is 10^8 times

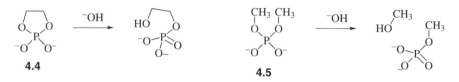

Scheme 4.12 ▶ Alkaline hydrolysis of phosphodiesters as an example of strain energy

greater than that for the corresponding acyclic phosphodiester **4.5**.[36] Therefore, if strain or distortion could be induced during enzyme catalysis, then the enzymatic reaction rate would be enhanced. This effect could be induced either in the enzyme, thereby converting it to a high-activity state, or in the substrate, thereby raising the ground state energy (destabilization) of the substrate and making it more reactive. In Chapter 3, Section 3.2.D.3, p. 139, the induced fit hypothesis of Koshland[37] was mentioned. This hypothesis suggests that the enzyme need not necessarily exist in the appropriate conformation required to bind the substrate. As the substrate approaches the enzyme, various groups on the substrate interact with particular active site functional groups, and this mutual interaction induces a conformational change in the active site of the enzyme. This can result in a change of the enzyme from a low-catalytic form to a high-catalytic form by destabilization of the enzyme (strain or distortion) or by inducing proper alignment of active site groups involved in catalysis, which could be responsible for the initiation of catalysis. Inspection of substrate binding sites in protein crystallographic databases indicates that most enzymes have at least a portion of the active site in a structure that is complementary to the substrate to permit binding on the first collisional event. The conformational change need not occur only in the enzyme; the substrate also could undergo deformation, which would lead to strain (destabilization; higher ground state energy) in the substrate. The enzyme was suggested to be elastic and could return to its original conformation after the product was released. This rationalizes how high-energy conformations of substrates are able to bind to enzymes.

According to Jencks,[38] strain or distortion of the bound substrate is essential for catalysis. Because ground state stabilization of the substrate occurs concomitant with transition state stabilization, the ΔG is no different from that of the uncatalyzed reaction, only displaced downward (Figure 4.4b). To lower the ΔG for the catalytic reaction, the E·S complex must be destabilized by strain, desolvation, or loss of entropy on binding, thereby raising the ΔG of the E·S and E·P complexes (Figure 4.4c). As the reaction proceeds, the ΔG can be lowered by release of strain energy or by other mechanisms described above.

4.2.G Example of the Mechanisms of Enzyme Catalysis

A very important bacterial enzyme in medicinal chemistry is the peptidoglycan transpeptidase, the enzyme that catalyzes the cross-linking of peptidoglycan strands to make the bacterial cell wall. This enzyme has not been purified, but studies with another penicillin-binding protein that is believed to be the *in vivo* transpeptidase which becomes uncoupled during its purification, namely, *D*-alanine carboxypeptidase, has been purified and has been useful in the elucidation of the mechanism of penicillin action.[39] Although a detailed mechanistic study of the transpeptidase has not yet been carried out, on the basis of the principles discussed above, the hypothetical mechanism shown in Scheme 4.13 can be proposed. The E·S complex formed from the two strands of peptidoglycan with transpeptidase would be stabilized

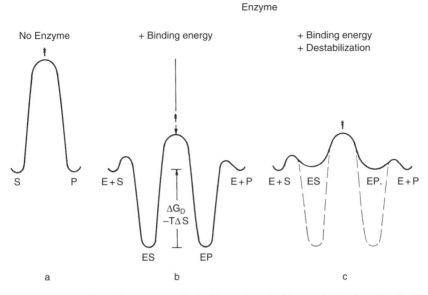

Figure 4.4 ► Energetic effect of enzyme catalysis. [Reproduced with permission from Jencks, W. P. (1987), *Cold Spring Harbor Symp. Quant. Biol. 52*, 65. Copyright ©1987 Cold Spring Harbor Laboratory Press.]

by the appropriate noncovalent binding interactions (see Chapter 3, Section 3.2.B, p. 124). These interactions could place the peptide carbonyl, which is ultimately the site of transpeptidation, very close (approximation) to the serine residue that is involved in covalent catalysis. Base catalysis (via a catalytic triad) could be utilized to activate the active site serine, and electrostatic catalysis (for example, by a positively charged arginine residue) could stabilize the oxyanion intermediate. Alternatively, an active site acidic group could donate a proton (acid catalysis) to the incipient oxyanion to lower the transition state energy for the formation of the tetrahedral intermediate. Breakdown of this intermediate could be facilitated by proton donation (acid catalysis) to the leaving *D*-alanine residue in addition to appropriate strain energy in the sp^3 tetrahedral carbon produced by a conformational change that favors sp^2 hybridization of the product (the covalent intermediate). If a proton-donating mechanism instead of electrostatic mechanism were used to activate the initial covalent reaction, then the active site conjugate base could remove the proton to facilitate tetrahedral intermediate breakdown. The third step, the cross-linking of the second peptidoglycan strand with the newly formed ester linkage of the activated initial peptidoglycan strand, could be catalyzed by approximation of the two reacting centers, by electrostatic catalysis as in the first step, and by another conformational change in the enzyme to distort the sp^2 ester carbonyl toward the second sp^3 tetrahedral intermediate. Breakdown of that tetrahedral intermediate could be catalyzed again by strain energy (sp^3 back to sp^2) as well as by base catalysis. Product release may occur more readily with the enzyme in the charged form shown. Proton transfer after product release would return the enzyme to its normal energy state.

It must be emphasized that the mechanism shown in Scheme 4.13 is hypothetical. However, on the basis of the mechanism of enzyme catalysis described above, it is not an unreasonable hypothesis.

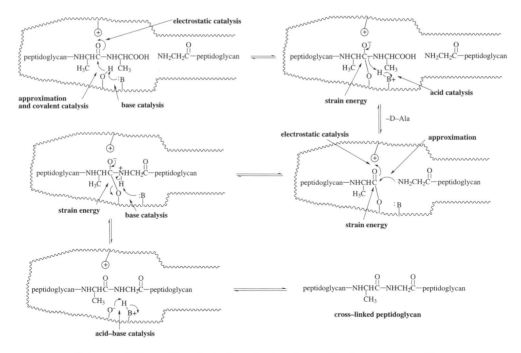

Scheme 4.13 ▶   Hypothetical mechanism for peptidoglycan transpeptidase

4.3   Coenzyme Catalysis

A *coenzyme*, or *cofactor*, is any organic molecule or metal ion that is essential for the catalytic action of the enzyme. The usual organic coenzymes are generally derived as products of the metabolism of vitamins that we consume (Table 4.2). Other organic molecules that are involved in essential enzyme functioning, but which are not derived from vitamins, include coenzyme A (**4.13b**), which is used to make the CoA thioesters of carboxylic acid substrates (and is not related to vitamin A); heme (**4.14**; protoporphyrin IX) and the tripeptide glutathione (**4.15**; GSH), which are very important to enzymes involved in drug metabolism (Chapter 7); adenosine triphosphate (**4.16**; ATP), which supplies the energy required to activate certain substrates during enzyme-catalyzed reactions; and lipoic acid (**4.17**) and ascorbic acid (**4.18**, vitamin C), which are involved in oxidation and reduction reactions, respectively.

Vitamins are, by definition, essential nutrients; human metabolism is incapable of producing them. Deficiency in a vitamin results in the shutting down of the catalytic activity of various enzymes that require the coenzyme made from the vitamin or results from the lack of other activities of the vitamins. This leads to certain disease states (Table 4.3).[40]

The remainder of this chapter is devoted to the chemistry of coenzyme catalysis. The discussion is limited to only those coenzymes whose chemistry of action will be important to the mechanisms of drug action that are described in later chapters.

4.3.A   Pyridoxal 5′-Phosphate (PLP)

Enzymes dependent on pyridoxal 5′-phosphate (PLP) catalyze several different reactions of amino acids that, at first glance, appear to be unrelated; however, when the mechanisms

TABLE 4.2 ▶ **Coenzymes Derived from Vitamins**

Vitamin	Structure	Coenzyme form	Structure	Coenzyme acronym
Vitamin B_1 (thiamin)	**4.6** (R = H)	Thiamin diphosphate	**4.6** (R $=^-O_2POPO_3^{2-}$)	TPP (TDP)
Vitamin B_2 (riboflavin)	**4.7** (R = H)	Flavin mononucleotide	**4.7** (R = PO_3^{2-})	FMN
		Flavin adenine dinucleotide	**4.7** (R $=^-O_2POPO_3^-$-5'-Ado)	FAD
Vitamin B_3 (niacin-amide)	**4.8a** (no R)	Nicotinamide adenine dinucleotide	**4.8a** (R' = H)	NAD^+
(niacin)	Corresponding carboxylic acid	Nicotinamide dinucleotide phosphate	**4.8a** (R' = PO_3^{2-})	$NADP^+$
		Reduced nicotinamide adenine dinucleotide	**4.8b** (R' = H)	NADH
		Reduced nicotinamide adenine dinucleotide phosphate	**4.8b** (R' = PO_3^{2-})	NADPH
Vitamin B_6 (pyridoxine)	**4.9** (R = CH_2OH, R' = H)	Pyridoxal 5'-phosphate	**4.9** (R = CHO, R' = PO_3^{2-})	PLP
		Pyridoxamine 5'-phosphate	**4.9** (R = CH_2NH_2, R' = PO_3^{2-})	PMP
Vitamin B_{12} (cyanoco-balamin)	**4.10** (R = CN)	Adenosylcobalamin	**4.10** (R = 5'-Ado)	CoB_{12}
Folic acid	**4.11a** (R = OH or poly-γ-glutamyl)	Tetrahydrofolate	**4.11b** (R = OH or poly-γ-glutamyl)	THF
Biotin	**4.12** (R = OH)	Covalently bound to enzyme as amide	**4.12** (R = lysine residue of enzyme)	—
Pantothenic acid	**4.13a**	Coenzyme A (no relation to vitamin A)	**4.13b**	CoASH

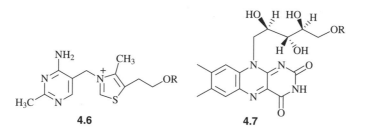

4.6   **4.7**

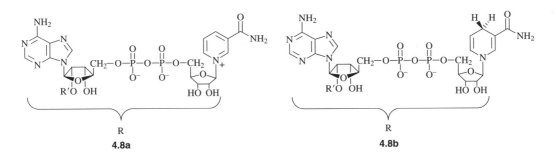

R
4.8a

R
4.8b

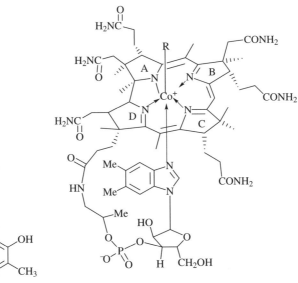

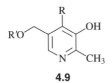

4.9

4.10

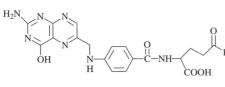

4.11a

4.11b

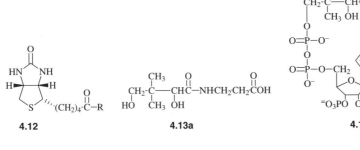

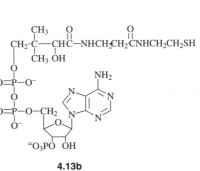

4.12

4.13a

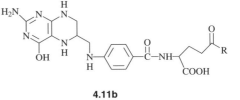

4.13b

TABLE 4.3 ▶ Symptoms of Vitamin Deficiency

Deficient vitamin	Disease state
Thiamin	Beriberi
Riboflavin	Dermatitis, anemia
Niacin	Pellagra
Pyridoxine	Dermatitis, convulsions
Vitamin B_{12}	Pernicious anemia
Folic acid	Pernicious anemia
Biotin	Dermatitis
Pantothenic acid	Neuromuscular effects
Ascorbic acid	Scurvy
Vitamin D	Rickets
Vitamin K	Hemorrhage
Vitamin A	Night blindness

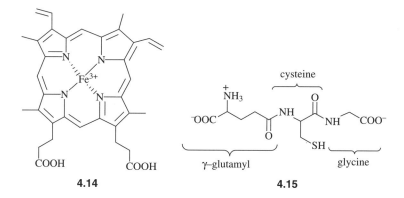

4.14   **4.15**

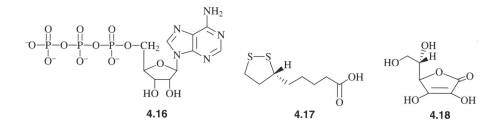

4.16   **4.17**   **4.18**

are discussed, the relationships will become apparent. The overall reactions are summarized in Table 4.4. PLP is the most versatile coenzyme, but all of these reactions are very specific depending on the enzyme to which the PLP is bound. Although these reactions can be catalyzed by PLP nonenzymatically, typically several of the possible reactions take place simultaneously.[41,42] For a given enzyme, only one reaction occurs almost exclusively.

Although several noncovalent binding interactions are responsible for holding the PLP in the active site, the major interaction is a covalent one (**4.19a**, Figure 4.5). To make the chemistry easier to visualize, this coenzyme will be abbreviated as shown in **4.19b** when the

TABLE 4.4 ▶ Reactions Catalyzed by Pyridoxal 5′-Phosphate-Dependent Enzymes

Substrate	Product	Reaction
R─CH(H)(COOH)─NH₂	R─CH(COOH)(H)─NH₂	racemization
R─CH(COOH)─NH₂	R─CH₂─NH₂ + CO_2	decarboxylation
R─CH(COOH)─NH₂	R─C(=O)─COOH + NH_4^+	transamination
R─CH(COOH)─NH₂	H_2N─CH₂─COOH + "R^+"	α-cleavage
X─CH₂─CH(COOH)─NH₂	CH₃─C(=O)─COOH + NH_4^+ + X^-	β-elimination
X─CH₂─CH(COOH)─NH₂	Y─CH₂─CH(COOH)─NH₂ + X^-	β-replacement
X─CH₂─CH₂─CH(COOH)─NH₂	CH₃─CH₂─C(=O)─COOH + NH_4^+ + X^-	γ-elimination
X─CH₂─CH₂─CH(COOH)─NH₂	Y─CH₂─CH₂─CH(COOH)─NH₂ + X^-	γ-replacement

chemistry is discussed. The aldehyde group of the PLP is held tightly at the active site by a Schiff base (iminium) linkage to a lysine residue. In addition to securing the PLP in the optimal position at the active site, Schiff base formation activates the carbonyl for nucleophilic attack. This is very important to the catalysis because the first step in *all* PLP-dependent enzyme transformations of amino acids (see Table 4.4) is a transimination reaction, i.e., the conversion of the lysine-PLP imine (**4.20**, Scheme 4.14) to the substrate PLP-imine (**4.21**). In Scheme 4.14 two different bases are shown to be involved in acid–base catalysis; a similar mechanism could be drawn with a single base. It is from **4.21** that all of the reactions shown in Table 4.4 occur. The property of **4.21** that links all of these reactions is that the pyridinium group can act as

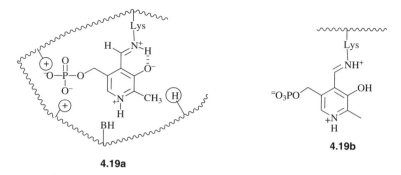

Figure 4.5 ▶ Pyridoxal 5′-phosphate covalently bound to the active site of an enzyme

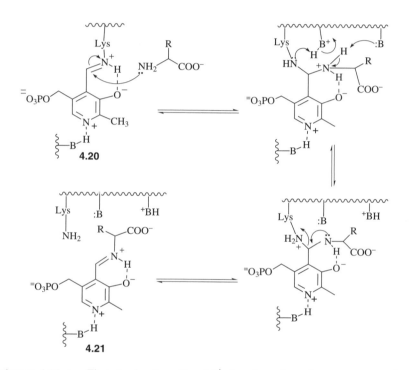

Scheme 4.14 ▶ First step in all pyridoxal 5′-phosphate-dependent enzyme reactions

an electron sink to stabilize electrons by resonance from the C–H, C–COO⁻, or C–R bonds. This could account for why all three of these bonds can be broken nonenzymatically. The important question is how can an enzyme catalyze the *regiospecific* cleavage of only one of these three bonds?

The bond that breaks must lie in a plane perpendicular to the plane of the PLP-imine π-electron system. In Figure 4.6 the C–H bond is the one perpendicular to the plane of the π-system, i.e., parallel with the p orbitals. This configuration results in maximum π-electron overlap (the sp^3 σ orbital of the C–H bond and p orbital of the aromatic system), and, therefore, minimizes the transition state energy for bond breakage of the C–H bond. The problem for the enzyme to solve, then, is how to control the conformation about the C_α–N bond so that only the bond that is to be cleaved is perpendicular to the plane of the π-system at the active site

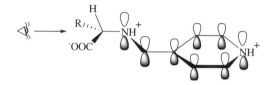

Figure 4.6 ▶ π-electron system of the PLP-imine

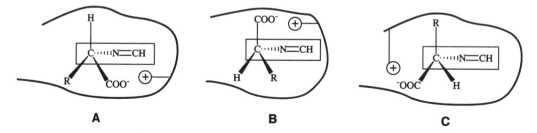

Figure 4.7 ▶ Dunathan hypothesis for PLP activation of the C_α−N bond. The rectangle represents the plane of the pyridine ring of the PLP. The angle of sight is that shown by the eye in Figure 4.6 [Adapted by permission of Dunathan, H. C. (1971). In *Advances in Enzymology*, Vol. 35, p. 79, Meister, A., Ed. Copyright © 1971 John Wiley & Sons, Inc.]

of the enzyme. The *Dunathan hypothesis*[43] gives a rational explanation for how an enzyme could control the C_α−N bond rotation (Figure 4.7). A positively charged residue at the active site could form a salt bridge with the carboxylate group of the amino acid bound to the PLP. This would make it possible for an enzyme to restrict rotation about the C_α−N bond and hold the H (Figure 4.7A), the COO$^-$ (Figure 4.7B), or the R (Figure 4.7C) group perpendicular to the plane of the aromatic system (the rectangles in Figure 4.7 are the pyridine ring systems). If the Dunathan hypothesis is accepted, then all of the PLP-dependent enzyme reactions can be readily understood. In fact, crystal structures of PLP-dependent enzymes generally show a group at the active site, often an arginine residue, positioned to bind to the substrate-bound carboxylate group.[44]

In the next four subsections mechanisms for only the classes of PLP-dependent enzymes that will be relevant to later chapters are described. The mechanism for α-cleavage will be discussed in Section 4.3.B.

A.1 Racemases

D-Amino acids are commonly found in bacteria for use in the assembly of their cell wall.[45] *D*-Serine and *D*-aspartate, however, have also been detected in mammalian brain;[46] *D*-serine activates glutamate/NMDA receptors in neurotransmission.[47] Because natural amino acids have the *L*-configuration, enzymes that convert the *L*-amino acids into their enantiomers must be available for the organisms to get the *D*-isomers. These enzymes are part of the family of enzymes known as *racemases*, which are highly specific for the amino acid used as its substrate; alanine racemase, for example, does not catalyze the racemization of gluta-mate. Also (and at least as fascinating as the substrate specificity), a decarboxylase does not also catalyze racemization of its substrate, so histidine decarboxylase converts *L*-histidine to histamine without formation of *D*-histidine, although there are some PLP-dependent enzymes that do catalyze a slow side reaction different from its normal function. Consider alanine racemase as an example of this type of reaction. The reaction catalyzed by this enzyme is the interconversion of *L*- and *D*-alanine with an equilibrium constant of 1, typical for racemases, so it does not matter which enantiomer is used as the substrate, the racemate is produced. The mechanism shown in Scheme 4.15 is typical for PLP-dependent racemases (for alanine racemase R $= CH_3$). Because the carbanion produced by α-proton removal from the PLP-imine is so highly delocalized, the pK_a of the α-proton is much lower than that in the parent amino acid. Deprotonation of the bound amino acid could occur with either one base for both enantiomers or two bases.[48] Alanine racemase uses two different active site bases to deprotonate bound *D*-alanine (Lys-39) or *L*-alanine (Tyr-265).[49] Note that all of the steps are reversible; therefore, these enzymes can accept either the *D*- or *L*-isomer and the equilibrium mixture will be obtained. Because of the importance of *D*-alanine to the construction of the bacterial cell wall, inhibition of this enzyme (and prevention of the formation of *D*-alanine) has been an approach for potential antibacterial drug design.

A.2 Decarboxylases

Decarboxylases catalyze the conversion of an amino acid to an amine and carbon dioxide (Scheme 4.16).[50] Because the loss of CO_2 is irreversible, the amine cannot be converted back to the amino acid. However, the last step in Scheme 4.16 is reversible, so it is possible to catalyze the exchange of the α-proton of the amine with solvent protons, for example, when the isolated enzyme reaction is carried out in ^2H_2O.

Specific decarboxylases are known for more than 10 of the common amino acids. *L*-Aromatic amino acid decarboxylase (sometimes referred to as dopa decarboxylase) plays a vital role in the conversion of the anti-parkinsonian drug *L*-dopa to dopamine in the brain; low dopamine levels are characteristic of parkinsonism. Many of the bacterial amino acid decarboxylases have low pH optima. It is possible that at least one of the functions of these decarboxylases, which generate amine bases from neutral amino acids, is to neutralize the acidic conditions in the cell. Another function may be to control the bacterial intracellular CO_2 pressure.

A.3 Aminotransferases (Formerly Transaminases)

Aminotransferases are the most complicated of the PLP-dependent reactions; they involve two substrates going to two products in two half reactions. Various PLP-dependent aminotransferases are known for α-, β-, and γ-amino acids. As the name implies, the amino group of the substrate amino acid is transferred to another molecule, an α-keto acid (Scheme 4.17). The labeling pattern given in Scheme 4.17 indicates that the amino group of **4.22** is transferred to the α-keto acid (**4.23**), which loses its oxygen as a molecule of water. The initial amino acid (**4.22**), therefore, is converted into an α-keto acid (**4.24**) and the starting α-keto acid (**4.23**) is converted into an amino acid (**4.25**). A mechanism consistent with these observations is given in Scheme 4.18 and Scheme 4.19. In Scheme 4.18 the ^{15}N substrate is transferred to the coenzyme, which is converted into ^{15}N-containing PMP (**4.28**). This is an internal redox reaction (a tautomerization): The amino acid substrate is oxidized to an α-keto acid at the expense of the reduction of the PLP to PMP. Note that the first two steps (up to the three resonance structures) are identical to the mechanism for PLP-dependent racemization (Scheme 4.15),

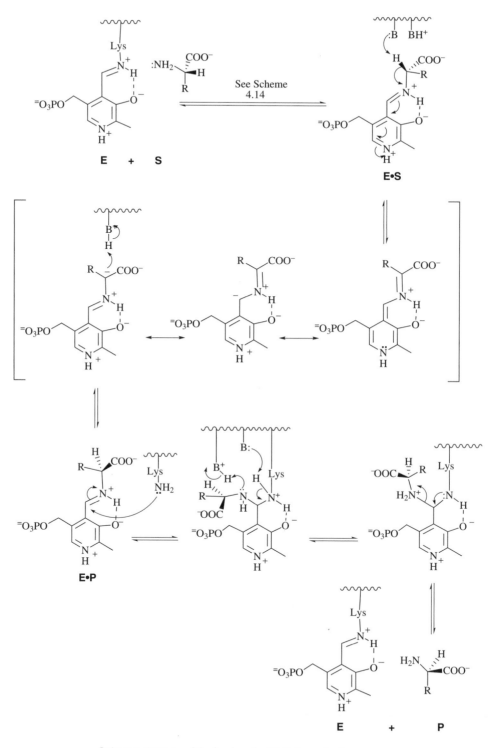

Scheme 4.15 ▶ Mechanism for PLP-dependent racemases

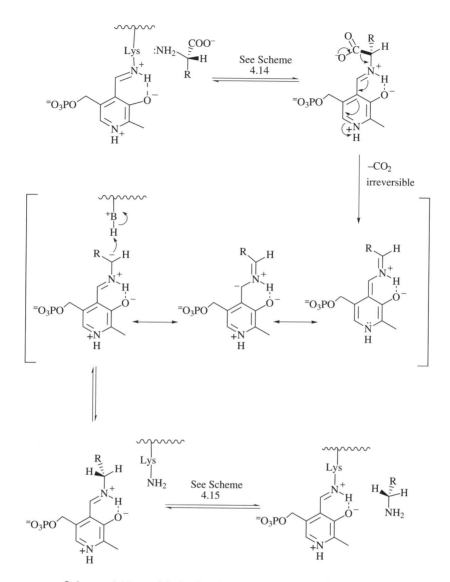

Scheme 4.16 ▶ Mechanism for PLP-dependent decarboxylases

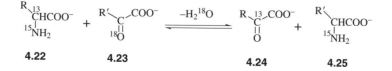

4.22   **4.23**   **4.24**   **4.25**

Scheme 4.17 ▶ Overall reaction catalyzed by PLP-dependent aminotransferases

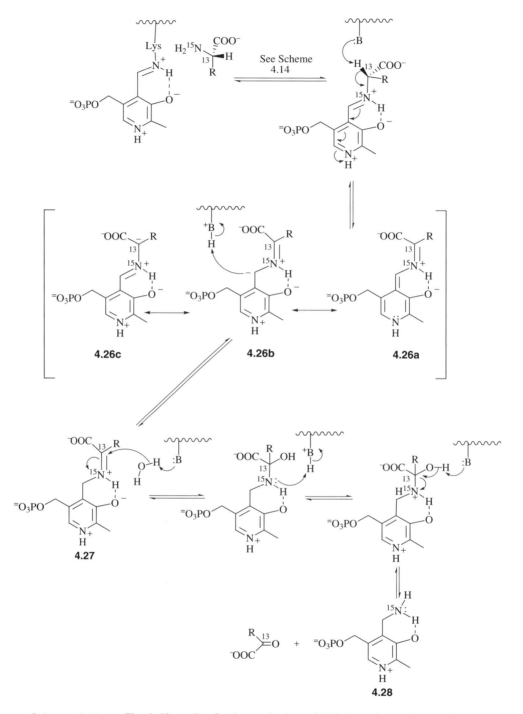

Scheme 4.18 ▶ First half-reaction for the mechanism of PLP-dependent aminotransferases

yet racemization does not occur. So how does the enzyme alter this racemization pathway in favor of transamination? Simply by the placement of an appropriate acidic residue closer to the PMP carbon (**4.26b**, Scheme 4.18) than to the substrate α-carbon (**4.26c**). This leads to the PMP-imine Schiff base (**4.27**), which can be hydrolyzed to PMP and the α-keto acid. At this point the coenzyme is no longer in the proper oxidation state (i.e., an imine) for reaction with another amino acid substrate molecule (remember, the first step is Schiff base formation with an amino acid).

If this were the end of the enzyme reaction, then the enzyme would be a reagent, not a catalyst, because only one turnover has taken place and the enzyme is inactive. To return the coenzyme in its PMP form to the requisite oxidation state of the PLP form for continued substrate turnover, a second substrate, another α-keto acid (not the product α-keto acid), binds to the active site of the enzyme and undergoes the exact reverse of the reaction shown in Scheme 4.18 (Scheme 4.19). In this process the amino group on the PMP, which originally came from the substrate amino acid, is then transferred to the second substrate (the α-keto acid), thereby producing a new amino acid and PLP. In the overall reaction (Schemes 4.18 and 4.19) the second half-reaction (Scheme 4.19) is only utilized to convert the enzyme back to the active form. For some aminotransferases, however, the converse is true, namely, that the first half reaction is to get the enzyme in the PMP form so that the intended substrate (an α-keto acid) can be converted to the product amino acid.

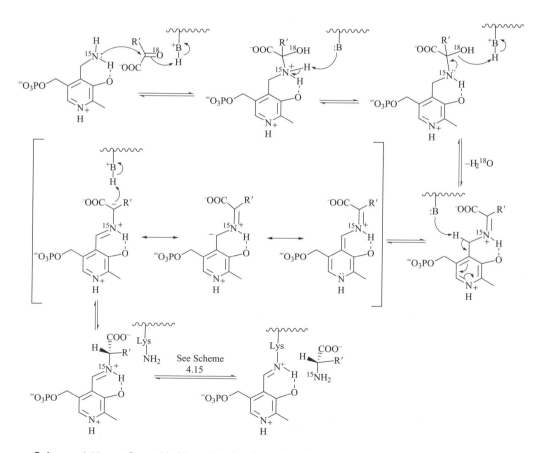

Scheme 4.19 ▶ Second half-reaction for the mechanism of PLP-dependent aminotransferases

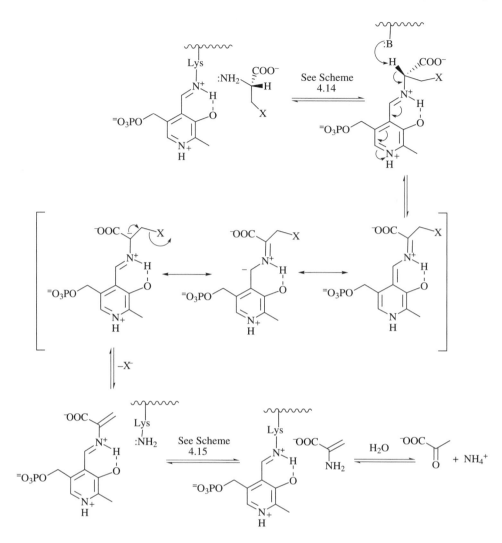

Scheme 4.20 ▶ Mechanism for PLP-dependent β-elimination reactions

A.4   PLP-Dependent β-Elimination

Some PLP-dependent enzymes catalyze the elimination of H–X from substrate molecules that contain a β-leaving group. Although these enzymes are not directly relevant to the later discussions, this sort of mechanism will be an important alternative pathway for certain inactivators of other PLP-dependent enzymes (see Chapter 5, Section 5.5.C.3, p. 287). The mechanism is shown in Scheme 4.20.

4.3.B   Tetrahydrofolate and Pyridine Nucleotides

The most important way for one-carbon-containing groups to be added to molecules involves enzymes that utilize coenzymes derived from the human vitamin folic acid (**4.29,**

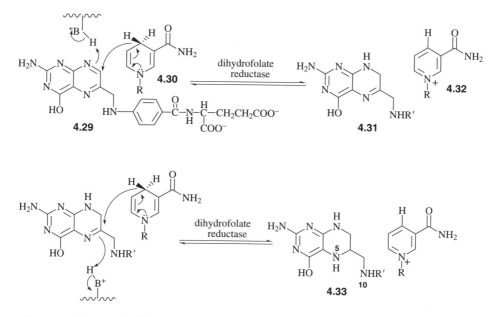

Scheme 4.21 ► Pyridine nucleotide-dependent reduction of folic acid to tetrahydrofolate

Scheme 4.21). Although folic acid is a monoglutamate, the coenzyme forms actually can contain an oligomer of as many as 12 glutamate residues, depending on the enzyme. The C–N double bonds in folic acid are reduced to tetrahydrofolate (**4.33**) by another coenzyme, reduced nicotinamide adenine dinucleotide (**4.30**; see **4.8b** for the entire structure; NADH), in the enzyme dihydrofolate reductase.[51] Because the sugar phosphate part of the pyridine nucleotide coenzymes is not involved in the chemistry, it is abbreviated as R; likewise, the part of folic acid not involved in the chemistry is abbreviated as R′. NADH and reduced nicotinamide adenine dinucleotide phosphate (NADPH) can be thought of as Mother Nature's sodium borohydride, a reagent used in organic chemistry to reduce active carbonyl and imine functional groups. As in the case of sodium borohydride, the reduced forms of the pyridine nucleotide coenzymes are believed to transfer their reducing equivalents as hydride ions[52] (Scheme 4.21).

The carbon atom at the C-4 position of NADH and NADPH is prochiral. An atom is *prochiral* if, by changing one of its substituents, it is converted from achiral to chiral. The C-4 carbon of NADH has two hydrogens attached to it; consequently, it is achiral. If one of the hydrogens is replaced by a deuterium, then the carbon becomes chiral. If the chiral center that is generated by replacement of hydrogen by deuterium has the *S*-configuration, then the hydrogen that was replaced is called the *pro-S* hydrogen; if the *R*-configuration is produced, then the hydrogen replaced is the *pro-R* hydrogen. This is demonstrated for ethanol and deuterated ethanol in Figure 4.8. The *pro-R* and *pro-S* hydrogens of the reduced pyridine nucleotide coenzymes are noted in **4.34**. As indicated in Section 4.1.B.1.a, p. 176, because enzymes bind molecules in specific orientations, any achiral compound bound to an enzyme can become prochiral, that is, its hydrogens can be differentiated by the enzyme. Some enzymes such as dihydrofolate reductase utilize the *pro-R* hydrogen (also called the A-side hydrogen) of the reduced pyridine nucleotides, others use the *pro-S*, or B-side, hydrogen. However, the reaction always is stereospecific; if an enzyme uses the *pro-R* hydrogen of NADH, then this is the only hydrogen that it transfers.

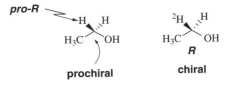

Figure 4.8 ▶   Determination of prochirality

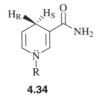

4.34

Although folate is converted into a reduced form (tetrahydrofolate), this coenzyme is generally not involved in redox reactions (one exception is thymidylate synthetase, which will be discussed in Chapter 5, Section 5.5.C.3.e, p. 298). Actually, tetrahydrofolate is not the full coenzyme; the complete coenzyme form contains an additional carbon atom between the N^5 and N^{10} positions (see **4.33** in Scheme 4.21 for nomenclature), which is transferred to other molecules. That additional carbon atom is derived from the methylene group of L-serine in a reaction catalyzed by the PLP-dependent enzyme serine hydroxymethyltransferase.[53] The reaction catalyzed by this PLP-dependent enzyme is one that we have not yet discussed, namely, α-cleavage (Scheme 4.22; the carbon atom that will be transferred to the tetrahydrofolate is marked with an asterisk). The hydroxymethyl group is held perpendicular to the plane of the PLP aromatic system (**4.35**), as in the case of the group cleaved during all of the other PLP-dependent reactions, so that deprotonation and, in this case, a carbon–carbon bond cleavage (a retroaldol reaction) can occur readily. This α-cleavage is an uncommon PLP-dependent reaction, but the cation formed from C–C bond cleavage is an oxygen-stabilizing one. The serine methylene is converted into formaldehyde as a result of this reaction.

Because of the potential toxicity of released formaldehyde (it is highly electrophilic and reacts readily with amino groups), serine hydroxymethyltransferase does not catalyze the degradation of serine (to an appreciable extent) until *after* the acceptor for formaldehyde, namely, tetrahydrofolate, is already bound at an adjacent site.[54] Once the tetrahydrofolate binds, the degradation of serine is triggered, and the formaldehyde generated reacts directly with the tetrahydrofolate to give, initially, the carbinolamine **4.36**, as shown in Scheme 4.23. Kallen and Jencks showed that the more basic nitrogen of tetrahydrofolate is the N^5 nitrogen,[55] which attacks the tetrahydrofolate first;[56] the N^{10} nitrogen is attached to the aromatic ring *para* to a carbonyl group, and, therefore, this nitrogen has amide-like character, and is not basic. Carbinolamine **4.36** is dehydrated to N^5-methylenetetrahydrofolate (**4.37**), which is in equilibrium with N^5,N^{10}-methylenetetrahydrofolate (**4.38**) and N^{10}-methylenetetrahydrofolate (**4.39**). The equilibrium in solution strongly favors the cyclic form (**4.38**) ($K_{eq} = 3.2 \times 10^4$ at pH 7.2).

The coenzyme forms of tetrahydrofolate are responsible for the transfer of one-carbon units in three different oxidation states: formate (transferred as a formyl group), formaldehyde (transferred as a hydroxymethyl group), and methanol (transferred as a methyl group).

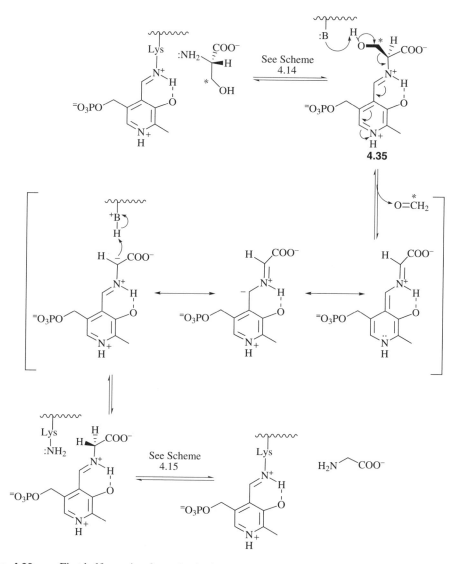

Scheme 4.22 ▶ First half-reaction for serine hydroxymethylase-catalyzed transfer of formaldehyde from serine to tetrahydrofolate

Methylenetetrahydrofolate (**4.37–4.39**) is in the correct oxidation state to transfer a hydroxymethyl group. What if the cell needs a one-carbon unit at the formate oxidation state? This is one oxidation state higher than the formaldehyde oxidation state. A $NADP^+$-dependent dehydrogenase (N^5,N^{10}-methylenetetrahydrofolate dehydrogenase) is involved in the oxidation of methylenetetrahydrofolate to give N^5,N^{10}-methenyltetrahydrofolate (**4.40**, Scheme 4.24). Hydrolysis of **4.40**, catalyzed by the enzyme N^5,N^{10}-methenyltetrahydrofolate cyclohydrolase, gives the carbinolamine **4.41**, which can break down to give either N^5-formyltetrahydrofolate (**4.42**, pathway a) or N^{10}-formyltetrahydrofolate (**4.43**, pathway b). Some enzymes utilize **4.42**, and others use **4.43** as their coenzyme.

If the cell needs a one-carbon unit at the methanol oxidation state, then N^5,N^{10}-methylenetetrahydrofolate (**4.37–4.39**) is reduced by N^5,N^{10}-methylenetetrahydrofolate reductase

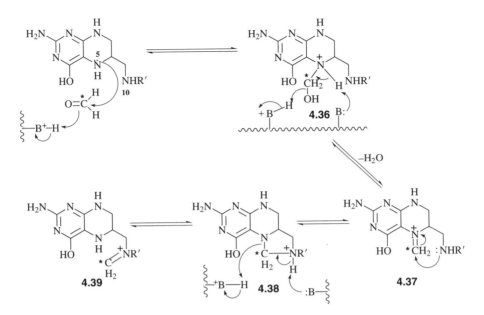

Scheme 4.23 ▶ Second half-reaction for serine hydroxymethylase-catalyzed transfer of formaldehyde from serine to tetrahydrofolate

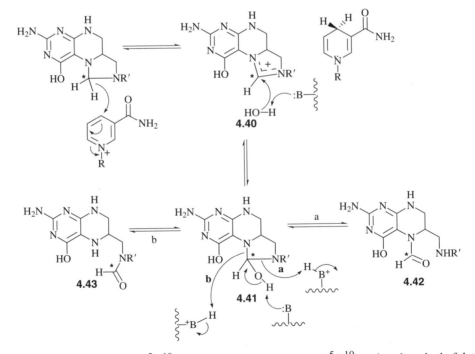

Scheme 4.24 ▶ Oxidation of N^5,N^{10}-methylenetetrahydrofolate to N^5,N^{10}-methenyltetrahydrofolate and to N^5- and N^{10}-formyltetrahydrofolate

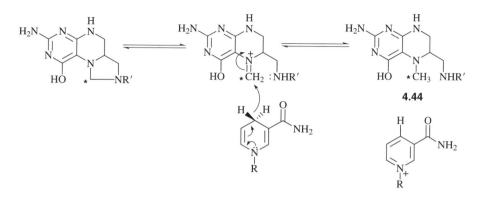

Scheme 4.25 ▶ Reduction of N^5,N^{10}-methylenetetrahydrofolate to N^5-methyltetrahydrofolate

to N^5-methyltetrahydrofolate (**4.44**, Scheme 4.25). This enzyme also requires another coenzyme, flavin adenine dinucleotide,[57] discussed in the next section.

All of these forms of the coenzyme are involved in enzyme-catalyzed transfers of one-carbon units at the formate (**4.40**, **4.42**, or **4.43**), formaldehyde (**4.37–4.39**), or methanol (**4.44**) oxidation states. An example of an enzyme that catalyzes the transfer of a one-carbon unit at the formate oxidation state (i.e., as a formyl group) is 5-aminoimidazole-4-carboxamide-5′-ribonucleotide transformylase (Scheme 4.26; AICAR).[58] This is the penultimate enzyme in the *de novo* biosynthetic pathway to purines, and it catalyzes the conversion of 5′-phosphoribosyl-4-carboxamide-5-aminoimidazole (**4.45**) to 5′-phosphoribosyl-4-carboxamide-5-formamidoimidazole (**4.47**), presumably via the tetrahedral intermediate **4.46**. The formyl group is transferred from 10-formyltetrahydrofolate, which is converted to tetrahydrofolate. The product is enzymatically dehydrated by inosine monophosphate cyclo-hydrolase (IMPCH) to inosine monophosphate (**4.48**), which is further converted to adenosine monophosphate (AMP) and guanosine monophosphate (GMP) by other enzymes. The carbon between the two nitrogens in the five-membered ring of purines was also derived from a methenyltetrahydrofolate-dependent reaction earlier in the biosynthetic sequence.

A one-carbon transfer at the formaldehyde oxidation state occurs in the biosynthesis of the pyrimidine DNA precursor, thymidylate, by the enzyme thymidylate synthase,[59] which is discussed in more detail in Chapter 5, Section 5.5.C.3.e, p. 298).

N^5-Methyltetrahydrofolate is involved in the enzyme-catalyzed transfer of a one-carbon unit at the methanol oxidation state of homocysteine, to give methionine.[60] Methyl transfer in general, however, is not carried out by methyltetrahydrofolate; another methyl transfer agent, *S*-adenosylmethionine, usually is implicated.

4.3.C Flavin

As in the case of the pyridine nucleotide coenzymes (see the section above), flavin coenzymes exist in several different forms.[61] All of the forms are derived from riboflavin (vitamin B$_2$, **4.49**), which is enzymatically converted to two other forms, flavin mononucleotide (FMN, **4.50**) and flavin adenine dinucleotide (FAD, **4.51**). Both **4.50** and **4.51** appear to be functionally equivalent, but some enzymes use one form, some the other, and some use one of each.

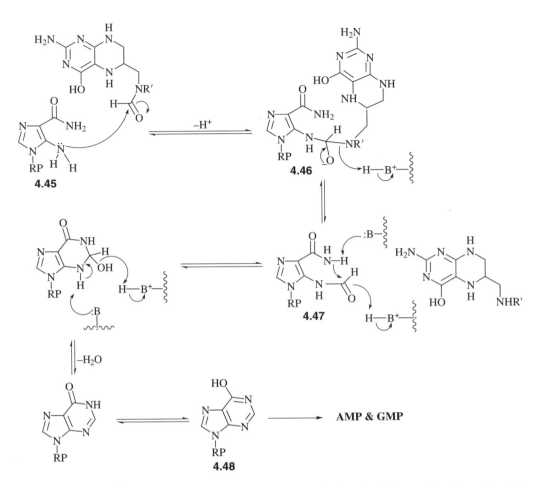

Scheme 4.26 ▶ N^{10}-formyltetrahydrofolate in the biosynthesis of purines. RP stands for ribose phosphate

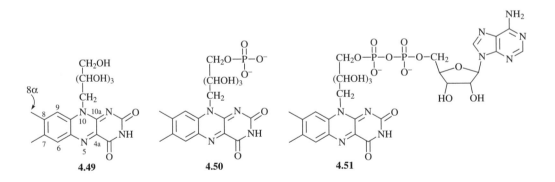

Unlike pyridoxal 5'-phosphate, which catalyzes reactions of amino acids, the flavin coenzymes catalyze a wide variety of *redox* and *monooxygenation reactions* on diverse classes of compounds (Table 4.5).

The highly conjugated isoalloxazine tricyclic ring system of the flavins is an excellent electron acceptor, and this is responsible for its strong redox properties. Flavins can accept either one electron at a time or two electrons simultaneously (Scheme 4.27); in many overall

TABLE 4.5 ▶ Reactions Catalyzed by Flavin-Dependent Enzymes

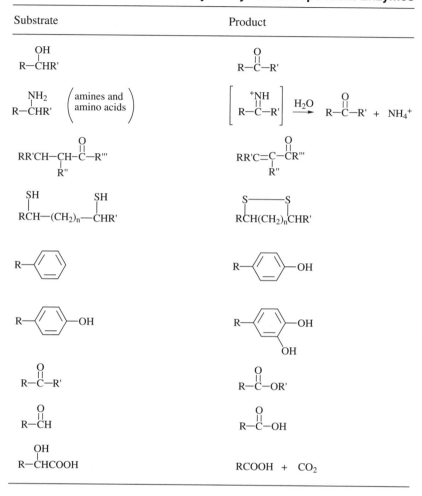

Substrate	Product

R—CHR' (OH) → R—C—R' (O)

R—CHR' (NH₂) (amines and amino acids) → [R—C—R' (⁺NH)] →(H₂O) R—C—R' (O) + NH₄⁺

RR'CH—CH—C—R''' (O, R'') → RR'C=C—CR''' (O, R'')

RCH—(CH₂)ₙ—CHR' (SH, SH) → RCH(CH₂)ₙCHR' (S——S)

R—⟨benzene⟩ → R—⟨ring⟩—OH

R—⟨ring⟩—OH → R—⟨ring⟩—OH (OH)

R—C—R' (O) → R—C—OR' (O)

R—CH (O) → R—C—OH (O)

R—CHCOOH (OH) → RCOOH + CO₂

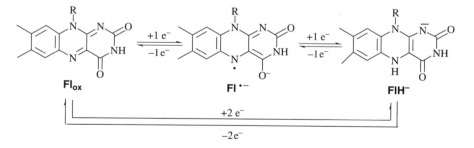

Scheme 4.27 ▶ One- and two-electron reductions of flavins

two-electron oxidations, it is not clear if the reaction proceeds by a single two-electron reaction or by two one-electron transfer steps. The three forms of the coenzyme are the oxidized form (Fl_{ox}), the semiquinone form ($Fl^{-\cdot}$), and the reduced form (FlH^{-}). Although most flavin-dependent enzymes (also called *flavoenzymes*) bind the flavin with noncovalent interactions, some enzymes have covalently bound flavins, in which the flavin is attached at its 8α-position or 6-position (see **4.49** for numbering) to either an active-site histidine[62] or cysteine[63] residue. Once the flavin has been reduced, the enzyme requires a second substrate to return the flavin to the oxidized form so that it can accept electrons from another substrate molecule. This is reminiscent of the requirement of the second substrate (an α-keto acid), which is required to return pyridoxamine 5′-phosphate to pyridoxal 5′-phosphate after PLP-dependent enzyme transamination. In the case of flavoenzymes, however, there are two mechanisms for conversion of reduced flavin back to oxidized flavin.

Some flavoenzymes are called *oxidases* and others *dehydrogenases*. The distinction between these names refers to the way in which the reduced form of the coenzyme is reoxidized so that the catalytic cycle can continue. Those enzymes that utilize electron transfer proteins, such as ubiquinone or cytochrome b_5, to accept electrons from the reduced flavin and proceed by two one-electron transfers (Scheme 4.28) are called *dehydrogenases*.[64] Oxidases use molecular oxygen to oxidize the coenzyme with concomitant formation of hydrogen peroxide; Scheme 4.29 shows possible mechanisms for this oxidation. Pathway a depicts the reaction with triplet oxygen, leading first to the caged radical pair of **4.52** and superoxide by electron transfer, which can either undergo radical combination via pathway c to give **4.53**,[65] or second electron transfer (pathway d) to go directly to the oxidized flavin. Pathway b shows the analogous reaction with singlet oxygen to give the flavin hydroperoxide (**4.53**) directly, which will only occur if there is a mechanism for spin inversion from the normal triplet oxygen, such as with a metal ion. Loss of hydrogen peroxide gives oxidized flavin.

Each of the reactions shown in Table 4.5 is specific for a particular enzyme. Some of the reactions are difficult to carry out nonenzymatically; therefore, although the flavin is essential for the redox reaction, the enzyme is responsible for catalyzing these reactions, and for both substrate and reaction specificity. Four types of mechanisms can be considered, one involving a carbanion intermediate, one with a carbanion and a radical intermediate, one with radical intermediates, and one with a hydride intermediate. As we will see, there is no definitive mechanism for flavin-dependent enzymes; each of these mechanisms may be applicable to different flavoenzymes and/or different substrates.

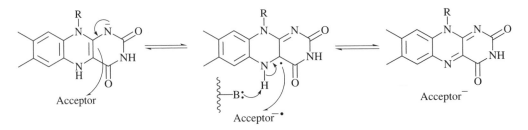

Scheme 4.28 ▶ Mechanism for dehydrogenase-catalyzed flavin oxidation

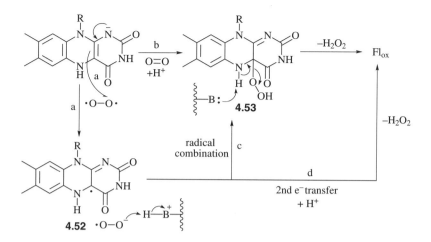

Scheme 4.29 ▶ Mechanism for oxidase-catalyzed flavin oxidation

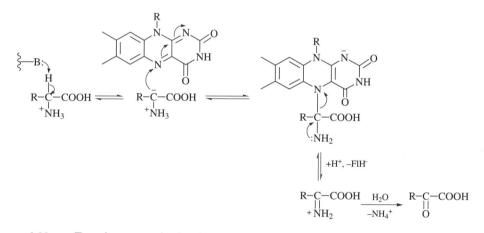

Scheme 4.30 ▶ Two-electron mechanism for flavin-dependent *D*-amino acid oxidase-catalyzed oxidation of *D*-amino acids

C.1 Two-Electron (Carbanion) Mechanism

Evidence for a carbanion mechanism has been provided for the flavoenzyme *D*-amino acid oxidase (DAAO),[66] which catalyzes the oxidation of *D*-amino acids to α-keto acids and ammonia (the second reaction in Table 4.5) (Scheme 4.30).

C.2 Carbanion Followed by Two One-Electron Transfers

Some enzymes may initiate a carbanion mechanism but then proceed by an electron transfer (radical) pathway. Enzymatic evidence for this kind of mechanism is now available from studies with general acyl-CoA dehydrogenase,[67] a family of enzymes that catalyzes the oxidation of fatty acid acyl-CoA derivatives (**4.54**, Scheme 4.31) to the corresponding α, β-unsaturated acyl-CoA compound (**4.55**).

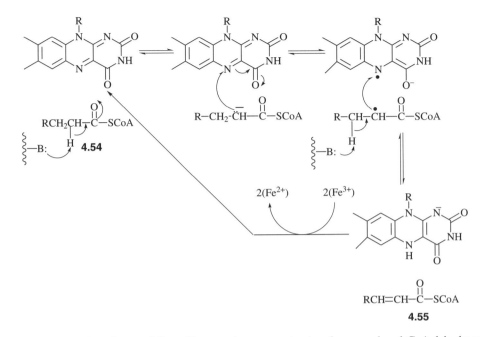

Scheme 4.31 ▶ Two-electron followed by one-electron mechanism for general acyl-CoA dehydrogenase

C.3 One-Electron Mechanism

Evidence for a one-electron (radical) flavin mechanism comes from a variety of experiments with monoamine oxidase,[68] a flavoenzyme important in medicinal chemistry (see Chapter 5, Section 5.5.C.2, p. 286, and 5.5.C.3.c/d, pp. 292 and 295). This is one of the enzymes responsible for the catabolism of various biogenic amine neurotransmitters, such as norepinephrine and dopamine. Possible mechanisms for monoamine oxidase (MAO), which exists in two isozymic forms called MAO A and MAO B, and which catalyzes the degradation of biogenic amine neurotransmitters, such as norepinephrine and dopamine (**4.56**), to their corresponding aldehydes (**4.57**) are summarized in Scheme 4.32.

C.4 Hydride Mechanism

The second step in bacterial cell wall peptidoglycan biosynthesis, catalyzed by uridine diphosphate-N-acetylenolpyruvylglucosamine reductase (or MurB), is the reduction of enolpyruvyl-uridine diphosphate-N-acetylglucosamine (**4.58**) to give uridine diphosphate-N-acetylmuramic acid (**4.59**) (Scheme 4.33). The proposed reduction mechanism[69] involves initial reduction of the FAD cofactor by NADPH (because a reduced flavin is highly prone to oxidation, it is typically formed at the active site of the enzyme by *in situ* reduction with a pyridine nucleotide cofactor at the time it is needed), followed by hydride transfer from reduced FAD via a Michael addition to the double bond of the α, β-unsaturated carboxylate (**4.58**).

Flavin monooxygenases are members of the class of liver microsomal mixed function oxygenases that are important in the oxygenation of *xenobiotics* (foreign substances) that enter the body, including drugs (see Chapter 7).[70] Unlike flavin oxidases and dehydrogenases, this class of enzymes incorporates an oxygen atom from molecular oxygen into the substrate. It is believed that a flavin C^{4a}-hydroperoxide is an important intermediate in this

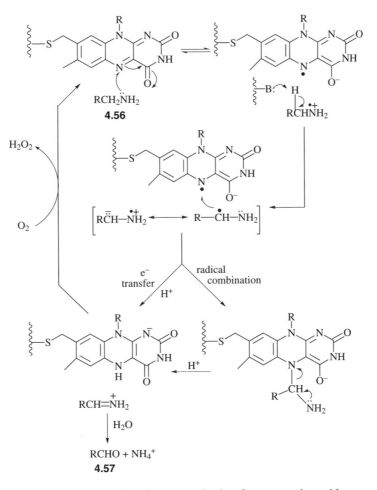

Scheme 4.32 ▶ One-electron mechanism for monoamine oxidase

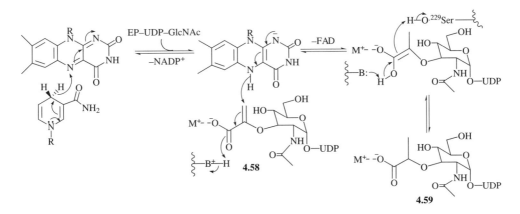

Scheme 4.33 ▶ Hydride reduction mechanism for uridine diphosphate-*N*-acetylenolpyruvylglucosamine reductase (MurB)

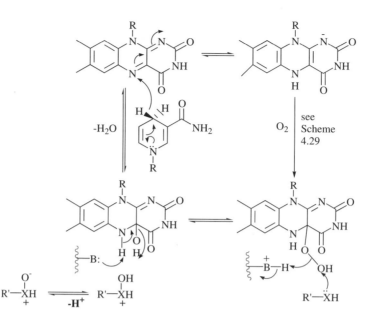

Scheme 4.34 ▶ Mechanism for oxygenation of flavin monooxygenase

process,[71] which can be formed by the same mechanisms suggested for the oxidation of reduced flavin (Scheme 4.29, pathways a/c and b). This then requires the flavin monooxygenase to have the flavin in its reduced form. As noted above (Section 4.3.C.4), the way this is accomplished is with a stable reducing agent, NADH or NADPH, which converts the oxidized flavin to its reduced form and initiates the oxygenation reaction (Scheme 4.34). On the basis of stopped-flow spectroscopic evidence for the formation and decay of C^{4a}-flavin hydroperoxide anion, C^{4a}-flavin hydroperoxide, and C^{4a}-flavin hydroxide intermediates in a flavoenzyme that converts a phenol into a catechol (sixth reaction in Table 4.5), the simplest mechanism that can be drawn involves nucleophilic attack by the substrate at the distal oxygen of the flavin hydroperoxide.[72]

4.3.D Heme

Heme, or protoporphyrin IX, is an iron(III)-containing porphyrin cofactor (**4.14**, p. 191) for a large number of liver microsomal mixed function oxygenases principally in the cytochrome P450 family of enzymes. These enzymes, like flavin monooxygenases, are important in the metabolism of xenobiotics, including drugs (see Chapter 7).[73] As in the case of the flavin monooxygenases, molecular oxygen binds to the heme cofactor (after reduction of the Fe^{3+} to Fe^{2+}), and is converted into a reactive form that is used in a variety of oxygenation reactions, especially hydroxylation and epoxidation reactions. The hydroxylation reactions often occur at seemingly unactivated carbon atoms. The mechanism for this class of enzymes is still in debate, but a high-energy iron-oxo species must be involved.[74] Formally, this species can be written as any one of the resonance structures **4.61a–d** (Scheme 4.35). The heme is abbreviated as **4.60**, where the peripheral nitrogens represent the four pyrrole nitrogens. The axial ligands in the case of cytochrome P450 are a cysteine thiolate from the protein and water. The electrons for reduction of the heme of cytochrome P450 (the second and fourth steps of Scheme 4.35) come from an enzyme complexed with cytochrome P450 called NADPH-cytochrome P450 reductase,[75] which contains NADPH and two different flavin coenzymes (FAD and FMN).

Heme-dependent enzymes catalyze reactions on a wide variety of substrates. Our discussion will be limited to reactions of cytochrome P450 because of its importance to drug metabolism (Chapter 7). Despite the substrate variability, there appears to be a common mechanism (with variations) for each. Below are given mechanisms for hydroxylation of alkanes (Scheme 4.36), epoxidation of alkenes (Scheme 4.37), and hydroxylation of molecules containing heteroatoms, such as sulfur (Scheme 4.38). In Scheme 4.36 the high-energy iron-oxo species (**4.61b**) abstracts a hydrogen atom from the alkane to give a very short-lived carbon radical (**4.62**),[76] which accepts a hydroxyl radical from the heme species (known as

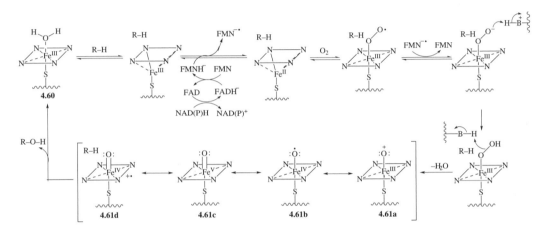

Scheme 4.35 ▶ Activation of heme for heme-dependent hydroxylation reactions

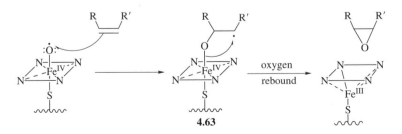

Scheme 4.36 ▶ Mechanism for heme-dependent hydroxylation reactions

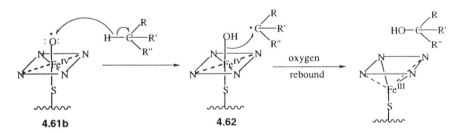

Scheme 4.37 ▶ Mechanism for heme-dependent epoxidation reactions

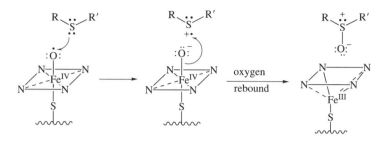

Scheme 4.38 ▶ Mechanism for heme-dependent heteroatom oxygenation reactions

$$PhCO_2H \; + \; NH_3 \; \longrightarrow \; PhCO_2^- \; NH_4^+$$

Scheme 4.39 ▶ Reaction of an amine with a carboxylic acid

oxygen rebound) to give the alcohol. Alkene epoxidation (Scheme 4.37) is similar to alkane hydroxylation, but instead of a hydrogen atom abstracted from the substrate, an electron is "abstracted" from the double bond to give the alkylated iron-oxo species (**4.63**) followed by Fe–O bond homolysis. Chemical model studies,[77] however, suggest that alkene epoxidation may involve an initial charge-transfer complex between the iron-oxo species and the alkene followed by a concerted process; if **4.63** is formed, it is very short lived (as in the case of alkane hydroxylation). Oxygenation of sulfides (Scheme 4.38) also is similar to hydroxylation of alkanes, except instead of initial hydrogen atom abstraction, the iron-oxo species "abstracts" an electron from the readily oxidizable nonbonded electron pair of the sulfur atom, followed by oxygen rebound. This electron-transfer mechanism occurs because of the low oxidation potential of sulfur.

4.3.E   Adenosine Triphosphate and Coenzyme A

Adenosine triphosphate (ATP) provides energy to those molecules that are not sufficiently reactive to allow them to undergo chemical reactions. Consider the nonenzymatic reaction of a carboxylic acid with ammonia. If you mix benzoic acid with a base such as ammonia, what are you going to get? You will not get an amide (unless you heat it excessively); you get the salt, ammonium benzoate (Scheme 4.39). What if you wanted to get the amide; how would you do it? There are numerous ways to make amides, but in general, the carboxylic acid must first be activated, because once the proton is removed, the carboxylate usually does not react with nucleophiles. Typically, dehydrating agents such as thionyl chloride or acetic anhydride are used first to activate the carboxylic acid to an acid chloride (**4.64**, Scheme 4.40A) or anhydride (**4.65**, Scheme 4.40B), respectively. Acid chlorides and anhydrides are very reactive toward nucleophiles, so ammonia can displace the good leaving groups chloride ion and acetate ion, respectively, and produce amides. That, in fact, is what enzymes do also, except thionyl chloride would be a bit harsh on our tissues, so instead, ATP is the substitute. Think of ATP, then, as the endogenous form of thionyl chloride or acetic anhydride.

The structure of ATP is shown in Figure 4.9. There are four electrophilic sites where nucleophiles can attack: at the γ-phosphate, the β-phosphate, the α-phosphate, and the $5'$-methylene group. Attacks at the γ- and α-positions are the most common. Attack at the

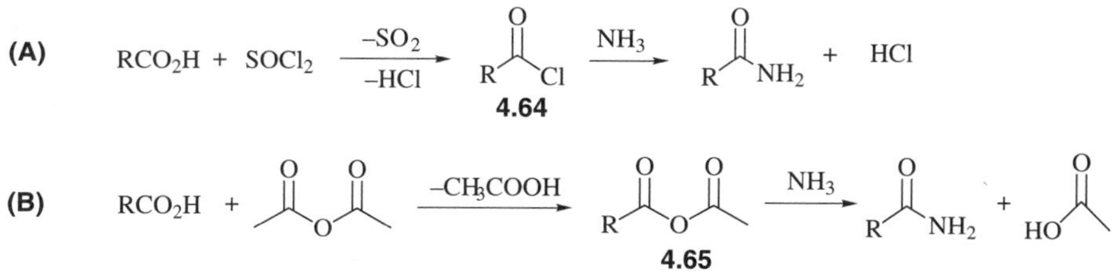

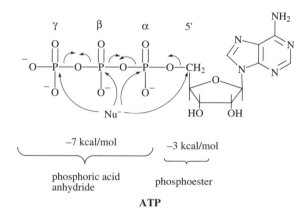

Scheme 4.40 ► Activation of carboxylic acids via acid chloride (A) and anhydride (B) intermediates

Figure 4.9 ► Electrophilic sites on ATP

5′-methylene is rare. For the most part, after the reaction, ATP is converted to ADP (adenosine diphosphate) + P_i (inorganic phosphate) or to AMP (adenosine monophosphate) + PP_i (inorganic diphosphate), depending on what site is attacked.

In the case of the conversion of fatty acids to the corresponding fatty acyl-coenzyme A derivatives (**4.67**), which are substrates for general acyl-CoA dehydrogenase (see Section 4.3.C.2, Scheme 4.31, p. 210), ATP is used to activate the carboxylic acid as an adenosine monophosphate ester (**4.66**) so that it is reactive enough for conversion to the CoA thioester (Scheme 4.41). The AMP ester intermediate undergoes rapid reaction with the highly nucleophilic thiol group of coenzyme A (**4.13b**).

Coenzyme A thioesters serve three important functions for different enzymes. With general acyl-CoA dehydrogenase the thioester group makes the α-proton much more acidic than that of a carboxylate, and therefore easier to remove. The pK_a of the α-proton of a thioester (about 21) is lower than that of an ester (about 25–26) and considerably lower than the α-proton of a carboxylate salt (about 33–34).[78] Secondly, thioesters are much more reactive toward acylation of nucleophilic substrates than are carboxylic acids and oxygen esters, but not so reactive, as in the case of the phosphate esters made from carboxylic acids with ATP, that they cannot be functional in aqueous media. Thirdly, coenzyme A esters are important in the transport of molecules and in their binding to enzymes.

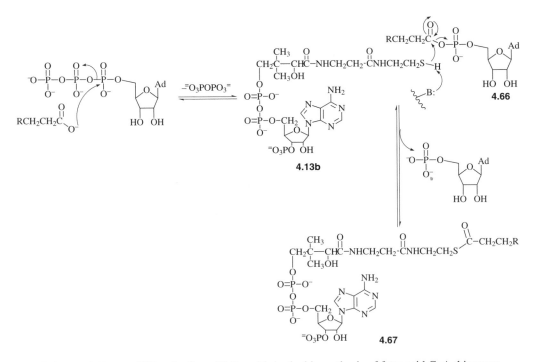

Scheme 4.41 ▶   ATP-activation of fatty acids in the biosynthesis of fatty acid CoA thioesters

4.4 Enzyme Therapy

Throughout this text small organic and some inorganic molecules are discussed in terms of their design and mechanism of action. Some enzymes, however, are themselves useful as drugs. For the most part, the enzymes that have therapeutic utility catalyze hydrolytic reactions. For example, amylase, ligase, cellulase, trypsin, papain, and pepsin are proteolytic or lipolytic digestive enzymes used for gastrointestinal disorders resulting from poor digestion. Trypsin also is used for degrading necrotic tissue from wounds. Collagenase hydrolyzes collagen in necrotic tissue, but does not attack collagen in healthy tissue. Lactase (β-D-galactosidase) is taken by those having low lactase activity for the hydrolysis of lactose into glucose and galactose. Deoxyribonuclease and fibrinolysin are used to dissolve the DNA and fibrinous material, respectively, in purulent exudates and in blood clots. Because malignant leukemia cells are dependent on an exogenous source of asparagine for survival, whereas normal cells are able to synthesize asparagine, the enzyme asparaginase (L-asparagine amidohydrolase), which hydrolyzes the exogenous asparagine to aspartate, has been successful in the treatment of acute lymphocytic leukemia. Three different enzymes, urokinase, streptokinase, and tissue plasminogen activator (tPA), convert the inactive protein plasminogen into plasmin, a proteolytic enzyme that digests fibrin clots; therefore, these enzymes are effective in the treatment of myocardial infarction, venous and arterial thrombosis, and pulmonary embolism.[79] Excessive bilirubin in the blood is the cause for neonatal jaundice. A blood filter containing immobilized bilirubin oxidase can degrade more than 90% of the bilirubin in the blood in a single pass through the filter.[80] The use of genetic engineering techniques to produce altered active enzymes that have increased stability should lead to increased use of enzymes as drugs. The major drawbacks to the use of enzymes in therapy are enzyme instability (other proteases degrade them) and allergic responses.

In the next chapter we examine the design and mechanism of action of enzyme inhibitors as drugs.

4.5 General References

Enzyme Catalysis

Boyer, P. D. (Ed.) *The Enzymes*, 3rd ed., Academic, New York, 1970–1987.

Fersht, A. *Enzyme Structure and Mechanism*, 2nd ed., W. H. Freeman, New York, 1985.

Jencks, W. P. *Catalysis in Chemistry and Enzymology*, McGraw-Hill, New York, 1969.

Jencks, W. P. *Adv. Enzymol.* **1975**, *43*, 219.

Page, M. I.; Williams, A. (Eds.) *Enzyme Mechanisms*, Royal Society of Chemistry, London, 1987.

Segel, I. H. *Enzyme Kinetics*, Wiley, New York, 1975.

Silverman, R. B. *The Organic Chemistry of Enzyme-Catalyzed Reactions*, Academic Press, San Diego, 2002.

Pyridoxal 5-Phosphate (PLP)

Dolphin, D.; Poulson, R.; Avramovic, O. (Eds.) *Vitamin B_6 Pyridoxal Phosphate*, Wiley, New York, 1986, Parts A and B.

Pyridine Nucleotides (NADH and NADPH)

Dolphin, D.; Poulson, R.; Avramovic O. (Eds.) *Pyridine Nucleotide Coenzymes*, Wiley, New York, 1987, Parts A and B.

Flavin

Ghisla, S.; Massey, V. *Eur. J. Biochem.* **1989**, 181, 1 and references therein.

Müller, F. *Top. Curr. Chem.* **1983**, *108*, 71.

Walsh, C. *Enzymatic Reaction Mechanisms*, Freeman, San Francisco, 1979.

Tetrahydrofolate

Blakley, R. L.; Benkovic, S. J. (Eds.) *Folates and Pterins*, Wiley, New York, 1984 and 1985 Vol. 1 and 2.

Heme

Guengerich, F. P. (Ed.) *Mammalian Cytochromes P-450*, CRC Press, Boca Raton, FL, 1987, Vols. I and II.

Guengerich, F. P.; MacDonald, T. L. *FASEB J.* **1990**, *4*, 2453.

Enzyme Therapy

Blohm, D.; Bollschweiler, C.; Hillen, H. *Angew. Chem. Int. Ed. Engl.* **1988**, *27*, 207.

Circhoke, A. J. *Enzymes and Enzyme Therapy*, Keats Publishing, Chicago, 2000, 2nd edition.

Holcenberg, J. S.; Roberts, J. (Eds.) *Enzymes as Drugs*, Wiley, New York, 1981.

4.6 PROBLEMS

(Answers can be found in Appendix at the end of the book.)

1. What are the two most important characteristics of an enzyme?

2. The $(3S, 4S)$-isomer of deuterated compound **1** is a substrate for a dephosphorylase that gives an anti-elimination to the Z-isomer **2**. The corresponding $(3R, 4S)$-isomer is not a substrate. How do you explain that?

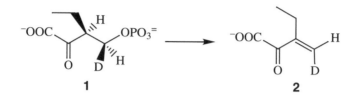

3. Compound **3** is a substrate for an enzyme that catalyzes the removal of the H_S proton exclusively. If compound **4** is used as the substrate, the deprotonation occurs mostly at H_S, but a small amount of products are observed from removal of H_R. This second reaction was shown to be catalyzed by the same enzyme as well. How could this be rationalized?

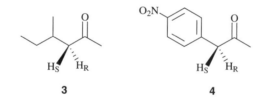

4. The following reaction is catalyzed by an enzyme at pH 7 and 25°C, whereas nonenzymatically, this reaction does not occur under these conditions. Explain how the enzyme can easily catalyze this reaction.

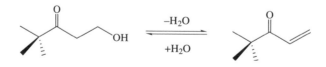

5. Porphobilinogen synthase is a zinc-dependent enzyme (contains two zinc ions) that catalyzes the condensation of two molecules of 5-aminolevulinic acid (**5**) to give porphobilinogen (**6**). A condensed mechanism is shown below. Indicate with an arrow every place possible that the enzyme catalyzes this reaction, and name the catalytic mechanisms.

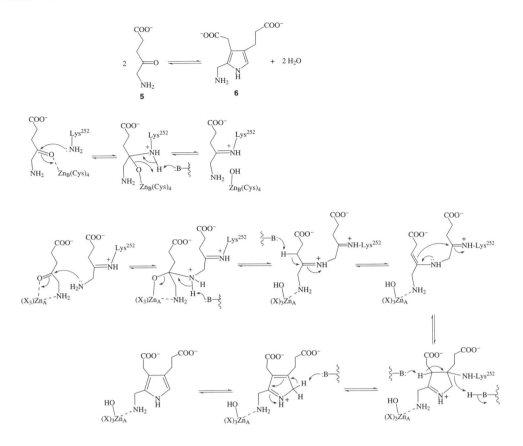

6. Draw a mechanism for the enzyme-catalyzed enolization of a ketone utilizing an electrostatic interaction.

7. Transition state stabilization accounts for a large portion of the stabilization energy of an enzyme-catalyzed reaction. Haloalkane dehalogenase catalyzes an S_N2 reaction of an active site carboxylate with an alkyl halide followed by hydrolysis to the alcohol. What transition state stabilization processes could the enzyme utilize? (Draw the transition state for guidance and note the processes.)

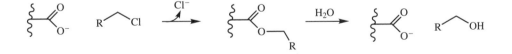

8. If you were using aspartate aminotransferase, and the enzyme stopped catalyzing the reaction, what two components of the enzyme reaction would you check?

9. Dopa (**7**, Ar = 3,4-dihydroxyphenyl) is converted into norepinephrine (**8**, Ar = 3,4-dihydroxyphenyl; R = H) in two enzyme-catalyzed reactions, both of which require coenzymes. If the first enzyme-catalyzed reaction is run in ^2H_2O, the norepinephrine is stereospecifically labeled with deuterium (R = ^2H). Draw mechanisms for the two enzyme-catalyzed reactions.

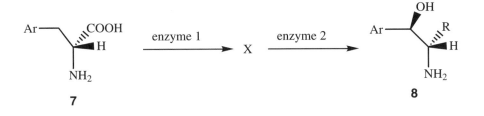

7   8

10. γ-Aminobutyric acid (GABA) aminotransferase catalyzes a PLP-dependent conversion of GABA to succinic semialdehyde (SSA).

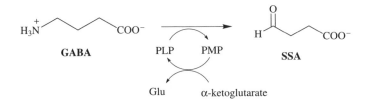

  A. Draw a mechanism for this reaction.

  B. Why is only one molecule of succinic semialdehyde formed in the absence of

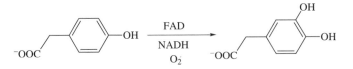

  C. If the reaction were carried out in $^2H_2{}^{18}O$, what would the products be?

11. *p*-Hydroxyphenylacetate 3-hydroxylase is a flavin adenine dinucleotide enzyme that catalyzes the reaction below. Draw a mechanism.

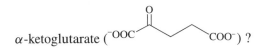

Hint: The mechanism is related to that in Scheme 4.34.

12. Draw a mechanism for the formation of N^5,N^{10}-methylenetetrahydrofolate and the transfer of a hydroxymethyl group to uridylate. (See next page for a hint.)

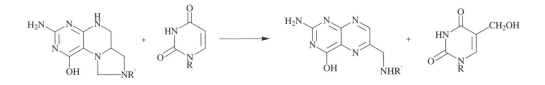

Hint: The first step in the transfer involves a Michael addition of an enzyme cysteine residue to uridylate.

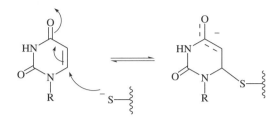

13. Cytochrome P450, a heme-dependent enzyme, catalyzes the oxidation of a wide variety of xenobiotics. Draw a mechanism for the conversion of propranolol (**9**) to **10** and acetone.

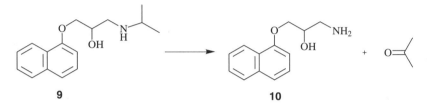

14. Vitamin B_6 (pyridoxine; **11**) is converted to pyridoxal $5'$-phosphate by a two-enzyme sequence, one that phosphylates the $5'$-hydroxyl group with ATP and one that oxidizes the $4'$-hydroxyl group with NAD^+. Draw mechanisms for the biosynthesis of PLP.

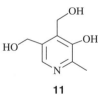

15. The formyl group (CHO) of *N*-formylmethionine (**13**) can be added to methionine (**12**) by incubation of **12** with three enzymes plus another amino acid and coenzymes. You need **13** with a ^{14}C label at the formyl carbon atom, which can be done biosynthetically starting from a ^{14}C-labeled amino acid.

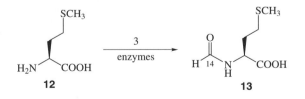

Show the starting amino acid with the ^{14}C label, the radioactive products of the first two enzymatic reactions (**13** is the product of the third enzyme), and any coenzymes involved in each enzyme reaction (indicate to which reactions the coenzymes belong).

4.7 References

1. (a) Joyce, G. F. *Proc. Natl. Acad. Sci. USA* **1998**, *95*, 5845. (b) Pan, T. *Curr. Opin. Chem. Biol.* **1997**, *1*, 17.

2. Sumner, J. B. *J. Biol. Chem.* **1926**, *69*, 435.

3. (a) Jabri, E.; Carr, M. B.; Hausinger, R. P.; Karplus, P. A. *Science* **1995**, *268*, 998. (b) Jabri, E.; Karplus, P. A. *Biochemistry* **1996**, *35*, 10616.

4. Hirs, C. H. W.; Moore, S.; Stein, W. H. *J. Biol. Chem.* **1960**, *235*, 633.

5. Radzicka, A.; Wolfenden, R. *Science* **1995**, *267*, 90.

6. Schramm, V. L. *Annu. Rev. Biochem.* **1998**, *67*, 693.

7. Page, M. I. In *Enzyme Mechanisms*, Page, M. I.; Williams, A. (Eds.), Royal Society of Chemistry, London, 1987, 1.

8. Brown, A. J. *Trans. Chem. Soc.* (*London*) **1902**, *81*, 373.

9. Henri, V. *Acad. Sci., Paris* **1902**, *135*, 916.

10. Fischer, E. *Berichte* **1894**, *27*, 2985.

11. (a) Koshland, D. E., Jr. *Proc. Natl. Acad. Sci. USA* **1958**, *44*, 98; (b) Koshland, D. E.; Neet, K. E. *Annu. Rev. Biochem.* **1968**, *37*, 359.

12. (a) Pauling, L. *Chem. Eng. News* **1946**, *24*, 1375. (b) Pauling, L. *Am Sci.* **1948**, *36*, 51.

13. Haldane, J. B. S. *Enzymes*, Longmans and Green, London, 1930 (reprinted by MIT Press, Cambridge, MA, 1965).

14. Polanyi, M. *Z. Elektrochem.* **1921**, *27*, 143.

15. Eyring, H. *J. Chem. Phys.* **1935**, *3*, 107.

16. Hackney, D. D. In *The Enzymes*, 3rd ed., Sigman, D. S.; Boyer, P. D. (Eds.), Academic Press, San Diego, 1990, Vol. 19, pp. 1–37.

17. Schowen, R. L. In *Transition States of Biochemical Processes*, Gandour, R. D.; Schowen, R. L. (Eds.), Plenum, New York, 1978, p. 77.

18. Richards, F. M. *Annu. Rev. Biophys. Bioeng.* **1977**, *6*, 151.

19. Miller, B. G.; Wolfenden, R. *Annu. Rev. Biochem.* **2002**, *71*, 847.

20. Guengerich, F. P. (Ed.) *Mammalian Cytochromes P-450*, CRC Press, Boca Raton, FL, 1987.

21. Jencks, W. P. *Adv. Enzymol.* **1975**, *43*, 219.

22. Wolfenden, R. *Acc. Chem. Res.* **1972**, *5*, 10.

23. Eigen, M.; Hammes, G. G. *Adv. Enzymol.* **1963**, *25*, 1.

24. Jencks, W. P. *Catalysis in Chemistry and Enzymology*, McGraw-Hill, New York, 1969, Chaps. 1, 2, 3, 5.

25. Jencks, W. P. *Adv. Enzymol.* **1975**, *43*, 219.

26. Wolfenden, R.; Frick, L. In *Enzyme Mechanisms*, (Page, M. I.; Williams, I. (Eds.), Royal Society of Chemistry, London, 1987, p. 97.

27. Bruice, T. C.; Pandit, U. K. *J. Am. Chem. Soc.* **1960**, *82*, 5858.

28. Bruice, T. C.; Pandit, U. K. *Proc. Natl. Acad. Sci. USA* **1960**, *46*, 402.

29. (a) Lightstone, F. C.; Bruice, T. C. *J. Am. Chem. Soc.* **1996**, *118*, 2595. (b) Bruice, T. C.; Lightstone, F. C. *Acc. Chem. Res.* **1999**, *32*, 127.

30. Kirby, A. J. *Adv. Phys. Org. Chem.* **1980**, *17*, 183.

31. March, J. *Advanced Organic Chemistry*, 3rd ed., Wiley, New York, 1985, p. 268.

32. Blow, D. M.; Birktoft, J.; Hartley, B. S. *Nature* **1969**, *221*, 337.

33. Kraut, J. *Annu. Rev. Biochem* **1977**, *46*, 331.

34. Bryan, P.; Pantoliano, M. W.; Quill, S. G.; Hsaio, H.-Y.; Poulos, T. *Proc. Natl. Acad. Sci. USA* **1986**, *83*, 3743.

35. Carter, P.; Wells, J. A. *Nature* **1988**, *332*, 564.

36. Covitz, T.; Westheimer, F. H. *J. Am. Chem. Soc.* **1963**, *85*, 1773.

37. (a) Koshland, D. E., Jr. *Proc. Natl. Acad. Sci. USA* **1958**, *44*, 98. (b) Koshland, D. E., Jr. *Biochem. Pharmacol.* **1961**, *8*, 57.(c) Koshland, D. E., Jr.; Neet, K. E. *Annu. Rev. Biochem.* **1968**, *37*, 359.

38. Jencks, W. P. *Cold Spring Harbor Symp. Quant. Biol.* **1987**, *52*, 65.

39. Waxman, D. J.; Strominger, J. L. *Annu. Rev. Biochem.* **1983**, *52*, 825.

40. Marcus, R.; Coulston, A. M. In *Goodman and Gilman's the Pharmacological Basis of Therapeutics*, 7th ed., Gilman, A. G.; Goodman, L. S.; Rall, T. W.; Murad, F. (Eds.), Macmillan, New York, 1985, p. 1551.

41. Martell, A. E. *Acc. Chem. Res.* **1989**, *22*, 115.

42. Leussing, D. L. In *Vitamin B$_6$ Pyridoxal Phosphate*, Dolphin, D.; Poulson, R.; Avramovic, O. (Eds.), Wiley, New York, 1986, Part A, p. 69.

43. Dunathan, H. C. *Adv. Enzymol.* **1971**, *35*, 79.

44. John, R. A. *Biochim. Biophys. Acta* **1995**, 1248, 81.

45. Walsh, C. T. *J. Biol. Chem.* **1989**, *264*, 2393.

46. (a) Hashimoto, A.; Nishikawa, T.; Oka, T.; Takahashi, K. *J. Neurochem.* **1993**, *60*, 783. (b) Hashimoto, A.; Oka, T. *Prog. Neurobiol.* **1997**, *52*, 325.

47. Wolosker, H.; Blackshaw, S.; Snyder, S. H. *Proc. Natl. Acad. Sci. USA* **1999**, *96*, 13409.

48. Soda, K.; Tanaka, H.; Tanizawa, K. In *Vitamin B6 Pyridoxal Phosphate*, Dolphin, D.; Poulson, R.; Avramovic, O. (Eds.), Wiley, New York, 1986, Part B, p. 223.

49. Watanabe, A.; Yoshimura, T.; Mikami, B.; Hayashi, H.; Kagamiyama, H.; Esaki, N. *J. Biol. Chem.* **2002**, *277*, 19166.

50. Sukhareva, B. S. In *Vitamin B6 Pyridoxal Phosphate*, Dolphin, D.; Poulson, R.; Avramovic, O. (Eds.), Wiley, New York, 1986, Part B, p. 325.

51. Bertino, J. R.; Booth, B. A.; Bieber, A. L.; Cashmore, A.; Sartorelli, A. C. *J. Biol. Chem.* **1964**, *239*, 479.

52. Westheimer, F. H. In *Pyridine Nucleotide Coenzymes*, Dolphin, D.; Poulson, R.; Avramovic, O. (Eds.), Wiley, New York, 1987, Part A, p. 253.

53. (a) Schirch, L. G. In *Folates and Pterins*, Blakely, R. L.; Benkovic, S. J. (Eds.), Wiley, New York, 1984, Vol. 1, p. 399. (b) Matthews, R. G.; Drummond, J. T. *Chem. Rev.* **1990**, *90*, 1275.

54. (a) Benkovic, S. J.; Bullard, W. P. In *Progress in Bioorganic Chemistry*, Kaiser, E. T.; Kedzy, F. (Eds.), Wiley, New York, 1973, Vol. 2, p. 133. (b) Jordan, P. M.; Akhtar, M. *Biochem. J.* **1970**, *116*, 277.

55. Kallen, R. G.; Jencks, W. P. *J. Biol. Chem.* **1966**, *241*, 5845.

56. Kallen, R. G.; Jencks, W. P. *J. Biol. Chem.* **1966**, *241*, 5851.

57. (a) Daubner, S. C.; Matthews, R. G. *J. Biol. Chem.* **1982**, *257*, 140. (b) Matthews, R. G.; Drummond, J. T. *Chem. Rev.* **1990**, *90*, 1275.

58. Wolan, D. W.; Greasley, S. E.; Beardsley, G. P.; Wilson, I. A. *Biochemistry* **2002**, *41*, 15505.

59. Carreras, C. W.; Santi, D. V. *Annu. Rev. Biochem.* **1995**, *64*, 721.

60. (a) Gonzalez, J. C.; Peariso, K.; Penner-Hahn, J. E.; Matthews, R. G. *Biochemistry* **1996**, *35*, 12228. (b) Matthews, R. G.; Goulding, C. W. *Current Opin. Chem. Biol.* **1997**, *1*, 332.

61. (a) Walsh, C. *Acc. Chem. Res.* **1980**, *13*, 148. (b) Bruice, T. C. *Acc. Chem. Res.* **1980**, *13*, 256. (c) Ghisla, S.; Massey, V. *Eur. J. Biochem.* **1989**, *181*, 1.

62. (a) Chlumsky, L. J.; Sturgess, A. W.; Nieves, E.; Jorns, M. S. *Biochemistry* **1998**, *37*, 2089. (b) Singer, T. P.; Edmondson, D. E. *Methods Enzymol.* **1980**, *66*, 253. (c) Edmondson, D. E.; Kenney, W. C.; Singer, T. P. *Methods Enzymol.* **1978**, *53*, 449.

63. (a) Kearney, E. B.; Salach, J. I.; Walker, W. H.; Seng, R. L.; Kenney, W.; Zeszotek, E.; Singer, T. P. *Eur. J. Biochem.* **1971**, *24*, 321. (b) Steenkamp, D. J.; Denney, W. C.; Singer, T. P. *J. Biol. Chem.* **1978**, *253*, 2812.

64. Massey, V.; Mller, F.; Feldberg, R.; Schuman, M.; Sullivan, P. A.; Howell, L. G.; Mayhew, S. G.; Matthews, R. G.; Foust, G. P. *J. Biol. Chem.* **1969**, *244*, 3999.

65. Bruice, T. C. *Israel J. Chem.* **1984**, *24*, 54.

66. (a) Neims, A. H.; DeLuca, D. C.; Hellerman, L. *Biochemistry* **1966**, *5*, 203. (b) Walsh, C. T.; Schonbrunn, A.; Abeles, R. H. *J. Biol. Chem.* **1971**, *246*, 6855. (c) Todone, F.; Vanoni, M. A.; Mozzarelli, A.; Bolognesi, M.; Coda, A.; Curti, B.; Mattevi, A. *Biochemistry* **1997**, *36*, 5853.

67. (a) Lenn, N. D.; Shih, Y.; Stankovich, M. T.; Liu, H.-W. *J. Am. Chem. Soc.* **1989**, *111*, 3065. (b) Lai, M.-T.; Liu, L.-D.; Liu, H.-W. *J. Am. Chem. Soc.* **1991**, *113*, 7388. (c) Lai, M.-T.; Oh, E.; Liu, H.-W. *J. Am. Chem. Soc.* **1993**, *115*, 1619.

68. (a) Silverman, R. B. In *Advances in Electron Transfer Chemistry*, Mariano, P. S. (Ed.), JAI Press, Greenwich, CT, 1992, Vol. 2, pp. 177–213. (b) Silverman, R. B. *Acc. Chem. Res.* **1995**, *28*, 335. (c) Lu, X.; Rodriguez, M.; Ji, H.; Silverman, R. B.; Vintém, A. P. B.; Ramsay, R. R. In *Flavins and Flavoproteins 2002*; Chapman, S. K.; Perham, R. N.; Scrutton, N. S. (Eds.), Rudolf Weber, Berlin, 2002, pp 817–830.

69. Benson, T. E.; Marquardt, J. L.; Marquardt, A. C.; Etzkorn, F. A.; Walsh, C. T. *Biochemistry* **1993**, *32*, 2024.

70. (a) Ziegler, D. M. In *Enzymatic Basis of Detoxification*, Jacoby, W. B. (Ed.), Academic, New York, 1980, Vol. 1, p. 201. (b) Poulsen, L. L. In *Reviews in Biochemical Toxicology*, Hodogson, E.; Bend, J. R.; Philpot, R. M. (Eds.), Elsevier, New York, 1981, Vol. 3, p. 33.

71. Ghisla, S.; Massey, V. *Eur. J. Biochem.* **1989**, *181*, 1.

72. Massey, V. *J. Biol. Chem.* **1994**, *269*, 22459.

73. Guengerich, F. P. (Ed.) *Mammalian Cytochromes P-450*, CRC Press, Boca Raton, FL, 1987; Vols. I and II.

74. Akhtar, M.; Wright, J. N. *Nat. Prod. Rep.* **1991**, *8*, 527.

75. (a) Kim, J.-J. P.; Roberts, D. L.; Djordjevic, S.; Wang, M.; Shea, T. M.; Masters, B. S. S. *Methods Enzymol.* **1996**, *272*, 368. (b) Strobel, H. W.; Dignam, J. D.; Gum, J. R. *Pharmacol. Ther.* **1980**, *8*, 525.

76. Newcomb, M.; Le Tadic-Biadatti, M.-H.; Chestney, D. L.; Roberts, E. S.; Hollenberg, P. F. *J. Am. Chem. Soc.* **1995**, 117, 12085.

77. (a) Ostović, D.; Bruice, T. C. *J. Am. Chem. Soc.* **1989**, *111*, 6511. (b) Ostović, D.; Bruice, T. C. *Acc. Chem. Res.* **1992**, *25*, 314.

78. Richard, J. P.; Williams, G.; O'Donoghue, A. C.; Amyes, T. L. *J. Am. Chem. Soc.* **2002**, *124*, 2957.

79. Haber, E.; Quertermous, T.; Matsueda, G. R.; Runge, M. S. *Science* **1989**, *243*, 51.

80. Lavin, A.; Sung, C.; Klibanov, A. M.; Langer, R. *Science* **1985**, *230*, 543.

Chapter 5

Enzyme Inhibition and Inactivation

5.1 Why Inhibit an Enzyme?

Many diseases, or at least the symptoms of diseases, arise from a deficiency or excess of a specific metabolite in the body, from an infestation of a foreign organism, or from aberrant cell growth. If the metabolite deficiency or excess can be normalized, and if the foreign organisms and aberrant cells can be destroyed, then these disease states will be remedied. These problems can be addressed by specific enzyme inhibition.

Any compound that slows down or blocks enzyme catalysis is an *enzyme inhibitor*. If the interaction with the *target enzyme* is irreversible (usually covalent), then the compound is a special type of enzyme inhibitor referred to as an *enzyme inactivator*. Of the 97 new drugs that went onto the world drug market during 1998, 1999, and 2000, almost one-third (31/97) were enzyme inhibitors; five (5% of the total 97 drugs for those years), cefoselis (**5.1**, Wincef; antibacterial), rabeprazole sodium (**5.2**, Aciphex; antiulcer), arglabin (**5.3**, anticancer), esomeprazole magnesium (**5.4**, Nexium; antiulcer), and exemestane (**5.5**, Aromasin; anticancer), were irreversible inhibitors.

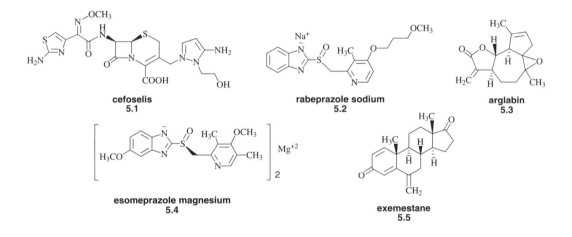

cefoselis
5.1

rabeprazole sodium
5.2

arglabin
5.3

esomeprazole magnesium
5.4

exemestane
5.5

Consider what happens when an enzyme activity is blocked. The substrates for that enzyme cannot be metabolized, and the metabolic products are not generated (that is, unless there is another enzyme that can metabolize the substrate, and unless there is another metabolic pathway that generates the same product). Why should these two outcomes be important to drug design? If a cell has a deficiency of the substrate for the target enzyme, and as a result of that deficiency, a disease state results, then inhibition of that enzyme would prevent the degradation of the substrate, thereby increasing its concentration. An example of this is the onset of seizures that arises from diminished γ-aminobutyric acid (GABA) levels in the brain. Inhibition of the enzyme that degrades GABA, namely, GABA aminotransferase, leads to an anticonvulsant effect. If there is an excess of a particular metabolite that produces a disease state, then inhibition of the enzyme that catalyzes the formation of that metabolite would diminish its concentration. Excess uric acid can lead to gout. Inhibition of xanthine oxidase, the enzyme that catalyzes the conversion of xanthine to uric acid, decreases the uric acid levels, and results in an antihyperuricemic effect. If the product of an enzyme reaction is required to carry out an important physiological function that the drug is supposed to block, then inhibition of that enzyme decreases the concentration of that product and can interfere with the physiological effect. Prostaglandins are important hormones that are involved in the pathogenesis of inflammation and fever. Inhibition of prostaglandin synthase results in anti-inflammatory, antipyretic, and analgesic effects. In the case of foreign organisms such as bacteria and parasites, or in the case of tumor cells, inhibition of one of their essential enzymes can prevent important metabolic processes from taking place, resulting in inhibition of growth or replication of the organism or aberrant cell. Inhibition of the bacterial alanine racemase, the enzyme that makes *D*-alanine for incorporation into peptidoglycan strands, for example, would prevent the biosynthesis of the peptidoglycan and, therefore, the biosynthesis of the bacterial cell wall. These compounds possess antibacterial activity. The use of drugs to combat foreign organisms and aberrant cells is called *chemotherapy*.

Enzyme inhibition is a promising approach for the rational discovery of new leads or drugs. Although numerous drugs exert their therapeutic action by inhibiting specific enzymes, the mechanisms of action of most of these drugs were determined subsequent to the discovery of the therapeutic properties of the drugs. Target enzymes selected for rational drug design are those whose inhibition *in vivo* would lead to the desired therapeutic effect. There are two general categories of target enzymes. In most cases a potential drug is designed for an enzyme whose inhibition is known to produce a specific pharmacological effect, but existing inhibitors have certain undesirable properties such as lack of potency or specificity or exhibit side effects. A more daring approach is to design inhibitors of enzymes whose inhibition has not yet been established to lead to a desired therapeutic effect. This category of enzyme targets requires knowledge of the pathophysiology of disease processes and the ability to identify important metabolites whose function or dysfunction results in a disease state. Not until an inhibitor is obtained will it be possible to determine the real effect of inhibition of that enzyme on the metabolism of the organism. Once an enzyme target is identified, lead compounds must be prepared that can inhibit it completely and specifically.

Of all the protein targets for potential therapeutic use, including hormone and neurotransmitter receptors and carrier proteins, enzymes are the most promising for rational inhibitor design. Enzyme purification is generally a much simpler task than receptor purification; a homogenous enzyme preparation can be obtained for preliminary screening purposes, and in some cases, may be used to elucidate the active site structure, which is useful for computer-based drug design approaches (see Chapter 2, Section 2.2.H, p. 78). Furthermore, whereas effective receptor antagonists often bear no structural similarity to agonists, enzyme inhibitors are often very similar in molecular structure to substrates or products of the target enzyme. Consequently, lead compounds are readily obtainable for enzyme targets. In addition, knowledge of enzyme mechanisms can be used in the design of transition state analogs and multisubstrate inhibitors (Section 5.4.C), slow tight-binding inhibitors (Section 5.4.D), and mechanism-based enzyme inactivators (Section 5.5.C).

To minimize side effects, there are certain properties that ideal enzyme inhibitors and/or enzyme targets should possess. An ideal enzyme inhibitor should be totally specific for the one target enzyme. Because this is rare, if attained at all, highly selective inhibition is a more realistic objective. By adjustment of the dose administered, essentially specific inhibition may be possible. In some cases, such as infectious diseases, enzyme targets can be identified because of biochemical differences in essential metabolic pathways between foreign organisms and their hosts.[1] In other instances, substrate specificity differences between enzymes from the two sources can be utilized in the design of selective enzyme inhibitor drugs. Unfortunately, when dealing with various organisms, and especially with tumor cells, the enzymes that are essential for their growth also are vital to human health. Inhibition of these enzymes can destroy human cells as well. Nonetheless, this approach is taken in various types of chemotherapy. The reason this approach is effective is that foreign organisms and tumor cells replicate at a much faster rate than do most *normal* human cells (those in the gut, the bone marrow, and the mucosa are exceptions). Consequently, rapidly proliferating cells have an elevated requirement for essential metabolites. *Antimetabolites*, compounds whose structures are similar to those of essential metabolites and which inhibit the metabolizing enzymes, are taken up by the rapidly replicating cells and, therefore, these cells are selectively inhibited. The *selective toxicity* in this case derives from a kinetic difference rather than a qualitative difference in the metabolism.

An ideal enzyme target in a foreign organism or aberrant cell would be one that is essential for its growth, but which is either nonessential for human health, or, even better, not even present in humans. This type of selective toxicity would destroy only the foreign organism or aberrant cell, and would not require the careful administration of drugs that is necessary when the inhibited enzyme is important to human metabolism as well. The penicillins, for example, which inhibit the bacterial peptidoglycan transpeptidase essential for the biosynthesis of the bacterial cell wall, have low toxicity in humans.

Enzyme inhibitors can be grouped into two general categories: reversible and irreversible inhibitors. As the name implies, inhibition of enzyme activity by a *reversible inhibitor* is reversible, suggesting that noncovalent interactions are involved. This is not strictly the case; there also can be reversible covalent interactions. An *irreversible enzyme inhibitor*, also called an *enzyme inactivator*, is one that prevents the return of enzyme activity for an extended period of time, suggesting the involvement of a covalent bond. This also is not strictly the case; it is possible for noncovalent interactions to be so effective that the enzyme-inhibitor complex is, for all intents and purposes, irreversibly formed. Prior to a more detailed discussion of each of these types of enzyme inhibitors, we turn our attention to two important concerns in drug design and drug action that will be referred to throughout this chapter: drug resistance and drug synergism.

5.2 Drug Resistance

5.2.A What Is Drug Resistance?

Drug resistance occurs when a formerly effective drug dose is no longer effective. This can be a natural resistance or an acquired resistance. Resistance arises mainly by *natural selection*. A drug destroys all of the organisms in a colony that are susceptible to the action of that drug; however, on average, 1 in 10 million organisms in a colony has one or more mutations that make it resistant to that drug.[2] Once all of the susceptible organisms have been killed, the few resistant ones replicate and eventually become the predominant species. Bacteria resist antibiotics as a result of chromosomal mutations or inductive expression of a latent chromosomal gene or by exchange of genetic material (gene transfer) through transformation, transduction, or conjugation by plasmids. Because mutagenic drugs generally are not used, resistance by drug-induced mutation seldom occurs.

Because of the remarkable ability of microorganisms to evolve and adapt, there is a need for new drugs with new mechanisms of action that are not susceptible to mechanisms of resistance. Hopefully, the elucidation of the sequences of genomes of various organisms will present new targets and new mechanisms. Resistance should not be confused with another term, tolerance, which is not related to microorganisms or cancer cell growth. *Tolerance* is when the body adapts to a particular drug and requires more of that drug to attain the same initial effect, typically leading to a decrease in the therapeutic index. This is what occurs with addiction to morphine. When morphine stimulates a receptor in the cell walls that respond to a transmitter, the cells adapt by increasing the number of receptors they produce so that some receptors are available to combine with the transmitter, even in the continued presence of morphine. Because morphine now activates only a fraction of the available receptors, its effects will be less than before, and a higher dose will be needed to achieve the same effect. It also is possible for tolerance to develop to the undesired effects of a drug, which leads to an *increase* in the therapeutic index (such as tolerance to sedation by phenobarbital).

5.2.B   Mechanisms of Drug Resistance

Antibiotic resistance has reached an alarming stage worldwide.[3] Many organisms today have acquired multiple systems to reduce or avoid the action of antibiotics.[4] The most threatening mechanisms of resistance involve changes in the target site for antibiotic interaction, because that confers resistance to all compounds with the same mechanism of action. It is most useful to be able to identify compounds that target many sites of action or have multiple mechanisms of action. *Exogenous resistance* occurs when new proteins are developed by the organism to protect it from drugs. *Endogenous resistance* occurs by mutation, even single-point mutations (one amino-acid change). Eight main mechanisms of drug resistance arise from natural selection, as discussed next.[5]

B.1   Altered Drug Uptake

One type of resistance involves the ability of the organism to exclude the drug from the site of action by preventing the uptake of the drug. The plasma membrane can adjust its net charge by varying its proportion of anionic (phosphatidylglycerol) to cationic (lysylphosphatidyl-glycerol) groups.[4] In this way a drug with the same charge can be repelled from the membrane. Aminoglycoside antibiotic resistance can arise from lack of quinones that mediate the drug transport or from lack of an electrical potential gradient required to drive the drug across the bacterial membrane.[6]

B.2   Overproduction of the Target Enzyme

Increased target enzyme production, by induction of extra copies of the gene encoding the enzyme, is another mechanism for drug resistance.[7] Resistance to inhibitors of the enzyme dihydrofolate reductase by malarial parasites[8] and by malignant white blood cells[9] has been shown to be the result of overproduction of that enzyme in an unaltered form.

B.3   Altered Target Enzyme (or Site of Action)

Mutation of the amino-acid residues in the active site of the target enzyme can result in poor binding of the drug to the active site.[10] Resistance to the antibiotic trimethoprim (**5.6**, Proloprim) derives from an altered dihydrofolate reductase, the enzyme inhibited by **5.6**.[11] The properties of the singly mutated enzyme differ somewhat from those of the normal enzyme, but it still binds dihydrofolate. Erythromycin resistance results from drug-induced formation of N^6,N^6-dimethyladenines in the 23S ribosomal RNA, the site of action of that antibiotic.[12] This reduces the affinity of erythromycin for the target RNA.

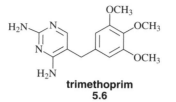

trimethoprim
5.6

The triphosphate nucleotides of 3'-azidothymidine (AZT, zidovudine, **5.7**, Retrovir) and (−)-2', 3'-dideoxy-3'-thiacytidine (3TC, lamivudine, **5.8**, Epivir) inhibit the reverse transcriptase of human immunodeficiency virus-type 1 (HIV-1) and are used in the treatment of AIDS. However, resistance can develop relatively rapidly.[13] Resistance to AZT arises from mutation of several residues in reverse transcriptase;[14] resistance to 3TC is conferred by a single mutation (different from the ones arising from AZT administration) in which a valine (preferentially) or isoleucine is substituted for Met-184.[15] However, a curious phenomenon occurs when both AZT and 3TC are administered in combination; a much longer delay in resistance occurs, even though Val-184 mutants rapidly emerge.[16] This results because the Val-184 mutant that develops from 3TC administration is much more susceptible to AZT, which therefore suppresses AZT resistance!

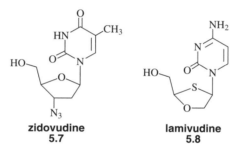

zidovudine
5.7

lamivudine
5.8

One way to minimize the effect of target enzyme mutation in drug resistance would be to design a drug that is very close in structure to that of the substrate for the target enzyme. If the organism mutates a residue in the active site to lower the binding affinity of the inhibitor, it will also lower the binding affinity of its substrate and will, therefore, auto-inhibit itself. For example, the structure of the anti-influenza drug zanamivir (see Chapter 2, **2.93**, Figure 2.18, Relenza) is very close to that of the substrate sialic acid (**2.94**), and it interacts with the same residues at the active site. This is an important strategy to protect the drug from resistance to an organism that tries to mutate its active site.

An interesting (terrifying!) variant of this mechanism for resistance was elucidated by Walsh and coworkers[17] for the antibiotic vancomycin (**5.9**, Vancocin), the drug of last defense against resistant streptococcal or staphylococcal organisms. The bacterial cell wall is constructed by a series of enzyme-catalyzed reactions, leading to the peptidoglycan, a branched polymer of alternating β-(D)-N-acetylglucosamine (NAG or GlcNAc) and β-(D)-N-acetylmuramic acid (NAM or MurNAc) residues. Attached to what was the carboxylic acid group of the lactyl group of MurNAc is a polypeptide chain that varies in structure according to the strain of bacteria. One example is shown in Scheme 5.1 with the mechanism for cross-linkage of the peptidoglycan. The terminal D-alanyl-D-alanine residues of the MurNAc side chain of the peptidoglycan bind to the transpeptidase, which initially acts as a serine protease, clipping the terminal peptide bond, releasing D-alanine, and making a serine ester. Cross-linkage of this ester with another peptidoglycan strand (**5.10**) builds the cell wall.

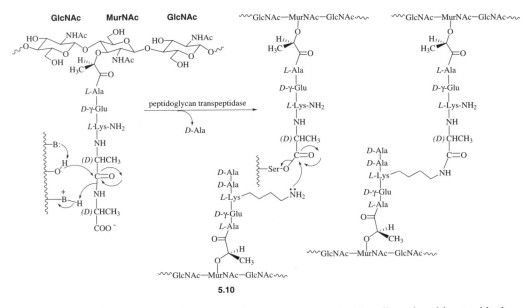

Scheme 5.1 ▶ Mechanism for the cross-linking of the bacterial cell wall catalyzed by peptidoglycan transpeptidase

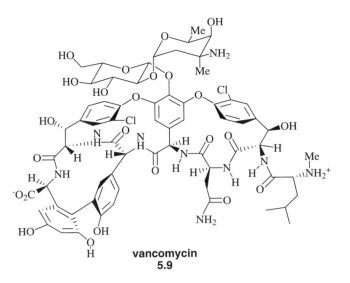

vancomycin
5.9

Vancomycin acts by forming a complex with the terminal *D*-alanyl-*D*-alanine of the peptidoglycan (Figure 5.1), thereby blocking the transglycosylation (the reaction that builds up the peptidoglycan by formation of the glycosyl linkage between GlcNAc and MurNAc) and transpeptidation steps. When bacterial cell wall biosynthesis is blocked, the high internal osmotic pressure (4–20 atmospheres) can no longer be sustained, and the bacteria burst (Figure 5.2).

A surprising resistance to vancomycin arises in bacteria by induction of five new genes (termed VanS, VanR, VanH, VanA, and VanX), used to construct an altered peptidoglycan in which the terminal *D*-alanyl-*D*-alanine is replaced by *D*-alanyl-*D*-lactate (**5.11**,

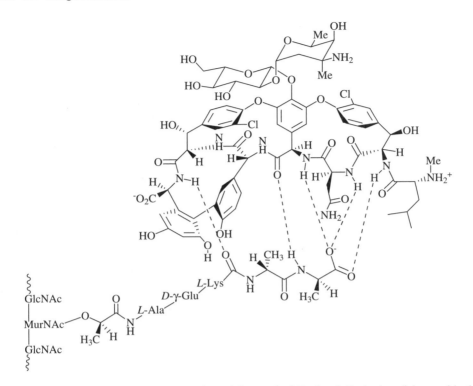

Figure 5.1 ▶ Complex between vancomycin and the terminal *D*-alanyl-*D*-alanine of the peptidoglycan

Figure 5.3). The VanS gene product is a transmembrane histidine kinase that initiates the signal transduction pathway, and the VanR gene product is a two-domain response regulator that accepts the phosphate group from phospho-VanS and activates VanH, VanA, and VanX transcription. The VanX gene product catalyzes the hydrolysis of *D*-Ala-*D*-Ala dipeptides so that only *D*-Ala-*D*-lactate is available for MurF-catalyzed condensation with UDP-muramyl-*L*-Ala-*D*-γ-Glu-*L*-Lys in the construction of the peptidoglycan strands (Figure 5.3). During transpeptidation (cross-linking of the peptidoglycan strands) in the antibiotic-sensitive organism, the terminal *D*-alanine residue is released (see Scheme 5.1); in the antibiotic-resistant organisms, *D*-lactate is released instead, but the same cross-linked product as in the wild-type organism is produced. However, substitution of *D*-alanine by *D*-lactate in the peptidoglycan leads to the deletion of one hydrogen bond to vancomycin and also produces a nonbonded electron repulsion between the lactate ester oxygen and the amide carbonyl of vancomycin (Figure 5.4). This produces a 1000-fold reduction in binding of the drug to the peptidoglycan substrate.[18] Therefore, in this case, the resistant organism has not mutated an essential enzyme, but has mutated the substrate for the enzyme, which is what forms the complex with vancomycin!

B.4 Production of a Drug-Destroying Enzyme

Resistance can occur by induction of genes that produce new enzymes to degrade the drug. In Section 5.5.B.2.a increased β-lactamase production is discussed as a mechanism for penicillin resistance.[19] Aminoglycoside antibiotics, such as kanamycins (**5.12**) and

(A)   (B)

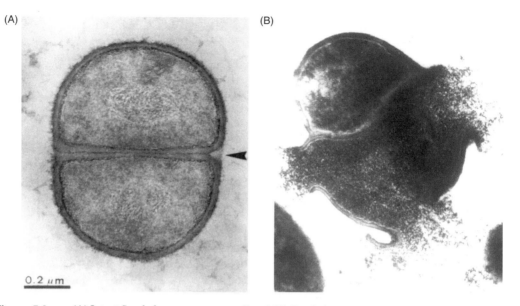

Figure 5.2 ▶ (A) Intact *Staphylococcus aureus* cell and (B) *Staphylococcus aureus* cell that has undergone lysis due to treatment with fosfomycin. [With permission of Dr. David Pompliano, Merck Laboratories, Rahway, NJ.]

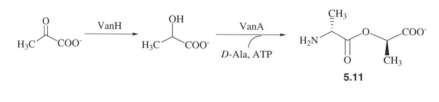

5.11

UDP-muramyl-*L*-Ala-*D*-γ-Glu-*L*-Lys   +   *D*-Ala-*D*-lactate

                          ⟋ ATP
            MurF │
                          ⟍ ADP + Pi

UDP-muramyl-*L*-Ala-*D*-γ-Glu-*L*-Lys-*D*-Ala-*D*-lactate

D-Ala-*D*-Ala $\xrightarrow{\text{VanX}}$ *D*-Ala + *D*-Ala

Figure 5.3 ▶ Biosynthesis of *D*-alanyl-*D*-lactate and incorporation into the peptidoglycan of vancomycin-resistant bacteria

neomycins (**5.13**, Neosporin), function by binding to the organism's ribosomal RNA, thereby inhibiting protein synthesis. Resistance develops because the bacteria acquire enzymes that catalyze the modification of hydroxyl and amino groups of these antibiotics by acetylation, adenylylation, and phosphorylation, thereby blocking their binding to the ribosomal RNA.[20] To avoid this type of resistance, analogs of the antibiotics can be designed that (1) bind poorly to the drug-modifying enzyme (aminoglycoside analogs with one less amino group had much diminished binding to and activity for the aminoglycoside 3′-phosphotransferase of resistant organisms, but had the same antibacterial activity as the unmodified compound);[21] (2) are less susceptible to modification, such as tobramycin (**5.14**,

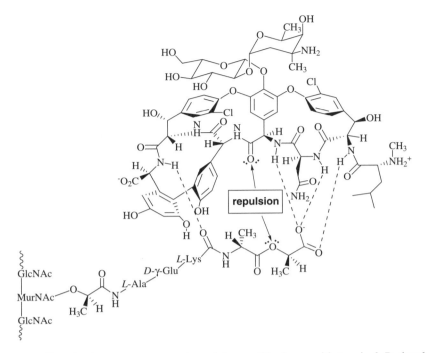

Figure 5.4 ▶ Complex between vancomycin and the peptidoglycan with terminal D-alanyl-D-lactate instead of D-alanyl-D-alanine in vancomycin-resistant bacteria

TobraDex), which lacks the 3′-hydroxyl group that is phosphorylated by resistant organisms; or (3) inhibit the enzymes responsible for destroying the drug. A clever approach to the design of an analog that is less susceptible to modification was reported by Mobashery and coworkers.[22] The 3′-hydroxyl group of kanamycin A (**5.12**) was converted to a carbonyl (**5.15**, Scheme 5.2), which is in equilibrium with the corresponding hydrate (**5.16**). The resistant organism can phosphorylate the hydrate to **5.17** (an ATP-dependent aminoglycoside 3′-phosphotransferase), but the phosphorylated product spontaneously decomposes back to **5.15**. Wong and coworkers[23] designed bifunctional dimeric neamine analogs (**5.18**) that both inhibit several of the aminoglycoside-modifying enzymes yet still bind to the organism's ribosomal RNA.

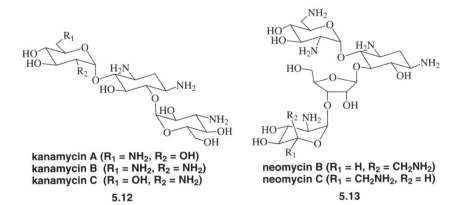

kanamycin A (R₁ = NH₂, R₂ = OH)
kanamycin B (R₁ = NH₂, R₂ = NH₂)
kanamycin C (R₁ = OH, R₂ = NH₂)

5.12

neomycin B (R₁ = H, R₂ = CH₂NH₂)
neomycin C (R₁ = CH₂NH₂, R₂ = H)

5.13

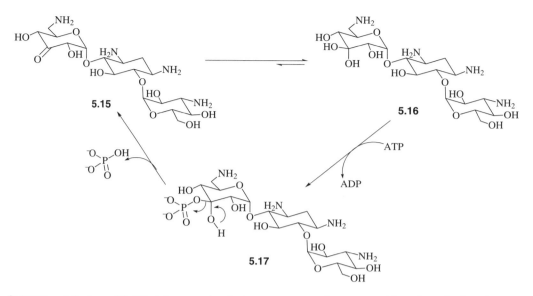

Scheme   5.2   ▶   Modified kanamycin A to overcome resistance resulting from hydroxyl group
phosphorylation

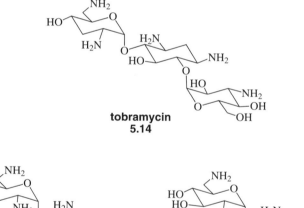

tobramycin
5.14

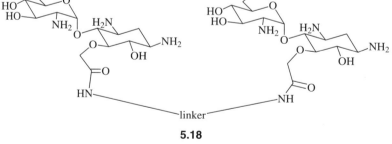

5.18

B.5   Deletion of a Prodrug-Activating Enzyme

Another form of resistance derives from the deletion of an enzyme required to convert
a prodrug, a compound that is enzymatically converted into a drug after its administration (see Chapter 8) into its active form. Tumor resistance to the antileukemia drug

6-mercaptopurine (**5.19**, R = H, Purinethol) is caused by deletion of hypoxanthine-guanine phosphoribosyltransferase,[24] the enzyme required to convert (**5.19**, R = H) into thioinosine monophosphate (**5.19**, R = ribosyl 5′-monophosphate), the active form of the drug.

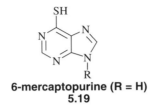

6-mercaptopurine (R = H)
5.19

B.6 Overproduction of the Substrate for the Target Enzyme

In Chapter 3, Section 3.2.C a competitive antagonist is described as a molecule that is in competition with the natural ligand for binding to a receptor. The same situation can arise between inhibitors and substrates of enzymes. Overproduction of the substrate for a target enzyme would competitively block the ability of the drug to bind at the active site (see Section 5.4.A). As mentioned in Section 5.4.B.2.d this is one mechanism of resistance to sulfa drugs.

B.7 New Pathway for Formation of the Product of the Target Enzyme

If the effect of a drug is to block production of a metabolite by enzyme inhibition, the organism could bypass the effect of the drug by inducing a new metabolic pathway that produces the same metabolite.

B.8 Efflux Pumps

Tumor cells and microorganisms can develop protein transporters that bind to drugs and carry them out of the cell before they exhibit their therapeutic effect.[25] Some transporters, known as *multidrug resistance pumps*, are more broad in their specificity and can efflux a variety of natural and synthetic drugs.[26]

5.3 Drug Synergism (Drug Combination)

5.3.A What Is Drug Synergism?

When drugs are given in combination, their effects can be antagonistic, subadditive, additive, or synergistic. *Drug synergism* arises when the therapeutic effect of two or more drugs used in combination is greater than the sum of the effects of the drugs administered individually.

5.3.B Mechanisms of Drug Synergism

B.1 Inhibition of a Drug-Destroying Enzyme

What if resistance to a drug occurs because a gene is induced that encodes for a new enzyme that destroys the drug? An important approach for synergism would be to design a new compound that inhibits the drug-destroying enzyme. This new compound has no real therapeutic effect; it only protects the drug from being destroyed. So, if that new compound is administered alone, it destroys the drug-destroying enzyme, but it has no therapeutic benefit. If the drug is administered alone, it gets destroyed, and it too has no therapeutic effect. But if both the drug and the new compound are administered together, then the new compound destroys the drug-destroying enzyme, which allows the drug to be effective again, a synergistic effect.

B.2   Sequential Blocking

A second mechanism for synergism is *sequential blocking*, the inhibition of two or more consecutive steps in a metabolic pathway. The reason this is effective is because it is difficult (particularly with a reversible inhibitor) to inhibit an enzyme 100%. If less than 100% of the enzyme activity is blocked, the metabolic pathway has not been shut down. With the combined use of inhibitors of two consecutive enzymes in the pathway, it is possible to block the metabolic pathway virtually completely. Because reversible enzyme inhibition (or receptor antagonism) is hyperbolic in nature, complete inhibition of an enzyme would require a large excess of the drug, and may be toxic. This approach becomes somewhat less important with an irreversible inhibitor which may inhibit an enzyme completely (see Section 5.5.C.3.b, p. 289, for an example of an irreversible inhibitor that does not shut down the target enzyme totally).

B.3   Inhibition of Enzymes in Different Metabolic Pathways

If the cause for resistance is the production of a new metabolic pathway that produces a particular metabolite, then the combination of drugs that inhibit enzymes in both metabolic pathways would again shut down the production of the undesirable metabolite. Use of only one of these drugs is not effective, but the combination of the two drugs becomes synergistic.

B.4   Efflux Pump Inhibitors

If an antimicrobial drug is being effluxed from the organism by a transporter protein, a compound can be designed that inhibits this efflux pump. The efflux pump inhibitor has no therapeutic activity, and without it, the drug is ineffective; however, the combination of the two is synergistic and will exhibit antimicrobial activity.[27]

B.5   Use of Multiple Drugs for the Same Target

A fifth mechanism for synergism is the use of two or more drugs to inhibit the growth of tumor cells or microbial mutants, because a mutant that is resistant to one drug does not easily undergo further mutation. Typically, only one in a culture of 10^7 bacteria may be resistant to a particular drug.[28] The chance of finding an organism resistant against two different drugs is one in 10^{14}, and against three drugs is one in 10^{21}, so the effect of multiple drugs is exponential, not additive, and the use of multiple antimicrobial agents greatly minimizes the opportunity for a mutant resistant organism to proliferate. For example, the antituberculosis drug, isoniazid (**5.20**, Rifamate), which inhibits the replication of the tubercle bacillus, is used in combination with other antimicrobial agents such as rifampin (**5.21**, Rifadin); these two drugs are sold in combination for the treatment of tuberculosis.

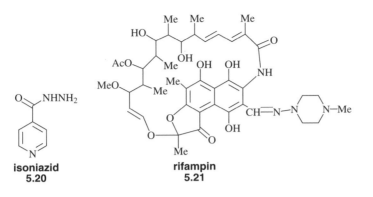

isoniazid
5.20

rifampin
5.21

5.4   Reversible Enzyme Inhibitors

5.4.A   Mechanism of Reversible Inhibition

The most common enzyme inhibitor drugs are the reversible type, particularly ones that compete with the substrate for active site binding. These are known as *competitive reversible inhibitors*, compounds that have structures similar to those of the substrates or products of the target enzymes and which bind at the substrate binding sites, thereby blocking substrate binding. Typically, these inhibitors establish their binding equilibria with the enzyme rapidly, so that inhibition is observed as soon as the enzyme is assayed for activity, although there are some cases (see Section 5.4.D, p. 262) in which inhibition can be relatively slow.

As in the case of the interaction of a substrate with an enzyme, an inhibitor (I) also can form a complex with an enzyme (E) (Scheme 5.3). The equilibrium constant K_i (k_{off}/k_{on}) is a *dissociation* constant for breakdown of the E·I complex; therefore, as discussed for the K_d of drug–receptor complexes (see Chapter 3, Section 3.2.A), the *smaller* the K_i value for I, the more potent the inhibitor. Another common measurement for inhibition, which is quicker to determine, but is more of an estimate, is the IC_{50}, the inhibitor concentration that produces 50% enzyme inhibition in the presence of substrate. The IC_{50} value can be converted into an approximate K_i value by Equation 5.1:[29]

$$IC_{50} = \left(1 + \frac{[S]}{K_m}\right) K_i \qquad (5.1)$$

When the inhibitor binds at the active site (the substrate binding site), then it is a competitive inhibitor. Formation of the E·I complex prevents the binding of substrate and, therefore, blocks the catalytic conversion of the substrate to product. The inhibitor, however, also may act as a substrate and may be converted to a metabolically useless product. In the context of drug

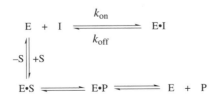

Scheme 5.3 ▶ Kinetic scheme for competitive enzyme inhibition

design, this is generally not a favorable process because the product formed may be toxic or may lead to other toxic metabolites. Nonetheless, there are drugs that function by this mechanism (see Section 5.4.B.2).

Interaction of the inhibitor with the enzyme can occur at a site other than the substrate binding site and still result in inhibition of substrate turnover. When this occurs, usually as a result of an inhibitor-induced conformational change in the enzyme to give a form of the enzyme that does not bind the substrate properly, then the inhibitor is a *noncompetitive reversible inhibitor*. Unless something is known about this *allosteric binding site* in the enzyme, it is not possible to design noncompetitive enzyme inhibitors. Consequently, the discussion in this chapter will be limited to the design and mechanism of action of competitive enzyme inhibitors.

The equilibrium shown in Scheme 5.3, and therefore the E·I concentration, will depend on the concentrations of the inhibitor and the substrate, as well as the K_i for the inhibitor and the K_m for the substrate. As the concentration of the inhibitor is increased, it drives the equilibrium toward the E·I complex. Because the substrate and the competitive inhibitor bind to the enzyme at the same site, they both cannot interact with the enzyme simultaneously. When the inhibitor concentration diminishes, the E·I complex concentration diminishes, and the effect of the inhibitor can be overcome by the substrate. (Increasing the substrate concentration also would displace the equilibrium from the E·I complex toward increased E·S complex formation, which was one of the mechanisms of drug resistance; see Section 5.2.B.6, p. 239.)

If the enzyme inhibitor is a drug, the maximal pharmacological effect will occur when the drug concentration is maintained at a saturating level at the target enzyme active site. As the drug is metabolized (see Chapter 7), and the concentration of I diminishes, repeated administration of the drug is required to maintain the integrity of the E·I complex. This accounts for why drugs often need to be taken two or more times a day. To increase the potency of reversible inhibitors and, thereby, reduce the dosage of the drug, the binding interactions with the target enzyme must be optimized (that is, the inhibitor should have a low K_i value).

When a drug is designed to be an enzyme inhibitor, it generally will be a competitive inhibitor because the lead compound often will be the substrate for the target enzyme. Because an enzyme is just a specific type of receptor, an analogy can be made between agonists, partial agonists, and antagonists with good substrates, poor substrates, and competitive inhibitors, respectively.

5.4.B   Selected Examples of Competitive Reversible Inhibitor Drugs

In this section we take a look at four different approaches to the design of competitive reversible inhibitors: simple competitive inhibition, alternative substrate inhibition, transition state analog inhibition, and slow, tight-binding inhibition.

B.1   Simple Competitive Inhibition: Captopril, Enalapril, Lisinopril, and Other Antihypertensive Drugs

The most common type of inhibition is simple competitive inhibition, which involves the design of molecules whose structure resembles that of the substrate for the target enzyme and which reversibly bind to the active site.

a. Humoral Mechanism for Hypertension

The elucidation of the molecular details of the renin-angiotensin system, one of the humoral mechanisms for blood pressure control, began more than 60 years ago.[30] Angiotensinogen, an α-globulin produced by the liver,[31] is hydrolyzed by the proteolytic enzyme renin to a decapeptide, angiotensin I (Scheme 5.4), which has little, if any, biological activity. The C-terminal histidylleucine dipeptide is cleaved from angiotensin I by angiotensin-converting enzyme (ACE or dipeptidyl carboxypeptidase I) mainly in the lungs and blood vessels to give the octapeptide angiotensin II. This peptide is responsible for the increase in blood pressure by acting as a very potent vasoconstrictor[32] and by triggering release of a steroid hormone, aldosterone (**5.23**), which regulates the electrolyte balance of body fluids by promoting excretion of potassium ions and retention of sodium ions and water. Both vasoconstriction and sodium ion/water retention lead to an increase in blood pressure. Angiotensin II is converted to another peptide hormone, angiotensin III, by aminopeptidase A;[33] angiotensin III also is made by aminopeptidase A hydrolysis of angiotensin I to **5.22** (Scheme 5.4), then by ACE-catalyzed hydrolysis of **5.22** to angiotensin III.[34] This hormone also stimulates the secretion of aldosterone and causes vasoconstriction.[35] Angiotensin II also is hydrolyzed by aminopeptidase N to a hexapeptide, angiotensin IV, which may be involved in memory retention and neuronal development, but it is unclear if it is involved in vasopressin release; angiotensin III also is converted to angiotensin IV by aminopeptidase N.[36] To make matters even worse, in addition to cleaving angiotensin I to angiotensin II, and **5.22** to angiotensin III, ACE also catalyzes the hydrolysis of the two C-terminal amino-acids from the potent vasodilator nonapeptide bradykinin (Arg-Pro-Pro-Gly-Phe-Ser-Pro-Phe-Arg), thereby destroying its vasodilation activity. Consequently, the action of ACE results in the generation of potent hypertensive agents (angiotensin II and angiotensin III), which also stimulate the release of another hypertensive agent (aldosterone), and destroys a potent antihypertensive agent (bradykinin). All of these outcomes of ACE action result in hypertension, an increase in blood pressure. Angiotensin-converting enzyme, therefore, is an important target for the design of antihypertensive agents; inhibition of ACE would shut down its three hypertensive mechanisms.

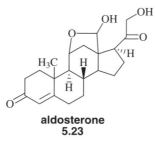

aldosterone
5.23

b. Lead Discovery

In 1965 Ferreira[37] reported that a mixture of peptides in the venom of the South American pit viper *Bothrops jararaca* potentiated the action of bradykinin by inhibition of some bradykininase activity. Bakhle and coworkers[38] subsequently showed that these peptides also inhibited the conversion of angiotensin I to angiotensin II. Nine active peptides were isolated from this venom; the structure of a pentapeptide (Pyr-Lys-Trp-Ala-Pro, where Pyr is *L*-pyroglutamate) was identified.[39] This peptide was shown to inhibit the conversion of angiotensin I to II and

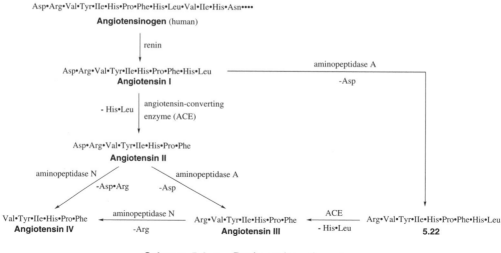

Asp•Arg•Val•Tyr•Ile•His•Pro•Phe•His•Leu•Val•Ile•His•Asn••••
Angiotensinogen (human)

Scheme 5.4 ▶   Renin-angiotensin system

bradykinin degradation *in vitro*[40] and *in vivo*.[41] The structures of six more of the peptides were determined by Ondetti and coworkers.[42] The peptide with the greatest *in vitro* activity was the pentapeptide,[43] but a nonapeptide (Pyr-Trp-Pro-Arg-Pro-Gln-Ile-Pro-Pro) called teprotide had the greatest *in vivo* potency[44] and was effective in lowering blood pressure.[45] Five other active peptides were isolated from the venom of the Japanese pit viper, *Agkistrodon halys blomhoffii*.[46] Because these compounds were peptides, they were not effective when administered orally, but they laid the foundation for the design of orally active peptidomimetic angiotensin-converting enzyme inhibitors.

c. Lead Modification and Mechanism of Action

The fact that *N*-acylated tripeptides are substrates of ACE indicated that it may be possible to prepare a small orally active ACE inhibitor. After testing numerous peptides as competitive inhibitors of ACE, it was concluded that proline was best in the C-terminal position and alanine was best in the penultimate position. An aromatic amino-acid is preferred in the antepenultimate position. When the search for a potent inhibitor of ACE was initiated at Squibb and Merck pharmaceutical companies, the enzyme had not yet been purified. Because the enzyme was inhibited by EDTA and other chelating agents, particularly bidentate ligands, it was believed to be a metalloenzyme. In fact, ACE purified to homogeneity from rabbit lung[47] was shown to contain 1 gram-atom of zinc ion per mole of protein. The zinc ion is believed to be a cofactor that assists in the catalytic hydrolysis of the peptide bond by both coordination to the carbonyl oxygen, making the carbonyl more electrophilic, and by coordination to a water molecule, making the water more nucleophilic. Coordination of both molecules to the zinc ion lowers the activation energy for attack of the water on the scissile peptide bond (Figure 5.5). Because the structure of the enzyme was not known, it was not obvious what peptide-like structures would be the best inhibitors. It was hypothesized that the mechanism and active site of ACE may resemble those of carboxypeptidase A, another zinc-containing peptidase whose X-ray structure was known.[48] Three important binding interactions between carboxypeptidase A and peptides are a carboxylate-binding group, a group that binds the C-terminal amino-acid side chain, and the zinc ion that coordinates to the carbonyl of the penultimate (the scissile)

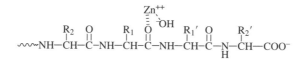

Figure 5.5 ► Function of the Zn(II) cofactor in angiotensin-converting enzyme catalysis

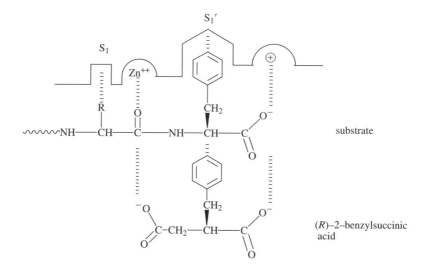

Figure 5.6 ► Hypothetical active site of carboxypeptidase A [Adapted with permission from Cushman, D. W., Cheung, H. S., Sabo, E. F., Ondetti, M. A. (1977). *Biochemistry* **16**, 5484. Copyright © 1977 American Chemical Society.]

peptide bond (Figure 5.6).[49] (*R*)-2-Benzylsuccinic acid, which can bind at all three of these sites, is a potent inhibitor of carboxypeptidase A.[50] The extreme potency of inhibition of carboxypeptidase A by (*R*)-2-benzylsuccinic acid was suggested to be derived from the resemblance of this inhibitor to the *collected products* (Figure 5.7) of hydrolysis of the substrate, and, therefore, it combines all of their individual binding characteristics into a single molecule. With this as a model, and the known effectiveness of a C-terminal proline for ACE inhibition, a series of peptidomimetic carboxyalkanoylproline derivatives (**5.24**) were tested as inhibitors of ACE. Note that to avoid having an orally unstable dipeptide, the N-terminal amino group was substituted by an isosteric CH_2 group to which the Zn(II)-coordinating carboxylate was attached. Although the results were encouraging, all of these compounds were only weak inhibitors of ACE. To increase the potency of the compounds, a better Zn(II)-coordinating ligand, a thiol group, was substituted for the carboxylate (**5.25**). These compounds were very potent inhibitors of ACE. Figure 5.8 shows a hypothesized depiction of the interaction of **5.24** and **5.25** with ACE. Note that carboxypeptidase A is a C-terminal *exopeptidase* (it cleaves the C-terminal amino-acid), whereas ACE is a C-terminal *endopeptidase* or, more precisely, a *dipeptidyl carboxypeptidase* (it cleaves a C-terminal dipeptide). Therefore, the active site of ACE (Figure 5.8) has two additional binding sites more than carboxypeptidase A has between the Zn(II) and the group that interacts with the C-terminal carboxylate group (Figure 5.6). The compound that had the best binding properties was **5.26** (captopril, Capoten), a competitive inhibitor of ACE with a K_i of 1.7×10^{-9} M. Furthermore, captopril is highly specific

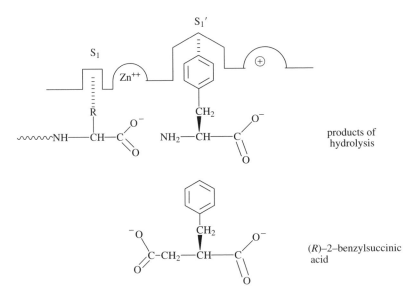

products of
hydrolysis

(R)–2–benzylsuccinic
acid

Figure 5.7 ▶   The collected products hypothesis of enzyme inhibition

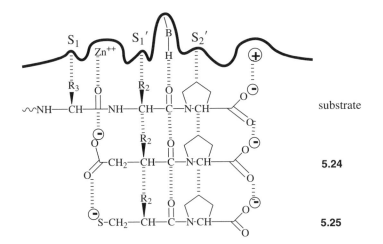

substrate

5.24

5.25

Figure 5.8 ▶   Hypothetical binding of carboxyalkylproline and mercaptoalkylproline derivatives to angiotensin-converting enzyme [Adapted with permission from Cushman, D. W., Cheung, H. S., Sabo, E. F., Ondetti, M. A. (1977). *Biochemistry* **16**, 5484. Copyright © 1977 American Chemical Society.]

for angiotensin-converting enzyme; the K_i values for captopril with carboxypeptidase A and carboxypeptidase B, two other Zn(II)-containing peptidases, are 6.2×10^{-4} M and 2.5×10^{-4} M, respectively.[51]

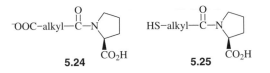

5.24   **5.25**

TABLE 5.1 ▶ Effect on K_i of Structural Modification of Captopril

Analog	Relative K_i
(captopril)	1.0
	12,500
	10
	12,000
	120
	120
	1,100

Presumably, the reason for the specificity is that there are many functional groups in **5.26** that can regio- and stereospecifically interact with groups at the active site of ACE, but they cannot interact with groups in other peptidases (compare Figures 5.6 and 5.8). The carboxylate group of the inhibitor can be stabilized by an electrostatic interaction with a cationic group on the enzyme, the amide carbonyl can be hydrogen bonded to a hydrogen donor group, the sulfhydryl can be liganded to the zinc ion, and the proline and (S)-methyl group can be involved in stereospecific hydrophobic and van der Waals interactions.

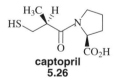

captopril
5.26

All of these interactions must be important because deletion or alteration of any of these groups raises the K_i considerably (Table 5.1). A myriad of analogs of this basic structure, including compounds with Zn(II)-coordinating ligands other than carboxylate and thiol groups, has been synthesized and tested as ACE inhibitors.

Captopril was the first ACE inhibitor on the drug market, and it was shown to be effective both for the treatment of hypertension and congestive heart failure. Given alone, captopril can normalize the blood pressure of about 50% of the hypertensive population. When given in combination with a diuretic, such as hydrochlorothiazide (**5.27**, Aldoril) (remember, angiotensin II releases aldosterone which causes water retention), this can be extended to 90% of the hypertensive population. In more severe cases, a *β-blocker*, an antagonist for the *β*-adrenergic receptor, which triggers vasodilation, may be used in a triple therapy with captopril and a diuretic.

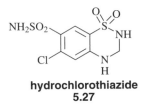

hydrochlorothiazide
5.27

Two side effects were observed in some patients during the early usage of captopril, namely, rashes and loss of taste. Both of these side effects were reversible on drug withdrawal or reduction of the dose.[52] Considering the potential lethality of hypertension, a minor rash or loss of taste would seem insignificant to the benefits of the therapy. However, hypertension is a disease without a symptom (that is, until it's too late); generally, a patient discovers he has this disease when his physician determines it by taking his blood pressure. Because of this lack of immediate discomfort, there may be difficulties getting the patient to comply with the therapy, especially if unpleasant side effects arise when the drug is taken. Consequently, the Merck group investigated the cause for the side effects. Because similar side effects arise when penicillamine is administered, it was hypothesized that the thiol group may be responsible.[53] Furthermore, deletion of this functional group should give inhibitors with greater metabolic stability because thiols undergo facile *in vivo* oxidation to disulfides. The approach taken was to attempt to increase the previously found weak potency of the carboxyalkanoylproline analogs by adding groups that can interact with additional sites on the enzyme, that is, to increase the pharmacophore.

If the carboxyalkanoylproline derivatives are collected product inhibitors (see Figure 5.7), then two features can be built into these analogs to make them look more product-like. One is to make them structurally more similar to dipeptides by substituting an NH for a CH_2 such as **5.28** (R = R' = H). Disappointingly, however, this compound had less than twice the potency of the isostere with a CH_2 in place of NH. The reason for this could be compensatory factors. An NH (or its protonated form) is much more hydrophilic than a CH_2 group. Therefore, an additional hydrophobic group should be added to counterbalance this hydrophilic effect. When a methyl group was appended (**5.28**, R = CH_3, R' = H), the potency increased about 55-fold. Because the other feature that could make these compounds structurally more similar to the products would be to append a group that might interact with the substrate S_1 subsite, the R group of **5.28** was modified further, and **5.28** (R = (*S*)-$PhCH_2CH_2$, R' = H), called enalaprilat, emerged as the viable drug candidate. The IC_{50} for enalaprilat is 19 times lower than that for captopril, suggesting that there are increased interactions of enalaprilat with ACE. These may be hydrophobic interactions of the phenylethyl group with the S_1 subsite

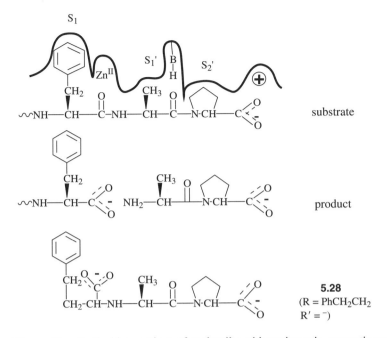

Figure 5.9 ▶ Hypothetical interactions of enalaprilat with angiotensin-converting enzyme

(see Figure 5.9). Also, see Section 5.4.C.2.a, p. 259, for an alternative explanation of the potency of enalaprilat.

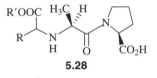

5.28

Enalaprilat, however, is poorly absorbed orally and, therefore, must be given by intravenous injection. This problem was remedied simply by conversion of the carboxyl group to an ethyl ester (**5.25**, R = (S)-PhCH$_2$CH$_2$, R' = CH$_3$CH$_2$), giving enalapril (Vasotec), which has excellent oral activity. Because the *in vitro* IC$_{50}$ for enalapril is 10^3 times higher than that for enalaprilat, the ethyl ester group must be hydrolyzed by esterases in the body to liberate the active form of the drug, namely, enalaprilat. Enalapril, then, is an example of a *prodrug*, a compound that requires metabolic activation for activity (see Chapter 8). The effect of esterification is to lower the pK_a of the NH group (pK_a 5.5 in enalapril, but 7.6 in enalaprilat),[54] and to remove the charge of the carboxylate, both of which would increase membrane transport. Another compound that was prepared by the Merck chemists as an alternative to enalaprilat was the lysylproline analog called lisinopril (**5.29**, Prinovil). Note the stereochemistry shown (S,S,S) is that found in the most potent isomer of lisinopril, and also is the stereochemistry of enalaprilat. Lisinopril is more slowly and less completely absorbed than enalapril, but its longer oral duration of action, and the fact that it does not require metabolic activation, make it an attractive alternative.

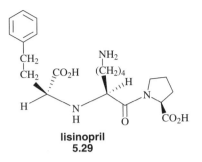

lisinopril
5.29

One approach for combination therapy with the "prils" stems from the discovery that a bacterium, *Helicobacter pylori*, is found in the stomach and is associated with peptic ulcers.[55] The organism protects itself from the acid in the stomach partly because it lives within the layer of mucous that the stomach secretes to protect itself against the acid, and partly because the bacterium produces the enzyme urease, which converts urea in the blood into ammonia to neutralize the acid.[56] The discovery of this bacterium was made in 1983 by Drs. Barry Marshall and Robin Warren at the Royal Perth Hospital in Australia, who were trying to grow mysterious cells taken from the stomach. The culture was left much longer to grow than normal because of the four-day Easter weekend that year, and on their return, they noticed the growth of a bacterium with spiral, helix-shaped cells, which they called *Helicobacter*. Although it took more than a decade to convince others that this bacterium was really living in the stomach and that it could cause ulcers, it is now widely accepted. Because there are many people who have this bacterium, but do not have ulcers, there must be additional factors, such as stress, that are needed for ulcer formation.[57] Treatment with antibacterial agents, such as metronidazole (**5.30**, Flagyl), can kill these bacteria, but generally other drugs that can lower stomach acid are needed in combination.[58]

metronidazole
5.30

d. Dual-Acting Drugs: Dual-Acting Enzyme Inhibitors

When there are two related enzymes whose inhibition leads to a synergistic effect, it may be beneficial to design a single inhibitor of both enzymes, a compound known as a *dual-acting enzyme inhibitor*. There are several advantages to the design of one compound that inhibits two different enzymes rather than two compounds, one for each enzyme: (1) with two drugs, two separate syntheses, two formulations, and two different metabolism studies (see Chapter 7) have to be developed; (2) two drugs will have different pharmacokinetic rates and metabolic profiles, making it difficult for both to be optimal in the same time frame; (3) the likelihood that both drugs would progress to the clinic at the same rate is small; (4) the cost for three sets of safety studies and three separate clinical trials (one for each drug plus one with the combination) would be enormous; and (5) the odds for a single drug just starting clinical trials to be approved for the drug market is 1 in 10; the odds for two drugs entering the market would be 1 in a 100! With a single, dual-acting drug, none of these problems exists.

The next generation of antihypertensive agents may be dual-acting enzyme inhibitors. Neutral endopeptidase (NEP) is another Zn^{2+}-containing endopeptidase. It degrades and deactivates atrial natriuretic peptide (ANP), a 28-amino-acid vasoactive peptide hormone produced by the heart that causes vasodilation and inhibition of the formation of aldosterone, the steroid hormone that regulates the electrolyte balance and leads to retention of sodium ions and water. Therefore, ANP acts to *lower* the blood pressure; the hormone actions of ANP and angiotensin II, therefore, are functionally opposite. If the formation of angiotensin II is blocked by the inhibition of ACE, and the concentration of ANP is increased by the inhibition of NEP, there should be a synergistic (or at least additive) antihypertensive effect.[59]

The lead compound devised at Bristol-Myers Squibb, which had a structure comprised of parts of captopril and enalapril (**5.31**), was found to have IC_{50} values for ACE and NEP of 30 and 400 nM, respectively, which is excellent for a lead compound.[60] The thiol group was retained because subsequent to the launch of captopril it was found that the lost of taste and rash side effects initially observed were related to high dosing, not to the thiol group; at a lower dose, captopril was just as effective, but without those side effects. Homologation gave a compound that was potent *in vitro* for both enzymes, but not very potent *in vivo* (**5.32**). The research group turned to conformationally restricted analogs to increase potency. Compound **5.33** was potent both *in vitro* and *in vivo*, and **5.34** had IC_{50} values of 5 and 17 nM for ACE and NEP, respectively, with oral activity greater than that of captopril in rats. A 7,6-fused bicyclic thiazepinone analog (**5.35**, omapatrilat, Vanlev) with IC_{50} values of 5 and 8 nM for ACE and NEP, respectively, was advanced to clinical trials.[61] The key advantages of the dual-acting enzyme inhibitor omapatrilat over the ACE inhibitors is its ability to lower both diastolic and systolic blood pressure better and its effectiveness in controlling subpopulations of patients, such as African-Americans and diabetics, where traditional drugs have been less effective.

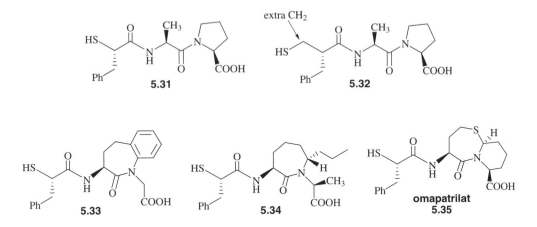

This dual-acting inhibitor approach for the design of new antihypertensive agents has been extended further; there is a third zinc metalloprotease, endothelin-converting enzyme (ECE), which displays high amino-acid sequence identity with NEP, especially in the active site,[62] that hydrolyzes a 38-amino-acid inactive peptide into endothelin-1, a 21-amino-acid peptide that is the most potent vasoconstrictor (raises the blood pressure) known.[63] Several classes of compounds that act as triple-acting enzyme inhibitors for ACE, NEP, and ECE have been identified.[64]

A *dual-acting drug* does not have to be limited to compounds that inhibit two different enzymes. It also could be a compound that inhibits one enzyme and acts as an antagonist for a receptor, or a compound that is an antagonist for two different receptors, or an agonist for two different receptors, or any combination thereof.

An example of a dual-acting drug that acts as an inhibitor of an enzyme and as an antagonist for a receptor is Z-350 (**5.36**),[65] an inhibitor of steroid 5α-reductase and an antagonist for the α₁-adrenoceptor.[66] Benign prostatic hyperplasia (BPH) is a progressive enlargement of the prostate gland leading to bladder outlet obstruction. This obstruction consists of a static component related to prostatic tissue mass and a dynamic component related to excessive contraction of the prostate and urethra.[67] Antagonists of the α₁-adrenoceptor, such as terazosin HCl (**5.37**, Hytrin), are used to relax the smooth muscle of the prostate and urethra.[68] Because dihydrotestosterone is known to be a dominant factor in prostatic growth, inhibitors of steroid 5α-reductase, the enzyme that converts testosterone into dihydrotestosterone, such as finasteride (**5.38**, Proscar), also are used to treat BPH. A dual-acting agent was designed[69] based on the α₁-adrenoceptor antagonist **5.39** (the structure is drawn in a conformation to resemble **5.36**) and the steroid 5α-reductase inhibitor **5.40** by combining features of both molecules into one structure (**5.41**), a common approach to the design of dual-acting agents. This compound was a potent antagonist for the α₁-adrenoceptor, but needed increased potency against steroid 5α-reductase. It was well established that the lipophilic part and the butanoic acid moiety are essential for steroid 5α-reductase activity, but the benzanilide moiety could be replaced by an acyl indole, so the next structures included **5.42**, which is as potent as **5.41** as an antagonist for α₁-adrenoceptor and more potent in steroid 5α-reductase inhibition.[70] A merging of the structures of **5.41** and **5.42** led to the design of **5.36**, which significantly reduces prostatic growth in rabbits and rats.

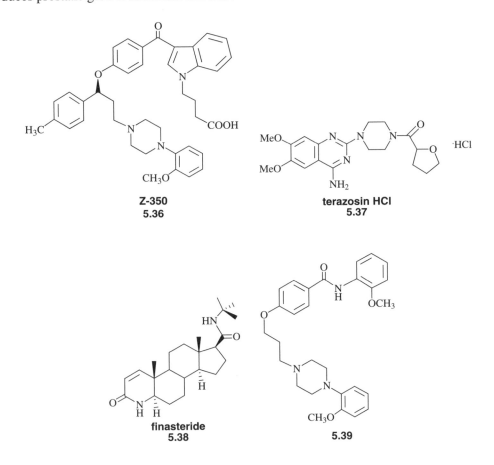

Z-350
5.36

terazosin HCl
5.37

finasteride
5.38

5.39

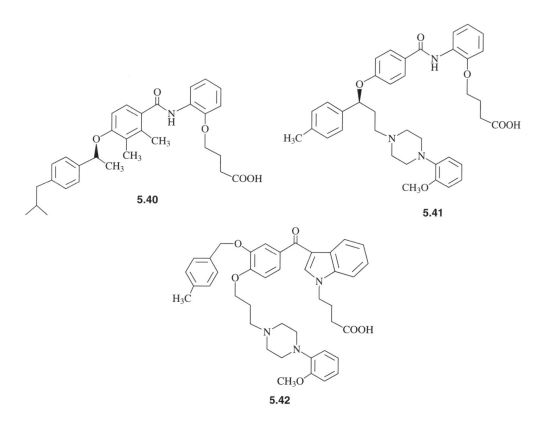

5.40

5.41

5.42

An example of a dual-acting receptor agonist is a compound that acts as an agonist for both the D_2-receptor and the β_2-adrenoceptor for the treatment of airway diseases, such as chronic obstructive pulmonary disease (COPD) and asthma. A D_2-receptor agonist reduces reflex bronchoconstriction, dyspnea, cough, and mucous production, but is less likely to diminish bronchoconstrictor activity of locally released mediators of bronchoconstriction.[71] β_2-Adrenoceptor agonists are the most commonly used antibronchoconstrictor agents,[72] but they have little effect on dyspnea, cough, and mucous production. A dual-acting agent for these receptors should combine all of the desired features of this class of drugs. The structures of a weak β_2-adrenoceptor agonist (**5.43**) and a potent D_2-receptor agonist (**5.44**) were hybridized to give **5.45**, which was improved by modification of the side chain to give **5.46**,[73] a clinical candidate for the treatment of the symptoms of COPD.

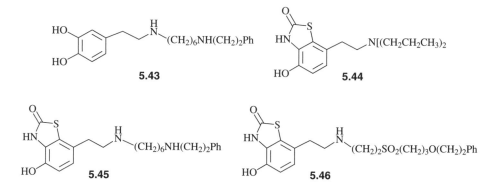

5.43

5.44

5.45

5.46

B.2 Alternative Substrate Inhibition: Sulfonamide Antibacterial Agents (Sulfa Drugs)

Competitive inhibition of an enzyme also can be attained with a molecule that not only binds to the enzyme, but also acts as a substrate. In this case, however, the product produced is not a compound that is useful to the organism. While the alternative substrate is being turned over, it is preventing the actual substrate from being converted to the product that the organism needs. The principal disadvantage to this approach is that the product generated could be toxic or cause an unwanted side effect. In the example below, this is not the case.

a. Lead Discovery

At the beginning of the 20th century Paul Ehrlich showed that various azo dyes were effective agents against trypanosomiasis in mice; however, none was effective in man. In the early 1930s Gerhard Domagk, head of bacteriological and pathological research at the Bayer Company in Germany, who was trying to find agents against streptococci, tested a variety of azo dyes. One of the dyes, prontosil (**5.47**), showed dramatically positive results and successfully protected mice against streptococcal infections.[74] Bayer was unwilling to move rapidly on getting prontosil onto the drug market. As Albert[75] tells it, when, in late 1935, Domagk's daughter cut her hand and was about to die of a streptococcal infection, her father gave her prontosil. Although she turned bright red from the dye, her recovery was rapid,[76] and the effectiveness of the drug became quite credible. In 1939, Domagk was awarded the Nobel Prize in Medicine for this achievement.

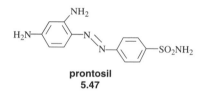

prontosil
5.47

An unexpected property of prontosil, however, was that it had no activity against bacteria *in vitro*. Tréfouël and coworkers[77] found that if a reducing agent was added to prontosil, then it was effective *in vitro*. They suggested that the reason for the lack of *in vitro* activity, but high *in vivo* activity, was that prontosil was converted by reduction to the active antibacterial agent, namely, *p*-aminobenzenesulfonamide (also called sulfanilamide) (**5.48**, AVC). Furthermore, they demonstrated that sulfanilamide was as effective as prontosil in protecting mice against streptococcal infections, and that it exerted a *bacteriostatic effect in vitro*. Unlike a *bacteriocidal* agent, which kills bacteria, a bacteriostatic drug inhibits further growth of the bacteria, thus allowing the host defenses to catch up in their fight against the bacteria. Because microorganisms replicate rapidly (with *E. coli*, for example, the number of cells can double every 20–30 minutes), a bacteriostatic agent will interrupt this rapid growth and allow the immune system to destroy the organism. Of course, with immunocompromised individuals, who are unable to contribute natural body defenses to fight their own disease, a bacteriocidal agent is necessary to prevent continuation of growth of the organism when the drug is withdrawn. Prontosil, then, is an early example of a *prodrug* (see Chapter 8), a compound that requires metabolic activation to be effective.

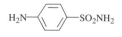

sulfanilamide
5.48

b. Lead Modification

The discovery of prontosil marks the beginning of modern chemotherapy. During the next decade thousands of sulfonamides were synthesized and tested as antibacterial agents. These were the first structure-activity relationship studies (see Chapter 2, Section 2.2.C, p. 21), and demonstrated the importance of molecular modification in drug design. Also, this was one of the first examples where new lead compounds for other diseases were revealed from side effects observed during pharmacological and clinical studies (see Section 2.1.B.4, p. 15). These early studies led to the development of new antidiabetic and diuretic agents. Another important scientific advance that was derived from work with sulfonamides was a simple method for the assay of these compounds in body fluids and tissues.[78] Furthermore, it was shown that the antibacterial effect of sulfanilamide was proportional to its concentration in the blood, and that at a given dose this varied from patient to patient. This was the beginning of the monitoring of blood drug levels during chemotherapy treatment, and led to the initiation of the routine use of *pharmacokinetics*, the study of the absorption, distribution, and excretion of drugs, in drug development programs. Proper drug dosage requirements could now be calculated.

c. Mechanism of Action

On the basis of the work by Stamp,[79] who showed that bacteria and other organisms contained a heat-stable substance that inhibited the antibacterial action of sulfonamides, Woods[80] in 1940 reported a breakthrough in the determination of the mechanism of action of this class of drugs. He hypothesized that because enzymes are inhibited by compounds whose structures resemble those of their substrates, the inhibitory substance should be a substrate for an essential enzyme, and it should have a structure similar to that of sulfanilamide. After various chemical tests, and a vague notion of the possible structure of this inhibitory substance, he deduced that it must be *p*-aminobenzoic acid (**5.49**), and proceeded to show that **5.49** was a potent inhibitor of sulfanilamide-induced bacteriostasis. The results of his experiments showed that sulfanilamide was competitive with *p*-aminobenzoic acid for microbial growth. To maintain growth with increasing concentrations of sulfanilamide, it is necessary also to increase the concentration of *p*-aminobenzoic acid. Selbie[81] found that coadministration of *p*-aminobenzoic acid and sulfanilamide into streptococcal-infected mice prevented the antibacterial action of the drug.

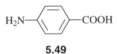

5.49

The observation of competitive inhibition by sulfanilamide was the basis for Fildes[82] to propose his theory of *antimetabolites*, compounds that block enzymes in metabolic pathways. He proposed a rational approach to chemotherapy, namely, enzyme inhibitor design, and suggested that the molecular basis for enzyme inhibition was that either the inhibitor combines with the enzyme and displaces its substrate or coenzyme or it combines directly with the substrate or coenzyme. In the mid-1940s Miller and coworkers[83] demonstrated that sulfanilamide inhibited folic acid biosynthesis, and in 1948 Nimmo-Smith *et al.*[84] showed that the inhibition of folic acid biosynthesis by sulfonamides was competitively reversed by *p*-aminobenzoic acid. Two enzymes from *Escherichia coli* were purified by Richey and Brown,[85] one that catalyzed the diphosphorylation of 2-amino-4-hydroxy-6-hydroxymethyl-7,8-dihydropteridine (**5.50**) and the other (dihydropteroate synthase) that catalyzed the synthesis of dihydrofolate (**5.52**) from diphosphate **5.51** and *p*-aminobenzoic

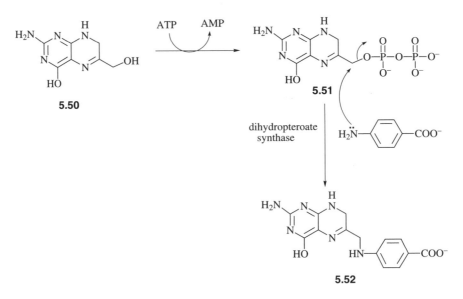

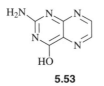

Scheme 5.5 ▶ Biosynthesis of bacterial dihydrofolic acid

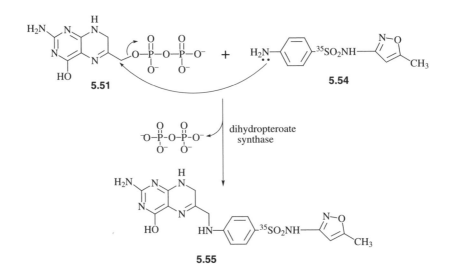

Scheme 5.6 ▶ Dihydropteroate synthase use of sulfamethoxazole in place of *para*-aminobenzoic acid

acid (Scheme 5.5). The name of the enzyme stems from the fact that folic acid is a derivative of the pterin ring system (**5.53**). Because of the structural similarity of sulfanilamide to *p*-aminobenzoic acid, it is a potent competitive inhibitor of the second enzyme. The reversibility of the inhibition was demonstrated by Weisman and Brown,[86] who suggested that sulfonamides were incorporated into the dihydrofolate. This was verified by Bock *et al.*,[87] who incubated dihydropteroate synthase with diphosphate **5.51** and [^{35}S]-sulfamethoxazole (**5.54**) and identified the product as **5.55** (Scheme 5.6). Therefore, this is an example of competitive reversible inhibition in which the inhibitor also is a substrate. However, the product (**5.55**) cannot produce dihydrofolate, and, therefore, the organism cannot get the tetrahydrofolate needed as a coenzyme to make purines (see Chapter 4, Section 4.3.B, p. 200), which are needed for DNA biosynthesis. This is why the sulfonamides are bacteriostatic, not bacteriocidal. Inhibition of tetrahydrofolate biosynthesis only inhibits replication; it does not kill the existing bacteria.

Inhibitors of dihydropteroate synthase, however, have no effect on humans, because we are incapable of biosynthesizing folic acid and, therefore, do not have that enzyme. Folic acid is a vitamin and must be eaten by humans. Furthermore, because bacteria biosynthesize their folic acid, they do not have a transport system for it.[88] Consequently, we can eat all the folic acid we want, and the bacteria cannot utilize it. This is another example of *selective toxicity*, inhibition of the growth of a foreign organism without affecting the host, and falls into the category of an ideal enzyme inhibitor (see Section 5.1). It is interesting to note that sulfonamides are not effective with pus-forming infections because pus contains many compounds that are the end products of tetrahydrofolate-dependent reactions, such as purines, methionine, and thymidine. Therefore, inhibition of folate biosynthesis is unimportant, and pus can contribute to bacterial sustenance.

Dihydropteroate synthase satisfies at least three of four important criteria for a good antimicrobial drug target: (1) The target is essential to the survival of the microorganism; (2) the target is unique to the microbe, so its inhibition does not harm humans; and (3) the structure and function of the target is highly conserved across a variety of species of that microbe so that inhibitors are broad spectrum agents.[89] The fourth criterion may be the most difficult to attain, namely, that resistance to inhibitors of the target not be easily acquired. Typically, it takes between 1 and 4 years for resistance to an antibacterial drug to emerge; in the case of the sulfonamides, it was almost 7 years.[90]

d. Drug Resistance

One major limitation to the use of sulfonamide antibacterial drugs is the development of drug resistance (see Section 5.2). There are principally three mechanisms of sulfonamide drug resistance. One mechanism is that organisms can overproduce *p*-aminobenzoic acid.[91] A second mechanism is the result of a plasmid-mediated synthesis of a less sensitive dihydropteroate synthase, one that binds *p*-aminobenzoic acid normally, but binds sulfonamides several thousand times less tightly than the normal enzyme can.[92] The third mechanism involves altered permeability to the sulfonamides.[93]

e. Drug Synergism

Combination therapy of sulfadoxine (**5.56**, Fansidar) with pyrimethamine (**5.57**, Daraprim) or sulfamethoxazole (**5.54**, Gantanol) with trimethoprim (**5.6**, Proloprim)[94] has been shown to be quite effective for the treatment of malaria and bacterial infections, respectively. Both of these combinations are examples of a sequential blocking mechanism[95] (see Section 5.3.B.2, p. 240). The sulfa drugs **5.54** and **5.56** inhibit dihydropteroate synthase, which catalyzes the synthesis of dihydrofolate, and **5.6** and **5.57** inhibit dihydrofolate reductase, which catalyzes the synthesis of tetrahydrofolate from dihydrofolate (see Section 4.3.B). Trimethoprim has the desirable property of being a very tight-binding inhibitor of bacterial dihydrofolate reductase, but a poor inhibitor of mammalian dihydrofolate reductase; the IC_{50} is 5 nM for the former and 2.6×10^5 nM for the latter.[96]

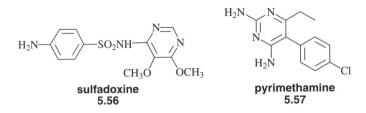

sulfadoxine
5.56

pyrimethamine
5.57

5.4.C Transition State Analogs and Multisubstrate Analogs

C.1 Theoretical Basis

As discussed in the last chapter (Section 4.1.B.2, p. 178), an enzyme accelerates the rate of a reaction by stabilizing the transition state, which lowers the free energy of activation. The enzyme achieves this rate enhancement by changing its conformation so that the strongest interactions occur between the substrate and enzyme active site *at the transition state* of the reaction. Some enzymes act by straining or distorting the substrate toward the transition state. The catalysis-by-strain hypothesis led to early observations that some enzyme inhibitors owe their effectiveness to a resemblance of the strained species. Bernhard and Orgel[97] theorized that inhibitor molecules resembling the transition state species would be much more tightly bound to the enzyme than would be the substrate; 11 years prior to that Pauling had mentioned that the best inhibitor of an enzyme would be one that resembled the "activated complex."[98] Therefore, why should we design inhibitors based on the ground state substrate structure? A potent enzyme inhibitor should be a stable compound whose structure resembles that of the substrate at a *postulated transition state* (or transient intermediate) of the reaction rather than that at the ground state. A compound of this type would bind much more tightly to the enzyme, and is called a *transition state analog inhibitor*. Jencks[99] was the first to suggest the existence of transition state analog inhibitors, and cited several possible literature examples; Wolfenden[100] and Lienhard[101] developed the concept further. Values for dissociation constants (K_i) of 10^{-15} M for enzyme–transition state complexes may not be unreasonable given the normal range of 10^{-3}–10^{-5} M for dissociation constants of enzyme–substrate complexes (K_m).

To design such an inhibitor, the mechanism of the enzyme reaction must be understood, so that a theoretical structure for the substrate at the transition state can be hypothesized. Because many enzyme-catalyzed reactions have similar transition states (for example, the different serine proteases), the basic structure of a transition state analog for one enzyme can be modified to meet the specificity requirements of another enzyme in the same mechanistic class, and, thereby, generate a transition state analog for the other enzyme. This modification may be as simple as changing an amino-acid in a peptidyl transition state analog inhibitor for one protease so that it conforms to the peptide specificity requirement of another protease. This is the approach that was taken to obtain serine protease specificity for the peptidyl trifluoromethyl ketone inhibitors of Abeles and coworkers[102] (see Section 5.4.D.4, p. 267). Christianson and Lipscomb[103] have renamed those reversible inhibitors that undergo a bond-forming reaction with the enzyme prior to the observation of the enzyme-inhibitor complex as *reaction coordinate analogs*. The peptidyl trifluoromethyl ketones (see Section 5.4.D.4,

p. 267) would be an example of this type of inhibitor. When more than one substrate is involved in the enzyme reaction, a single stable compound can be designed that has a structure similar to that of the two or more substrates at the transition state of the reaction. This special case of a transition state analog is termed a *multisubstrate analog inhibitor*. Because these compounds are a combination of two or more substrates, their structures are unique, and they are often highly specific. Their great binding affinity for the target enzyme arises because the free energy of binding is roughly the product of the free energies of each independent substrate that it mimics.

C.2 Transition State Analogs

a. Enalaprilat

In Section 5.4.B.1 enalaprilat was shown to be a very potent competitive reversible inhibitor of angiotensin-converting enzyme. The rationalization for its binding effectiveness was that it had multiple binding interactions with the substrate and product binding sites. The resemblance of enalaprilat to the substrate and products of the enzyme reaction (Figure 5.9, p. 249) supported this notion. Both of these rationalizations are ground state arguments, but transition state theory suggests that the most effective interactions occur at the transition state of the reaction. With that in mind, let's consider a potential mechanism for angiotensin-converting enzyme-catalyzed substrate hydrolysis (Scheme 5.7) and see if the transition state structure is relevant. It is not known if a general base mechanism, as shown in Scheme 5.7, or a covalent catalytic mechanism is involved. Enalaprilat has been drawn beneath transition states 1 and 2 ($\ddagger_1$ and $\ddagger_2$) to show how the structures are related. An enzyme conformational change at the transition state (shown as a blocked enzyme instead of a rounded enzyme in Scheme 5.7) could increase the binding interactions. The resemblance of enalaprilat to the transition state structure also could account for the observation that it is a slow, tight-binding inhibitor of the enzyme (see Section 5.4.D.2, p. 263).

b. Pentostatin

2'-Deoxycoformycin (pentostatin, **5.58**, Nipent), an antineoplastic agent isolated from fermentation broths of the bacterium *Streptomyces antibioticus*, is a potent inhibitor of the enzyme adenosine deaminase (adenosine aminohydrolase).[104] The K_i is 2.5×10^{-12} M, which is 10^7 times lower than the K_m for adenosine! As in the case of enalaprilat (Section 5.4.B.1 and 5.4.C.2.a, pp. 242 and 259), pentostatin is a slow, tight-binding inhibitor (see Section 5.4.D).[105] The k_{on} with human erythrocyte adenosine deaminase is 2.6×10^6 M^{-1} s^{-1} and the k_{off} is 6.6×10^{-6} s^{-1}. The very small k_{off} value is reflected in the very low K_i value.

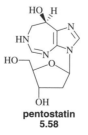

pentostatin
5.58

Pentostatin is an analog of the natural nucleoside 2'-deoxyinosine in which the purine is modified to contain a seven-membered ring with two sp^3 carbon atoms. It is believed

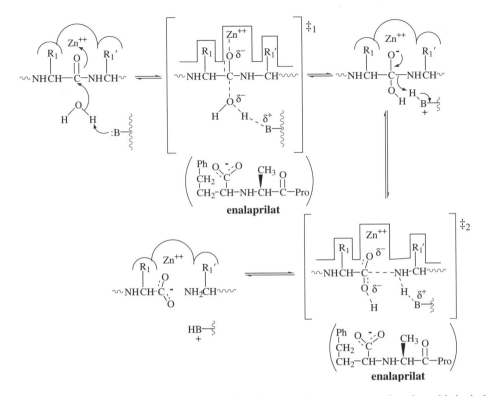

Scheme 5.7 ► Hypothetical mechanism for angiotensin-converting enzyme-catalyzed peptide hydrolysis

that this compound mimics the transition state structure of the substrates adenosine and 2′-deoxyadenosine (**5.59**, Scheme 5.8) during their hydrolysis to inosine and 2′-deoxyinosine (**5.62**), respectively, by adenosine deaminase. A crystal structure to 2.4-Å resolution of the transition state analog 6*R*-hydroxy-1,6-dihydropurine ribonucleoside (**5.63**), complexed to adenosine deaminase, revealed a zinc ion cofactor and suggested a possible mechanism for the enzyme.[106] In this case the resemblance of pentostatin (**5.58**) to the intermediate **5.61** (Scheme 5.8) is clearer than its similarity to the transition state shown (**5.60**); however, a late transition state would look more like **5.61**.

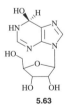

5.63

It is not clear why inhibition of this enzyme should result in selective lymphotoxicity. One hypothesis is that 2′-deoxyadenosine accumulates, which, in turn, is an inhibitor of ribonucleotide reductase and *S*-adenosylhomocysteine hydrolase.[107] Ribonucleotide reductase, which catalyzes the conversion of ribonucleotides to the corresponding 2′-deoxyribonucleotides, is essential for DNA biosynthesis. Inhibition of this enzyme leads to inhibition of DNA biosynthesis. *S*-Adenosylhomocysteine competitively inhibits most of the methyltransferases that utilize *S*-adenosylmethionine as the methyl donating agent. This, apparently, is a mechanism for the regulation of these methyltransferases. Inhibition of

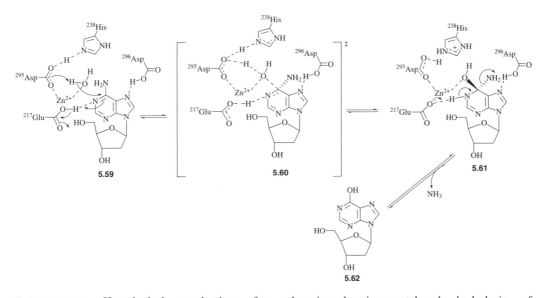

Scheme 5.8 ▶ Hypothetical mechanism for adenosine deaminase-catalyzed hydrolysis of 2′-deoxyadenosine

S-adenosylhomocysteine hydrolase, the enzyme that degrades *S*-adenosylhomocysteine, results in an accumulation of *S*-adenosylhomocysteine, which inhibits the growth and replication of various tumors (and viruses), particularly those requiring a methylated 5′-cap structure on their messenger ribonucleic acids (mRNAs). Furthermore, various lymphocytic functions are suppressed by the accumulation of extracellular adenosine.

A major obstacle to the success of *antipurines*, enzyme inhibitors that mimic purines and block their metabolism, is acquired resistance. Unlike adenosine and various other 2′-deoxyribonucleosides, which are converted directly to the corresponding 5′-monophosphates by nucleoside kinases, a similar reaction does not occur with inosine or 2′-deoxyinosine. Instead, they are converted by purine nucleoside phosphorylase to hypoxanthine (2′-deoxyinosine without the sugar moiety), which is transformed into the corresponding nucleotide by hypoxanthine-guanine phosphoribosyltransferase (HGPRT). The most common mechanism for antipurine drug resistance is a lack of the enzyme HGPRT or is the result of an altered HGPRT that binds the substrates poorly.

Animals treated with pentostatin show marked immunosuppression. Synergistic effects are observed when pentostatin is used in combination with other antipurines, especially the antiviral drug vidarabine (ara-A; **5.64**) which is degraded by adenosine deaminase. Although pentostatin has induced remissions of nodular lymphomas and lymphocytic leukemia, some unexplained deaths occurred in clinical trials; consequently, further study has been restricted.[108]

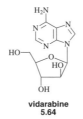

vidarabine
5.64

c. Multisubstrate Analogs

One of the first steps in the *de novo* biosynthesis of pyrimidines is the condensation of carbamoyl phosphate (**5.65**) and *L*-aspartic acid, catalyzed by aspartate transcarbamylase, which produces *N*-carbamoyl-*L*-aspartate (**5.66**, Scheme 5.9). Below the transition state structure is drawn *N*-phosphonoacetyl-*L*-aspartate (**5.67**, PALA), which is a stable compound (the isosteric exchange of a CH_2 for the O prohibits loss of the PO_3^{-2} moiety) that resembles the transition state for condensation of the two substrates.[109]

PALA was ineffective in clinical trials as a result of tumor resistance.[110] When tumor cells acquired the ability to utilize preformed circulating pyrimidine nucleosides, they no longer needed to have a *de novo* metabolic pathway for pyrimidines. Other mechanisms of resistance are increased carbamoyl phosphate or aspartate transcarbamylase production. To overcome resistance to PALA, nitrobenzylthioinosine, which inhibits the diffusion of nucleoside transport and should block the uptake of the preformed pyrimidines, was tested for its synergistic effect with PALA. These two compounds were synergistic *in vitro*, but were too toxic at effective dosages to be used *in vivo*.

5.4.D Slow, Tight-Binding Inhibitors

D.1 Theoretical Basis

As indicated above, the equilibrium between an enzyme and a reversible inhibitor is typically established rapidly (generally, diffusion controlled). With *slow-binding inhibitors*, however, the equilibrium between enzyme and inhibitor is reached slowly, and inhibition is time dependent, reminiscent of the kinetics for irreversible inhibition (see Section 5.5.B.1, p. 275). *Tight-binding inhibitors* are those inhibitors for which substantial inhibition occurs when the concentrations of inhibitor and enzyme are comparable.[111] *Slow, tight-binding inhibitors* have both properties. These inhibitors can bind noncovalently[112] or covalently;[113] when a covalent bond is formed, a slowly reversible adduct may be involved. Noncovalent slow, tight-binding inhibitors are the bridge between rapidly reversible and covalent, irreversible inhibitors. Depending on the tightness of binding, these inhibitors can become functionally equivalent to covalent, irreversible inhibitors with half-lives (time for half of the E·I complex to break down) of hours, days, or even months!

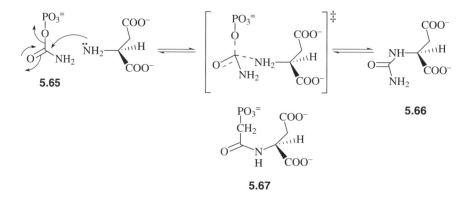

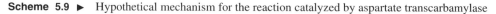

Scheme 5.9 ▶ Hypothetical mechanism for the reaction catalyzed by aspartate transcarbamylase

The reason for slow-binding inhibition is not known. One possibility is that these inhibitors are such good analogs of the substrate that they induce a conformational change in the enzyme which resembles that associated with the transition state;[114] typically these compounds are transition-state analogs (see Section 5.4.C, p. 258). If this is the case, then inhibitor binding would be slow because it does not have all of the essential structural features of the substrate transition state geometry. The dissociation would be even slower because the dissociation rate is not enhanced by product formation. The conformational change may result from a change in the protonation state of the enzyme[115] or from the displacement of an essential water molecule by the inhibitor.[116] To differentiate simple competitive inhibition from slow-binding inhibition, it is necessary to carry out kinetic studies. Many simple competitive inhibitors may turn out to be slow, tight-binding inhibitors after the kinetic analysis.

D.2 Enalaprilat

The interaction of enalaprilat with purified rabbit lung angiotensin-converting enzyme was studied, and the kinetics indicate that it is a slow, tight-binding inhibitor.[117] The rate constant for the formation of the E·I complex (k_{on}) at pH 7.5 was determined to be 2×10^6 $M^{-1}sec^{-1}$, which is at least two orders of magnitude smaller than expected for a diffusion-controlled reaction. Steady-state kinetics gave the value of the K_i as 1.8×10^{-10} M. Since $K_i = k_{off}/k_{on}$, the k_{off} should be 3.6×10^{-4} sec^{-1}, which is in satisfactory agreement with the measured k_{off} value of 1.6×10^{-4} sec^{-1}. The small k_{off} value for this noncovalent E·I complex emphasizes the strong affinity of enalaprilat for angiotensin-converting enzyme.

D.3 Lovastatin and Simvastatin, Antihypercholesterolemic Drugs

An example of noncovalent slow, tight-binding inhibitors is a family of drugs used to treat high cholesterol levels.

a. Cholesterol and Its Effects

Coronary heart disease is the leading cause of death in the United States and other Western countries; about one-half of all deaths in the United States can be attributed to atherosclerosis,[118] which results from the buildup of fatty deposits called plaque on the inner walls of arteries. The major component of atherosclerotic plaque is cholesterol. In humans more than one-half of the total body cholesterol is derived from its *de novo* biosynthesis in the liver.[119] Cholesterol biosynthesis requires more than 20 enzymatic steps starting from acetyl CoA. The rate-determining step is the conversion of 3-hydroxy-3-methylglutaryl coenzyme A (HMG-CoA; **5.68**) to mevalonic acid (**5.69**), catalyzed by HMG-CoA reductase (Scheme 5.10). For the structure of coenzyme A, see Chapter 4, **4.13b**. Because hypercholesterolemia is a primary risk factor for coronary heart disease,[120] and the overall rate of cholesterol biosynthesis is a function of this enzyme, efforts were initiated to inhibit HMG-CoA reductase as a means of lowering plasma cholesterol levels.

b. Lead Discovery

Endo and coworkers[121] at the Sankyo Company in Tokyo tested 8000 strains of microorganisms for metabolites that inhibited sterol biosynthesis *in vitro*, and discovered three active

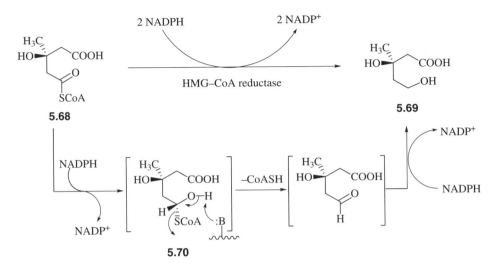

Scheme 5.10 ▶ HMG-CoA reductase, the rate-determining enzyme in *de novo* cholesterol biosynthesis

compounds in the culture broths of the fungus *Penicillium citrinum*. The most potent compound, called mevastatin, also was isolated from broths of *Penicillium brevicompactum* by Brown and coworkers[122] at Beecham Pharmaceuticals in England, who named it compactin (**5.71**, R = H). A second, more potent, compound was isolated by Endo[123] from the fungus *Monascus ruber*, which he named monacolin K; the same compound was isolated by a group at Merck from *Aspergillus terreus*,[124] which they named mevinolin (**5.71**, R = CH$_3$). Mevinolin is now known as lovastatin (Mevacor). Several related metabolites also were isolated from cultures of these fungi,[125] including dihydrocompactin from *P. citrinum* (**5.72**, R = H, R′ = (S)-CH$_3$CH$_2$CH(CH$_3$)CO$_2^-$), dihydromevinolin from *A. terreus* (**5.72**, R = CH$_3$, R′ = (S)-CH$_3$CH$_2$CH(CH$_3$)CO$_2^-$), and dihydromonacolin L from a mutant strain of *M. ruber* (**5.72**, R = CH$_3$, R′ = H).

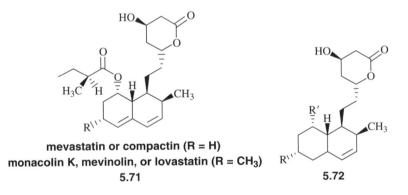

mevastatin or compactin (R = H)
monacolin K, mevinolin, or lovastatin (R = CH$_3$)
5.71

5.72

c. Mechanism of Action

Compactin[126] and lovastatin are potent competitive reversible inhibitors of HMG-CoA reductase. The K_i for compactin is 1.4×10^{-9} M and for lovastatin is 6.4×10^{-10} M (rat liver enzyme); for comparison, the K_m for HMG-CoA is about 10^{-5} M. Therefore, the affinity of

HMG-CoA reductase for compactin and lovastatin is 7,140 and 16,700 times, respectively, *greater* than for its substrate. Compactin and lovastatin do not affect any other enzyme in cholesterol synthesis except HMG-CoA reductase.

It may not be immediately obvious why lovastatin is a competitive reversible inhibitor of HMG-CoA reductase, given it should resemble the structure of the substrate or product to be competitive, and it does not appear to mimic either. The reason is that the active form is not that shown in **5.71**, but rather, the hydrolysis product, that is, the open chain 3,5-dihydroxyvaleric acid form (**5.73**). This form mimics the structure of the proposed intermediate **5.70** (Scheme 5.10) in the reduction of HMG-CoA by HMG-CoA reductase, that is, if the structure of CoA (see Chapter 4, **4.13b**) is taken into account. Enzyme studies with compactin and analogs indicate that there are two important binding domains at the active site, the hydroxymethylglutaryl (HMG) binding domain to which the upper part of **5.73** binds, and a hydrophobic pocket located adjacent to the active site to which CoA and the decalin (lower) part of **5.73** bind.[127] The high affinity of compactin and its analogs to HMG-CoA reductase derives from the simultaneous interactions of the two parts of these inhibitors with the two binding domains on the enzyme. This is the same phenomenon that was responsible for the increased binding of two inhibitors attached by a linker in SAR by NMR (see Section 2.2.E.6, p. 44). As a result of the interactions in two adjacent binding pockets, dissociation of the E·I complex is very slow. A kinetic analysis of the on and off rate constants (k_{on} and k_{off}, respectively; see Scheme 5.3) for HMG-CoA and compactin (that is, the rate constants for the binding and dissociation of compactin to HMG-CoA reductase) with yeast HMG-CoA reductase were 1.9×10^5 M^{-1} sec^{-1} and 0.11 sec^{-1} for HMG-CoA and 2.7×10^7 M^{-1} sec^{-1} and 6.5×10^{-3} sec^{-1} for compactin, respectively. Therefore, compactin binds faster and dissociates slower than does HMG-CoA, which accounts for the difference in their K_m and K_i (k_{off}/k_{on}) values. It, also, may be possible to classify these inhibitors as transition state analogs (see Section 5.4.C, p. 258).

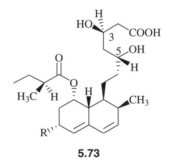

5.73

d. Lead Modification

Numerous structural modifications were made on compactin and lovastatin to determine the importance of the lactone moiety and its stereochemistry, the ability of the lactone moiety to be opened to the dihydroxy acid, the optimal length and structure of the moiety bridging the lactone and the lipophilic groups, and the size and shape of the lipophilic group (i.e., the SAR). It was found that potency was greatly reduced[128] unless a carboxylate anion could be formed and the hydroxyl groups were left unsubstituted in an erythro relationship. Insertion of a bridging unit other than ethyl or (*E*)-ethenyl between the 5-carbinol moiety and the lipophilic moiety also diminishes the potency. Modifications of the lower (lipophilic) part of compactin in most cases led to compounds with considerably lower potencies, except for certain substituted biphenyl analogs.[129] If the substituted biphenyl rings were constrained as fluorenylidine moieties, the corresponding potencies decreased.[130] When the hydroxyl group in the lactone ring was replaced by an amino or thiol group, diminished potencies were observed.[131]

Modification of the 2(S)-methylbutyryl ester side chain ($CH_3CH_2CH(CH_3)CO_2^-$) of lovastatin indicated that introduction of an additional aliphatic group on the carbon α to the ester carbonyl group increased the potency of lovastatin.[132] To block ester hydrolysis and produce a compound with a much longer plasma half-life, a second methyl group was added to the side chain of the lower half of lovastatin (**5.73**, but with ester side chain $CH_3CH_2C(CH_3)_2CO_2^-$) called simvastatin (Zocor), which also has a potency about 2.5 times greater than that of lovastatin. The lactone epimer of lovastatin (the epimer of the carbon adjacent to the lactone oxygen in **5.71**) has less than 10^{-4} times the potency of lovastatin.[133] Modifications in the 3,5-dihydroxyvaleric acid moiety of analogs of **5.73** resulted in lower potencies, except when the 5-hydroxyl group was replaced by a 5-keto group, in which case potencies comparable to the parent compounds were observed. Presumably, the 5-keto group becomes reduced, but it is not known if that occurs by HMG-CoA reductase and, if it does, whether it occurs prior to inhibition or whether it is the cause for inhibition.

The generic names of this family of drugs end in the suffix -*statin*, so they are known collectively as the *statins*. The early analogs, such as compactin, lovastatin, and simvastatin, are classified as *type 1 statins*; the newer analogs, known as *type 2 statins*, retain the HMG mimic of the molecule, but the CoA mimic has been modified. Cerivastatin (**5.74**, Baycol), a very potent anticholesterol drug (0.1–0.15 mg taken once a day!), was launched by Bayer in 1999 but had to be recalled from the drug market in 2001 because of its association with deaths of some patients taking it who developed rhabdomyolysis, a side effect involving muscular weakness.

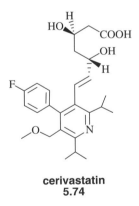

**cerivastatin
5.74**

Although side effects generally have a negative connotation, another "side effect" of both the type 1 and type 2 statins is an enhancement in new bone formation in rodents.[134] This new bone growth is associated with an increased expression of the bone morphogenetic protein-2 gene in bone cells, which may have implications in the treatment of osteoporosis. The activity of these compounds was identified from a random screen of a collection of 30,000 natural product compounds.

The X-ray crystal structures of several of the statins bound to HMG-CoA reductase confirm the biochemical hypothesis, and show that they occupy a portion of the HMG-CoA binding site which then blocks access of the substrate to the active site.[135]

D.4 Peptidyl Trifluoromethyl Ketone Inhibitors of Human Leukocyte Elastase

An example of a covalent, slow, tight-binding inhibitor is a family of trifluoromethyl ketones that inhibit human leukocyte elastase. Human leukocyte elastase and cathepsin G are serine proteases that are released normally by the immune system neutrophils in the lungs to digest dead lung tissue and destroy invading bacteria. Natural inhibitors of these enzymes (α_1-protease inhibitor[136] and bronchial mucous inhibitor)[137] also are released to prevent these enzymes from destroying the key structural protein component of the lung, elastin, and lung connective tissue. It is hypothesized that an imbalance in the protease and protease inhibitor concentrations (the *protease/antiprotease hypothesis*)[138] may be the cause for pulmonary emphysema,[139] cystic fibrosis,[140] and chronic bronchitis.[141] An imbalance could arise because of a genetic deficiency in α_1-protease inhibitor or from inhalation of cigarette smoke which oxidizes Met-358 at the active site of the inhibitor.[142]

A drug design approach that would return the imbalance to normal would be to discover an inhibitor of leukocyte elastase and cathepsin G as a mimic of the action of their natural antiproteases.[143] Chemists at AstraZeneca Pharmaceuticals developed a class of peptidyl trifluoromethyl ketones with the structures X-Val-CF$_3$, X-Pro-Val-CF$_3$, X-Val-Pro-Val-CF$_3$, and X-Lys(Z)-Val-Pro-Val-CF$_3$, where X is N-(methoxysuccinyl) and Z is N-(carbobenzoxy).[144] The most potent analog, Z-Lys(Z)-Val-Pro-Val-CF$_3$, had a $K_i < 10^{-10}$ M. All of these compounds were shown to be competitive slow, tight-binding inhibitors of human leukocyte elastase. The kinetic constants for the most potent analog were $k_{on} = 8 \times 10^4$ M^{-1} s^{-1} and $k_{off} = <10^{-5}$ s^{-1} (obtained from the equation $k_{off} = k_{on}K_i$). Numerous orally active and bioavailable analogs of these trifluoromethyl ketones were synthesized,[145] and **5.75** went to clinical trials, but was later abandoned.

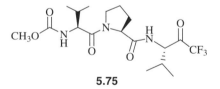

5.75

The design of peptidyl trifluoromethyl ketones as inhibitors of serine proteases stems from the work of Abeles and coworkers,[146] who suggested that because trifluoromethyl ketones exist almost exclusively as the hydrate in water and because they enhance nucleophilic addition to the carbonyl, then fluoroketones may form a reasonably stable, covalent, but reversible, hemiketal with the active site serine of a serine protease (Scheme 5.11). This inhibition mechanism allows for various possible conformational changes in the enzyme that may occur when the enzyme is in different protonation forms. The (E·I)$'$ complex may involve a reorientation of the inhibitor in the active site to allow for more efficient attack of the serine at the trifluoromethyl carbonyl. According to this hypothetical mechanism, the covalently bound inhibitor only dissociates when the active site imidazole is protonated. This could account for the small k_{off}. These inhibitors are an example of *covalent* reversible inhibition. Before we discuss covalent irreversible inhibitors, let's look at a case history for the design of a competitive reversible inhibitor.

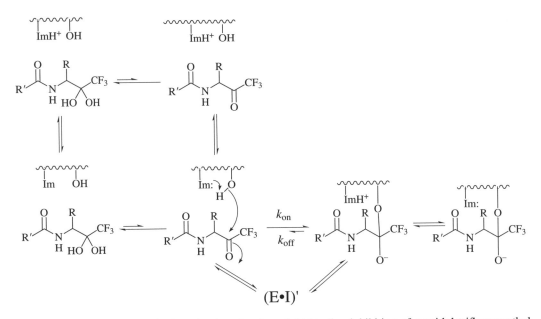

Scheme 5.11 ▶ Hypothetical mechanism for slow, tight-binding inhibition of peptidyl trifluoromethyl ketones with serine proteases. Im is the imidazole of histidine

5.4.E Case History of Rational Drug Design of an Enzyme Inhibitor: Ritonavir

The genome of the human immunodeficiency virus-1 (HIV-1) encodes an aspartate protease (HIV-1 protease), which proteolytically processes the *gag* and *gag-pol* gene products into mature, functional proteins.[147] If these processing steps are blocked, the progeny virions are immature and noninfectious. Consequently, HIV-1 protease should be an important target for design of inhibitors to act as potential anti-AIDS drugs. Several companies discovered very potent and effective inhibitors by various lead discovery/modification approaches. What follows is the approach taken at Abbott Laboratories leading to ritonavir, a potent HIV-1 protease inhibitor with high oral bioavailability that was shown to be an effective drug for the treatment of AIDS. Similar approaches were taken by other pharmaceutical companies as well.

The general approach is first to identify molecules that have good potency (preferably in the nanomolar range), then use those molecules as starting points for solving pharmacokinetic problems. The potency must be monitored constantly while the pharmacokinetic issues are being dealt with.

E.1 Lead Discovery

HIV protease is an unusual enzyme because the *homodimer* (two identical polypeptides come together to form the active enzyme) has C_2 symmetry (a 180° rotation about an axis through the center gives the same structure). This symmetry element was used as a key starting point in the design of novel inhibitor structures.[148] The initial plan was to design C_2-symmetrical compounds that should show selectivity for HIV protease over other mammalian aspartate

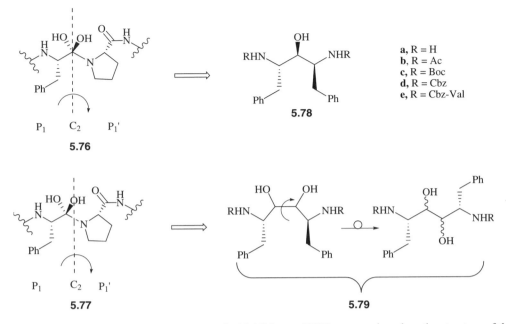

Figure 5.10 ▶ Lead design for C_2-symmetrical inhibitors of HIV protease based on the structure of the tetrahedral intermediate during hydrolysis of HIV protease

proteases because of the lack of symmetry with other aspartate proteases. The other design element was to make a transition state analog (or, more accurately, an intermediate analog). The proposed tetrahedral intermediate structure for the hydrolysis of a good asymmetric substrate, such as -Phe-Pro-, was bisected either through the scissile carbon (**5.76**, Figure 5.10) or adjacent to it (**5.77**). However, to make a C_2 symmetric compound, the two amino acid residues must be the same. Because the *P region* (the *N*-terminal side of the scissile bond) had been shown to be more important than the *P′ region* (the *C*-terminal side of the scissile bond), the *P′ region* was deleted, and the proline was substituted by phenylalanine. A C_2 symmetry operation was performed on the remainder of the substrate, generating two possible lead compounds, **5.78a** and **5.79a** (Figure 5.10), respectively. The amino groups were acylated with a variety of *N*-protecting groups to increase lipophilicity, including Ac (**b**), Boc (**c**), and Cbz (**d**). Compound **5.78b** was very weakly inhibitory; the stereoisomers of **5.79c**, however, were good inhibitors.

E.2   Lead Modification

To increase the pharmacophore of these inhibitors, P_2/P_2' residues were added. The Cbz-Val analogs, **5.78e** and **5.79e**, were low nanomolar and subnanomolar inhibitors, respectively; those analogs with structure **5.79** were generally 10 times more potent than **5.78** analogs. Varying the stereochemistry of the hydroxyl groups in **5.79e** had little or no effect on inhibition. Compounds **5.79e** (different stereochemistries) were potent *in vitro* inhibitors of HIV-1 protease in H9 cells (IC$_{50}$ values 20–150 nM). The therapeutic indices were in the range 500–5000, which is much better than could be hoped for when designing agents for life-threatening diseases.

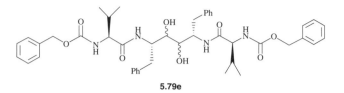

5.79e

At this point in the lead modification process, relatively potent inhibitors have been identified, so efforts can concentrate on pharmacokinetic problems. Because of generally poor pharmacokinetic behavior of peptides, peptidomimetics are generally sought (see Section 2.2.E.7, p. 47). However, peptidomimetics also can suffer from low oral/intestinal absorption and rapid hepatic elimination.[149] The major causes for poor peptidomimetic pharmacokinetic profiles are high molecular weights, low aqueous solubility, susceptibility to proteolytic degradation, hepatic metabolism, and biliary extraction.

The aqueous solubility of **5.78e** and **5.79e** was very poor. A crystal structure of HIV-1 protease with **5.78e** bound[150] indicated that the Cbz groups were exposed; that is, they were not interacting with the protein as part of the pharmacophore, but they also were not interfering with protein binding. As discussed in Section 2.2.A, these are ideal groups to modify, because it is likely that modification will not affect potency. The crystal structure showed that there was room for structural modification, so the terminal Cbz phenyl groups were modified with polar, heterocyclic bioisosteres, such as pyridinyl (at the 2-, 3-, and 4-positions of the pyridinyl ring) and thiazolyl (at the 2- and 4-positions) groups.[151] The P_1 phenyl (from Phe) and P_2 isopropyl (from Val) groups were embedded in lipophilic pockets, so modification of those groups would not be fruitful. Conversion of one of the Cbz phenyls to a pyridinyl group increased the aqueous solubility by 20-fold without a change in the inhibitory potency, but the water solubility was still less than 1 μg/mL. More basic nonaromatic heterocycles had much greater water solubilities, but the potencies dropped. Substitution of both Cbz phenyl groups by heteroaromatic groups had little effect on the inhibitory potencies, but gave dramatic increases in water solubilities. Some of the compounds showed oral bioavailabilities in the range of 10–20%, but these bioavailabilities did not exceed concentrations required for effective anti-HIV activity *in vitro*. Compound **5.80** had the best combination of inhibitory potency and aqueous solubility; unfortunately, it showed no oral bioavailability in rats. In general, the compounds with the **5.78** framework, although less potent than the compounds with the **5.79** framework, had consistently superior oral bioavailabilities in animals. To take advantage of both of these properties, the next series of analogs investigated had general structure **5.81**, which combined the extended framework of **5.79** with the mono alcohol structure of **5.78**.[152] Although these were still poorly bioavailable in rats, the **5.81** series was more potent than the **5.79** series (IC_{50} values were generally subnanomolar). With all of these modifications and still no oral bioavailability, it is time to consider Lipinski's Rule of Five (see Section 2.2.F.4, p. 65). One of his rules for oral bioavailability is that the molecular weight should not exceed about 500; the molecular weight of **5.80** is 794. To increase oral bioavailability, then, it was necessary to decrease the size of the molecules and the number of rotatable bonds. To do that, and retain the C_2 symmetry, both ends of **5.80** would have to be deleted, but that brings the molecule back to the **5.78** and **5.79** series. That was when the C_2-symmetry design had to be put aside, and unsymmetrically substituted derivatives (different A groups at the N and C termini) of **5.79** and **5.81** were investigated.[153] Several SAR observations were made that were incorporated into the later designs: (1) Incorporation of a carbamate

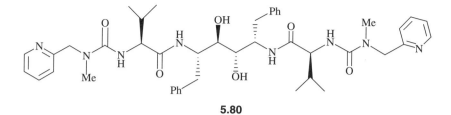

5.80

linkage (ROCONHR′) resulted in improved potency over the use of an *N*-alkylurea linkage (RNR′CONHR″); (2) an *N*-ethylurea linkage was less tolerated than *N*-methylurea; (3) there was no difference between 2-pyridinyl and 3-pyridinyl groups at P_3; and (4) methyl substitution on the P_3 pyridinyl group did not diminish potency and often enhanced it. With regard to the pharmacokinetics, there was little correlation between aqueous solubility and oral bioavailability! Compounds with an *N*-methylurea linkage between the pyridinyl group and the P_2 aminoacyl residue generally showed greater oral bioavailability and solubility than the ones with a carbamate linkage. Also, compounds in the **5.81** series exhibited greater oral bioavailability than those in the **5.79** series. It is not clear if this is because of absorption or biliary excretion; the additional hydrogen bonding of the diols results in lower intestinal absorption.

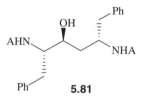

5.81

A good measure of the overall potential of HIV-1 protease inhibitors is the ratio of the plasma levels achieved *in vivo* (C_{max}) to the concentration required for anti-HIV activity *in vitro* (ED_{50}). The compound that emerged from this study, **5.82**, has a C_{max}/ED_{50} of 21.6 (4.11 μM/0.19 μM). The aqueous solubility is 3.2 μg/mL, the oral bioavailability is 32%, and the plasma half-life after a 5 mg/kg dose i.v. was 2.3 h. This molecule is appreciably smaller than those in the **5.78** and **5.79** series (but the molecular weight is 653, still quite high), yet it maintained the submicromolar *in vivo* antiviral activity. However, the relatively short plasma half-life prohibits the maintenance of the plasma levels sufficiently in excess of the ED_{95} for viral replication observed *in vitro*. A pharmacokinetic study of the metabolism of **5.82** indicated that the probable cause for the short plasma half-life was the production of three metabolites: the *N*-oxide of the 2-pyridinyl group, the *N*-oxide of the 3-pyridinyl group, and the bis(pyridine *N*-oxide). Using ^{14}C-labeled **5.82**, these metabolites corresponded to 92–95% of the total bile radioactivity. Obviously, the next modification had to be designed to minimize this metabolism. Attempts to hinder oxidative metabolism sterically by the addition of substituents at the 6-position of the pyridinyl group generally yielded compounds that showed lower C_{max} values and oral bioavailability, although they did have greater potency than **5.82**. Attempts to modify the electronic nature of the P_3 pyridinyl group by the addition of electron-donating groups, such as methoxyl or amino, led to more potent and more soluble analogs, but they had poorer pharmacokinetic profiles.

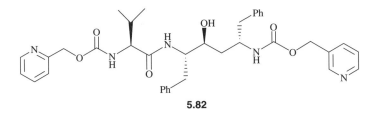

5.82

By diminishing the oxidation potential of the electron-rich pyridinyl groups, drug metabolism should be diminished; consequently, the pyridinyl groups were replaced with other heteroaromatic bioisosteres.[154] The more electron-deficient (less basic) heterocycles, however, also had lower aqueous solubility and oral absorption. 5-Pyrimidinyl substitution gave inhibitors of equal potencies to the pyridinyl compounds, but with lower bioavailability; the furanyl analogs were more potent, but showed even lower bioavailability. However, 5-thiazolyl analogs, which generally increase aqueous solubility, showed excellent pharmacokinetic properties, although their solubilities were still low.

Alkyl substitution on the P_3 heterocycle led to increased antiviral potency, but substitution on the P_2' heterocycle gave decreased potency. Later, the crystal structure of ritonavir bound to the enzyme showed that the alkyl group participated in a hydrophobic interaction with Val-82; this interaction was optimized with an isopropyl group. As noted above, though, an N-methylurea linkage between the P_3 and P_2' groups provided much higher solubility than the corresponding carbamate linkages. Consequently, P_3 N-methylurea analogs were made. Furthermore, it was found that in the N-methylurea series, but not the carbamate series, the regioisomeric position of the hydroxyl group became significant; compounds in which the hydroxyl group was distal to the P_2 valine residue (e.g., **5.83**) were 10-fold more potent than the corresponding analogs with the hydroxyl group proximal to the P_2 valine (**5.84**). By combining all of these characteristics, the best analog was found to be **5.85** (ritonavir, Norvir), which has a solubility of 6.9 μg/mL at pH 4, a C_{max}/ED_{50} of 105 (2.62 μM/0.025 μM), and an oral bioavailability of 78%. The in vitro K_i for HIV-1 protease is 15 pM! The metabolic reactivity of pyridinyl groups versus thiazolyl groups was elucidated by determining the metabolism of **5.82** versus **5.83** versus **5.85**. The relative rates of metabolism are **5.82** (1.0) > **5.83** (0.2) > **5.85** (0.05), as predicted. Oxazolyl and isoxazolyl analogs had profiles similar to that of ritonavir, both in potency and pharmacokinetic behavior, but plasma concentrations were maintained for shorter periods of time than ritonavir. It was later found that one additional reason for the improved pharmacokinetic properties of ritonavir is that it is a potent inhibitor of the 3A4 isozyme of cytochrome P450, the enzyme responsible for the oxidative metabolism of ritonavir.[155] Inhibition results from the binding of the unhindered nitrogen atom on the P_2' 5-thiazolyl group to the heme cofactor in the active site of CYP3A4.[156]

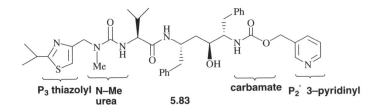

P₃ thiazolyl   N–Me urea   **5.83**   carbamate   P₂' 3–pyridinyl

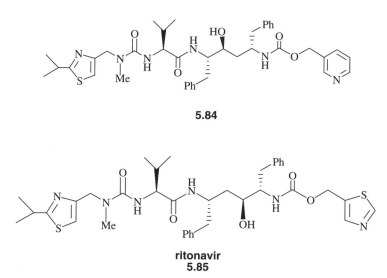

5.84

ritonavir
5.85

Because of the potency of inhibition of cytochrome P450 by ritonavir, it was found that combination of other HIV-1 protease inhibitors with ritonavir was synergistic. However, resistance to ritonavir developed, which was shown to be the result of a single-point mutation in HIV-1 protease, namely, Val-82 was mutated to Thr, Ala, or Phe, which interferes with the interaction of the isopropyl group attached to the thiazolyl group at P_3. To avoid this repulsive interaction with the mutant, the (3-isopropylthiazolyl)methyl group of ritonavir was excised (**5.86**). This inhibited the mutant enzyme, but only poorly. The pharmacophore was increased by a ring-chain transformation (**5.87**), and to increase potency further, the other thiazolyl group was replaced by a 2,6-dimethylphenyl group, giving lopinavir (**5.88**, Kaletra). Of course, you realize that removing the P2′ 5-thiazolyl group destroys the cytochrome P450 inhibition activity, and, in fact, the plasma half-life for lopinavir is low. To get around that problem, it was found that ritonavir could be added as an inhibitor of cytochrome P450, which protects lopinavir from metabolic degradation. A combination of lopinavir (to inhibit HIV-1 protease in the resistant strains) and ritonavir (to inhibit HIV-1 protease in susceptible strains plus to inhibit cytochrome P450), known as Kaletra, is the commercial product. In clinical trials with this combination, no HIV was detectable in the blood of the patients for 3 years.

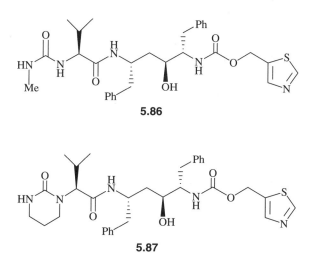

5.86

5.87

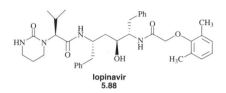

lopinavir
5.88

The general approach taken is to start by modifying the lead to increase potency for binding to the appropriate receptor (pharmacodynamics). Once this is accomplished, though, modifications to improve the pharmacokinetic behavior need to be carried out in conjunction with studies to determine if these modifications affect the pharmacodynamics. The discovery of ritonavir demonstrates these approaches work very well[157] and also shows that peptidomimetic analogs can be capable of high absorption, oral bioavailability, and slow hepatic clearance.

5.5   Irreversible Enzyme Inhibitors

5.5.A   Potential of Irreversible Inhibition

A reversible enzyme inhibitor was effective as long as a suitable concentration of the inhibitor was present to drive the equilibrium $E + I \rightleftharpoons E \cdot I$ to the right (see Section 5.4.A, p. 241). Therefore, a reversible inhibitor drug is effective only while the drug concentration is maintained at a high enough level to sustain the enzyme–drug complex. Because of drug metabolism and excretion (see Chapter 7), repetitive administration of the drug is required.

A competitive *irreversible enzyme inhibitor*, also known as an *active-site directed irreversible inhibitor* or an *enzyme inactivator*, is a compound whose structure is similar to that of the substrate or product of the target enzyme and which generally forms a covalent bond to an active site residue. (A slow, tight-binding inhibitor, however, often is a noncovalent inhibitor that can be *functionally* irreversibly bound.) In the case of irreversible inhibition it is not necessary to sustain the inhibitor concentration to retain the enzyme–inhibitor interaction. Because this is an irreversible reaction, once the target enzyme has reacted with the irreversible inhibitor, the complex cannot dissociate (there are exceptions), and, therefore, the enzyme remains inactive, even in the absence of additional inhibitor. This effect could translate into the requirement for smaller and fewer doses of the drug. Even though the target enzyme is destroyed by the irreversible inhibitor, it does not mean that only one dose of the drug would be sufficient to destroy the enzyme permanently. Yes, it destroys that copy of the enzyme permanently, but our genes are constantly encoding more copies of proteins that diminish in concentration. As the enzyme loses activity, additional copies of the enzyme are synthesized, but this process can take hours or even days.[158] In some cases, however, particularly where genetic translation of the target enzyme is slow, it may be safer to design reversible inhibitors whose effects can be controlled more effectively by termination of their administration.

One may wonder what the effect on metabolism would be if a particular enzyme activity were completely inhibited for an extended period of time. Consider the case of aspirin, an irreversible inhibitor of prostaglandin synthase (see Section 5.5.B.2.b, p. 280). If the quantity of aspirin consumed in the United States (about 55 tons a day) were averaged over the entire population, then every man, woman, and child would be taking about 180 mg of aspirin every day, enough to shut down human prostaglandin biosynthesis for the entire country

$$E + I \underset{}{\overset{K_I}{\rightleftharpoons}} E \cdot I \xrightarrow{k_{inact}} E \text{-} I$$

Scheme 5.12 ▶ Basic kinetic scheme for an affinity labeling agent

permanently! Suffice it to say that there are many irreversible enzyme inhibitor drugs in medical use.

The term *irreversible* is a loose one; either a very stable covalent bond or a labile bond may be formed between the drug and the enzyme active site. As pointed out earlier, some tight-binding reversible inhibitors also are functionally irreversible. As long as the enzyme remains nonfunctional long enough to produce the desired pharmacological effect, it is considered irreversibly inhibited. The two principal types of enzyme inactivators are reactive compounds called *affinity labeling agents* and unreactive compounds that are activated by the target enzyme known as *mechanism-based enzyme inactivators*.

5.5.B Affinity Labeling Agents

B.1 Mechanism of Action

An affinity labeling agent is a reactive compound that has a structure similar to that of the substrate for a target enzyme. Subsequent to reversible E·I complex formation, it reacts with active site nucleophiles (amino acid side chains), generally by acylation or alkylation (S_N2) mechanisms, thereby forming a stable covalent bond to the enzyme (Scheme 5.12). Note that this reaction scheme is similar to that for the conversion of a substrate to a product (see Section 4.1.B.1, p. 176); instead of a k_{cat}, the catalytic rate constant for product formation, there is a k_{inact}, the inactivation rate constant for enzyme inactivation. On the assumption that the equilibrium for reversible E·I complex formation (K_I) is rapid and the rate of dissociation of the E·I complex (k_{off}) is fast relative to that of the covalent bond forming reaction, then k_{inact} will be the rate-determining step. In this case, unlike simple reversible inhibition, there will be a time-dependent loss of enzyme activity (as is the case with slow, tight-binding reversible inhibitors because of the relatively small k_{off}).

The rate of inactivation is proportional to low concentrations of inhibitor, but becomes independent at high concentrations. As is the case with substrates, the inhibitor also can reach *enzyme saturation* when the k_{inact} is slow relative to the k_{off}. Once all of the enzyme molecules are tied up in an E·I complex, the addition of more inhibitor will have no effect on the rate of inactivation.

Because an affinity labeling agent contains a reactive functional group, not only can it react with the active site of the target enzyme, but it also can react with thousands of nucleophiles associated with many other enzymes and biomolecules in the body. Consequently, these inactivators are potentially quite toxic. In fact, many cancer chemotherapy drugs (see Chapter 6) are affinity labeling agents, and they are quite toxic. Therefore, they are not as common in drug design as other types of enzyme inhibitors.

There are several principal reasons why these reactive molecules, nonetheless, can be effective drugs. First, once the inactivator forms an E·I complex, a unimolecular reaction ensues (the E·I complex is now a single molecule), which can be many orders of magnitude (10^8 times; see Section 4.2.A, p. 179) more rapid than nonspecific bimolecular reactions with nucleophiles on other proteins. Furthermore, the inactivator may form an E·I complex with other enzymes, but if there is no nucleophile near the reactive functional group, no reaction will take place. Third, in the case of antitumor agents, mimics of DNA precursors are rapidly transported to the appropriate site and, therefore, they are preferentially concentrated at the desired tumor target.

The key to the effective design of affinity labeling agents as drugs is specificity of binding. If the molecule has a very low K_i for the target enzyme, then E·I complex formation with the target enzyme will be favored, and the selective reactivity will be enhanced. Another approach to increase the selectivity of this class of inactivators is to modulate the reactivity of the active functional group. The effectiveness of this approach can be seen by comparing the relatively moderate reactivity of the functional groups in the nontoxic affinity labeling agents described in Section 5.5.B.2 with the highly reactive functional groups in some of the cancer chemotherapeutic drugs in Section 6.3.B, p. 371.

One approach that takes advantage of modulated reactivity of an affinity labeling agent has been termed by Krantz[159] *quiescent affinity labeling*. In this case the inactivator reactivity is so low that reactions with nucleophiles in solution at physiological pH and temperature are exceedingly slow or nonexistent. However, because of the exceptional nucleophilicity of groups in some enzymes that use covalent catalysis as a catalytic mechanism (see Section 4.2.B, p. 181), the poorly electrophilic sites in the quiescent affinity labeling agents become reactive enough for nucleophilic reaction, but only at the active site of the enzyme. High selectivity for inactivation of a particular enzyme can be built into a molecule if the structure of the inactivator is designed so that it binds selectively to the target enzyme. For example, peptidyl acyloxymethyl ketones (**5.89**, Scheme 5.13) have low chemical reactivity (the acyloxyl group is a weak leaving group), but are potent and highly selective inactivators of cathepsin B,[160] a cysteine protease that has been implicated in osteoclastic bone resorption,[161] tumor metastasis,[162] and muscle wasting in Duchenne muscular dystrophy.[163]

A third feature that would increase the potential effectiveness of an affinity labeling agent can be built into the molecule if something is known about the location of the active site nucleophiles. In cases where a nucleophile is known to be at a particular position relative to the bound substrate, the reactive functional group can be incorporated into the affinity labeling agent so that it is near that site when the inactivator is bound to the target enzyme. This increases the probability for reaction with the target enzyme by approximation (see Section 4.2.A, p. 179).

Because many enzymes involved in DNA biosynthesis utilize substrates with similar structures, high concentrations of reversible inhibitors may block multiple enzymes. A lower concentration of an irreversible inhibitor may be more selective if it only reacts with those enzymes having appropriately juxtaposed active site nucleophiles.

Even when all of these design factors are taken into consideration, nonspecific reactions still can take place that may result in side effects. Judicious use of low electrophilic moieties can result in highly effective potential drug candidates.

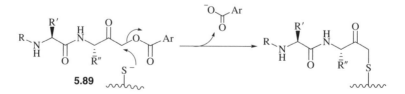

Scheme 5.13 ▶ Peptidyl acyloxymethyl ketones as an example of a quiescent affinity labeling agent

B.2 Selected Affinity Labeling Agents

a. Penicillins and Cephalosporins/Cephamycins

Penicillins have the general structure **5.90**; for example, **5.90a** is penicillin G (Bicillin), **5.90b** is penicillin V (PenicillinV), **5.90c** (R' = H) is oxacillin, **5.90c** (R' = Cl, Bactocill) is cloxacillin, **5.90d** (R' = H) is ampicillin (Ampicillin), and **5.90d** (R' = OH) is amoxicillin (Amoxil). The differences in these derivatives (other than structure) are related to absorption properties, resistance to penicillinases, and specificity for organisms for which they are most effective.[164]

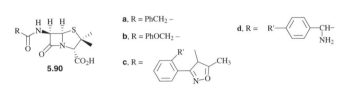

The structures of some cephalosporins and cephamycins are shown in **5.91**; for example, **5.91a** is cefazolin (injectable; Ancef), **5.91b** is cefoxitin (injectable; Mefoxin), **5.91c** is cefaclor (oral; Ceclor), and **5.91d** is ceftizoxime (injectable; Cefizox). The analogs where X = H are cephalosporins and those where X = OCH_3, such as cefoxitin, **5.91b**, are cephamycins. The penicillins, cephalosporins, and cephamycins all have in common the β-lactam ring and are known collectively as *β-lactam antibiotics*. Cephalosporins and cephamycins are classified by generations that are based on general features of antimicrobial activity.[165] Cefazolin is a first-generation cephalosporin, cefoxitin and cefaclor are second-generation cephalosporins, and ceftizoxime is a third-generation cephalosporin. Modifications in the structure of these antibiotics have been extensive,[166] so much so that essentially every atom excluding the lactam nitrogen has been replaced or modified in the search for improved antibiotics.

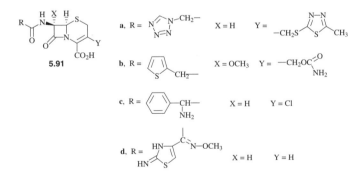

The discovery of penicillin was described in Section 2.1.A.1, p. 9.[167] Penicillins and cephalosporins/cephamycins are *bacteriocidal* (they kill existing bacteria), unlike the sulfonamides (see Section 5.4.B.2, p. 254), which are bacteriostatic. They are ideal drugs in that they inactivate an enzyme that is essential for bacterial growth, but which does not exist in animals, namely, the peptidoglycan transpeptidase. This enzyme catalyzes the cross-linking of the peptidoglycan to form the bacterial cell wall and was discussed in connection with the

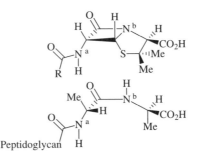

Figure 5.11 ▶ Comparison of the structure of penicillins with acyl D-alanyl-D-alanine

drug vancomycin (see Scheme 5.1 in Section 5.2.B.3, p. 234). Because animal cells do not have cell walls, there is no need for this enzyme in animals.

The transpeptidase has not been purified, but d-alanine carboxypeptidase, an enzyme believed to be the actual *in vivo* transpeptidase that became uncoupled during purification, has been purified.[168] This enzyme acts freely as a serine protease and carries out the first half of the transpeptidase reaction, that is, the formation of the serine ester. The acceptor molecule, then, is water instead of the amino group of another peptidoglycan chain. This leads to the hydrolyzed product rather than the cross-linked one. An active site serine has been identified as the residue involved in catalysis.[169]

By comparison of a molecular model of penicillin with that of D-alanyl-D-alanine (the terminal dipeptide of the peptidoglycan side chain), Tipper and Strominger[170] suggested that penicillin could mimic the structures of the terminus of the peptidoglycan and bind at the active site of transpeptidase (Figure 5.11). The N^a to N^b distances (3.3 Å) and the N^b to carboxylate carbon distances (2.5 Å) in both molecules are identical. The N^a to carboxylate carbon distance is 5.4 Å in the penicillins and 5.7 Å in D-alanyl-D-alanine. The β-lactam carbonyl may be further activated by torsional effects of the thiazolidine ring in the penicillins. This carbonyl corresponds to the carbonyl in the acyl D-alanyl-D-alanine that acylates the active site serine and, therefore, penicillins also could acylate the transpeptidase serine residue[171] (Scheme 5.14). This hypothesis was supported many years later in a crystal structure obtained to 1.2-Å resolution of a cephalosporin bound to a bifunctional serine-type D-alanyl-D-alanine carboxypeptidase/transpeptidase.[172] The bulk of the penicillin molecule, when it is attached to the active site, precludes hydrolysis or transamidation either for steric reasons or because it induces a conformational change in the enzyme that prevents these processes from occurring.[173] Covalent binding at the active site prevents the substrate from binding. Similar arguments could be made for cephalosporins. The double bond in the dihydrothiazine ring also may activate the β-lactam carbonyl (Scheme 5.15).[174]

On the basis of the structural similarity of penicillins to acyl D-alanyl-D-alanine (see Figure 5.11), Tipper and Strominger predicted that 6-methylpenicillin (a methyl on the sp^3 carbon adjacent to N^a) would be a more potent inhibitor than the parent molecule. However, both 6-methylpenicillin and 7-methylcephalosporin (the numbering is different for cephalosporins because it has one more carbon in the ring, but the methyl in 7-methylcephalosporin corresponds to that in 6-methylpenicillin) were synthesized and were shown to be inactive.[175] Because the corresponding 7-methoxycephalosporins (that is, cephamycins) are better inhibitors than the parent cephalosporins, it is not clear why the methyl analogs are poor inhibitors.

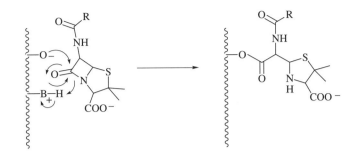

Scheme 5.14 ▶ Acylation of peptidoglycan transpeptidase by penicillins

Scheme 5.15 ▶ Activation of the β-lactam carbonyl of cephalosporins

The beauty of the penicillins (and cephalosporins) is that they are not exceedingly reactive; consequently, few nonspecific acylation reactions occur. Their modulated reactivity and nontoxicity make them ideal drugs. If it were not for allergic responses and problems associated with drug resistance and digestion, penicillins might be considered nutritious foods, comprised of various carboxylic acid derivatives (the RCO side chains), cysteine, and the essential amino acid valine!

Because of the discovery of highly effective antibiotics, such as the penicillins and the sulfa drugs, and the realization of even more effective synthetic analogs of these natural products, Nobel laureate immunologist F. Macfarlane Burnet noted in 1962 that by the late 20th century "the virtual elimination of infectious disease as a significant factor in social life" should be anticipated.[176] William Stewart, the former surgeon general of the United States, testified before Congress in 1967 that it was time to "close the book on infectious diseases." Apparently, rampant resistance to these and all antibacterial drugs was not foreseen.

Although penicillins are "wonder drugs" in their activity against a variety of bacteria, many strains of bacteria have become resistant. For example, within 10 years of the introduction of penicillin, half of the strains of *Staphylococcus aureus* had become resistant, and by the 1990s 90% of these strains were resistant.[177] The principal cause for resistance, as a result of gene transfer and recombination, is the excretion of the enzyme β-lactamase. This enzyme catalyzes the hydrolysis of β-lactams, presumably by having the ability to hydrolyze the acylated serine residue, a process that the susceptible strains cannot carry out.[178] Other important causes for resistance are that the transpeptidase becomes less susceptible to acylation by penicillins, and there is a change in the outer membrane permeability of the penicillins into the periplasm.

Because the major cause for resistance is the excretion of β-lactamases, an obvious approach to drug synergism would be the combination of a penicillin with a β-lactamase inhibitor. In the 1970s certain naturally occurring β-lactams that did not have the general penicillin or cephalosporin structure were isolated from various organisms[179] and were found to be potent mechanism-based inactivators (see Section 5.5.C) of β-lactamases.[180] These compounds are used in combination with penicillins to destroy penicillin-resistant strains of bacteria, for example, the combination of amoxicillin (**5.90d**, R' = OH) with clavulanate (**5.92**) is sold as Augmentin, and ampicillin (**5.90d**, R' = H) plus sulbactam (**5.93**) are in Unasyn (in unison, get it?). The β-lactamase inhibitors have no antibiotic activity, but they protect the penicillin from destruction so that it can interfere with cell wall biosynthesis. This is an example drug synergism (see Section 5.3.B.1, p. 239).

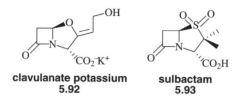

clavulanate potassium
5.92

sulbactam
5.93

b. Aspirin

It is stated in the *Papyrus Ebens* (circa 1550 B.C.) that dried leaves of myrtle provide a remedy for rheumatic pain. Hippocrates (460–377 B.C.) recommended the use of willow bark for pain during childbirth.[181] A boiled vinegar extract of willow leaves was suggested by Aulus Cornelius Celsus (30 A.D.) for relief of pain.[182] In 1763 Reverend Edmund Stone of England announced his findings that the bark of the willow (*Salix alba vulgaris*) provided an excellent substitute for Peruvian bark (*Cinchona* bark, a source of quinine) in the treatment of fevers.[183] The connection between the two barks was discovered by his tasting them and making the observation that they both had a similar bitter taste. The bitter active ingredient with antipyretic activity in the willow bark was called salicin, which was first isolated in 1829 by Leroux. On hydrolysis, salicin produced glucose and salicylic alcohol, which was metabolized to salicylic acid. Sodium salicylate was first used for the treatment of rheumatic fever and as an antipyretic agent in 1875. Toward the end of the 19th century Felix Hoffmann, a chemist employed by Bayer Company, had a father who suffered from severe rheumatoid arthritis and pleaded with his son to search for a less irritating drug than the sodium salicylate he was using (salicylic acid not only tastes awful, but it causes ulcerations of the mouth and stomach linings with prolonged use). Many salicylate derivatives were synthesized, and acetylsalicylate (aspirin, **5.94**) was the best. He gave it to his father, who responded well, and in 1899 Bayer introduced aspirin as an antipyretic (fevers), anti-inflammatory (arthritis), and analgesic (pain) agent.[184] Aspirin became the first drug ever to be tested in clinical trials before registration. Also, because of its insolubility in water, it had to be sold in solid form and, therefore, became the first major medicine to be sold in the form of tablets. The trade name *aspirin* was coined by adding an "a" for acetyl to spirin for *Spiraea*, the plant species from which salicylic acid was once prepared. In the first 100 years since its introduction on the market, it has been estimated that one trillion (10^{12}) aspirin tablets have been consumed.[185]

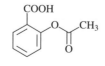

acetylsalicylate
5.94

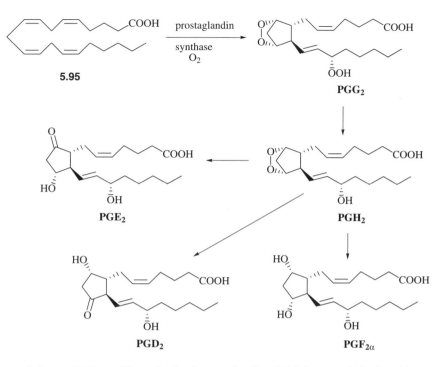

Scheme 5.16 ▶ Biosynthesis of prostaglandins (PGs) from arachidonic acid

The mechanism of action was initially reported by Vane,[186] who shared the 1982 Nobel Prize in Medicine (with Sune Bergström and Bengt Samuelsson) for this discovery, and by Smith and Willis[187] to be the result of inhibition of prostaglandin biosynthesis. Prostaglandins (PGs) are derived from arachidonic acid (**5.95**) by the action of prostaglandin synthase (also known as cyclooxygenase) (Scheme 5.16). At the time of the initial report there was evidence that various prostaglandins were involved in the pathogenesis of inflammation and fever. It is now known that all mammalian cells (except erythrocytes) have microsomal enzymes that catalyze the biosynthesis of prostaglandins, and that prostaglandins are always released when cells are damaged, and they are in increased concentrations in inflammatory exudates. When prostaglandins are injected into animals, the effects are reminiscent of those observed during inflammatory responses, namely, redness of the skin (erythema) and increased local blood flow. A long-lasting vasodilatory action also is prevalent. Prostaglandins can cause headache and vascular pain when infused in man. Elevation of body temperature during infection also is mediated by the release of prostaglandins.

From the above discussion it is apparent that inhibition of prostaglandin synthase, the enzyme responsible for the biosynthesis of all of the prostaglandins and related compounds [PGH_2 also is converted by prostacyclin synthase to prostacyclin (**5.96**) and by thromboxane synthase to thromboxane A_2 (**5.97**)], would be a desirable approach for the design of anti-inflammatory, analgesic, and antipyretic drugs. Inhibition of platelet cyclooxygenase is particularly effective at blocking prostaglandin biosynthesis because, unlike most other cells, platelets cannot regenerate the enzyme, because they have little or no capacity for protein biosynthesis. Therefore, a single dose of 40 mg per day of aspirin is sufficient to destroy the cyclooxygenase for the life of the platelet! When prostaglandin synthesis is blocked, there

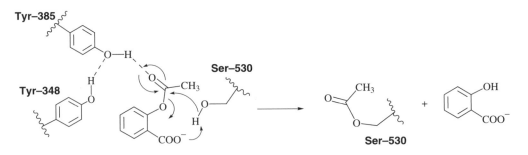

Scheme 5.17 ▶ Hypothetical mechanism for acetylation of prostaglandin synthase by aspirin

is less stimulation of the pain-sensitive nerve endings, resulting in less aching of muscles and joints, as well as less relaxation of blood vessels in the head, so fewer headaches; fewer prostaglandins in the hypothalamus means that the body temperature set point is not changed, and no fever is induced.

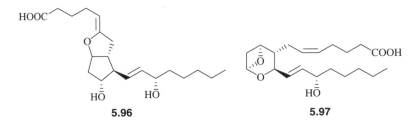

5.96   5.97

Sheep vesicular gland prostaglandin synthase was shown to be irreversibly inactivated by aspirin.[188] When microsomes of sheep seminal vesicles were treated with aspirin tritiated in the methyl of the acetyl group, acetylation of a single protein was observed. The same experiment carried out with aspirin tritiated in the benzene ring resulted in no tritium incorporation, suggesting that acetylation was occurring. Incubation of purified prostaglandin synthase with [^{3}H-acetyl] aspirin led to irreversible inactivation with incorporation of one acetyl group per enzyme molecule.[189] Pepsin digestion of the tritiated enzyme gave a 22-amino-acid labeled peptide; tryptic digestion of this peptide gave a tritiated decapeptide in which the serine residue was acetylated.[190] Thermolysin digestion of the tritiated enzyme gave the labeled dipeptide Phe-Ser, where the hydroxyl group of Ser-530 had become acetylated.[191] The most straightforward mechanism for acetylation would be a transesterification mechanism by aspirin acting as an affinity labeling agent; based on site-directed mutagenesis studies, it was proposed by Marnett and coworkers[192] that Tyr-385 and Tyr-348 in the active site hydrogen bond to the acetyl carbonyl of aspirin, which directs it specifically to Ser-530 for acetylation (Scheme 5.17).

The principal side effect with aspirin and all nonsteroidal anti-inflammatory agents (NSAIDs) is that their chronic use leads to ulceration of the stomach lining. In the late 1980s it was found that cyclooxygenase (COX) activity was increased in certain inflammatory states and could be induced by inflammatory cytokines.[193] This suggested the existence of two different forms (isozymes) of COX, now known as COX-1 and COX-2. COX-1, the *constitutive form* of the enzyme (that is, present in the cells all the time), is responsible for the physiological production of prostaglandins and is important in maintaining tissues in the stomach lining and kidneys. COX-2,[194] induced by cytokines in inflammatory cells, is responsible for the elevated production of prostaglandins during inflammation and is associated with inflammation, pain, and fevers. The discovery of a second COX enzyme suggested that aspirin and other NSAIDs may inhibit both COX-1 and COX-2, leading to stomach irritation (and, in some cases, ulcers); ideally, inhibition of only COX-2 would produce the desired anti-inflammatory effect without stomach lining irritation.[195]

Both G. D. Searle and Merck initiated programs to discover a COX-2 selective (reversible) inhibitor. At Searle, more than 2500 compounds were synthesized and almost 2000 screened of which 280 were potent. Almost 350 compounds were tested for oral activity of which seven candidates were selected. Compound **5.98a** has an IC_{50} of 60 nM and a selectivity of >1700 against COX-2 versus COX-1.[196] Series **5.98b** has a range of IC_{50} values of 10–100 nM and *in vitro* selectivities of 10^3–10^4 for COX-2 versus COX-1.[197] However, celecoxib (**5.99**, Celebrex), with an IC_{50} of 40 nM and *in vitro* selectivity of only 375 in favor of COX-2 inhibition, emerged as the commercial compound because of its overall favorable properties.[198] Its *in vivo* selectivity (inflammatory effect/stomach irritation; a therapeutic index) was >1000. Merck also discovered a COX-2 selective inhibitor, rofecoxib (**5.100**, Vioxx),[199] that is about 5.5 times more selective than celecoxib and is a once-a-day treatment for osteoarthritis and rheumatoid arthritis. Searle then discovered valdecoxib (**5.101**, Bextra), having an IC_{50} for COX-1 of 140 μM and for COX-2 of 5 nM or 28,000-fold selectivity for COX-2 over COX-1.[200]

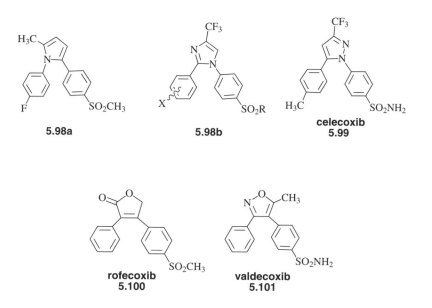

A third isozyme of cyclooxygenase called COX-3 (actually a splice variant of COX-1), found principally in the cerebral cortex, was shown to be selectively inhibited by NSAIDs that have good analgesic and antipyretic activity, but low anti-inflammatory activity, such as acetaminophen (**5.102a**, Tylenol).[201] Analgesic/antipyretic drugs penetrate the blood–brain barrier well and may accumulate in the brain where they inhibit COX-3. NSAIDs that contain a carboxylate group, such as aspirin and ibuprofen (**5.102b**, Advil) cross the blood–brain barrier poorly, but are more potent inhibitors of COX-3, so they also exhibit analgesic and antipyretic activities. COX-2 selective inhibitor drugs inhibit inflammatory pain,[202] whereas COX-1-selective inhibitor drugs, such as aspirin and ibuprofen,[203] are superior to COX-2 inhibitors against chemical pain stimulators.[204] The analgesic effect of COX-1 selective inhibitors occurs at lower doses than those needed to inhibit inflammation.[205]

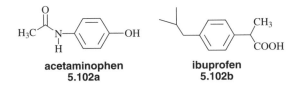

acetaminophen
5.102a   **ibuprofen**
                    5.102b

X-ray crystal structures of COX-1 and COX-2 are known, and there is very little difference in the active sites of these two isozymes. COX-1[206] has an active site isoleucine at position 523 (Ile-523) and COX-2[207] has a valine (Val-523) at that position. Because isoleucine is larger than valine (by one methyl group), when COX-2 selective inhibitors bind to COX-1, a substituent on the inhibitor has a repulsive interaction with the Ile, but not with the smaller Val in COX-2, resulting in preferential binding to COX-2. Site-directed mutagenesis of these amino acids demonstrated that they were important to the selective inhibitor binding.[208] This small difference in size between Ile and Val is not sufficient to use in structure-based drug design approaches (see Section 2.2.H, p. 78), and in fact the COX-2 selective inhibitors were discovered without the aid of crystal structures. Note that these COX-2 selective inhibitors are reversible inhibitors. They are being discussed in this section on irreversible inhibitors only because they follow from our discussion of aspirin.

A series of acetoxybenzenes ("aspirin" analogs) substituted in the *ortho*-position with alkyl sulfides instead of the carboxylate were synthesized, and the analog with a 2-heptynylsulfide substituent (**5.103**) was 21-fold selective for COX-2 (IC$_{50}$ 0.8 μM versus 17 μM for COX-1) while still retaining irreversible inhibition.[209] Interestingly, if the sulfide is changed to an ether, or the acetyl changed to propionyl, all selectivity is lost and the potency is decreased with both isozymes. When [1-^{14}C-acetyl]-**5.103** was used to inactivate COX-2, the radioactive label was transferred to Ser-516 (this corresponds to Ser-530 in COX-1).

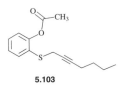

5.103

Both penicillin and aspirin are examples of affinity labeling agents that involve acylation mechanisms. Other affinity labeling agents function by alkylation or arylation mechanisms. An interesting example involves selective, covalent modification of β-tubulin, leading to disruption of microtubule polymerization and cytotoxicity against multidrug-resistant tumors.[210] The pentafluorobenzene analog **5.104** produces time-dependent binding to β-tubulin; tritium-labeled **5.104** incorporates tritium into a protein that comigrates on sodium dodecyl sulfate (SDS)/polyacrylimide gel electrophoresis (PAGE) with β-tubulin. Covalent modification occurs at a conserved cysteine residue (Cys-239) shared by β_1, β_2, and β_4-tubulin isoforms. Replacement of the *para*-fluorine with other halogens causes a substantial decrease in potency; replacement with a hydrogen abolishes the cytotoxicity of the compound.[211]

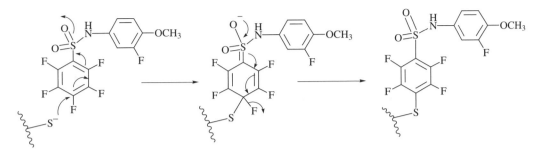

Scheme 5.18 ▶ Hypothetical nucleophilic aromatic substitution mechanism for arylation of β-tubulin

These results are consistent with a nucleophilic aromatic substitution mechanism for covalent inactivation (Scheme 5.18).

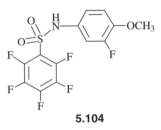

5.104

5.5.C   Mechanism-Based Enzyme Inactivators

C.1   Theoretical Aspects

A *mechanism-based enzyme inactivator* is an unreactive compound that bears a structural similarity to the substrate or product for a specific enzyme. Once the inactivator binds to the active site, the target enzyme, via its normal catalytic mechanism, converts the compound into a product that inactivates the enzyme prior to escape from the active site. Therefore, these inactivators are acting initially as substrates for the target enzyme. Although the product generally forms a covalent bond to the target enzyme, it is not essential (a tight-binding inhibitor may form). However, the target enzyme must transform the inactivator into the actual inactivating species, and inactivation must occur prior to the release of this species from the active site. Because there is an additional step in the inactivation process relative to that for an affinity labeling agent (see Section 5.5.B.1, p. 275), the kinetic scheme for mechanism-based inactivation differs from that for affinity labeling (Scheme 5.19). Provided k_4 is a fast step and the equilibrium k_1/k_{-1} is set up rapidly, then k_2 is the inactivation rate constant (k_{inact}) that determines the rate of the inactivation process. The two key features of this type of inactivator that differentiate it from an affinity labeling agent are its initial unreactivity and the requirement for the enzyme to catalyze a reaction on it, thereby converting it into a product, that is, the actual inactivator species. Often this converted inactivator species is quite reactive, and therefore acts as an affinity labeling agent which already is at the active site of the target enzyme. Inactivation (k_4 in Scheme 5.19) does not necessarily occur every time the inactivator is transformed into the inactivating species; sometimes it escapes from the active

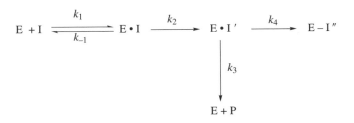

Scheme 5.19 ▶ Kinetic scheme for simple mechanism-based enzyme inactivation

site (k_3 in Scheme 5.19). The ratio of the number of turnovers that gives a released product per inactivation event, k_3/k_4, is called the *partition ratio*.

C.2 Potential Advantages in Drug Design Relative to Affinity Labeling Agents

Because of the generally high reactivity of affinity labeling agents, they can react with enzymes and biomolecules other than the target enzyme. When this occurs, toxicity and side effects can arise. However, mechanism-based enzyme inactivators are unreactive compounds, and this is the key feature that makes them so amenable to drug design. Consequently, nonspecific alkylations and acylations of other proteins should not be a problem. In the ideal case only the target enzyme will be capable of catalyzing the appropriate chemistry for conversion of the inactivator to the activated species, and inactivation will result with every turnover, that is, the partition ratio will be zero (no metabolites formed per inactivation event). This may be quite important for potential drug use. If the partition ratio is greater than zero, the released activated species may react with other proteins, possibly resulting in a toxic effect. In this case the inactivator is called a *metabolically activated inactivator*.[212] Alternatively, the released species may be hydrolyzed by the aqueous medium prior to reaction with other biomolecules, but the product formed may be toxic or may be metabolized further to other toxic substances. Under the ideal conditions mentioned above, the inactivator would be a strong drug candidate because it would be highly enzyme specific and low in toxicity. In fact, α-difluoromethylornithine (eflornithine), a specific mechanism-based inactivator of ornithine decarboxylase (see 5.5.C.3.b, p. 290), has been administered to patients in amounts of 30 g day for several weeks with only minor side effects![213]

Relatively few drugs in current medical use are mechanism-based inactivators, and most were determined *ex post facto*, rather than being designed to be mechanism-based inactivators. Some known mechanism-based inactivators not discussed in the section below include the antidepressant drug phenelzine sulfate (**5.105**, Nardil) and the antihypertensive drug, hydralazine hydrochloride (**5.106**, Apo-Hydral), both of which inactivate monoamine oxidase; clavulanic acid (**5.92**, Section 5.5.B.2.a, p. 277), a compound used to protect penicillins and cephalosporins against bacterial degradation (inactivates β-lactamases); the antiviral agent, trifluridine (**5.107**, Viroptic), which inactivates thymidylate synthase; gemcitabine HCl (**5.108**, Gemzar), an antitumor drug that inactivates ribonucleotide reductase; the antihyperuricemic agent, allopurinol (**5.109**, Aloprim), which inactivates xanthine oxidase; and the antithyroid drug, methimazole (**5.110**, Tapazole), which inactivates thyroid peroxidase. There are a few mechanism-based inactivators of cytochrome P450, however, those inactivators do not derive their pharmacological effects as a result of that inactivation.

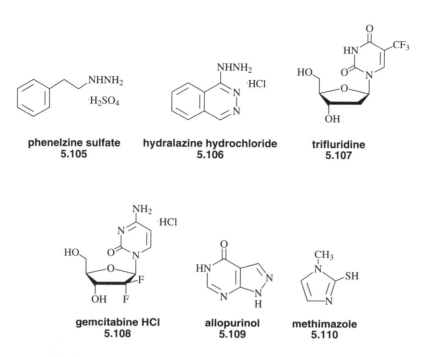

phenelzine sulfate
5.105

hydralazine hydrochloride
5.106

trifluridine
5.107

gemcitabine HCl
5.108

allopurinol
5.109

methimazole
5.110

Because the activation of mechanism-based inactivators depends on the catalytic mechanism of their target enzyme, these inactivators can be designed by a rational organic mechanistic approach. However, this approach to drug design is now infrequently used for several reasons. First, knowledge of the catalytic mechanism of the target enzyme is required to use the mechanism in the design of the inactivator. Also, this approach requires that a specific molecule be synthesized, which may be time consuming to do, just to test the mechanistic hypothesis. Even if the compound inactivates the enzyme, issues of potency, specificity, and pharmacokinetics have to be managed while maintaining the core structure for catalytic turnover. Because of all of these complications, relatively few mechanism-based inactivator drugs are commercial. Furthermore, by the last decade of the 20th century, high-throughput screening (see Section 2.1.B, p. 11) and combinatorial chemistry (see Section 2.2.E.5, p. 34) made random and focused screens popular again, leaving the more rational approaches behind. In the next section several mechanism-based inactivators are discussed, first in terms of the medicinal relevance of their target enzyme, then from a mechanistic point of view. The reason for the relatively large number of examples of this type of enzyme inhibitor discussed below, despite the lesser prominence of this approach in drug design, derives from my interest in organic mechanisms, mechanism-based inactivation, and the history of these drugs.

C.3 Selected Examples of Mechanism-Based Enzyme Inactivators

a. Vigabatrin, an Anticonvulsant Drug

Epilepsy is a disease that was described more than 4000 years ago in early Babylonian and Hebrew writings. Full clinical descriptions were written in Hippocrates' monograph *On the Sacred Disease* in about 400 B.C.[214] Epilepsy is not a single disease, but a family of central nervous system diseases characterized by recurring convulsive seizures. By that definition then 1–2% of the world population has some form of epilepsy.[215] The disease is categorized as *primary* or *idiopathic* when no cause for the seizure is known, and *secondary* or *symptomatic*

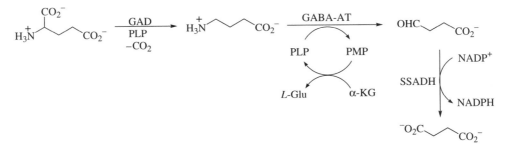

Scheme 5.20 ▶   Metabolism of L-glutamic acid

when the etiology has been identified. Symptomatic epilepsy can result from specific physiological phenomena such as brain tumors, syphilis, cerebral arteriosclerosis, multiple sclerosis,
Buerger's disease, Pick's disease, Alzheimer's disease, sunstroke or heat stroke, acute intox-
ication, lead poisoning, head trauma, vitamin B_6 deficiency, hypoglycemia, and labor. The
biochemical mechanism leading to central nervous system electrical discharges and epilepsy
are unknown, but there may be multiple mechanisms involved. However, it has been shown
that convulsions arise when there is an imbalance in two principal neurotransmitters in the
brain, L-glutamic acid, an excitatory neurotransmitter, and γ-aminobutyric acid (GABA), an
inhibitory neurotransmitter. The concentration of GABA is regulated by two PLP-dependent
enzymes, L-glutamic acid decarboxylase (GAD), which converts glutamate to GABA, and
GABA aminotransferase (GABA-AT), which degrades GABA to succinic semialdehyde with
the regeneration of glutamate (Scheme 5.20). Although succinic semialdehyde is toxic to
cells, there is no buildup of this metabolite, because it is efficiently oxidized to succinic acid
by the enzyme succinic semialdehyde dehydrogenase (SSADH). GABA system dysfunction
has been implicated in the symptoms associated with epilepsy, Huntington's disease, Parkinson's disease, and tardive dyskinesia. When the concentration of GABA diminishes below a
threshold level in the brain, convulsions begin. If a convulsion is induced in an animal, and
GABA is injected directly into the brain, the convulsions cease. It would seem, then, that an
ideal anticonvulsant agent would be GABA; however, peripheral administration of GABA
produces no anticonvulsant effect. This was shown to be the result of the failure of GABA,
under normal circumstances, to cross the *blood–brain barrier*, a membrane that surrounds
the capillaries of the circulatory system in the brain and protects it from passive diffusion of
undesirable (generally hydrophilic) chemicals from the bloodstream. Another approach for
increasing the brain GABA concentration, however, would be to design a compound capable
of permeating the blood–brain barrier that subsequently inactivated GABA-AT, the enzyme
that catalyzes the degradation of GABA. Provided that GAD also is not inhibited, GABA
concentrations should rise. This, in fact, has been shown to be an effective approach to the
design of anticonvulsant agents.[216] Compounds that both cross the blood–brain barrier and
inhibit GABA-AT *in vitro* have been reported to increase whole brain GABA levels *in vivo*
and possess anticonvulsant activity. The anticonvulsant effect does not correlate with whole
brain GABA levels, but it does correlate with an increase in the GABA concentration at the
nerve terminals of the *substantia nigra*.[217]

The mechanism for the PLP-dependent aminotransferase reactions was discussed earlier (see Section 4.3.A.3, p. 195). On the basis of this mechanism researchers at the former
Merrell Dow Pharmaceuticals (now Aventis)[218] designed 4-amino-5-hexenoic acid (vigabatrin, **5.111**, Scheme 5.21, Sabril). This is the first rationally designed mechanism-based

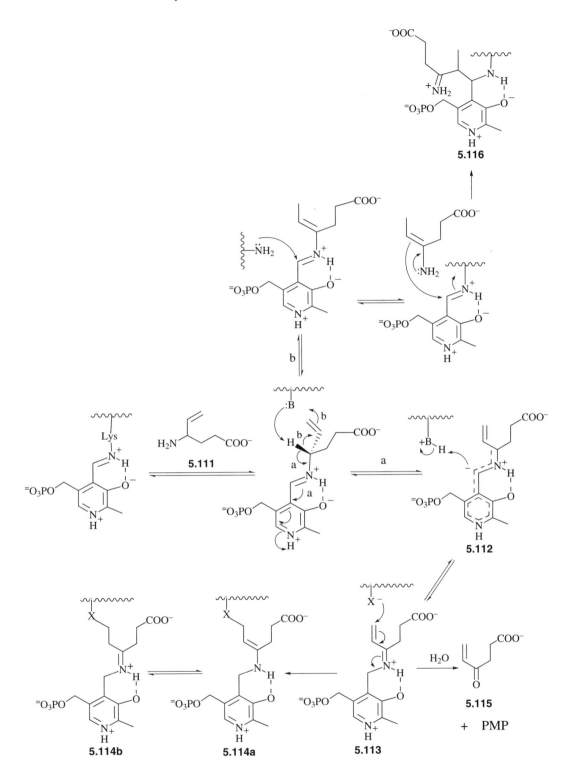

Scheme 5.21 ► Hypothetical mechanism for the inactivation of GABA aminotransferase by vigabatrin

inactivator drug (currently, it is in medical use everywhere in the world, except in the United States). The proposed inactivation mechanism is shown in Scheme 5.21. By comparison of Scheme 5.21 with Scheme 4.18 in Chapter 4 (p. 198; in Scheme 4.18 an α-amino acid is used, but here a γ-amino acid is the example), it is apparent that identical mechanisms are proposed up to compound **5.113** (Scheme 5.21) and compound **4.27** (Scheme 4.18). In the case of normal substrate turnover, hydrolysis of **4.27** gives pyridoxamine 5′-phosphate (PMP) and the keto acid (Scheme 4.18). The same hydrolysis could occur with **5.113** to give the corresponding products, PMP and **5.115**. However, **5.113** is a potent electrophile, a Michael acceptor, which can undergo conjugate addition by an active site nucleophile (X^-) and produce inactivated enzyme (**5.114a** or **5.114b**). The active site nucleophile was identified as Lys-329, the lysine residue that holds the PLP at the active site.[219] The mechanism in Scheme 5.21 (pathway a) appears to be relevant for about 70% of the inactivation.[220] The other 30% of the inactivation is accounted for by an allylic isomerization and enamine rearrangement leading to **5.116** (pathway b). Note that vigabatrin is an unreactive compound that is converted by the normal catalytic mechanism of the target enzyme into a reactive compound (**5.113**) that attaches to the enzyme. This is the typical course of events for a mechanism-based inactivator.

It may seem strange that GABA does not cross the blood–brain barrier, but vigabatrin, which also is a small charged molecule, *can* diffuse through that lipophilic membrane. The attachment of a vinyl substituent to GABA, apparently, has two effects that permit this compound to cross the blood–brain barrier (albeit poorly). First, the vinyl substituent increases the lipophilicity of the molecule. Second, it is an electron-withdrawing substituent which would have the effect of lowering the pK_a of the amino group. This would increase the concentration of the nonzwitterionic form (**5.117b**, Scheme 5.22), which is more lipophilic than the zwitterionic form (**5.117a**) because it is uncharged.

b. Eflornithine, an Antiprotozoal Drug and Beyond

The *polyamines*, spermidine (**5.118**) and spermine (**5.119**), and their precursor, putrescine (**5.120**), are important regulators of cell growth, division, and differentiation. The mechanisms by which they do this are unclear, but they appear to be required for DNA synthesis. Rapidly growing cells have much higher levels of polyamines (and ornithine decarboxylase; see below) than do slowly growing or quiescent cells. When quiescent cells are stimulated, the polyamine and ornithine decarboxylase levels increase prior to an increase in the levels of DNA, RNA, and protein. The polycationic nature of the polyamines may be responsible for their interaction with cellular structures that have negatively charged groups such as DNA.

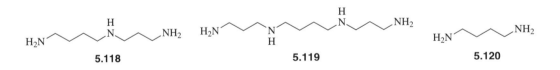

5.118   5.119   5.120

Ornithine decarboxylase, the enzyme that catalyzes the conversion of ornithine to putrescine (**5.120**) is the rate-limiting step in polyamine biosynthesis (Scheme 5.23). Spermidine (**5.118**) is produced by the spermidine synthase-catalyzed reaction of putrescine with *S*-adenosylhomocysteamine (**5.122**), which is derived from the decarboxylation of *S*-adenosylmethionine (**5.121**, SAM) in a reaction catalyzed by *S*-adenosylmethionine decarboxylase. Another aminopropyltransferase, namely, spermine synthase, catalyzes the reaction of spermidine with **5.122** to produce spermine (**5.119**).

Scheme 5.22 ▶ Zwitterionic and nonzwitterionic forms of vigabatrin

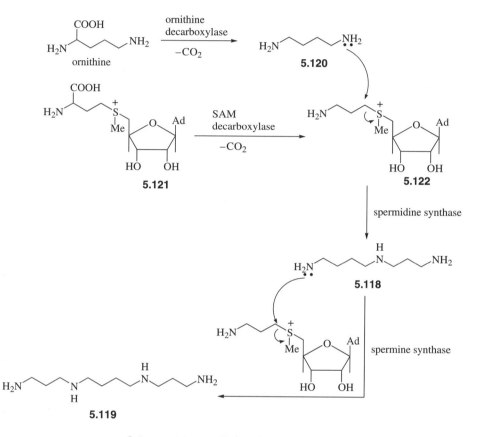

Scheme 5.23 ▶ Polyamine biosynthesis

Because polyamines are important for rapid cell growth, inhibition of polyamine biosynthesis should be an effective approach for the design of antitumor and antimicrobial agents. In fact, inactivation of ornithine decarboxylase by the potent mechanism-based inactivator eflornithine (α-difluoromethylornithine, **5.123**), results in virtually complete reduction in the putrescine and spermidine content; however, the spermine concentration is only slightly affected. There are at least two reasons for this latter observation. Inactivation of ornithine decarboxylase activity by eflornithine *in vivo* may not be complete because gene synthesis of ornithine decarboxylase has a half-life of only 30 minutes, and, also, the endogenous ornithine that is present competitively protects the enzyme from the inactivator. Furthermore, inactivation of ornithine decarboxylase induces an increase in the levels of *S*-adenosylmethionine decarboxylase, which leads to higher *S*-adenosylhomocysteamine levels. This can be efficiently used to convert the existing putrescine and spermidine to spermine.[221] The inability of eflornithine to shut down all of polyamine biosynthesis may be responsible for its discouragingly poor antitumor effects observed in clinical trials.[222] However, eflornithine has been found to have great value in the treatment of certain protozoal infestations such as *Trypanosoma brucei gambiense* (the more virulent strain, *Trypanosoma brucei rhodesiense*, is resistant) which causes African sleeping sickness (300,000 deaths per year)[223] and *Pneumocystis carinii*, the microorganism that produces pneumonia in AIDS patients.[224] Unfortunately, large amounts of the compound (150 mg/kg) are needed every 4–6 hours, which makes it very expensive, especially in Third World countries where the disease is prevalent.

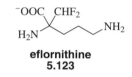

eflornithine
5.123

More recently, another indication has been discovered for eflornithine, namely, as a topical cream for the reduction of unwanted facial hair in women (Vaniqa). The mechanism of action is not known, but it could involve inhibition of polyamine biosynthesis important to hair growth.

The mechanism for the PLP-dependent decarboxylases was discussed in Chapter 4 (see Section 4.3.A.2, p. 195). Ornithine decarboxylase catalyzes the decarboxylation of eflornithine (**5.123**, Vaniqa), producing a reactive product that inactivates the enzyme; a possible inactivation mechanism is shown in Scheme 5.24.

According to the mechanism for PLP-dependent decarboxylases (see Chapter 4, Scheme 4.16), the only irreversible step is the one where CO_2 is released. Therefore, if you incubate the decarboxylase with the product amine, it catalyzes the reverse reaction up to the loss of CO_2, that is, the removal of the α-proton. This is the microscopic reverse of the reaction that occurs once the CO_2 is released. The *principle of microscopic reversibility* states that for a reversible reaction the same mechanistic pathway will be followed in the forward and reverse reactions. Therefore, a mechanism-based inactivator that has a structure similar to that of the product of the ornithine decarboxylase reaction (the amine) should undergo catalytic α-deprotonation, which then produces the same set of resonance structures produced by decarboxylation of the ornithine analog. α-Difluoromethylputrescine (**5.125**, Scheme 5.25) was shown to inactivate ornithine decarboxylase,[225] presumably because deprotonation gives the same intermediate that is obtained by decarboxylation of eflornithine (compare **5.124** in Scheme 5.24 with **5.124** in Scheme 5.25).

For any target enzyme that is reversible, mechanism-based inactivators should be designed for both the forward (substrate-like) and backward (product-like) reactions. Depending on the metabolic pathway involved, enzyme selectivity may be more favorable for one over the other.

c. Tranylcypromine, an Antidepressant Drug

The modern era of therapeutics for the treatment of depression began in the late 1950s with the introduction of both the monoamine oxidase (MAO) inhibitors and the tricyclic antidepressants. The first MAO inhibitor was iproniazid (**5.126**, Marsilid), which initially was used as an antituberculosis drug until it was observed that patients taking it exhibited excitement and

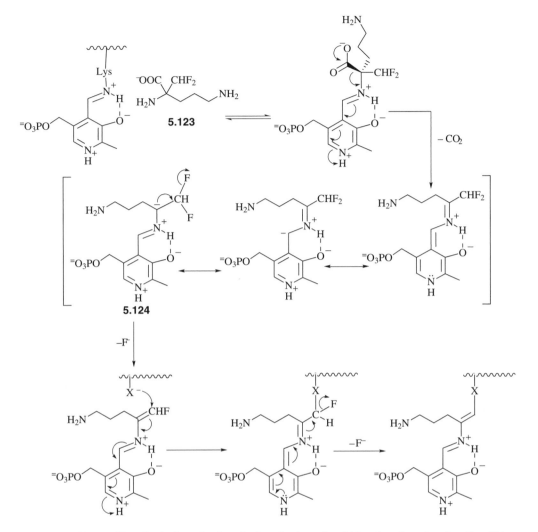

Scheme 5.24 ▶ Hypothetical mechanism for inactivation of ornithine decarboxylase by eflornithine

euphoria.[226] In 1952 Zeller *et al.*[227] showed that iproniazid was a potent inhibitor of MAO, and clinical studies got under way in the late 1950s.[228]

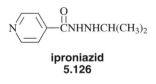

iproniazid
5.126

The brain concentrations of various biogenic (pressor) amines such as norepinephrine, serotonin, and dopamine were found to be depleted in chronically depressed individuals. A correlation was observed between an increase in the concentrations of these brain biogenic amines and the onset of an antidepressant effect.[229] This was believed to be the result of MAO inhibition, because MAO is one of the enzymes responsible for the catabolism of these

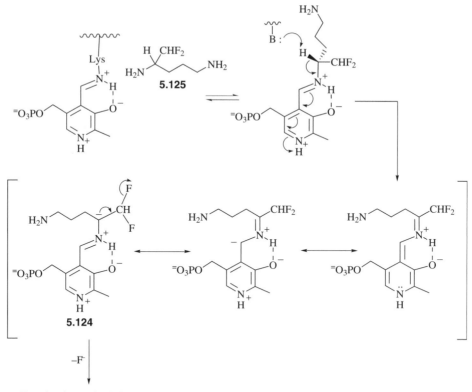

Inactivation as in Scheme 5.24

Scheme 5.25 ▶ Hypothetical mechanism for inactivation of ornithine decarboxylase by
α-difluoromethylputrescine

biogenic amines. By the early 1960s several MAO inactivators were being used clinically
for the treatment of depression. Unfortunately, it was found that in some cases there was
a cardiovascular side effect that led to the deaths of several patients. Consequently, these
drugs were removed from the drug market until the cause of death could be ascertained.
Within a few months the problem was understood. It was determined that all of those who
died while taking an MAO inhibitor had two things in common: They had all died from
a hypertensive crisis and, prior to their deaths, they had eaten foods containing a high tyramine content (e.g., cheese, wine, beer, and yeast products). The connection between these
observations is that the ingested tyramine triggers the release of norepinephrine, a potent
vasoconstrictor, which raises the blood pressure. Under normal conditions the excess norepinephrine is degraded by MAO and catechol O-methyltransferase. If the MAO is inactivated,
then the norepinephrine does not get degraded fast enough, the blood pressure keeps ris-
ing, and this can lead to a hypertensive crisis. This series of events has been termed the
cheese effect because of the high tyramine content found in certain cheeses. Because the
MAO inhibitors were not toxic, except when taken with certain foods, these drugs were
allowed to be returned to the drug market, but they were prescribed with strict dietary regulations. Because of this inconvenience, and the discovery of the tricyclic antidepressants
(which block the reuptake of biogenic amines at the nerve terminals), MAO inhibitors are

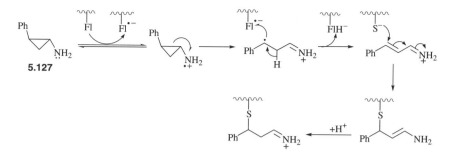

Scheme 5.26 ▶ Hypothetical mechanism for the inactivation of monoamine oxidase by tranylcypromine

not the drugs of choice, except in those types of depression that do not respond to tricyclic antidepressants or when treating phobic-anxiety disorders that respond well to MAO inhibitors.

The resurgence of interest by the pharmaceutical industry in MAO inhibitors in the late 1980s and 1990s[230] came about because MAO exists in at least two isozymic forms,[231] termed MAO A and MAO B. The main difference in these two isozymes is their selectivity for the oxidation of the various biogenic amines. Because the antidepressant effect is related to increased concentrations of brain serotonin and norepinephrine, both of which are MAO A substrates, compounds that selectively inhibit MAO A possess antidepressant activity; selective inhibitors of MAO B show potent antiparkinsonian properties.[232] To have an antidepressant drug without the cheese effect, however, it is necessary to inhibit brain MAO A selectively without inhibition of peripheral MAO A, particularly MAO A in the gastrointestinal tract and sympathetic nerve terminals. Inhibition of brain MAO A increases the brain serotonin and norepinephrine concentrations, which leads to the antidepressant effect; the peripheral MAO A must remain active to degrade the peripheral tyramine and norepinephrine that cause the undesirable cardiovascular effects.

Tranylcypromine (**5.127**, Parnate), a nonselective MAO A/B inactivator that exhibits a cheese effect, was one of the first MAO inactivators approved for clinical use; many other cyclopropylamine analogs show antidepressant activity.[233] The one-electron mechanism for MAO was discussed in Chapter 4 (Section 4.3.C.3; Scheme 4.32). If tranylcypromine acts as a substrate, and one-electron transfer from the amino group to the flavin occurs, the resulting cyclopropylaminyl radical will undergo rapid cleavage,[234] further oxidation, and attachment to a cysteine residue [235,236] (Scheme 5.26).

tranylcypromine
5.127

d. Selegiline (*L*-Deprenyl), an Antiparkinsonian Drug

Parkinson's disease, the second most common neurodegenerative disease and afflicting more than one-half million people in the United States, is characterized by chronic, progressive motor dysfunction resulting in severe tremors, rigidity, and akinesia. The symptoms of Parkinson's disease arise from the degeneration of dopaminergic neurons in the *substantia nigra* and a marked reduction in the concentration of the pyridoxal 5′-phosphate-dependent aromatic *L*-amino acid decarboxylase, the enzyme that catalyzes the conversion of *L*-dopa to the inhibitory neurotransmitter dopamine. Because dopamine is metabolized primarily by monoamine oxidase B in man (see Section 5.5.C.3.c, p. 292), and Parkinson's disease is characterized by a reduction in the brain dopamine concentration (see Chapter 8, Section 8.2.B.10, p. 544), selective inactivation of MAO B has been shown to be an effective approach to increase the dopamine concentration and, thereby, treat this disease. Actually, a MAO B-selective inactivator is used in combination with the antiparkinsonian drug *L*-dopa (Dopar). Selective inhibition of MAO B does not interfere with the MAO A-catalyzed degradation of tyramine and norepinephrine; therefore, no cardiovascular side effects (the cheese effect; see Section 5.5.C.3.c, p. 292) are observed with the use of a MAO B-selective inactivator. The earliest MAO B-selective inactivator, selegiline (**5.128**, Eldepryl),[237] was approved in 1989 by the Food and Drug Administration for the treatment of Parkinson's disease in the United States.

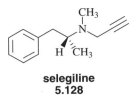

selegiline
5.128

Little was known about the cause for this disease until 1977 when a previously healthy 23-year-old man, who was a chronic street drug user, was referred to the National Institutes for Mental Health for investigation of symptoms of what appeared to be Parkinson's disease.[238] Although the patient responded favorably to the usual treatment for Parkinson's disease (*L*-dopa/carbidopa; see Chapter 8, Section 8.2.B.10, p. 544), the speed with which, and the age at which, the symptoms developed were inconsistent with the paradigm of Parkinson's disease as a regressive geriatric disease. When this man later died of a drug overdose, an autopsy showed the same extensive destruction of the *substantia nigra* that is found in idiopathic parkinsonism. In 1982 four young Californians who had tried some "synthetic heroin" also developed symptoms of an advanced case of Parkinson's disease, including near total immobility.[239] It was found that the drug they had been taking was a *designer drug*, a synthetic narcotic that has a structure designed to be a variation of an existing illegal narcotic; at that time, because the new structure was not listed as a controlled substance, it was not an illegal drug (this law has since been changed).

    In this case the "designers" of the street drug were using the illegal analgesic meperidine (**5.129**, Demerol) as the basis for their structure modification. The compound synthesized, 1-methyl-4-phenyl-4-propionoxypiperidine (**5.130**), was referred to as a "reverse ester" of meperidine (note the ester oxygen in **5.130** is on the opposite side of the carbonyl from that in **5.129**). However, by "reversing" the ester, the drug designers converted a stable ethoxycarbonyl group of meperidine to a propionoxyl group, a good leaving group, in **5.130**. In fact, **5.130** decomposes on heating or in the presence of acids with elimination of propionate to give 1-methyl-4-phenyl-1,2,5,6-tetrahydropyridine (5.131, MPTP). Analysis of several samples of the designer drug (called "new heroin") revealed the MPTP contamination. This same contamination was identified in the samples of drugs taken by the 23-year-old man in 1977, but the drug samples at that time were found to exhibit no neurotoxicity in rats, so it was thought that MPTP was not responsible for the neurological effects. However, when the young California drug addicts were observed to have similar symptoms, tests in primates[240] and mice[241] showed that MPTP produced the same neurological symptoms and histological changes in the *substantia nigra* as those observed with idiopathic parkinsonism. Rats, however, it is now known, are remarkably resistant to the neurotoxic effects of MPTP, which explains why the earlier tests in rats were negative. The observation that an industrial chemist developed Parkinson's disease after synthesizing large amounts of MPTP as a starting material led to the suggestion that cutaneous absorption or vapor inhalation may be significant pathways for introduction of the neurotoxin. Because of this, it can be hypothesized that Parkinson's disease may actually be an environmental disease arising from long-term slow degeneration of dopaminergic neurons by ingested or inhaled neurotoxins similar to MPTP. Because the symptoms of Parkinson's disease do not appear until 60–80% of the dopaminergic neurons are destroyed, it is reasonable that this disease is associated with the elderly. Opponents of this hypothesis note that the interregional and subregional patterns of striated dopamine loss by MPTP differ from those of idiopathic Parkinson's disease.[242] However, results of surveys taken throughout the world suggest that environmental toxins may have an important role in the etiology of Parkinson's disease.[243] Furthermore, it is interesting that many frequently used medicines have structures related to MPTP; some, particularly neuroleptics, produce parkinsonian side effects.[244] Also, genetic factors do not appear to be important in most cases of Parkinson's disease.

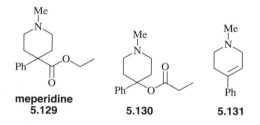

meperidine
5.129   **5.130**   **5.131**

Once a connection was made between MPTP and Parkinson's disease, a vast amount of research with MPTP was initiated, and it soon was realized that the neurotoxic agent was actually a metabolite of MPTP. Pretreatment of animals with the MAO B-selective inactivator selegiline was shown to protect the animals from the neurotoxic effects (both the disease symptoms and the damage to the *substantia nigra*) of MPTP, whereas the MAO-A selective agent clorgyline (**5.132**) did not.[245] This indicated that the neurotoxic metabolite was generated by MAO B oxidation of MPTP. The two metabolites produced by MAO B are 1-methyl-4-phenyl-2,3-dihydropyridinium ion (**5.133**, MPDP$^+$) and 1-methyl-4-phenylpyridinium ion (**5.134**, MPP$^+$).[246] The latter compound (**5.134**) accumulates in selected areas in the brain and, therefore, is believed to be the actual neurotoxic agent.[247] Because MPTP is a neutral molecule, it can cross the blood–brain barrier and enter the brain; once it is oxidized by MAO B, the pyridinium ion (**5.134**) cannot diffuse out of the brain. Selective toxicity of MPTP appears to be the result of the transport of MPP$^+$, but not MPTP, into dopamine neurons via an amine uptake system.[248] These studies suggest that Parkinson's disease may be more than just a geriatric disease; it may be caused by molecules in the environment, possibly pesticides, that are consumed and metabolized to molecules that are neurotoxic.[249]

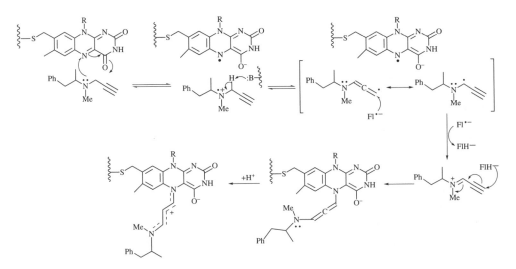

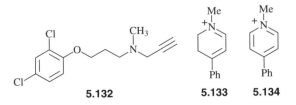

Scheme 5.27 ▶   Hypothetical mechanism for the inactivation of monoamine oxidase by selegiline

5.132   **5.133**   **5.134**

It is apparent, then, that, by inactivation of MAO B, selegiline is important both to the prevention of the oxidation of neurotoxin precursors such as MPTP and to the degradation of dopamine. Although the mechanism of inactivation of MAO by selegiline has not been studied, that of an analog, 3-dimethylamino-1-propyne (**5.135**), has; this compound becomes attached to the N^5 position of the flavin.[250] Because of the evidence for a one-electron mechanism for MAO-catalyzed oxidations (see Section 4.3.C.3), the inactivation mechanisms by selegiline shown in Scheme 5.27 seem most reasonable.

5.135

e. 5-Fluoro-2′-Deoxyuridylate, Floxuridine, and 5-Fluorouracil, Antitumor Drugs

Cancer is a family of diseases characterized by abnormal and uncontrolled cell division. Neither the etiology nor the way in which it causes death is understood in most cases. One important approach to *antineoplastic* (antitumor) agents is the design of an antimetabolite (see Section 5.4.B.2.c, p. 255) whose structure is related to those of pyrimidines and purines involved in the biosynthesis of DNA. These compounds interfere with the formation or utilization of one of these essential normal cellular metabolites. This interference generally results from the inhibition of an enzyme in the biosynthetic pathway of the metabolite or from incorporation, as a false building block, into vital macromolecules such as proteins and polynucleotides. Antimetabolites usually are obtained by making a small structural change in the metabolite, such as a bioisosteric interchange (see Chapter 2, Section 2.2.E.4, p. 29).

5-Fluorouracil (**5.136**, one trade name is Adrucil), its 2′-deoxyribonucleoside, floxuridine (**5.137**, FUDR), and its 2′-deoxyribonucleotide, 5-fluoro-2′-deoxyuridylate (**5.138**), are potent antimetabolites of uracil and its congeners and are also potent antineoplastic agents. 5-Fluorouracil itself is not active, but it is converted *in vivo* to the 2′-deoxynucleotide (**5.138**), which is the active form. Because the van der Waals radius of fluorine (1.35 Å) is similar to that of hydrogen (1.20 Å), 5-fluorouracil and its metabolites are recognized by enzymes that act on uracil and its metabolites. There are several pathways for this *in vivo* activation. A minor pathway to the intermediate 5-fluorouridylate (**5.140**) begins with the conversion of 5-fluorouracil (**5.136**) to 5-fluorouridine (**5.139**), catalyzed by ribose-1-phosphate uridine phosphorylase, followed by further conversion of **5.139** to **5.140**, catalyzed by uridine kinase (Scheme 5.28). The major pathway to **5.140** is the direct conversion of 5-fluorouracil, which is catalyzed by orotate phosphoribosyltransferase. 5-Fluoro-2′-deoxyuridylate (**5.138**) is produced from **5.140** by the circuitous route shown in Scheme 5.28 or by direct conversion of **5.136** to its 2′-deoxyribonucleoside, floxuridine (**5.137**), catalyzed by uridine phosphorylase, followed by 5′-phosphorylation, which is catalyzed by thymidine kinase (Scheme 5.28). However, when **5.137** is administered rapidly, it is converted back to **5.136** faster than it is phosphorylated to **5.138**. Under these circumstances attempts to use floxuridine to bypass the long metabolic route for conversion of **5.136** to **5.138** are unsuccessful. Continuous intra-arterial infusion of floxuridine, however, enhances the direct conversion of **5.137** to **5.138**.

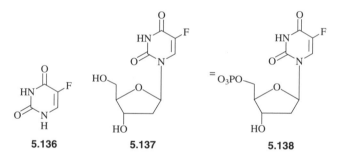

    5.136   **5.137**   **5.138**

The principal site of action of **5.138** is thymidylate synthase, the enzyme that catalyzes the last step in *de novo* biosynthesis of thymidylate, namely, the conversion of 2′-deoxyuridylate to 2′-deoxythymidylate (referred to as just thymidylate). The reaction catalyzed by thymidylate synthase is the only *de novo* source of thymidylate, which is an essential constituent of the DNA. Therefore, inhibition of thymidylate synthase in tumor cells inhibits DNA biosynthesis and produces what is known as *thymineless death* of the cell.[251] Unfortunately, normal cells also require thymidylate synthase for *de novo* synthesis of their thymidylate. Nonetheless, inhibitors of thymidylate synthase are effective antineoplastic agents. There are several reasons for this selective toxicity against tumor cells; all are related to the difference in the rates of cell division for normal and abnormal cells. Because aberrant cells replicate much more rapidly than do most normal cells, the rapidly proliferating tumor cells have a higher requirement for its DNA than do the slower proliferating normal cells. This means that the activity of thymidylate synthase is elevated in tumor cells relative to normal cells. Because uracil is one of the precursors of thymidylate, it *and* 5-fluorouracil are taken up into tumor cells much more efficiently than into normal cells. Finally, and possibly most importantly, enzymes that degrade uracil in normal cells also degrade 5-fluorouracil, and these degradation processes do not take

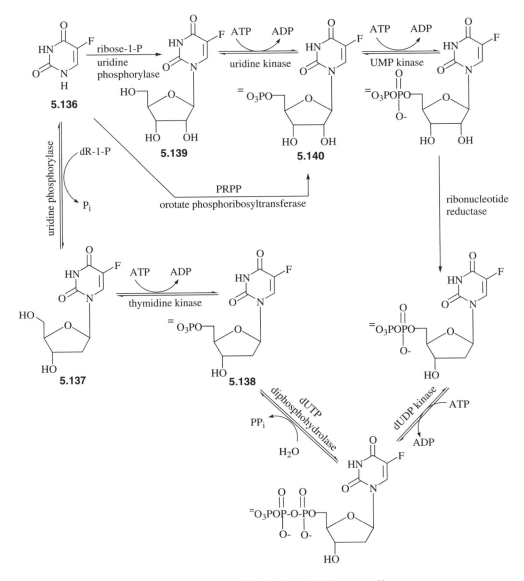

Scheme 5.28 ▶ Metabolism of 5-fluorouracil

place in cancer cells.[252] The adverse side effects accompanying the use of 5-fluorouracil in humans generally arise from the inhibition of thymidylate synthase and destruction of the rapidly proliferating normal cells of the intestines, the bone marrow, and the mucosa. The effects of anticancer drugs are discussed in more detail in Chapter 6.

Unlike other tetrahydrofolate-dependent enzymes (see Section 4.3.B), thymidylate synthase utilizes methylenetetrahydrofolate as both a one-carbon donor *and* as a reducing agent (Scheme 5.29).[253] An active site cysteine residue undergoes Michael addition to the 6-position of 2′-deoxyuridylate (**5.141**, dRP is deoxyribose phosphate) to give an enolate (**5.142**) that attacks N^5,N^{10}-methylenetetrahydrofolate (more likely, in one of the more reactive open, iminium forms; see Chapter 4, Scheme 4.23, p. 204), and forms a ternary complex

(**5.143**) of the enzyme, the substrate, and the coenzyme. Enzyme-catalyzed removal of the C-5 proton leads to β-elimination of tetrahydrofolate (**5.143** to **5.144**). Oxidation of the tetrahydrofolate (a hydride mechanism is shown, but a one-electron mechanism, that is, first transfer of an N-5 nitrogen nonbonded electron to the alkene followed by hydrogen atom transfer, is possible) gives dihydrofolate (**5.145**) and the enzyme-bound thymidylate enolate (**5.146**). Reversal of the first step releases the active site cysteine residue and produces thymidylate (**5.147**). This reaction changes the oxidation state of the coenzyme. Because of this, another enzyme, dihydrofolate reductase, is required to reduce the dihydrofolate back to tetrahydrofolate (see Chapter 4, Scheme 4.21).

5-Fluoro-2′-deoxyuridylate (**5.138**) inactivates thymidylate synthase because once the ternary complex (**5.148**) forms, there is no C-5 proton that the enzyme can remove to eliminate the tetrahydrofolate (Scheme 5.30).[254] Consequently, the enzyme remains as the ternary complex. Note that in this mechanism the inactivator is not converted into a reactive compound that attaches to the enzyme. Instead, it first attaches to the enzyme, then requires condensation with N^5,N^{10}-methylenetetrahydrofolate to generate a stable complex.

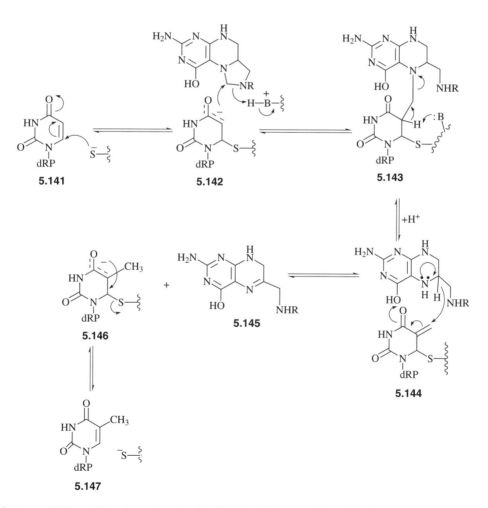

Scheme 5.29 ▶ Hypothetical mechanism for thymidylate synthase (dRP is deoxyribose phosphate)

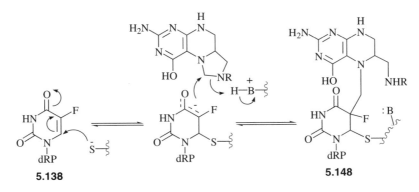

Scheme 5.30 ► Hypothetical mechanism for the inactivation of thymidylate synthase by 5-fluoro-2′-deoxyuridylate

A mechanism-based inactivator, therefore, does not necessarily require formation of a reactive species. It only requires the enzyme to catalyze a reaction on it that leads to inactivation prior to release of the product. If the enzyme were inactivated without the requirement of N^5,N^{10}-methylenetetrahydrofolate, that is, by simple Michael addition of the active site cysteine to the C-6 position of the inactivator, then **5.138** would be an affinity labeling agent.

Because dihydrofolate reductase is essential for the regeneration of tetrahydrofolate from the dihydrofolate produced in the thymidylate synthase reaction, drug synergism occurs when inhibitors of dihydrofolate reductase (such as methotrexate, **5.149**, Rheumatrex) and thymidy-late synthase are used in combination. This is an example of a sequential blocking synergistic mechanism (see Section 5.3.B.2, p. 240).

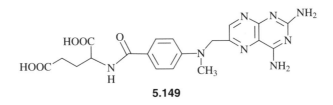

Next, in Chapter 6, we will consider drug interactions with another type of receptor, DNA.

◩ 5.6   General References

Drug Resistance and Synergism; Chemotherapy

Albert, A. *Selective Toxicity*, 7th ed., Chapman and Hall, London, 1985.
Lewis K.; Salvers, A. A.; Taber, H. W.; Wax, R. G. (Eds.), *Bacterial Resistance to Antimicro-bials*, Marcel Dekker, New York, 2002.
Walsh, C. *Antibiotics: Actions, Origins, Resistance*, ASM Press, Washington, DC, 2003.

Slow, Tight-Binding Inhibitors

Morrison, J. F.; Walsh, C. T. *Adv. Enzymol.* **1988**, *61*, 201.
Schloss, J. V. *Acc. Chem. Res.* **1988**, *21*, 348.

Sculley, M. J.; Morrison, J. F. *Biochim. Biophys. Acta* **1986**, *874*, 44.
Szedlacsek, S. E.; Duggleby, R. G. *Meth. Enzymol.* **1995**, *249* (Part D), 144.

Transition State Analogs and Multisubstrate Analogs
Andrews, P. R.; Winkler, D. A. In *Drug Design: Fact or Fantasy?* Jolles, G.; Wooldridge, K. R. H. (Eds.), Academic, London, 1984, p. 145.
Radzicka, A.; Wolfenden, R. *Meth. Enzymol.* **1995**, *249* (Part D), 284.

Various Classes of Drugs
Hardman, J. G.; Limbird, L. E.; Gilman, A. G. (Eds.) *Goodman and Gilman's The Pharmacological Basis of Therapeutics*, 10th ed., McGraw-Hill, New York, 2001.
Wolff, M. E. (Ed.) *Burger's Medicinal Chemistry and Drug Discovery*, John Wiley & Sons, New York, 1995–1997, Vols. 1–6.

Mechanism-Based Enzyme Inactivators
(a) Silverman, R. B. *Mechanism-Based Enzyme Inactivation: Chemistry and Enzymology*, CRC, Boca Raton, FL, **1988**, Vols. I and II. (b) Silverman, R. B. *Meth. Enzymol.* **1995**, *249* (Part D), 240.

Polyamines
Johnson, L. R.; McCormack, S. A. *J. Physiol. Pharmacol.* **2001**, *52*, 327.
McCann, P. P.; Pegg, A. E.; Sjoerdsma, A. (Eds.) *Inhibition of Polyamine Metabolism. Biological Significance and Basis for New Therapies*, Academic, Orlando, 1987.
Pegg., A. E. *Cancer Res.* **1988**, *48*, 759.
Tabor, C. W.; Tabor, H. *Annu. Rev. Biochem.* **1984**, *53*, 749.

5.7 PROBLEMS

(Answers can be found in Appendix at the end of the book.)

1. Why would you want to design a drug that is an enzyme inhibitor?

2. If you wanted to inhibit an enzyme in a microorganism that is also in humans, what approaches would you take in your research?

3. *S*-Adenosylmethionine (SAM) is biosynthesized from methionine and ATP, catalyzed by methionine adenosyl transferase. The mechanism is shown below.

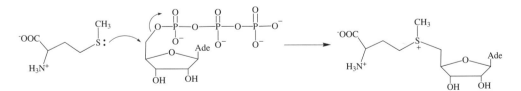

   A. Design a competitive reversible inhibitor for this enzyme.

   B. Design a multisubstrate analog inhibitor and show the basis for your design.

4. What advantage does a slow, tight-bonding inhibitor have over a simple reversible inhibitor?

5. Resistance to your new potent antibacterial drug (**1**) was shown to be the result of a single-point genetic mutation in the target enzyme such that an important active site lysine residue was mutated to an aspartate residue. Suggest a simple way to proceed toward the design of a new antibacterial drug against the resistant strain.

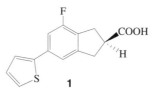

6. Thromboxane A_2 (TXA$_2$) is biosynthesized in humans from prostaglandin H_2 (PGH$_2$) by the enzyme thromboxane A_2 synthase. Binding of TXA$_2$ to a receptor causes vasoconstriction (raises the blood pressure) and platelet aggregation; therefore, TXA$_2$ has been implicated as a causative factor in a number of cardiovascular and renal diseases.

     A. Without looking at part B of this question, briefly describe three approaches you could take regarding TXA$_2$ to design a new antihypertensive agent.

     B. Thromboxane A_2 synthase inhibitors, such as **2**, have not been successful, possibly because of the buildup of PGH$_2$, the substrate for thromboxane A_2 synthase, which also is a potent agonist at the TXA$_2$ receptor. PGH$_2$ is converted to PGI$_2$, a potent vasodilator and platelet inhibitory agent, by PGI$_2$ synthase. TXA$_2$ receptor antagonists, such as **3**, block the binding of both PGH$_2$ and TXA$_2$ to TXA$_2$ receptor, but because TXA$_2$ synthase is not inhibited, there is no buildup of PGH$_2$ and therefore no conversion to PGI$_2$. What approach would you take to get around these problems?

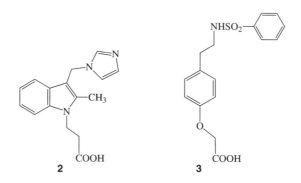

     C. Design a molecule that uses the principle you proposed in part B.

7. Show the transition state for the reaction below, and draw a reasonable transition state analog inhibitor.

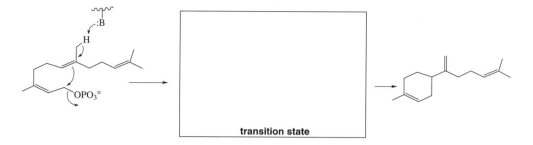

transition state

8. 5-Fluoro-2′-deoxyuridylate inactivates thymidylate synthase and is an antitumor agent. What would be a good choice for drug combination with 5-FdUMP? Why?

9. 5-Lipoxygenase requires a ferric ion as a cofactor. Show how **4** acts as an inhibitor of 5-lipoxygenase.

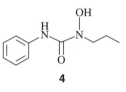

4

10. AG 7088 was designed to inhibit a picornavirus protease and is a potent antiviral agent. Why do you think moieties A–E were incorporated into the structure?

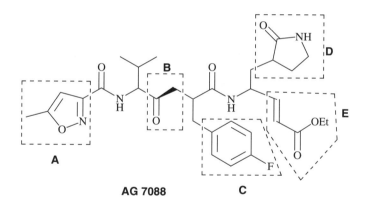

AG 7088

11. Two isoforms of an enzyme were discovered; isoform-1 produces a hormone that causes muscle spasms and isoform-2 makes another hormone from the same substrate that lowers cholesterol levels.

   A. What would you do to prevent muscle spasms without raising cholesterol levels?

   B. If the active sites of isoform-1 and -2 are the same except isoform-1 has a cysteine residue and isoform-2 has a phenylalanine residue at that same position, what two approaches would you take for a muscle spasm drug without a cholesterol-increase side effect?

12. Fosfomycin (**5**) is a potent antibacterial agent that interferes with cell wall biosynthesis at the enzyme called MurA. MurA contains an active site cysteine residue.

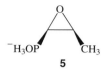

5

    A. Draw a potential chemical mechanism for how fosfomycin acts.

    B. Give two reasonable mechanisms for resistance to fosfomycin that do not involve the bacterial cell membrane.

13. GABA aminotransferase catalyzes a PLP-dependent conversion of GABA to succinic semialdehyde (see Section 5.5.C.3.a).

    A. Draw a mechanism for how **6** inactivates this enzyme.

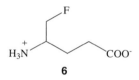

6

    B. A hypothetical anticonvulsant drug that inhibits GABA aminotransferase was given to a patient in overdose quantities. Not only did the patient stop convulsing, but he went into a coma. If the problem was that the GABA concentration became too high, mention two possible solutions to the problem.

14. An excess of androgenic hormones such as testosterone can cause benign prostatic hypertrophy (enlarged prostate). Most of the androgenic activity appears to be caused by a metabolite of testosterone (**7**), namely, 5α-dihydrotestosterone (**8**), produced from testosterone in a NADPH-dependent reaction catalyzed by steroid 5α-reductase. The mechanism for this reductase is shown below.

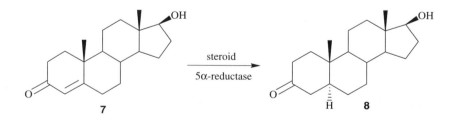

7   **8**

Finasteride (**9**, Proscar) is a potent inhibitor of steroid 5α-reductase. It has been proposed to be a mechanism-based inhibitor. Draw a mechanism consistent with this proposal.

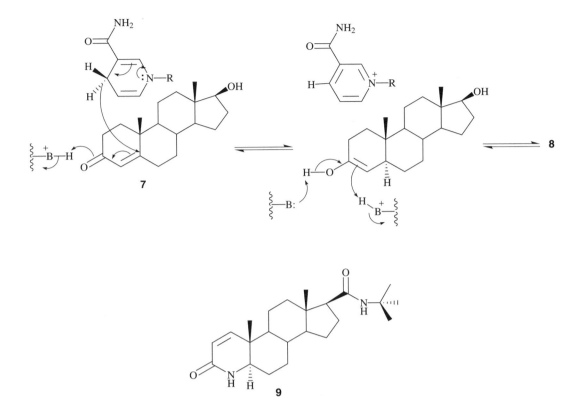

5.8 References

1. Cohen, S. S. *Science*, **1979**, *205*, 964.

2. Mobashery, S.; Azucena, E. In *Encyclopedia Life Science*, Nature Publishing Group, United Kingdom, 2002, Vol. 2, pp. 472–477.

3. (a) Neu, H. C. *Science* **1992**, *257*, 1064. (b) Tomasz, A. *N. Engl. J. Med.* **1994**, *330*, 1247.

4. (a) Gold, H. S.; Moellering, R. C. *N. Engl. J. Med.* **1996**, *335*, 1445. (b) Domagala, J. M.; Sanchez, J. P. *Annu. Rep. Med. Chem.* **1997**, *32*, 111. (c) Levy, S. B. *Sci. Am.* **1998**, *278*, 46.

5. (a) Albert, A. *Selective Toxicity*, 7th ed., Chapman and Hall, London, **1985**, p. 256. (b) Lowe, J. A. III *Annu. Rep. Med. Chem.* **1982**, *17*, 119.

6. (a) Haest, C. W. M.; De Gier, J.; Op den Kamp, J. A. F.; Bartels, P.; Van Deenen, L. L. M. *Biochim. Biophys. Acta* **1972**, *255*, 720. (b) Bryan, C. E.; Kwan, S. *J. Antimicrob. Chemother.* **1981**, *8* (Suppl. D), 1.

7. Alt, F. W.; Kellems, R. E.; Bertino, J. R.; Schimke, R. T. *J. Biol. Chem.* **1978**, *253*, 1357.

8. Kan, S.; Siddiqui, W. *J. Protozool.* **1979**, *26*, 660.

9. Bertino, J.; Cashmore, A.; Fink, N.; Calabresi, P.; Lefkowitz, E. *Clin. Pharmacol. Ther.* **1965**, *6*, 763.

10. Spratt, B. G. *Science* **1994**, *264*, 388.

11. Then, R. L.; Hermann, F. *Chemotherapy*, **1981**, *27*, 192.

12. Weisblum, B. In *Microbiology—1974*; Schlessinger, D. (Ed.), ASM, Washington, DC, 1975, p. 199.

13. (a) Larder, B. A.; Darby, G.; Richman, D. D. *Science* **1989**, *243*, 1731. (b) Schuurman, R.; Nijhuis, M.; VanLeeuwen, R.; Schipper, P.; DeJong, D.; Collis, P.; Danner, S.A.; Mulder, J.; Loveday, C.; Christopherson, C.; Kwok, S.; Sninsky, J.; Boucher, C. A. B. *J. Infect. Dis.* **1995**, *171*, 1411.

14. Kellam, P.; Boucher, C. A. B.; Larder, B. A. *Proc. Natl. Acad. Sci. USA* **1992**, *89*, 1934.

15. (a) Gao, Q.; Gu, Z.; Parniak, M. A.; Cameron, J.; Cammack, N.; Boucher, C.; Wainberg, M. A. *Antimicrob. Agents Chemother.* **1993**, *37*, 1390. (b) Tisdale, M.; Kemp, S. D.; Parry, N. R.; Larder, B. A. *Proc. Natl. Acad. Sci. USA* **1993**, *90*, 5653.

16. Larder, B. A.; Kemp, S. D.; Harrigan, P. R. *Science* **1995**, *269*, 696.

17. (a) Bugg, T. D. H.; Walsh, C. T. *Nat. Prod. Rep.* **1992**, *9*, 199. (b) Walsh, C. T. *Science* **1993**, *261*, 308. (c) Walsh, C. T.; Fisher, S. L.; Park, I.-S.; Prahalad, M.; Wu, Z. *Chem. Biol.* **1996**, *3*, 21.

18. Bugg, T. D. H.; Wright, G. D.; Dutka-Malen, S.; Arthur, M.; Courvalin, P.; Walsh, C. T. *Biochemistry* **1991**, *30*, 10408.

19. (a) Kotra, L. P.; Samama, J. P.; Mobashery, S. In *Bacterial Resistance to Antimicrobials, Mechanisms, Genetics, Medical Practice and Public Health*; Lewis, K.; Salyers, A. A.; Haber, H. W.; Wax, R. G. (Eds.), Marcel Dekker, New York, 2002, pp. 123–159. (b) Bush, K. *Curr. Pharm. Des.* **1999**, *5*, 839. (c) Rice, L. B.; Bonomo, R. A. *Drug Resist. Updates* **2000**, *3*, 178.

20. (a) Wright, G. D.; Berghuis, A. M.; Mobashery, S. In *Aminoglycoside Antibiotics: Structures, Functions and Resistance*; Rosen, B. P.; Mobashery, S. (Eds.), Plenum Press, New York, 1998, pp. 27–69. (b) Kondo, S.; Hotta, K. *J. Infect. Chemother.* **1999**, *5*, 1. (c) Mingeot-Leclerco, M.-P.; Glupczynski, Y.; Tulkens, P. M. *Antimicrob. Agents Chemother.* **1999**, *43*, 727.

21. Roestanmadji, J.; Grapsas, I.; Mobashery, S. *J. Am. Chem. Soc.* **1995**, *117*, 11060.

22. Haddad, J.; Vakulenko, S.; Mobashery, S. *J. Am. Chem. Soc.* **1999**, *121*, 11922.

23. Sucheck, S. J.; Wong, A. L.; Koeller, K. M.; Boehr, D. D.; Draker, K.-a.; Sears, P.; Wright, G. D.; Wong, C.-H. *J. Am. Chem. Soc.* **2000**, *122*, 5230.

24. Harrap, K. In *Scientific Foundations of Oncology*; Symington, T.; Carter, R. (Eds.), Heinemann, London, 1976, p. 641.

25. (a) Lawrence, L. E.; Barrett, J. F. *Exp. Opin. Invest. Drugs* **1998**, *7*, 199. (b) Marshall, N. J.; Piddock, L. J. *Microbiologia* **1997**, *13*, 285. (c) Nikaido, H. *Clin. Infect. Dis.* **1998**, *27*, S32.

26. (a) Nikaido, H. *Curr. Opin. Microbiol.* **1998**, *1*, 516. (b) Paulsen, I. T.; Brown, M. H.; Skurray, R. A. *Microbiol. Rev.* **1996**, *60*, 575. (c) Nikaido, H. *Science* **1994**, *264*, 382.

27. (a) Renau, T. E.; Leger, R.; Flamme, E. M.; Sangalang, J.; She, M. W.; Yen, R.; Gannon, C. L.; Griffith, D.; Chamberland, S.; Lomovskya, O.; Hecker, S. J.; Lee, V. J.; Ohta, T.; Nakayama, K. *J. Med. Chem.* **1999**, *42*, 4928. (b) Stermitz, F. R.; Lorenz, P.; Tawara, J. N.; Zenewicz, L. A.; Lewis, K. *Proc. Natl. Acad. Sci. USA* **2000**, *97*, 1433.

28. Albert, A. *Selective Toxicity*, 7th ed., Chapman and Hall, London, 1985, p. 256.

29. (a) Segel, I. H. *Enzyme Kinetics*, John Wiley & Sons, New York, 1975, p. 106. (b) Burlingham, B. T.; Widlanski, T. S. *J. Chem. Educ.* **2003**, *80*, 214.

30. Espiner, E. A.; Nicholls, M. G. In *The Renin-Angiotensin System*; Robertson, J. I. S.; Nicholls, M. G. (Eds.), Gower Medical Publishing, London, 1993, pp. 33.1–33.24.

31. Tewksbury, D. A.; Dart, R. A.; Travis, J. *Biochem. Biophys. Res. Commun.* **1981**, *99*, 1311; Tewksbury, D. In *Biochemical Regulation of Blood Pressure*; Soffer, R. L. (Ed.), Wiley, New York, 1981, p. 95.

32. Moeller, I.; Allen, A. M.; Chai, S.-Y.; Zhuo, J.; Mendelsohn, F. A. O. *J. Human Hypertension* **1998**, *12*, 289.

33. Wilk, D.; Healy, D. P. *Adv. Neuroimmunol.* **1993**, *3*, 195.

34. Larner, A.; Vaughan, E. D., Jr.; Tsai, B.-S.; Peach, M. J. *Proc. Soc. Exp. Biol. Med.* **1976**, *152*, 631.

35. (a) Zini, S.; Fournie-Zaluski, M.-C.; Chauvel, E.; Roques, B. P.; Corvol, P.; Llorens-Cortes, C. *Proc. Natl. Acad. Sci. USA* **1996**, *93*, 11968. (b) Blair-West, J. R.; Coghlan, J. P.; Denton, D. A.; Funder, J. W.; Scoggins, B. A.; Wright, R. D. *J. Clin. Endocrinol. Metab.* **1971**, *32*, 575. (c) Caldicott, W. J. H.; Taub, K. J.; Hollenberg, N. K. *Life Sci.* **1977**, *20*, 517.

36. Palmieri, F. E.; Bausback, H. H.; Ward, P. E. *Biochem. Pharmacol.* **1989**, *38*, 173.

37. Ferreira, S. H. *Brit. J. Pharmacol. Chemother.* **1965**, *24*, 163.

38. (a) Bakhle, Y. S. *Nature (London)* **1968**, *220*, 919. (b) Bakhle, Y. S.; Reynard, A. M.; Vane, J. R. *Nature (London)* **1969**, *222*, 956.

39. Ferreira, S. H.; Bartelt, D. C.; Greene, L. J. *Biochemistry* **1970**, *9*, 2583.

40. Ferreira, S. H.; Greene, L. J.; Alabaster, V. A.; Bakhle, Y. S.; Vane, J. R. *Nature (London)* **1970**, *225*, 379.

41. Stewart, J. M.; Ferreira, S. H.; Greene, L. J. *Biochem. Pharmacol.* **1971**, *20*, 1557.

42. Ondetti, M. A.; Williams, N. J.; Sabo, E. F.; Pluscec, J.; Weaver, E. R.; Kocy, O. *Biochemistry* **1971**, *10*, 4033.

43. Cheung, H. S.; Cushman, D. W. *Biochim. Biophys. Acta* **1973**, *293*, 451.

44. Cushman, D. W.; Cheung, H. S. In *Hypertension*; Genest, J.; Koiw, E. (Eds.), Springer, Berlin, 1972, p. 532.

45. Ondetti, M. A.; Cushman, D. W. In *Biochemical Regulation of Blood Pressure*; Soffer, R. L. (Ed.), Wiley, New York, 1981, p. 165.

46. Kato, H.; Suzuki, T. *Biochemistry* **1971**, *10*, 972.

47. Das, M.; Soffer, R. L. *J. Biol. Chem.* **1975**, *250*, 6762.

48. Quiocho, F. A.; Lipscomb, W. N. *Adv. Protein Chem.* **1971**, *25*, 1.

49. Cushman, D. W.; Cheung, H. S.; Sabo, E. F.; Ondetti, M. A. *Biochemistry*, **1977**, *16*, 5484.

50. Byers, L. D.; Wolfenden, R. *J. Biol. Chem.* **1972**, *247*, 606; Byers, L. D.; Wolfenden, R. *Biochemistry* **1973**, *12*, 2070.

51. Ondetti, M. A.; Cushman, D. W.; Sabo, E. F.; Cheung, H. S. In *Drug Action and Design: Mechanism-Based Enzyme Inhibitors*; Kalman, T. I. (Ed.), Elsevier/North Holland, New York, 1979, p. 271.

52. Atkinson, A. B.; Robertson, J. I. S. *Lancet* **1979**, *ii*, 836.

53. Patchett, A. A.; Harris, E.; Tristram, E. W.; Wyvratt, M. J.; Wu, M. T.; Taub, D.; Peterson, E. R.; Ikeler, T. J.; ten Broeke, J.; Payne, L. G.; Ondeyka, D. L.; Thorsett, E. D.; Greenlee, W. J.; Lohr, N. S.; Hoffsommer, R. D.; Joshua, H.; Ruyle, W. V.; Rothrock, J. W.; Aster, S. D.; Maycock, A. L.; Robinson, F. M.; Hirschmann, R.; Sweet, C. S.; Ulm, E. H.; Gross, D. M.; Vassil, T. C.; Stone, C. A. *Nature (London)* **1980**, *288*, 280.

54. Wyvratt, M. J.; Patchett, A. A. *Med. Res. Rev.* **1985**, *4*, 483.

55. Marshall, B. J. In *Helicobacter pylori*, Mobley, H. L. T.; Mendz, G. L.; Hazell, S. L. (Eds.), ASM Press, Herndon, VA, 2001, pp. 19–24.

56. Montecucco, C.; Rappuoli, R. *Nat. Rev. Mol. Cell Biol.* **2001**, *2*, 457.

57. Israel, D. A.; Peek, R. M. *Aliment. Pharmacol. Ther.* **2001**, *15*, 1271.

58. (a) Xia, H. H.-X.; Wong, B. C. Y.; Talley, N. J.; Lam, S. K. *Expert Opin. Pharmacother.* **2001**, *2*, 253. (b) Williamson, J. S. *Curr. Pharm. Des.* **2001**, *7*, 355.

59. (a) De Lombaert, S.; Chatelain, R. E.; Fink, C. A.; Trapani, A. J. *Curr. Pharm. Des.* **1996**, *2*, 443. (b) Fink, C. A. *Expert Opin. Ther. Pat.* **1996**, *6*, 1147. (c) Seymour, A. A.; Asaad, M. M.; Lanoce, V. M.; Langenbacher, K. M.; Fennell, S. A.; Rogers, W. L. *J. Pharmacol. Exp. Ther.* **1993**, *266*, 872. (d) Pham, I.; Gonzalez, W.; El Amraani, A. I.; Fournie-Zaluski, M. C.; Philippe, M.; Laboulandine, I.; Roques, B. P.; Michel, J. B. *J. Pharmacol. Exp. Ther.* **1993**, *265*, 1339.

60. Robl, J. A.; Cimarusti, M. P.; Simpkins, L. M.; Brown, B.; Ryono, D. E.; Bird, J. E.; Asaad, M. M.; Schaeffer, T. R.; Trippodo, N. C. *J. Med. Chem.* **1996**, *39*, 494.

61. Robl, J. A.; Sun, C. Q.; Stevenson, J.; Ryono, D. E.; Simpkins, L. M.; Cimarusti, M. P.; Dejneka, T.; Slusarchyk, W. A.; Chao, S.; Stratton, L.; Misra, R. N.; Bednarz, M. S.; Asaad, M. M.; Cheung, H. S.; AbboaOffei, B. E.; Smith, P. L.; Mathers, P. D.; Fox, M.; Schaeffer, T. R.; Seymour, A. A.; Trippodo, N. C. *J. Med. Chem.* **1997**, *40*, 1570.

62. Turner, A. J.; Tanzawa, K. *FASEB J.* **1997**, *11*, 355.

63. (a) Rubanyi, G. M.; Polokoff, M. A. *Pharmacol. Rev.* **1994**, *46*, 325. (b) Patel, T. R. *CNS Drugs* **1996**, *5*, 293. (c) Benigni, A.; Remuzzi, G. *Exp. Opin. Ther. Patents* **1997**, *7*, 139.

64. (a) Loffler, B.-M. *J. Cardiovasc. Pharmacol.* **2000**, *35* (*Suppl. 2*), S79. (b) Vemulapalli, S.; Chintala, M.; Stamford, A.; Watkins, R.; Chiu, P.; Sybertz, E.; Fawzi, A. B. *Cardiovasc. Drug Rev.* **1997**, *15*, 260.

65. Fukuda, Y.; Fukuta, Y.; Higashino, R.; Ogishima, M.; Yoshida, K.; Tamaki, H.; Takei, M. *Naunyn-Schmiedeberg's Arch. Pharmacol.* **1999**, *359*, 433.

66. (a) Furuta, S.; Fukuda, Y.; Sugimoto, T.; Miyahara, H.; Kamada, E.; Sano, H.; Fukuta, Y.; Takei, M.; Kurimoto, T. *Eur. J. Pharmacol.* **2001**, *426*, 105. (b) Fukuta, Y.; Fukuda, Y.; Higashino, R.; Yoshida, K.; Ogishima, M.; Tamaki, H.; Takei, M. *J. Pharmacol. Exp. Ther.* **1999**, *290*, 1013.

67. Kenny, B.; Ballard, S.; Blagg, J.; Fox, D. *J. Med. Chem.* **1997**, *40*, 1293.

68. Lepor, H.; Knap-Maloney, G.; Sunshine, H.; *J. Urol.* **1990**, *144*, 1393.

69. Yoshida, K.; Horikoshi, Y.; Eta, M.; Chikazawa, J.; Ogishima, M.; Fukuda, Y.; Sato, H. *Bioorg. Med. Chem. Lett.* **1998**, *8*, 2967.

70. Sato, H.; Kitagawa, O.; Aida, Y.; Chikazawa, J.; Kurimoto, T.; Takei, M.; Fukuta, Y.; Yoshida, K. *Bioorg. Med. Chem. Lett.* **1999**, *9*, 1553.

71. (a) Missale, C.; Nash, S. R.; Robinson, S. W.; Jaber, M.; Caron, M. G. *Physiol. Rev.* **1998**, *78*, 189. (b) Strange, P. G. In *Advances in Drug Research*, Testa, B., Meyer, U. A. (Eds.), Academic Press, London, 1996, Vol. 28, pp. 313–352.

72. Andersen, G. P. In *New Drugs for Asthma Therapy. Agents and Actions Supplement 34*; Anderson, G. P., Chapman, I. D., Morley, J. (Eds.), Birkhauser Verlag, Basel, 1991, pp. 97–115.

73. Bonnert, R. V.; Brown, R. C.; Chapman, D.; Cheshire, D. R.; Dixon, J.; Ince, F.; Kinchin, E. C.; Lyons, A. J.; Davis, A. M.; Hallam, C.; Harper, S. T.; Unitt, J. F.; Dougall, I. G.; Jackson, D. M.; McKechnie, K.; Young, A.; Simpson, W. T. *J. Med. Chem.* **1998**, *41*, 4915.

74. Domagk, G. *Deut. Med. Wochenschr.* **1935**, *61*, 250.

75. Albert, A. *Selective Toxicity*; 7th ed., Chapman and Hall, London, 1985, p. 220.

76. Domagk, G. *Klin. Wochenschr.* **1936**, *15*, 1585.

77. Tréfouël, J.; Tréfouël, Mme. J.; Nitti, F.; Bovet, D. *Compt. Rend. Soc. Biol., Paris* **1935**, *120*, 756.

78. Marshall, E. K., Jr. *J. Biol. Chem.* **1937**, *122*, 263; Bratton, A. C.; Marshall, E. K., Jr. *J. Biol. Chem.* **1939**, *128*, 537.

79. Stamp, T. C. *Lancet* **1939**, *ii*, 10.

80. Woods, D. D. *Brit. J. Exper. Pathol.* **1940**, *21*, 74.

81. Selbie, F. R. *Brit. J. Exper. Pathol.* **1940**, *21*, 90.

82. Fildes, P. *Lancet*, **1940**, *i*, 955.

83. Miller, A. K. *Proc. Soc. Exper. Pathol. Med.* **1944**, *57*, 151.; Miller, A. K.; Bruno, P.; Berglund, R. M. *J. Bacteriol.* **1947**, *54*, 9 (G20).

84. Nimmo-Smith, R. H.; Lascelles, J.; Woods, D. D. *Brit. J. Exper. Pathol.* **1948**, *29*, 264.

85. Richey, D. P.; Brown, G. M. *J. Biol. Chem.* **1969**, *244*, 1582.

86. Weisman, R. A.; Brown, G. M. *J. Biol. Chem.* **1964**, *239*, 326.

87. Bock, L.; Miller, G. H.; Schaper, K.-J.; Seydel, J. K. *J. Med. Chem.* **1974**, *17*, 23.

88. Wood, R. C.; Ferone, R.; Hitchings, G. H. *Biochem. Pharmacol.* **1961**, *6*, 113.

89. Wong, K. K.; Pompliano, D. L. In *Resolving the Antibiotic Paradox*, Rosen, B. P.; Mobashery, S. (Eds.), Plenum Press, New York, 1998, pp. 197–217.

90. Davies, J. E. In *Antibiotic Resistance: Origins, Evolution and Spread*; Chadwick, D. J.; Goode, J. (Eds.), Wiley, New York, 1997.

91. Landy, M.; Gerstung, R. B. *J. Bacteriol.* **1944**, *47*, 448.

92. (a) Wise, E. M., Jr.; Abou-Donia, M. M. *Proc. Natl. Acad. Sci. U.S.A.* **1975**, *72*, 2621. (b) Skold, O. *Antimicrob. Agents Chemother.* **1976**, *9*, 49.

93. (a) Swedberg, G.; Skold, O. *J. Bacteriol.* **1980**, 142, 1. (b) Nagate, T.; Inoue, M.; Inoue, K.; Mitsuhashi, S. *Microbiol. Immun.* **1978**, 22, 367.

94. Wormser, G. P.; Keusch, G. T.; Rennie, C. H. *Drugs* **1982**, *24*, 459.

95. Hitchings, G. H.; Burchall, J. J. *Adv. Enzymol.* **1965**, *27*, 417; Anand, N. In *Inhibition of Folate Metabolism in Chemotherapy;* Hitchings, G. H. (Ed.), Springer-Verlag, Berlin, 1983, p. 25.

96. Ferone, R.; Burchall, J. J.; Hitchings, G. H. *Mol. Pharmacol.* **1969**, *5*, 49.

97. Bernhard, S. A.; Orgel, L. E. *Science* **1959**, *130*, 625.

98. Pauling, L. *Am. Scientist* **1948**, *36*, 51.

99. Jencks, W. P. In *Current Aspects of Biochemical Energetics*, Kennedy, E. P. (Ed.), Academic, New York, 1966, p. 273.

100. Wolfenden, R. *Annu. Rev. Biophys. Bioeng.* **1976**, *5*, 271.; Wolfenden, R. *Nature* **1969**, *223*, 704; Wolfenden, R. *Meth. Enzymol.* **1977**, *46*, 15.

101. Lienhard, G. E. *Science* **1973**, *180*, 149; Lienhard, G. E. *Annu. Repts. Med. Chem.* **1972**, *7*, 249.

102. (a) Imperiali, B.; Abeles, R. H. *Biochemistry* **1986**, *25*, 3760. (b) Gelb, M. H.; Svaren, J. P.; Abeles, R. H. *Biochemistry* **1985**, *24*, 1813.

103. Christianson, D. W.; Lipscomb, W. N. *Acc. Chem. Res.* **1989**, *22*, 62.

104. (a) Loo, T. L.; Nelson, J. A. In *Cancer Medicine*, 2nd ed., Holland, J. F.; Frei, E. III (Eds.), Lea & Febiger, Philadelphia, 1982, p. 790, (b) McCormack, J. J.;

Johns, D. G. In *Pharmacologic* Principles of Cancer Treatment, Chabner, B. A. (Ed.), W. B. Saunders, Philadelphia, 1982, p. 213.

105. Agarwal, R. P.; Spector, T.; Parks, R. E., Jr. *Biochem. Pharmacol.* **1977**, *26*, 359.

106. Wilson, D. K.; Rudolph, F. B.; Quiocho, F. A. *Science* **1991**, 252, 1278.

107. Berne, R. M.; Rall, T. W.; Rubio, R. (Eds.), *Regulatory Functions of Adenosine*, Martinus Nijhoff, Boston, 1983.

108. Tritsch, G. L. (Ed.) *Adenosine Deaminase in Disorders of Purine Metabolism and in Immune Deficiency*; New York Academy of Sciences, New York, 1985, Vol. 451.

109. (a) Stark, G. R.; Bartlett, P. A. *Pharmacol. Ther.* **1983**, *23*, 45. (b) Collins, K. D.; Stark, G. R. *J. Biol. Chem.* **1971**, *246*, 6599.

110. Erlichman, C.; Vidgen, D. *Biochem. Pharmacol.* **1984**, *33*, 3177.

111. (a) Waley, S. G. *Biochem. J.* **1993**, *294*, 195. (b) Sculley, M. J.; Morrison, J. F. *Biochim. Biophys. Acta* **1986**, *874*, 44. (c) Schloss, J. V. *Acc. Chem. Res.* **1988**, *21*, 348. (d) Morrison, J. F.; Walsh, C. T. *Adv. Enzymol.* **1988**, *61*, 201.

112. (a) Rich, D. H.; *J. Med. Chem.* **1985**, *28*, 263. (b) Morrison, J. F.; Walsh, C. T. *Adv. Enzymol.* **1988**, *61*, 201.

113. (a) Imperiali, B.; Abeles, R. H. *Biochemistry* **1986**, *25*, 3760. (b) Stein, R. L.; Strimpler, A. M.; Edwards, P. D.; Lewis, J. J.; Mauger, R. C.; Schwartz, J. A.; Stein, M. M.; Trainor, D. A.; Wildonger, R. A.; Zottola, M. A. *Biochemistry* **1987**, *26*, 2682.

114. Morrison, J. F. *Trends Biochem. Sci.* **1982**, *7*, 102.

115. Bartlett, P. A.; Marlowe, C. K. *Science* **1987**, *235*, 569.

116. Rich, D. H.; *J. Med. Chem.* **1985**, *28*, 263.

117. Bull, H. G.; Thornberry, N. A.; Cordes, M. H. J.; Patchett, A. A.; Cordes, E. H. *J. Biol. Chem.* **1985**, *260*, 2952.

118. Witztum, J. L. In *Goodman and Gilman's The Pharmacological Basis of Therapeutics*, 9th ed., Hardman, J. G.; Limbird, L. E.; Molinoff, P. B.; Ruddon, R. W.; Gilman, A. G. (Eds.), McGraw-Hill, New York, 1996, p. 875.

119. Grundy, S. M. *West. J. Med.* **1978**, *128*, 13.

120. (a) Stamler, J.; *Arch. Surg.* **1978**, *113*, 21. (b) Havel, R. J.; Goldstein, J. L.; Brown, M. S. In *Metabolic Control and Disease*; Bundy, P. K.; Rosenberg, L. E. (Eds.), W. B. Saunders, Philadelphia, 1980, p. 393.

121. (a) Endo, A.; Kuroda, M.; Tsujita, Y. *J. Antibiot.* **1976**, *29*, 1346. (b) Endo, A.; Tsujita, Y.; Kuroda, M.; Tanzawa, K. *Eur. J. Biochem.* **1977**, *77*, 31.

122. Brown, A. G.; Smale, T. C.; King, T. J.; Hasenkamp, R.; Thompson, R. H. *J. Chem. Soc. Perkin Trans. I* **1976**, 1165.

123. (a) Endo, A. *J. Antibiot.* **1979**, *32*, 852. (b) Endo, A. *J. Antibiot.* **1980**, *33*, 334.

124. Alberts, A. W.; Chen, J.; Kuron, G.; Hunt, V.; Huff, J.; Hoffman, C.; Rothrock, J.; Lopez, M.; Joshua, H.; Harris, E.; Patchett, A.; Monaghan, R.; Currie, S.; Stapley, E.; Albers-Schonberg, G.; Hensens, O.; Hirschfield, J.; Hoogsteen, K.; Liesch, J.; Springer, J. *Proc. Natl. Acad. Sci. USA* **1980**, *77*, 3957.

125. Endo, A. *J. Med. Chem.* **1985**, *28*, 401.

126. Tanzawa, K.; Endo, A. *Eur. J. Biochem.* **1979**, *98*, 195.

127. Nakamura, C. E.; Abeles, R. H. *Biochemistry* **1985**, *24*, 1364.

128. Stokker, G. E.; Hoffman, W. F.; Alberts, A. W.; Cragoe, E. J., Jr.; Deana, A. A.; Gilfillan, J. L.; Huff, J. W.; Novello, F. C.; Prugh, J. D.; Smith, R. L.; Willard, A. K. *J. Med. Chem.* **1985**, *28*, 347.

129. (a) Hoffman, W. F.; Alberts, A. W.; Cragoe, E. J., Jr.; Deana, A. A.; Evans, B. E.; Gilfillan, J. L.; Gould, N. P.; Huff, J. W.; Novello, F. C.; Prugh, J. D.; Rittle, K. E.; Smith, R. L.; Stokker, G. E.; Willard, A. K. *J. Med. Chem.* **1986**, *29*, 159. (b) Stokker, G. E.; Alberts, A. W.; Anderson, P. S.; Cragoe, E. J., Jr.; Deana, A. A.; Gilfillan, J. L.; Hirschfield, J.; Holtz, W. J.; Hoffman, W. F.; Huff, J. W.; Lee, T. J.; Novello, F. C.; Prugh, J. D.; Rooney, C. S.; Smith, R. L.; Willard, A. K. *J. Med. Chem.* **1986**, *29*, 170.

130. Stokker, G. E.; Alberts, A. W.; Gilfillan, J. L.; Huff, J. W.; Smith, R. L. *J. Med. Chem.* **1986**, *29*, 852.

131. Bartmann, W.; Beck, G.; Granzer, E.; Jendralla, H.; Kerekjarto, B. V.; Wess, G. *Tetrahedron Lett.* **1986**, *27*, 4709.

132. Hoffman, W. F.; Alberts, A. W.; Anderson, P. S.; Chen, J. S.; Smith, R. L.; Willard, A. K. *J. Med. Chem.* **1986**, *29*, 849.

133. (a) Stokker, G. E.; Rooney, C. S.; Wiggins, J. M.; Hirschfield, J. *J. Org. Chem.* **1986**, *51*, 4931. (b) Heathcock, C. H.; Hadley, C. R.; Rosen, T.; Theisen, P. D.; Hecker, S. J. *J. Med. Chem.* **1987**, *30*, 1858.

134. Mundy, G.; Garrett, R.; Harris, S.; Chan, J.; Chen, D.; Rossini, G.; Boyce, B.; Zhao, M.; Gutierrez, G. *Science* **1999**, *286*, 1946.

135. Istvan, E. S.; Deisenhofer, J. *Science* **2001**, *292*, 1160.

136. Travis, J.; Salvesen, G. S. *Annu. Rev. Biochem.* **1983**, *52*, 655.

137. Stockley, R. A.; Morrison, H. M.; Smith, S.; Tetley, T. *Hoppe-Seyler's Z. Physiol. Chem.* **1984**, *365*, 587.

138. (a) Weinbaum, G.; Groutas, W. C. In *Pulmonary Pharmacology and Toxicology*; Hollinger, M. A. (Ed.), CRC Press, Boca Raton, FL, 1991, Vol. II, pp. 153–181. (b) Mittman, C.; Taylor, J. C. (Eds.), *Pulmonary Emphysema and Proteolysis*, Academic Press, New York, 1988.

139. Eriksson, S. *Eur. Respir. J.* **1991**, *4*, 1041.

140. Nadel, J. A. *Am. Rev. Respir. Dis.* **1991**, *144*, S48.

141. Weinbaum, G.; Giles, R. E.; Krell, R. D. (Eds.), *Ann. N.Y. Acad. Sci.* **1991**, *624*, 1–370.

142. (a) Johnson, D.; Travis, J. *J. Biol. Chem.* **1978**, *253*, 7142. (b) Beatty, K.; Matheson, N.; Travis, J. *Hoppe Seyler's Z. Physiol. Chem.* **1984**, *365*, 731.

143. Groutas, W. C. *Med. Res. Rev.* **1987**, *7*, 227.

144. Stein, R. L.; Strimpler, A. M.; Edwards, P. D.; Lewis, J. J.; Mauger, R. C.; Schwartz, J. A.; Stein, M. M.; Trainor, D. A.; Wildonger, R. A.; Zottola, M. A. *Biochemistry* **1987**, *26*, 2682.

145. (a) Veale, C. A.; Bernstein, P. R.; Bohnert, C. M.; Brown, F. J.; Bryant, C.; Damewood, J. R., Jr.; Earley, R.; Feeney, S. W.; Edwards, P. D.; Gomes, B.; Hulsizer, J. M.; Kosmider, B. J.; Krell, R. D.; Moore, G.; Salcedo, T. W.; Shaw, A.; Silberstein, D. S.; Steelman, G. B.; Stein, M.; Strimpler, A.; Thomas, R. M.; Vacek, E. P.; Williams, J. C.; Wolanin, D. J.; Woolson, S. *J. Med. Chem.* **1997**, *40*, 3173. (b) Edwards, P. D.; Andisik, D. W.; Bryant, C. A.; Ewing, B.; Gomes, B.; Lewis, J. J.; Rakiewicz, D.; Steelman, G.; Strimpler, A.; Trainor, D. A.; Tuthill, P. A.; Mauger, R. C.; Veale, C. A.; Wildonger, R. A.; Williams, J. C.; Wolanin, D. J.; Zottola, M. *J. Med. Chem.* **1997**, *40*, 1876.

146. (a) Imperiali, B.; Abeles, R. H. *Biochemistry* **1986**, *25*, 3760. (b) Gelb, M. H.; Svaren, J. P.; Abeles, R. H. *Biochemistry* **1985**, *24*, 1813.

147. (a) Kramer, R. A.; Schaber, M. S.; Skalka, A. M.; Ganguly, K.; Wong-Staal, F.; Reedy, E. P. *Science* **1986**, *231*, 1580. (b) Debouck, C.; Gorniak, J. G.; Strickler, J. E.; Meek, T. D.; Metcalf, B.W.; Rosenberg, M. *Proc. Natl. Acad. Sci. USA* **1987**, *84*, 8903. (c) Kohl, N. E.; Emini, E. A.; Schleif, W. A.; Davis, L. J.; Heimbach, J. C.; Dixon, R. A. F.; Scolnick, E. M.; Sigal, I. S. *Proc. Natl. Acad. Sci. USA* **1988**, *85*, 4686.

148. Kempf, D. J.; Norbeck, D. W.; Codacovi, L. M.; Wang, X. C.; Kohlbrenner, W. E.; Wideburg, N. E.; Paul, D. A.; Knigge, M. F.; Vasavanonda, S.; Craigkennard, A.; Saldivar, A.; Rosenbrook, W.; Clement, J. J.; Plattner, J. J.; Erickson, J. *J. Med. Chem.* **1990**, *33*, 2687.

149. Plattner, J. J.; Norbeck, D. W. In *Drug Discovery Technologies*; Clark, R.; Moos, W. H. (Eds.), Ellis Horwood Ltd, Chichester, 1990, pp. 92–126.

150. Erickson, J.; Neidhart, D. J.; Vandrie, J.; Kempf, D. J.; Wang, X. C.; Norbeck, D. W.; Plattner, J. J.; Rittenhouse, J. W.; Turon, M.; Wideburg, N.; Kohlbrenner, W. E.; Simmer, R.; Helfrich, R.; Paul, D. A.; Knigge, M. *Science* **1990**, *249*, 527.

151. Kempf, D. J.; Codacovi, L.; Wang, X. C.; Kohlbrenner, W. E.; Wideburg, N. E.; Saldivar, A.; Vasavanonda, S.; Marsh, K. C.; Bryant, P.; Sham, H. L.; Green, B. E.; Betebenner, D. A.; Erickson, J.; Norbeck, D. W. *J. Med. Chem.* **1993**, *36*, 320.

152. Kempf, D. J.; Norbeck, D. W.; Codacovi, L.; Wang, X. C.; Kohlbrenner, W. F.; Wideburg, N. E.; Saldivar, A.; Craig-Kennard, A.; Vasavanonda, S.; Clement, J. J.; Erickson, J.; In *Recent Advances in the Chemistry of Anti-Infective Agents*; Bentley, P. H.; Ponsford, R. (Eds.), Royal Society of Chemistry, Cambridge, 1993, pp. 297.

153. Kempf, D. J.; Marsh, K. C.; Fino, L. C.; Bryant, P.; Craig-Kennard, A.; Sham, H. L.; Zhao, C.; Vasavanonda, S.; Kohlbrenner, W. E. *Bioorg. Med. Chem.* **1994**, *2*, 847.

154. Kempf, D. J.; Marsh, K. C.; Denissen, J. F.; McDonald, E.; Vasavanonda, S.; Flentge, C. A.; Green, B. E.; Fino, L.; Park, C. H. *Proc. Natl. Acad. Sci. USA* **1995**, *92*, 2484.

155. Kumar, G. N.; Gravowski, B.; Lee, R.; Denissen, J. F. *Drug Metab. Dispos.* **1996**, *24*, 615.

156. Kempf, D. J.; Marsh, K. C.; Kumar, G.; Rodrigues, A. D.; Denissen, J. F.; McDonald, E.; Kukulka, M. J.; Hsu, A.; Granneman, G. R.; Baroldi, P. A.; Sun, E.; Pizzuti, D.; Plattner, J. J.; Norbeck, D. W.; Leonard, J. M. *Antimicrob. Agents Chemother.* **1997**, *41*, 654.

157. Kempf, D. J.; Sham, H. L.; Marsch, K. C.; Flentge, C. A.; Betebenner, D.; Green, B. E.; McDonald, E.; Vasavanonda, S.; Saldivar, A.; Wideburg, N. E.; Kati, W. M.; Ruiz, L.; Zhao, C.; Fino, L.; Patterson, J.; Molla, A.; Plattner, J. J.; Norbeck, D. W. *J. Med. Chem.* **1998**, *41*, 602.

158. Lippert, B.; Jung, M. J.; Metcalf, B. W. *Brain Res. Bull.* **1980**, *5*(Supp. 2), 375.

159. Krantz, A. In *Advances in Medicinal Chemistry*, JAI Press, London, 1992, Vol. 1, pp. 235–261.

160. Smith, R. A.; Copp, L. J.; Coles, P. J.; Pauls, H. W.; Robinson, V. J.; Spencer, R. W.; Heard, S. B.; Krantz, A. *J. Am. Chem. Soc.* **1988**, *110*, 4429.

161. Kominami, E.; Tsukahara, T.; Bando, Y.; Katunuma, N. *J. Biochem.* **1985**, *98*, 87.

162. Sloane, B. F.; Lah, T. T.; Day, N. A.; Rozhin, J.; Bando, Y.; Honn, K. V. In *Cysteine Proteinases and Their Inhibitors*; Turk, V. (Ed.), Walter de Gruyter, New York, 1986, pp. 729–749.

163. Prous, J. R. (Ed.), *Drugs Future* **1986**, *11*, 927.

164. Neuhaus, F. C.; Georgopapadakou, N. H. In *Emerging Targets for Antibacterial and Antifungal Chemotherapy*; Sutcliffe, J.; Georgopapadakou, N. H. (Eds.), Chapman and Hall, New York, 1992, pp. 206–273.

165. Mandell, G. L. In *Principles and Practice of Infectious Diseases*, 2nd ed., Mandell, G. L.; Douglas, R. G., Jr.; Bennett, J. E. (Eds.), Wiley, New York, 1985, p. 180.

166. (a) Sammes, P. G. (Ed.) *Topics in Antibiotic Chemistry*, Ellis Horwood, Chichester, 1980, Vol. 4. (b) Brown, A. G.; Roberts, S. M. (Eds.), *Recent Advances in the Chemistry of β-Lactam Antibiotics*, Royal Society of Chemistry, London, 1985.

167. Bush, K.; Mobashery, S. *Adv. Exp. Med. Biol.* **1998**, *456*, 71.

168. Waxman, D. J.; Strominger, J. L. *Annu. Rev. Biochem.* **1983**, *52*, 825.

169. Yocum, R. R.; Rasmussen, J. R.; Strominger, J. L. *J. Biol. Chem.* **1980**, *255*, 3977.

170. Tipper, D. J.; Strominger, J. L. *Proc. Natl. Acad. Sci. USA* **1965**, *54*, 1133.

171. Izaki, K.; Matsuhashi, M.; Strominger, J. L. *J. Biol. Chem.* **1968**, *243*, 3180.

172. Lee, W.; McDonough, M. A.; Kotra, L. P.; Li, Z.-H.; Silvaggi, N. R.; Takeda, Y.; Kelly, J. A.; Mobashery, S. *Proc. Natl. Acad. Sci. USA* **2001**, *98*, 1427.

173. Kuzin, A.; Liu, H., Kelly, J. A.; Knox, J. R. *Biochemistry* **1995**, *34*, 9532.

174. Sweet, R. M.; Dahl, K. F. *J. Am. Chem. Soc.* **1970**, *92*, 5489.

175. Böhme, E. H. W.; Applegate, H. E.; Toeplitz, B.; Dolfini, J. E.; Gougoutas, J. Z. *J. Am. Chem. Soc.* **1971**, *93*, 4324.

176. Kotra, L. P.; Golemi, D.; Vakulenko, S.; Mobashery, S. *Chem. Ind.* **2000** (May 22), 341.

177. Jones, R. N. *Am. J. Med.* **1996**, *100*, 3S.

178. (a) Knowles, J. R. *Acc. Chem. Res.* **1985**, *18*, 97. (b) Kelly, J. A.; Dideberg, O.; Charlier, P.; Wery, J. P.; Libert, M.; Moews, P. C.; Knox, J. R.; Duez, C.; Fraipoint, Cl.; Joris, B.; Dusart, J.; Frère, J. M.; Ghuysen, J. M. *Science* **1986**, *231*, 1429.

179. (a) Brown, A. G.; Butterworth, D.; Cole, M.; Hanscomb, G.; Hood, J. D.; Reading, C.; Rolinson, G. N. *J. Antibiot.* **1976**, *29*, 668. (b) English, A. R.; Retsema, J. A.; Girard, J. A.; Lynch, J. E.; Barth, W. E. *Antimicrob. Agents Chemother.* **1978**, *14*, 414.

180. (a) Silverman, R. B. *Mechanism-Based Enzyme Inactivation: Chemistry and Enzymology*, CRC, Boca Raton, FL, 1988, Vol. I, p. 135. (b) Cartwright, S. J.; Waley, S. G. *Med. Res. Rev.* **1983**, *3*, 341. (c) Charnas, R. L.; Knowles, J. R. *Biochemistry* **1981**, *20*, 3214. (d) Brenner, D. G.; Knowles, J. R. *Biochemistry* **1984**, *23*, 5833.

181. Gross, M.; Greenberg, L. A. *The Salicylates. A Critical Bibliographic Review*, Hillhouse, New Haven, CT, 1948.

182. Margotta, R. In *An Illustrated History of Medicine*, Lewis, P. (Ed.), Paul Hamlyn, London, 1968.

183. Stone, E. *Phil. Trans. Roy. Soc. London* **1963**, *53*, 195.

184. Martin, B. K. In *Salicylates, An International Symposium*; Dixon, A. St. J.; Martin, B. K.; Smith, M. V. H.; Wood, R. H. N. (Eds.), Little, Brown and Company, Boston, 1963, p. 6.

185. Jourdier, S. *Chem. Brit.* **1999**, *35*, 33.

186. Vane, J. R. *Nature (New Biol.)* **1971**, *231*, 232.

187. Smith, J. B.; Willis, A. L. *Nature (New Biol.)* **1971**, *231*, 235.

188. Roth, G. J.; Stanford, N.; Majerus, P. W. *Proc. Natl. Acad. Sci. USA* **1975**, *72*, 3073.

189. (a) Hemler, M.; Lands, W. E. M.; Smith, W. L. *J. Biol. Chem.* **1976**, *251*, 5575. (b) Van der Ouderaa, F. J.; Buytenhek, M.; Nugteren, D. H.; Van Dorp, D. A. *Eur. J. Biochem.* **1980**, *109*, 1.

190. Roth, G. J.; Machuga, E. T.; Ozols, J. *Biochemistry* **1983**, *22*, 4672.

191. (a) Van der Ouderaa, F. J.; Buytenhek, M.; Nugteren, D. H.; Van Dorp, D. A. *Eur. J. Biochem.* **1980**, *109*, 1. (b) DeWitt, D. L.; El-Harith, E. A.; Kraemer, S. A.; Andrews, M. J. Yao, E. F.; Armstrong, R. L.; Smith, W. L. *J. Biol. Chem.* **1990**, *265*, 5192.

192. Hochgesang, G. P., Jr.; Rowlinson, S. W.; Marnett, L. J. *J. Am. Chem. Soc.* **2000**, *122*, 6514.

193. (a) Raz, A.; Wyche, A.; Siegel, N.; Needleman, P. *J. Biol. Chem.* **1988**, *263*, 3022. (b) Masferrer, J. L.; Zweifel, B. S.; Seibert, K.; Needleman, P. *J. Clin. Invest.* **1992**, *86*, 1375.

194. (a) Xie, W.; Chipman, J. G.; Robertson, D. L.; Erikson, R. L.; Simmons, D. L. *Proc. Natl. Acad. Sci. USA* **1991**, *88*, 2692. (b) Kujubu, D. A.; Fletcher, B. S.; Varnum, C. R.; Lim, W.; Herschman, H. *J. Biol. Chem.* **1991**, *266*, 12866.

195. (a) Tally, J. J. *Exp. Opin. Ther. Patents* **1997**, *7*, 55. (b) Bjorkman, D. J. *Am. J. Med.* **1996**, *101*(Suppl. A), 25S. (c) Seibert, K.; Zhang, Y.; Leahy, K.; Hauser, S.; Masferrer, J.; Perkins, W.; Lee, L.; Isakson, P. *Proc. Natl. Acad. Sci. USA* **1994**, *91*, 12013.

196. Khanna, I. K.; Weier, R. M.; Yu, Y.; Collins, P. W.; Miyashiro, J. M.; Koboldt, C. M.; Veenhuizen, A. W.; Currie, J. L.; Seibert, K.; Isakson, P. C. *J. Med. Chem.* **1997**, *40*, 1619.

197. Khanna, I. K.; Weier, R. M.; Yu, Y.; Xu, X. D.; Koszyk, F. J.; Collins, P. W.; Koboldt, C. M.; Veenhuizen, A. W.; Perkins, W. E.; Casler, J. J.; Masferrer, J. L.; Zhang, Y. Y.; Gregory, S. A.; Seibert, K.; Isakson, P. C. *J. Med. Chem.* **1997**, *40*, 1634.

198. Penning, T.D.; Talley, J. J.; Bertenshaw, S. R.; Carter, J. S.; Collins, P. W.; Docter, S.; Graneto, M. J.; Lee, L. F.; Malecha, J. W.; Miyashiro, J. M.; Rogers, R. S.; Rogier, D. J.; Yu, S. S.; Anderson, G. D.; Burton, E. G.; Cogburn, J. N.; Gregory, S. A.; Koboldt, C. M.; Perkins, W. E.; Seibert, K.; Veenhuizen, A. W.; Zhang, Y. Y.; Isakson, P. C. *J. Med. Chem.* **1997**, *40*, 1347.

199. (a) Prasit, P.; Wang, Z.; Brideau, C.; Chan, C. C.; Charleson, S.; Cromlish, W.; Ethier, D.; Evans, J. F.; Ford-Hutchinson, A. W.; Gauthier, J. Y.; Gordon, R.; Guay, J.; Gresser, M.; Kargman, S.; Kennedy, B.; Leblanc, Y.; Leger, S.; Mancini, J.; O'Neill, G. P.; Ouellet, M.; Percival, M. D.; Perrier, H.; Riendeau, D.; Rodger, I.; Tagari, P.; Therien, M.; Vickers, P.; Wong, E.; Xu, L. J.; Young, R. N.; Zamboni, R.; Boyce, S.; Rupniak, N.; Forrest, N.; Visco, D.; Patrick, D. *Bioorg. Med. Chem. Lett.* **1999**, *9*, 1773. (b) Chan, C.-C.; Boyce, S.; Brideau, C.; Charleson, S.; Cromlish, W.; Ethier, D.; Evans, J.; Ford-Hutchinson, A. W.; Forrest, M. J.; Gauthier, J. Y.; Gordon, R.; Gresser, M.; Guay, J.; Kargman, S.; Kennedy, B.; Leblanc, Y.; Leger, S.; Mancini, J.; O'Neill, G. P.; Ouellet, M.; Patrick, D.; Percival, M. D.; Perrier, H.; Prasit, P.; Rodger, I.; Tagari, P.; Therien, M.; Vickers, P.; Visco, D.; Wang, Z.; Webb, J.; Wong, E.; Xu, L.-J.; Young, R. N.; Zamboni, R.; Riendeau, D. *J. Pharmacol. Exp. Ther.* **1999**, *290*, 551.

200. Talley, J. J.; Brown, D. L.; Carter, J. S.; Graneto, M. J.; Koboldt, C. M.; Masferrer, J. L.; Perkins, W. E.; Rogers, R. S.; Shaffer, A. F.; Zhang, Y. Y.; Zweifel, B. S.; Seibert, K. *J. Med. Chem.* **2000**, *43*, 775.

201. (a) Chandrasekharan, N. V.; Dai, H.; Turepu Roos, K. L.; Evanson, N. K.; Tomsik, J.; Elton, T. S.; Simmons, D. L. *Proc. Natl. Acad. Sci. USA* **2002**, *99*, 13926. (b) Botting, R. M. *Clin. Infect. Dis.* **2000**, *31(Suppl. 5)*, S202. (c) Willoughby, D. A.; Moore, A. R.; Colville-Nash, P. R. *Lancet* **2000**, *355*, 646.

202. Matheson, A. J.; Figgitt, D. P. *Drugs* **2001**, *61*, *833*.

203. Warner, T. D.; Giuliano, F.; Vojnovic, I.; Bukasa, A.; Mitchell, J. A.; Vane, J. R. *Proc. Natl. Acad. Sci. USA* **1999**, *96*, 7563.

204. Ochi, T.; Motoyama, Y.; Goto, T. *Eur. J. Pharmacol.* **2000**, *391*, 49.

205. Buckley, M. M.; Brogden, R. N. *Drugs* **1990**, *39*, 86.

206. Picot, D.; Loll, P. J.; Garavito, R. M. *Nature* **1994**, *367*, 243.

207. Kurumbail, R. G.; Stevens, A. M.; Gierse, J. K.; McDonald, J. J.; Stegeman, R. A.; Pak, J. Y.; Gildehaus, D.; Miyashiro, J. M.; Penning, T. D.; Seibert, K.; Isakson, P. C.; Stallings, W. C.; *Nature* **1996**, *384*, 644.

208. Gierse, J. K.; McDonald, J. J.; Hauser, S. D.; Rangwala, S. H.; Koboldt, C. M.; Seibert, K. *J. Biol. Chem.* **1996**, *271*, 15810.

209. (a) Kalgutkar, A. S.; Crews, B. C.; Rowlinson, S. W.; Garner, C; Seibert, K.; Marnett, L. J. *Science* **1998**, *280*, 1268. (b) Kalgutkar, A. S.; Kozak, K. R.; Crews, B. C.; Hochgesang, G. P., Jr.; Marnett, L. J. *J. Med. Chem.* **1998**, *41*, 4800.

210. Shan, B.; Medina, J. C.; Santha, E.; Frankmoelle, W. P.; Chou, T. C.; Learned, R. M.; Narbut, M. R.; Stott, D.; Wu, P. G.; Jaen, J. C.; Rosen, T.; Timmermans, P. B. M. W. M.; Beckman, H. *Proc. Natl. Acad. Sci. USA* **1999**, *96*, 5686.

211. Medina, J. C.; Roche, D.; Shan, B.; Learned, R. M.; Frankmoelle, W. P.; Clark, D. L.; Rosen, T.; Jaen, J. C. *Bioorg. Med. Chem. Lett.* **1999**, *9*, 1843.

212. Nelson, S. D. *J. Med. Chem.* **1982**, *25*, 753.

213. Schechter, P. J.; Barlow, J. L. R.; Sjoerdsma, A. In *Inhibition of Polyamine Metabolism. Biological Significance and Basis for New Therapies*, McCann, P. P.; Pegg, A. E.; Sjoerdsma, A. (Eds.), Academic, Orlando, FL, 1987, p. 345.

214. Isaacson, E. I.; Delgado, J. N. In *Burger's Medicinal Chemistry*, 4th ed., Wolff, M. E. (Ed.), Wiley, New York, 1981, Part III, p. 829.

215. Houser, W. A. In *Epilepsy. A Comprehensive Textbook*, Engel, J.; Pedley, T. A. (Eds.), Lippincott-Raven, Philadelphia, 1997, Vol. 1, Sec. 1.

216. Nanavati, S. M.; Silverman, R. B. *J. Med. Chem.* **1989**, *32*, 2413.

217. Iadarola, M. J.; Gale, K. *Science* **1982**, *218*, 1237.

218. Lippert, B.; Metcalf, B. W.; Jung, M. J.; Casara, P. *Eur. J. Biochem.* **1977**, *74*, 441.

219. De Biase, D.; Barra, D.; Bossa, F.; Pucci, P.; John, R. A. *J. Biol. Chem.* **1991**, *266*, 20056.

220. Nanavati, S. M.; Silverman, R. B. *J. Am. Chem. Soc.* **1991**, *113*, 9341.

221. Pegg, A. E. *Cancer Res.* **1988**, *48*, 759.

222. Schechter, P. J.; Barlow, J. L. R.; Sjoerdsma, A. In *Inhibition of Polyamine Metabolism. Biological Significance and Basis for New Therapies*, McCann, P. P.; Pegg, A. E.; Sjoerdsma, A. (Eds.), Academic, Orlando, FL, 1987, p. 345.

223. (a) Kuzoe, F. A. S. *Acta Trop.* **1993**, *54*, 153. (b) Pegg, A. E.; Shantz, L. M.; Coleman, C. S. *J. Cell Biol.* **1995**, *22*, 132. (c) Wang, C. C. In *Burger's Medicinal Chemistry and Drug Discovery*, 5th ed., Wolff, M. E. (Ed.), John Wiley & Sons, New York, 1997, Vol. 4, p. 459.

224. McCann, P. P.; Pegg, A. E. *Pharmacol. Ther.* **1992**, *54*, 195.

225. Danzin, C.; Bey, P.; Schirlin, D.; Claverie, N. *Biochem. Pharmacol.* **1982**, *31*, 3871.

226. Selikoff, I. J.; Robitzek, E. H.; Ornstein, G. G. *Quart. Bull. Seaview Hosp.* **1952**, *13*, 17 and 27.

227. Zeller, E. A.; Barsky, J.; Fouts, J. P.; Kirchheimer, W. F.; Van Orden, L. S. *Experientia* **1952**, *8*, 349.

228. (a) Kline, N. S. *J. Clin. Exp. Psychopathol. Quart. Rev. Psychiat. Neurol.* **1958**, *19*(Suppl. 1), 72. (b) Zeller, E. A. (Ed.), *Ann. N.Y. Acad. Sci.* **1959**, *80*, 551.

229. Ganrot, P. O.; Rosengren, E.; Gottfries, C. G. *Experientia* **1962**, *18*, 260.

230. Dostert, P. L.; Strolin Benedetti, M.; Tipton, K. F. *Med. Res. Rev.* **1989**, *9*, 45.

231. (a) Squires, R. F.; Lassen, J. B. *Biochem. Pharmacol.* **1968**, *17*, 369. (b) Squires, R. F. *Biochem. Pharmacol.* **1968**, *17*, 1401. (c) Johnston, J. P. *Biochem. Pharmacol.* **1968**, *17*, 1285.

232. (a) Palfreyman, M. G.; McDonald, I. A.; Bey, P.; Schechter, P. J.; Sjoerdsma, A. *Prog. Neuro-Psychopharmacol. & Biol. Psychiat.* **1988**, *12*, 967 (b) McDonald, I. A.; Bey, P.; Palfreyman, M. G. In *Design of Enzyme Inhibitors as Drugs*, Sandler, M.; Smith, H. J. (Eds.), Oxford University Press, Oxford, 1989, p. 227.

233. Green, L. D.; Dawkins, K. In *Burger's Medicinal Chemistry and Drug Discovery*, 5th ed., Wolff, M. E. (Ed.), Wiley, New York, 1997, Vol. 5, p. 121.

234. Maeda, Y.; Ingold, K. U. *J. Am. Chem. Soc.* **1980**, *102*, 328.

235. (a) Silverman, R. B. *J. Biol. Chem.* **1983**, *258*, 14766 (b) Lu, X.; Rodriguez, M.; Ji, H; Silverman, R. B.; Vintém, A. P. B.; Ramsay, R. R. In *Flavins and Flavoproteins 2002*, Chapman, S. K.; Perham, R. N.; Scrutton, N. S. (Eds.), Rudolf Weber, Berlin, 2002, p. 817–830.

236. Paech, C.; Salach, J. I.; Singer, T. P. *J. Biol. Chem.* **1980**, *255*, 2700.

237. Riederer, P.; Przuntek, H. (Eds.), *MAO-B Inhibitor Selegiline (R-(−)-Deprenyl)*, Springer-Verlag, Wein, 1987.

238. Davis, G. C.; Williams, A. C.; Markey, S. P.; Ebert, M. H.; Caine, E. D.; Reichert, C. M.; Kopin, I. *J. Psychiatry Res.* **1979**, *1*, 249.

239. Langston, J. W.; Ballard, P.; Tetrud, J. W.; Irwin, I. *Science* **1983**, *219*, 979.

240. (a) Burns, R. S.; Chiueh, C. C.; Markey, S. P.; Ebert, M. H.; Jacobowitz, D. M.; Kopin, I. J. *Proc. Natl. Acad. Sci. USA* **1983**, *80*, 4546. (b) Langston, J. W.; Forno, L. S.; Robert, C. J.; Irwin, I. *Brain Res.* **1984**, *292*, 390.

241. Heikkila, R. E.; Hess, A.; Duvoisin, R. C. *Science* **1984**, *224*, 1451.

242. Hornykiewicz, O. *Prog. Neuro-Psychopharmacol. Biol. Psychiatry* **1989**, *13*, 319.

243. Tanner, C. M. *Trends Neurosci.* **1989**, *12*, 49.

244. Markey, S. P.; Schmuff, N. R. *Med. Res. Rev.* **1986**, *6*, 389.

245. (a) Langston, W. B.; Irwin, I.; Langston, E. B.; Forno, L. S. *Science* **1984**, *225*, 1480. (b) Heikkila, R. E.; Manzino, L.; Cabbat, F. S.; Duvoisin, R. C. *Nature* **1984**, 311, 467.

246. Chiba, K.; Trevor, A.; Castagnoli, N. Jr. *Biochem. Biophys. Res. Commun.* **1984**, *120*, 574.

247. Markey, S. P.; Johannessen, J. N.; Chiueh, C. C.; Burns, R. S.; Herkenham, M. A. *Nature* **1984**, *311*, 464.

248. Javitch, J. A.; D'Amato, R. J.; Strittmater, S. M.; Snyder, S. H. *Proc. Natl. Acad. Sci. USA* **1985**, *82*, 2173.

249. (a) Priyadarshi, A.; Khuder, S. A.; Schaub, E. A.; Priyadarshi, S. S. *Environ. Res.* **2001**, *86*, 122. (b) Priyadarshi, A.; Khuder, S. A.; Schaub, E. A.; Shrivastava, S. *Neurotoxicology* **2000**, *21*, 435. (c) Couteur, D. G. Le; McLean, A. J.; Taylor, M. C.; Woodham, B. L.; Board, P. G. *Biomed. Pharmacother.* **1999**, *53*, 122.

250. Maycock, A. L.; Abeles, R. H.; Salach, J. I.; Singer, T. P. *Biochemistry* **1976**, *15*, 114.

251. Cohen, S. S. *Ann. N.Y. Acad. Sci.* **1971**, *186*, 292.

252. Mukherjee, K. L.; Heidelberger, C. *J. Biol. Chem.* **1960**, *235*, 433.

253. (a) Douglas, K. T. *Med. Res. Rev.* **1987**, *4*, 441. (b) Benkovic, S. J. *Annu. Rev. Biochem.* **1980**, *49*, 227.

254. (a) Santi, D. V.; McHenry, C. S.; Raines, R. T.; Ivanetich, K. M. *Biochemistry* **1987**, *26*, 8606. (b) Silverman, R. B. *Mechanism-Based Enzyme Inactivation: Chemistry and Enzymology*, CRC, Boca Raton, FL, 1988, Vol. I, p. 59.

DNA-Interactive Agents

6.1   Introduction

6.1.A   Basis for DNA-Interactive Drugs

Another receptor (broadly defined) with which drugs can interact is deoxyribonucleic acid or DNA, the polynucleotide that carries the genetic information in cells. Because this receptor is so vital to human functioning, and from the perspective of a medicinal chemist the overall shape and chemical structure of DNA found in normal and abnormal cells is nearly indistinguishable, drugs that interact with this receptor (*DNA-interactive drugs*) are generally very toxic to normal cells. Therefore, these drugs are reserved only for life-threatening diseases such as cancers. Because the medical term for cancer is *neoplasm*, anticancer drugs may be referred to as *antineoplastic agents*. Unlike the design of drugs that act on enzymes in a foreign organism, there is little that is useful to direct the design of selective agents against abnormal DNA. One feature of cancer cells that differentiates them from that of most normal cells is that cancer cells undergo a rapid, abnormal, and uncontrolled cell division. Genes coding for differentiation in cancer cells appear to be shut off or inadequately expressed, while genes coding for cell proliferation are expressed when they should not be. Because these cells are continually undergoing mitosis, there is a constant need for rapid production of DNA (and its precursors). One difference, then, is quantitative rather than qualitative. Because of the correspondence of normal and abnormal DNA, a compound that reacts with a cancer cell will react with a normal cell as well. However, because of rapid cell division, cancer cell mitosis can be halted preferentially to that found in normal cells where there is sufficient time for the triggering of repair mechanisms.[1] This quantitative difference is not the only difference. DNA damage in a cell is sensed by several as yet poorly defined mechanisms involving a number of proteins, especially p53.[2] Activation of p53 in response to DNA damage in normal cells can result in several other possible cellular responses, including upregulation of DNA repair systems, cell cycle arrest (to allow time for DNA repair to occur),[3] or programmed cell death (*apoptosis*). Tumor cells, however, are defective in their ability to undergo cell cycle arrest or apoptosis in response to DNA damage.[4] Cancer cells that cannot undergo cell cycle arrest are sensitive to DNA damaging agents.[5]

Because DNA is constantly becoming damaged, which leads to 80–90% of human cancers,[6] these DNA lesions must be excised, generally by DNA repair enzymes.[7] Repair systems in humans protect the genome by a variety of mechanisms that repair modified bases, DNA adducts, cross-links, and double-strand breaks, such as direct reversal, base excision, nucleotide excision, and recombination. Some nucleotides may be altered by UV light; enzymes known as *photolyases* can reverse the lesions in the presence of visible light.[8] Base excision repair, discovered in 1964,[9] eliminates single damaged base residues by the action of various DNA glycosylases, and then the abasic sugar is excised by *AP endonucleases*. Nucleotide excision repair enzymes hydrolyze two phosphodiester bonds, one on either side of the lesion. They can excise damage within oligomers that are 25 to 32 nucleotides long as well as other types of modified nucleotides. *DNA polymerases* and *ligases* are used to reinsert nucleotides and complete the repairs. Double-strand breaks in DNA are repaired by mechanisms that involve *DNA protein kinases* and recombination enzymes.

In general, anticancer drugs are most effective against malignant tumors with a large proportion of rapidly dividing cells, such as leukemias and lymphomas. Unfortunately, the most common tumors are solid tumors, which have a small proportion of rapidly dividing cells.

This is not a chapter on antitumor agents, but rather on the organic chemistry of DNA-interactive drugs and the ways in which DNA damage relates to cancer chemotherapy. Therefore, only relatively few drugs have been selected as representative examples to demonstrate the organic chemistry involved. Some of the principles of antitumor drug design were discussed in Chapter 5, Section 5.5.C.3.e, p. 298, and will also be discussed in Chapter 8 (Sections 8.2.B.5.b–d, 6.b, 7, pp. 530–535, 538, and 540).

6.1.B Toxicity of DNA-Interactive Drugs

The toxicity associated with cancer drugs usually is observed in those parts of the body where rapid cell division normally occurs, such as in the bone marrow, the gastrointestinal (GI) tract, the mucosa, and the hair. Chemotherapy-induced hair loss is believed to arise because the proliferating epithelium is ablated, and normal maturation of precursor epithelial cells to the hair strand is blocked.[10] Inhibition of cyclin-dependent kinase 2 (CDK2), a positive regulator of cell cycle progression,[11] may prevent chemotherapy-induced hair loss by temporarily arresting the hair follicle cell cycle, rendering them less susceptible to anticancer agent attack. A group at Glaxo Wellcome (now GlaxoSmithKline) used structure-based methods to design potent inhibitors of CDK2.[12] Starting from a series of 3-(benzylidene)indoline-2-ones (**6.1**), known inhibitors of receptor tyrosine kinases,[13] **6.2** was synthesized as an isosteric homolog; **6.2** was found to inhibit CDK2 with an IC_{50} of 60 nM. A crystal structure of **6.2** bound to CDK2 (Figure 6.1A) showed that position 5 of the inhibitor was adjacent to Lys-33 in the enzyme, suggesting that a hydrogen bond acceptor at that position may enhance binding. Likewise, the proximity of Val-18 indicated that a hydrophobic group added to position 4 may provide a hydrophobic interaction to that residue. Furthermore, the sulfonamide functionality was in

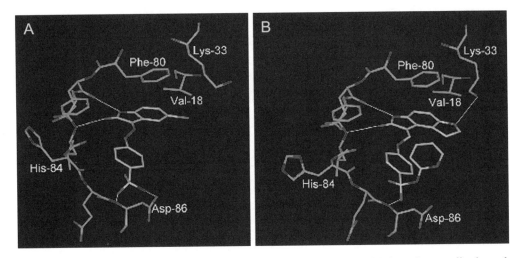

Figure 6.1 ► Crystal structure of (A) **6.2** (beige) and (B) **6.3** (purple) bound to cyclin-dependent kinase 2 [Reprinted with permission from *Science* 291, 134. Copyright © 2001 American Association for the Advancement of Science.]. Reproduced in color between pages 172 and 173

an opening in the active site, suggesting that substituents added to the sulfonamide should not interfere with binding; those substituents could be added to enhance pharmacokinetic and solubility properties of the molecule without affecting enzyme binding as discussed in Chapter 2, Section 2.2.A, p. 17. A series of analogs was prepared and screened, and **6.3** emerged as a potent (IC_{50} 10 nM) and selective inhibitor of CDK2. A crystal structure of **6.3** bound to CDK2 (Figure 6.1B) confirmed the desired binding interactions: The thiazole nitrogen atom at position 5 is within hydrogen-bond distance to Lys-33; the thiazole sulfur atom could provide a hydrophobic interaction with Val-18; and the pyridine substituent on the sulfonamide group does not interfere with binding. Topical application of **6.3** to a neonatal rat model reduced hair loss at the site of application in up to 50% of the animals.

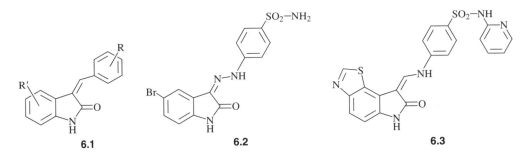

The clinical effectiveness of a cancer drug requires that it generally be administered at doses in the toxic range so that it kills tumor cells but allows enough normal cells in the critical tissues, such as the bone marrow and GI tract, to survive, thereby allowing recovery to be possible. There is some evidence that the nausea and vomiting that often occurs from these toxic agents are triggered by the central nervous system rather than as a result of destruction of cells in the GI tract.[14]

Even though cancer drugs are very cytotoxic, they must be administered repeatedly over a relatively long period of time to be assured that all of the malignant cells have been eradicated.

According to the *fractional cell kill hypothesis*,[15] a given drug concentration that is applied for a defined time period will kill a constant fraction of the cell population, independent of the absolute number of cells. Therefore, each cycle of treatment will kill a specific fraction of the remaining cells, and the effectiveness of the treatment is a direct function of the dose of the drug administered and the frequency of repetition. Furthermore, it is now known that single-drug treatments are only partially effective and produce responses of short duration. When complete remission is obtained with these drugs, it is only short lived, and relapse is associated with resistance to the original drug. Because of this, combination chemotherapy was adopted.

6.1.C Combination Chemotherapy

The introduction of cyclic *combination chemotherapy* for acute childhood lymphatic leukemia in the late 1950s marked the turning point in effective treatment of neoplastic disease. The improved effectiveness of combination chemotherapy compared to single-agent treatment is derived from a variety of reasons: Initial resistance to any single agent is frequent; initially responsive tumors rapidly acquire resistance after drug administration, probably because of selection of the preexisting resistant tumor cells in the cell population; anticancer drugs themselves increase the rate of mutation of cells into resistant forms; multiple drugs having different mechanisms of action allow independent cell killing by each agent; cells resistant to one drug may be sensitive to another; if drugs have nonoverlapping toxicities, each can be used at full dosage and the effectiveness of each drug will be maintained in combination. Also, unlike enzymes, which require gene-encoded resynthesis to restore activity after inactivation, covalent modification of DNA can be reversed by repair enzymes. In repair-proficient tumor cells it is possible to potentiate the cytotoxic effects of DNA-reactive drugs with a combination of alkylating drugs and inhibitors of DNA repair.[16]

6.1.D Drug Interactions

The most significant problem associated with the use of combination chemotherapy is drug interactions; overlapping toxicities are of primary concern. For example, drugs that cause renal cytotoxicity must be used cautiously or not at all with other drugs that depend on renal elimination as their primary mechanism of excretion. The order of administration also is important. An example (unrelated to DNA-interactive drugs) is the synergistic effects that are obtained when methotrexate (an inhibitor of dihydrofolate reductase) precedes 5-fluorouracil (an inhibitor of thymidylate synthase; see Chapter 5, Section 5.5.C.3.e, p. 298),[17] probably because of increased activation of 5-fluorouracil to its nucleotide form. The opposite order of administration leads to initial inactivation of thymidylate synthase so that the intracellular stores of tetrahydrofolate are not consumed (remember, thymidylate synthase consumes tetrahydrofolate in the form of methylenetetrahydrofolate when it converts deoxyuridylate to thymidylate). That being the case, inhibition of dihydrofolate reductase, the enzyme that catalyzes the resynthesis of tetrahydrofolate, then becomes inconsequential.

6.1.E Drug Resistance

As indicated earlier, the prime reason for the utilization of combination chemotherapy is to avoid *drug resistance*, which generally arises because of one or more of the following reasons: selection of cells that have increased expression of membrane glycoproteins, increases in levels of cytoplasmic thiols, increases in deactivating enzymes or decreases in activating enzymes (see Chapter 8) by changes in specific gene sequences, and increases in DNA repair. All of these mechanisms of resistance involve gene alterations. Membrane glycoproteins (or *P-glycoproteins*) are responsible for the efflux of drugs from cells and represent a type of *multidrug resistance* (MDR).[18] These P-glycoproteins bind and extrude drugs from tumor cells. By increasing pools of cytoplasmic thiols, such as glutathione, the cell increases its ability to destroy reactive electrophilic anticancer drugs (see Chapter 7, Section 7.4.C.5, p. 464).[19] More specifically, the gene encoding the family of glutathione *S*-transferases, which catalyze the reaction of glutathione with electrophilic compounds, may be altered so that these enzymes are overproduced (*gene amplification*). As described in more detail in Chapter 8, Section 8.2.B, p. 527, many drugs that covalently bind to DNA require enzymatic activation (*prodrugs*). The gene encoding these enzymes may be altered so that certain tumor cells no longer produce sufficient quantities of the activating enzymes to allow the drugs to be effective. Finally, once the DNA has been modified, a resistant cell could produce DNA repair enzymes, to excise the mutation in the DNA and repair the polynucleotide strands.

Specific mechanisms of resistance will be discussed for the different classes of drugs that interact with DNA in the appropriate sections. Before we discuss these DNA-interactive drugs, we need to consider the structure and properties of DNA.

6.2   DNA Structure and Properties

6.2.A   Basis for the Structure of DNA

The elucidation of the structure of DNA by Watson and Crick[20] was the culmination and synthesis of experimental results reported by a large number of scientists over several years.[21] Todd and coworkers[22] established that the four deoxyribonucleotides, containing the two purine bases, adenine (A) and guanine (G), and the two pyrimidine bases, cytosine (C) and thymine (T), are linked by bonds joining the 5′-phosphate group of one nucleotide to a 3′-hydroxyl group on the sugar of the adjacent nucleotide to form 3′,5′-phosphodiester linkages (**6.4**). The phosphodiester bonds are stable because they are negatively charged, thereby repelling nucleophilic attack. Chargaff and coworkers[23] showed that for any duplex DNA molecule the ratios of A/T and G/C are always equal to one regardless of the base composition of the DNA. They also noted that the number of adenines and thymines relative to the number of guanines and cytosines is characteristic of a given species but varies from species to species (in humans, for example, 60.4% of DNA is comprised of adenine and thymine bases). Astbury reported the first X-ray photographs of fibrous DNA, which exhibited a very strong meridional reflection at 3.4-Å distance, suggesting that the bases are stacked on each other.[24] On the basis of electrotitrimetric studies, Gulland[25] concluded that the nucleotide bases were linked by hydrogen bonding. X-ray data by Wilkins[26] and Franklin[27] indicated that DNA was a helical molecule that was able to adopt a variety of conformations. All of these data were digested by Watson and Crick, who then proposed that two strands of DNA are intertwined into a helical duplex which is held together by specific hydrogen bonding between base pairs of adenine with thymine (**6.5**) and guanine with cytosine

(**6.6**) to explain the results of Gulland and Chargaff and that these base pairs were stacked at 3.4-Å distance, as observed in the X-ray photographs. Furthermore, right-handed rotation between adjacent base pairs by about 36° produces a double helix with 10 base pairs per turn. A model of the helix was constructed using dimensions and conformations of the individual nucleotides based on the structure of cytidine.[28] The bases are located along the axis of the helix with sugar phosphate backbones winding in an antiparallel orientation along the periphery (Figure 6.2). Because the sugar and phosphate groups are always linked by 3′,5′-phosphodiester bonds, this part of DNA is very regular; however, the order (or sequence) of the nucleotides along the chain varies from one DNA molecule to another. The purine and pyrimidine bases are flat and tend to stack above each other approximately perpendicular to the helical axis; this base stacking is stabilized mainly by London dispersion forces[29] and by hydrophobic effects.[30] The two chains of the double helix are held together by hydrogen bonds between the bases.

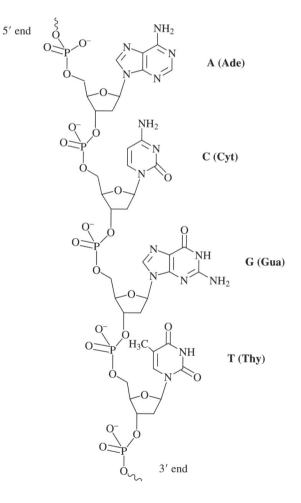

6.4

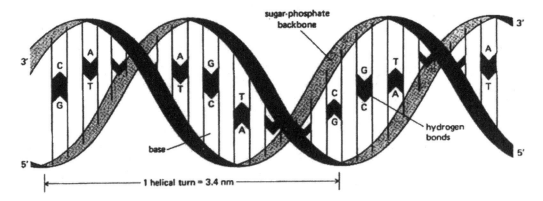

Figure 6.2 ▶   DNA structure. [Reproduced with permission from Alberts, B., Bray, D., Lewis, J., Raff, M., Roberts, K., and Watson, J. D. (1989). *Molecular Biology of the Cell*, 2nd ed., p. 99. Garland Publishing, New York. Copyright 1989 Garland Publishing.] Reproduced in color between pages 172 and 173

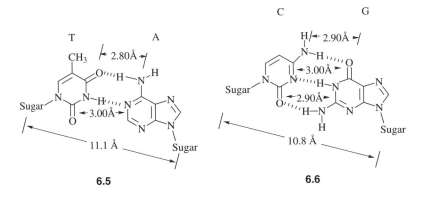

All of the bases of the DNA are on the inside of the double helix, and the sugar phosphates are on the outside; therefore, the bases on one strand are close to those on the other. Because of this fit, specific base pairings between a large purine base (either A or G) on one chain and a smaller pyrimidine base (T or C) on the other chain are essential. Base pairing between two purines would occupy too much space to allow a regular helix, and base pairing between two pyrimidines would occupy too little space. In fact, hydrogen bonds between guanine and cytosine or adenine and thymine are more effective than any other combination. Therefore, *complementary base pairs* (also called *Watson-Crick base pairs*) form between guanines and cytosines or adenines and thymines only, resulting in a complementary relation between sequences of bases on the two polynucleotide strands of the double helix. For example, if one strand has the sequence 5'-TGCATG-3', then the complementary strand must have the sequence 3'-ACGTAC-5' (note that the chains are antiparallel). As you might predict, because there are three hydrogen bonds between G and C base pairs and only two hydrogen bonds between A and T base pairs, the former are more stable.

The two glycosidic bonds that connect the base pair to its sugar rings are not directly opposite each other, and, therefore, the two sugar-phosphate backbones of the double helix are not equally spaced along the helical axis (Figure 6.3). As a result, the grooves that are formed between the backbones are not of equal size; the larger groove is called the *major groove* and the smaller one is called the *minor groove* (Figures 6.3 and 6.4). One side of every

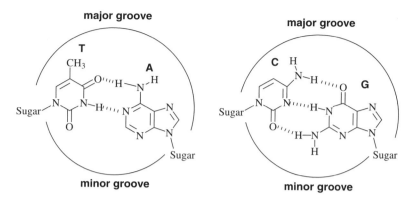

Figure 6.3 ▶ Characteristic of DNA base pairs that causes formation of major and minor grooves

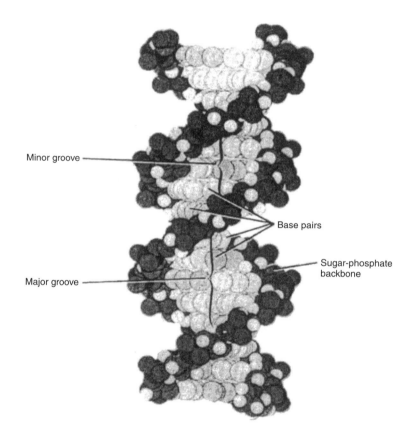

Figure 6.4 ▶ Major and minor grooves of DNA [With permission from Kornberg, A. (1980). From *DNA Replication* by Arthur Kornberg. Copyright ©1980 by W. H. Freeman and Company. Used with permission.]

base pair faces into the major groove, and the other side faces into the minor groove. The floor of the major groove is filled with base pair nitrogen and oxygen atoms that project inward from their sugar phosphate backbones toward the center of the DNA. The floor of the minor groove is filled with nitrogen and oxygen atoms of base pairs that project outward from their sugar phosphate backbones toward the outer edge of the DNA.

6.2.B   Base Tautomerization

Because of the importance of hydrogen bonding to the structure of DNA, we need to consider the tautomerism of the different heterocyclic bases (Figure 6.5), which depends largely on the dielectric constant of the medium and on the pK_a of the respective heteroatoms.[31] As shown in Figure 6.5, a change in the tautomeric form would have disastrous consequences with regard

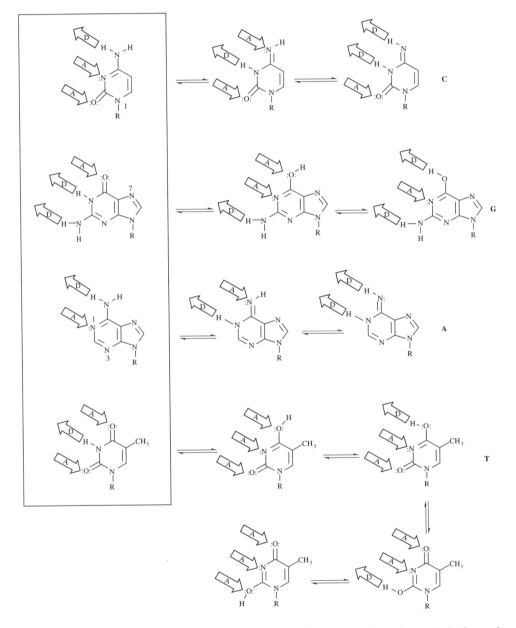

Figure 6.5 ▶   Hydrogen bonding sites of the DNA bases. D, hydrogen bond donor; A, hydrogen bond acceptor [Adapted with permission from Watson, J. D., Hopkins, N. H., Roberts, J. W., Steitz, J. A., and Weiner, A. M. (1987). *Molecular Biology of the Gene*, 4th ed., Vol. 1, p. 243. Benjamin/Cummings Publishing, Menlo Park, California. Copyright 1987 Benjamin/Cummings Publishing Company.]

to hydrogen bonding, because groups that are hydrogen bond donors in one tautomeric form become hydrogen bond acceptors in another form, and protons are moved to different positions on the heterocyclic ring. At physiological pH the more stable tautomeric form for the bases having an amino substituent (A, C, and G) is, by far (>99.99%), the amino form, not the imino form; the oxygen atoms of guanine and thymine also strongly prefer (>99.99%) to be in the keto form rather than the enol form.[32] Apparently, these four bases are ideal for maximizing the population of the appropriate tautomeric forms for complementary base recognition. Note that the donor and acceptor arrows for the tautomers in the box are complementary for C and G and for A and T, but not for C and A or G and T. Simple modifications of the bases, such as replacement of the carbonyl group in purines by a thiocarbonyl group, increases the enol population to about 7%;[33] this would have a significant effect on base pairing.

It is becoming clear that hydrogen bonding is not the only factor that controls the specificity of base pairing; shape may also play a key role. Kool has synthesized several nonpolar nucleoside isosteres that lack hydrogen bonding functionality to determine the importance of hydrogen bonding for DNA (Figure 6.6).[34] The difluorotoluene isostere nucleoside (**6.7**, a thymidine isostere; log $P = +1.39$) makes a nearly perfect mimic of thymidine (log $P = -1.27$) in the crystalline state and in solution.[35] When substituted in a DNA in which it is paired opposite adenine, it adopts a structure identical to that of a T-A base pair.[36] The benzimidazole isostere of deoxyadenosine (**6.8**) is less perfect in shape but is still a good adenine mimic in DNAs.[37] When **6.7** was incorporated into a template strand of DNA, common polymerases could selectively insert adenine opposite it,[38] and the efficiency was similar to that of a natural base pair. This suggests that Watson-Crick hydrogen bonds are not necessary to replicate a base pair with high efficiency and selectivity and that steric and geometric factors may be at least as important in the polymerase active site.[39] The nucleoside triphosphate derivative of **6.7** was made, and it was shown to insert selectively opposite an A in the template strand.[40] Isostere **6.8** also was a substrate for polymerases, and a pair between **6.7** and **6.8** also was replicated well.[41] The most important property of bases for

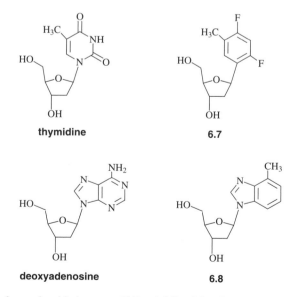

Figure 6.6 ▶ Nonpolar nucleoside isosteres (**6.7** and **6.8**) of thymine and adenine, respectively, that base pair by nonhydrogen bond interactions

successful base pairing may be that they fit as snugly into the tight, rigid active site of DNA polymerase as would a normal base pair, and, therefore, complementarity of size and shape may be more important than its hydrogen bond ability.[42]

6.2.C   DNA Shapes

DNA exists in a variety of sizes and shapes. The length of the DNA that an organism contains varies from micrometers to several centimeters in size. In the nucleus of human somatic cells each of the 46 chromosomes consists of a single DNA duplex molecule about 4 cm long. If the chromosomes in each somatic cell were placed end to end, the DNA would stretch almost 2 m long! How is it possible for such a large quantity of DNA to be crammed into each nucleus of a cell, given that the nucleus is only 5 μm in diameter? It is accomplished by the packaging of DNA into *chromatin* (Figures 6.7 and 6.8). Formation of chromatin

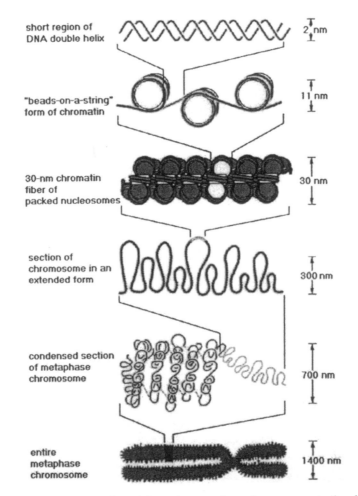

Figure 6.7 ► Stages in the formation of the entire metaphase chromosome starting from duplex DNA [With permission from Alberts, B. (1994). Copyright 1994 From *Molecular Biology of the Cell, 3rd Ed.* by Bruce Alberts, Dennis Bray, Julian Lewis, Martin Raff, Keith Roberts, and James D. Watson. Reproduced by permission of Routledge, Inc., part of The Taylor & Francis Group.]. Reproduced in color between pages 172 and 173

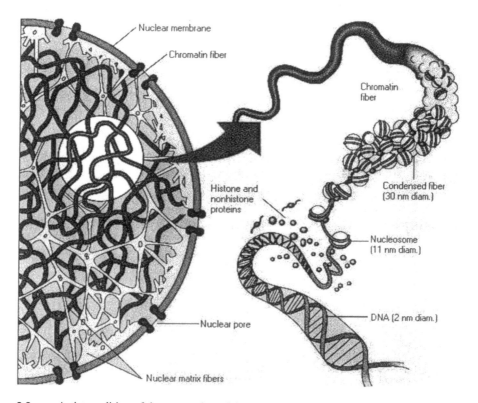

Figure 6.8 ▶ Artist rendition of the conversion of duplex DNA into chromatin fiber. Reproduced in color between pages 172 and 173

starts with *nucleosomes*, particles of DNA coiled around small, richly basic proteins called *histones* at regular intervals of about 200 base pairs.[43] The nucleosomes, held together by the electrostatic interactions between the positively charged lysine and arginine residues of the histone protein and the negatively charged phosphates of DNA, are then packed into chromatin fibers, which then associate with the chromosome scaffold, and are packed further to form the metaphase chromosome. The latter steps in this process are not clear, but Figures 6.7 and 6.8 give an artist's rendition of what they could be.

Some DNA is single stranded or triple stranded (*triplex*), but mostly it is in the double stranded (duplex) form. Some DNA molecules are linear and others (in bacteria) are circular (known as *plasmids*). Linear DNA can freely rotate until the ends become covalently linked to form circular DNA; then, the absolute number of times the DNA chains twist about each other (called the *linkage number*) cannot change. To accommodate further changes in the number of base pairs per turn of the duplex DNA, the circular DNA must twist, like when a rubber band is twisted, into *supercoiled DNA* (Figure 6.9). Untwisting of the double helix prior to rejoining the ends in circular DNA usually leads to *negative supercoiling* (left-handed direction); overtwisting results in *positive supercoiling* (right-handed direction). Virtually all duplex DNA within cells exists as chromatin in the negative supercoiled state, which is the direction opposite that of the twist of the double helix (see Section 6.2.D, p. 341). Because supercoiled DNA is a higher energy state than uncoiled DNA, the cutting (called *nicking*) of one of the DNA strands of supercoiled DNA converts it into relaxed DNA. Likewise, in mammalian DNA the superhelical stress produced during replication must be released.

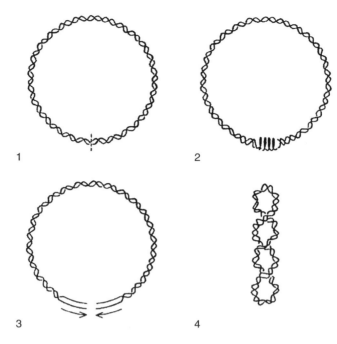

Figure 6.9 ▶ Conversion of duplex DNA into supercoiled DNA [Reproduced with permission from Watson, J. D., Hopkins, N. H., Roberts, J. W., Steitz, J. A., and Weiner, A. M. (1987). *Molecular Biology of the Gene*, 4th ed., Vol. 1, p. 258. Benjamin/Cummings Publishing, Menlo Park, California. Copyright 1987 Benjamin/Cummings Publishing Company.]

Nicking of DNA is catalyzed by a family of enzymes called *DNA topoisomerases*.[44] These ubiquitous nuclear enzymes catalyze the conversion of one topological isomer of DNA into another and also function to resolve topological problems in DNA such as overwinding and underwinding, knotting and unknotting, and catenation and decatenation (Figures 6.10 and 6.11), which normally arise during replication, transcription, recombination, repair, and other DNA processes.[45] They also are required for maintenance of proper chromosome structure. There are a number of ways for a long DNA molecule to lose or gain a few turns of twist; these excess or deficient turns of twist need to be corrected by topoisomerases.[46] DNA also has to untwist during several of its normal functions, for example, when it is copied into the messenger RNA, which is responsible for making proteins in the cell (*transcription*), near start sites of all genes so that RNA polymerase can construct new RNA strands, and when DNA is copied into another DNA strand by DNA polymerase just before a single cell divides into two cells (*replication*). During DNA replication the two strands of the DNA must be unlinked by topoisomerases, and during transcription, the translocating RNA polymerase generates supercoiling tension in the DNA that must be relaxed. The association of DNA with histones and other proteins also introduces supercoiling that requires relaxation by topoisomerases. Transcription from some promoters in bacteria requires a minimal level of negative supercoiling, but too much supercoiling of either sign is disastrous. In all cells completely replicated chromosomes must be untangled by DNA topoisomerases before partitioning and cell division can occur. Topoisomerases are known that relax only negative supercoils, that relax supercoils of both signs, or that introduce either negative (bacterial *DNA gyrase*) or positive supercoils into the DNA (*reverse gyrase*).

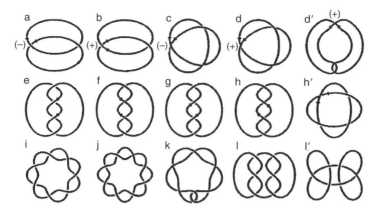

Figure 6.10 ▶ Catenane and knot catalog. Arrows indicate the orientation of the DNA primary sequence: a and b, singly-linked catenanes; c and d, simplest knot, the trefoil; e–h, multiply interwound torus catenanes; i, right-handed torus knot with 7 nodes; j, right-handed torus catenane with 8 nodes; k, right-handed twist knot with 7 nodes; l, 6-noded knots composed of 2 trefoils [From Wasserman, S. A. and Cozzarelli, N. R. (1972). Reprinted with permission from *Science* 232, 95. Copyright 1986 American Association for the Advancement of Science.]

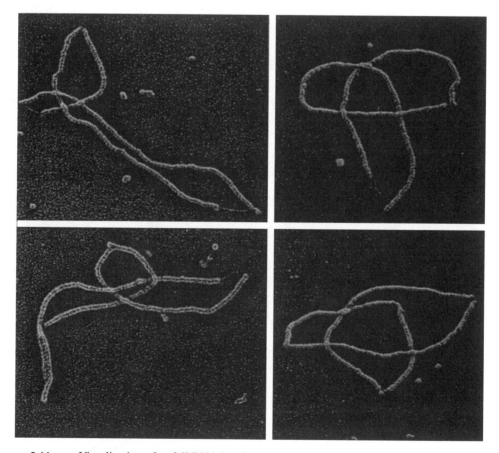

Figure 6.11 ▶ Visualization of trefoil DNA by electron microscopy [Reproduced with permission from Griffith, J. D., Nash, H. A., *Proc. Natl. Acad. Sci. USA* **1985**, *82*, 3124.]

There are two general types of topoisomerases into which at least six different topoisomerases have been classified;[47] these topoisomerases regulate the state of super-coiling of intracellular DNA. One type, known as *DNA topoisomerase I*, removes positive and negative supercoils by catalyzing a transient break of one strand of duplex DNA and allowing the unbroken, complementary strand to pass through the enzyme-linked strand, thereby result-ing in DNA relaxation by one positive turn. The other type is called *DNA topoisomerase II* (or, in the case of the bacterial enzyme, DNA gyrase), which catalyzes the transient breakage of both strands of the duplex DNA, with a four base pair stagger between the nicks. This generates a gate through which another region of DNA can be passed prior to resealing the strands. The outcome of this process is the supercoiling of the DNA in the negative direction or relaxation of positively supercoiled DNA, which changes the linkage number by -2. The type I enzymes are further classified into either the type IA subfamily if the enzyme link is to a $5'$ phosphate (formerly called type I-5$'$) or the type IB subfamily if the enzyme is attached to a $3'$ phosphate (formerly called type I-3$'$). Type II topoisomerases also are divided into type IIA and type IIB subfamilies by the same criteria. DNA topoisomerase I has no requirement for an energy cofactor to complete the rejoining process, but DNA topoisomerase II requires ATP and Mg(II) for the "strand passing" activity.[48] *Topoisomerase III* (which actually is a member of the type IA subfamily of topoisomerases) catalyzes the removal of negative, but not positive, supercoils.[49] *Topoisomerase IV* (subfamily IIA), a target for quinolone antimi-crobial agents, is required for the terminal stages of unlinking of DNA during replication; its inhibition causes accumulation of replication catenanes.[50] DNA *topoisomerase V* (subfamily IB) relaxes both negatively and positively supercoiled DNA in hyperthermophilic organisms in the temperature range from 60 to 122°C and salt concentrations from 0 to 0.65 M KCl or NaCl or 0 to 3.1 M of potassium glutamate.[51] The discovery of DNA topoisomerase VI in hyperthermophilic archaea was responsible for the subdivision of type II topoisomerases into type IIA and IIB subfamilies.[52] *Topoisomerase VI*, a type IIB topoisomerase, is able to decatenate intertwined DNA and to relax either positively or negatively supercoiled DNA in the presence of ATP and divalent cations.

The mechanisms for DNA strand cleavage by DNA topoisomerase I and II are different, but a common feature is the involvement of an active-site tyrosine residue that catalyzes a covalent catalytic cleavage mechanism. Scheme 6.1 shows general mechanisms for topoi-somerase IA and IB.[53] A tyrosyl group on the enzymes attacks the phosphodiester bond of DNA, giving one of two possible covalent adducts, known as *cleavable complexes*, depend-ing on whether the enzyme is in subfamily type IA (Scheme 6.1, pathway a) or type IB (Scheme 6.1, pathway b). Figure 6.12 depicts a possible mechanism for a topoisomerase I reaction. According to this mechanism, a single strand of DNA is cleaved by attack of an active site tyrosine residue (Figure 6.12, structure a), which becomes covalently bound to the $5'$-phosphate (this is a topoisomerase type IA) on one end of the cleaved strand (Figure 6.12, structure b). The enzyme holds onto both ends at the site of the break (the released $3'$-end is held noncovalently), producing a bridge across the gap through which the intact strand passes (structure c). The enzyme religates the two ends (structure d), and releases the relaxed DNA.[54] Figure 6.13 depicts a cartoon view of how topoisomerase IA may relax super-coiled DNA (Figure 6.13A) or catalyze decatenation (Figure 6.13B).[55] A crystal structure of a fragment of the *E. coli* enzyme[56] supports the mechanism in Figures 6.12 and 6.13. In the case of a topoisomerase type IB, nucleophilic attack by the tyrosine residue breaks the DNA strand to generate a phosphodiester link between the tyrosine and the $3'$ phosphate, releasing a $5'$-hydroxyl end (Scheme 6.1, pathway b).

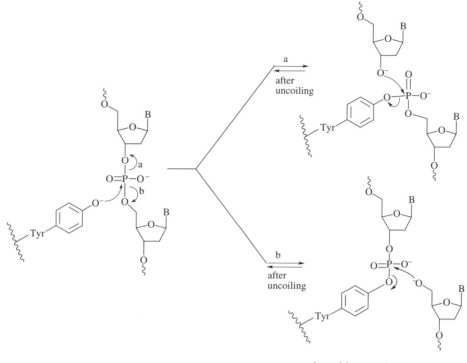

cleavable complexes

Scheme 6.1 ▶ DNA topoisomerase-catalyzed strand cleavage to cleavable complexes

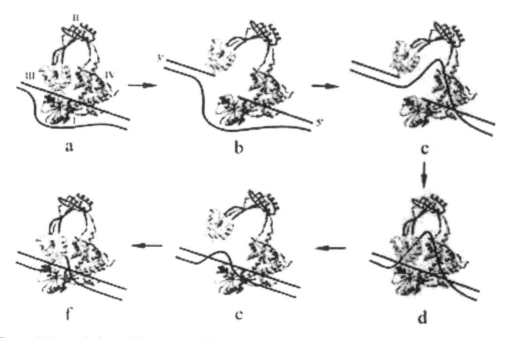

Figure 6.12 ▶ Artist rendition of a possible mechanism for a topoisomerase I reaction [With permission from Champoux, J. J. (2001). With permission from the *Annual Review of Biochemistry*, Volume 70 © 2001 by Annual Reviews www.annualreviews.org.]. Reproduced in color between pages 172 and 173

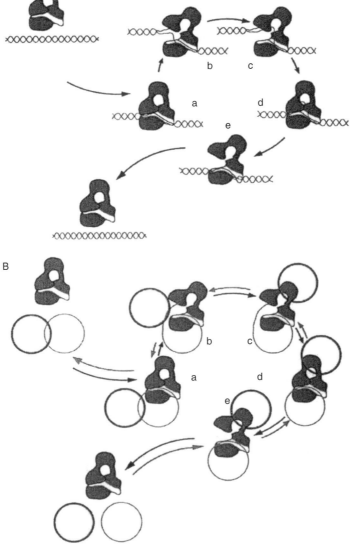

Figure 6.13 ▶ Artist rendition of possible mechanisms of topoisomerase IA-catalyzed relaxation of (A) supercoiled DNA and (B) decatenation of a DNA catenane [Reproduced with permission from Li, Z.; Mondragón, A.; DiGate, R. J., *Mol. Cell* **2001**, *7*, 301.]. Reproduced in color between pages 172 and 173

In the case of topoisomerase II enzymes, two tyrosine residues cleave phosphodiester bonds on each of the DNA strands to form cleavable complexes. The mechanism for how these enzymes coil and uncoil DNA is even more in debate than the topoisomerase I enzymes and would be difficult to show in a pictorial form anyway, so no mechanism is given. As we will see (Section 6.3.A), one important mechanism of action of antitumor and antibacterial agents is the stabilization of the cleavable complex after it forms so that religation of the two DNA fragments does not occur, leaving the DNA cleaved.[57] Likewise, resistance to some DNA-interactive agents arises from a reduction in cleavable complex formation.[58]

6.2.D DNA Conformations

There are three general helical conformations of DNA; two are right-handed DNA (A-DNA and B-DNA) and one is left-handed DNA (Z-DNA). Each conformation involves a helix made up of two antiparallel polynucleotide strands with the bases paired through Watson-Crick hydrogen bonding, but the overall shapes of the helices are quite different (Figure 6.14).

The right-handed forms differ in the distance required to make a complete helical turn (called the *pitch*), differ in the way their sugar groups are bent or puckered, differ in the angle

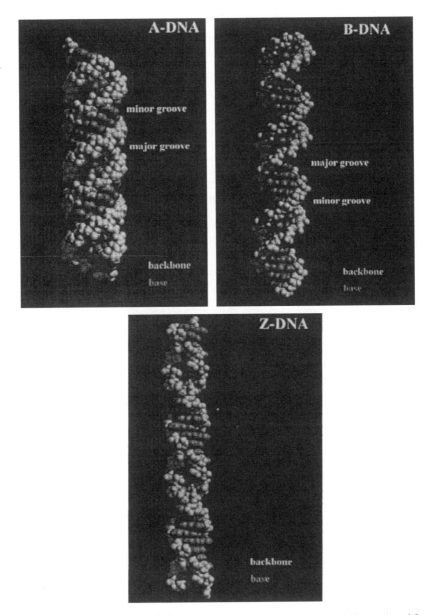

Figure 6.14 ▶ Computer graphics depictions of A-DNA, B-DNA, and Z-DNA [Reproduced from the IMB Jena Image Library of Biological Macromolecules; http://www.imb-jena.de/IMAGE.html.]. Reproduced in color between pages 172 and 173

of tilt that the base pairs make with the helical axis, and differ in the dimensions of the grooves. The predominant form by far is B-DNA, but in environments with low hydration, A-DNA occurs. In contrast to B-DNA, individual residues in A-DNA display uniform structural features; nucleotides in A-DNA have more narrowly confined conformations. Whereas there are 11 nucleotides in one helical turn in A-DNA, there are only 10 base pairs per pitch in B-DNA. Therefore, A-DNA is shorter and squatter than B-DNA.

Z-DNA, a minor component of the DNA of a cell, is a left-handed double helix having 12 base pairs per helical turn. It has only a minor groove because the major groove is filled with cytosine C-5 and guanine N-7 and C-8 atoms. In A- and B-DNA the glycosyl bond is always oriented *anti* (**6.9**). In Z-DNA the glycosyl bond connecting the base to the deoxyribose group is oriented *anti* at the pyrimidine residues but *syn* at the purine residues (**6.10**). This alternating *anti-syn* configuration gives the backbone (a line connecting phosphorus atoms) an overall zigzag appearance, hence, the name Z-DNA.

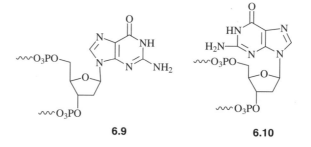

                    6.9   **6.10**

With this brief introduction to the structure of DNA we can now explore the different mechanisms by which drugs interact with DNA.

6.3 Classes of Drugs That Interact with DNA

In general, there are three major classes of clinically important DNA-interactive drugs: *reversible binders*, which interact with DNA through the reversible formation of noncovalent interactions; *alkylators*, which react covalently with DNA bases; and *DNA strand breakers*, which generate reactive radicals that produce cleavage of the polynucleotide strands. The ideal DNA interactive drug may turn out to be a nonpeptide molecule that is targeted for a specific sequence and site size.[59] However, it is not yet clear what DNA sequences (genes) should be targeted. Also, in traditional cancer chemotherapy significant amounts of DNA damage (rather than small amounts of sequence-selective DNA damage) are required to elicit the cell killing that is necessary for effective anticancer drugs. Proteins are examples of molecules that exhibit unambiguous DNA sequence recognition. The primary sequence recognition by proteins results from complementary hydrogen bonding between amino acid residues on the protein and nucleic acid bases in the major and minor grooves of DNA.[60] Proteins generally use major groove interactions with B-DNA because there are more donor and acceptor sites for hydrogen bonding than in the minor groove.[61]

You may be wondering how drugs can interact with DNA at all, given that essentially all of the DNA in the cell is packed as chromatin (see Section 6.2.C, p. 334). A close-up view

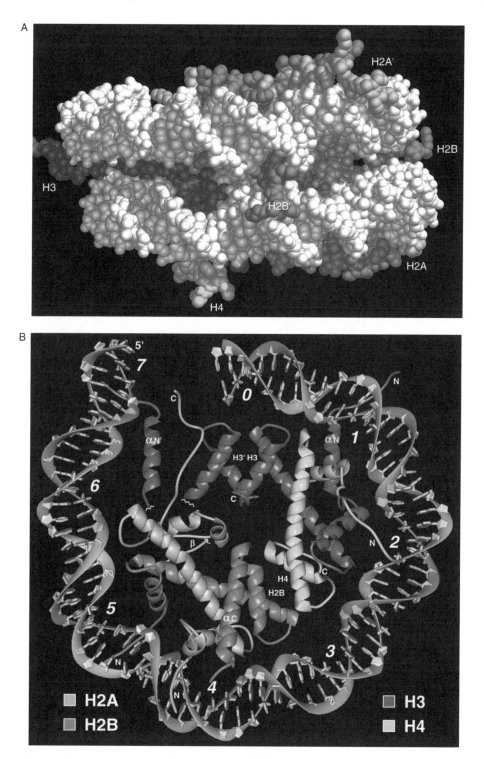

Figure 6.15 ▶ (A) Molecular model of a nucleosome. (B) Cutaway view of the nucleosome with the histones in the center and duplex DNA wrapped around them [With permission from Luger, K (1997). Reprinted with permission from *Nature* 398, 251–260. Copyright 1993 Macmillan Publishers Limited.]. Reproduced in color between pages 172 and 173

of a computer model of a nucleosome (Figure 6.15)[62] shows that the outer surface of the DNA is directly accessible to small molecules. Larger molecules, however, also can interact with DNA because *in vitro* studies have demonstrated that the nucleosomes are in dynamic equilibrium with uncoiled DNA (Figure 6.16),[63] so the drug can bind after uncoiling, which interferes with the binding of the DNA to the histone.

6.3.A   Reversible DNA Binders

Nucleic acids inside the cell interact with a variety of small molecules, including water, metal cations, small organic molecules, and proteins, all of which are essential for stabilization of the nucleic acid structure.[64] Interference with these interactions can disrupt the DNA structure. There are three important ways small molecules can reversibly bind to duplex DNA and lead to interference of DNA function: (1) by electrostatic binding along the exterior of the helix, (2) by interaction with the edges of the base pairs in either the major or minor groove, and (3) by intercalation between the base pairs (Figure 6.17).

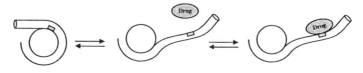

Figure 6.16 ▶   Schematic of how a drug could bind to DNA wrapped around histones in the nucleosome [Reproduced with permission from Polach K. J., Widom J. *J. Mol. Biol.* **1995**, *254*, 130.]. Reproduced in color between pages 172 and 173

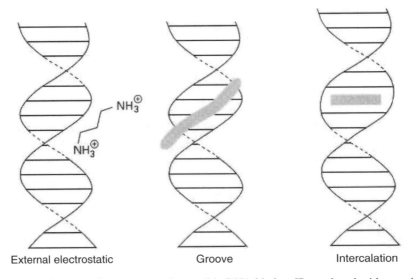

Figure 6.17 ▶   Schematic of three types of reversible DNA binders [Reproduced with permission from Blackburn G. M., Gait M. J., Eds. *Nucleic Acids in Chemistry and Biology*, 2nd ed., p. 332. Oxford University Press, Oxford; Copyright 1996 Oxford University Press.]. Reproduced in color between pages 172 and 173

A.1 External Electrostatic Binding

Duplex DNA contains a negatively charged sugar phosphate backbone; this polyanionic nature strongly affects both the structure and function of DNA. Cations and water molecules bind to DNA and allow it to exist in various secondary structures described in Section 6.2.D, p. 341. Release of phosphate counterions on binding of specific cation ligands can provide both favorable (increased entropy) and unfavorable (decreased enthalpy because of loss of specific ionic interactions) contributions to the overall free energy, leading to disruption of the DNA structure. These types of interactions are generally not dependent on DNA sequence.

A.2 Groove Binding

The major and minor grooves have significant differences in their electrostatic potentials, hydrogen bonding characteristics, steric effects, and degree of hydration. Proteins exhibit binding specificity primarily through major groove interactions, but small molecules prefer minor groove binding. Minor groove binding molecules generally have aromatic rings connected by single bonds that allow for torsional rotation in order to fit into the helical curvature of the groove with displacement of water molecules. The minor groove is generally not as wide in A-T-rich regions relative to G-C-rich regions; therefore, A-T regions may be more amenable to flat aromatic molecule binding than are G-C regions. The more narrow A-T regions produce a more snug fit of molecules into the minor groove and lead to van der Waals interactions with the DNA functional groups that define the groove. Binding also arises from interactions with the edges of the base pairs on the floor of the groove. Groove binders do not significantly unwind DNA base pairs. Hydrogen bonding from the C-2 carbonyl oxygen of T or the N-3 nitrogen of A to minor groove binders is very important. Similar groups also are present in the G-C base pairs, but the amino group of G sterically hinders hydrogen bond formation at N-3 of G and at the C-2 carbonyl oxygen of C. Also, the hydrogen bonds between the amino groups of G and carbonyl oxygens of C in G-C base pairs lie in the minor groove, and these sterically inhibit penetration of molecules into G-C-rich regions of this groove. Because of greater negative electrostatic potential in the A-T regions of the minor groove relative to the G-C regions,[65] there is a higher selectivity for cationic molecules in A-T regions; it is possible to enhance G-C region binding by the design of molecules that can accept hydrogen bonds from the amino group of G. Groove binders can be elongated to extend the interactions within the groove, which leads to high sequence-specific recognition by these molecules.

Molecules that bind in the A-T regions of the minor groove typically are crescent shaped with hydrogen bonding NH groups on the interior of the crescent. The NH groups hydrogen bond with the A-T base pairs in the minor groove, but are excluded from these interactions with G-C base pairs by the amino group of G. Cationic groups undergo electrostatic interactions with the negative electrostatic potential in the minor groove. A typical minor groove binder is the antitumor agent netropsin (**6.11**).[66] The refined 2.2 Å crystal structure of **6.11** bound to a B-DNA dodecamer shows that the drug displaces the water molecules so it is centered in the AATT region of the minor groove and forms three good bifurcated hydrogen bonds with N-3 of adenine and the C-2 carbonyl oxygen of thymine along the floor of the groove. The pyrrole rings of netropsin are packed against the C-2 positions of adenines, which leaves no room for the amine group of guanine, thereby providing a structural rationale for the A-T specificity of netropsin. Binding of netropsin neither unwinds nor elongates the double helix, but it causes a widening of the minor groove (0.5–2.0 Å) in the AATT region and a bending of the helix axis (8°) away from the site of binding.[67]

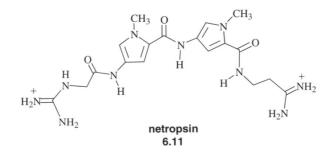

netropsin
6.11

A.3 Intercalation and Topoisomerase-Induced DNA Damage

Flat, generally aromatic or heteroaromatic molecules bind to DNA by inserting (i.e., *intercalating*) and stacking between the base pairs of the double helix. The principal driving forces for intercalation are stacking and charge-transfer interactions, but hydrogen bonding and electrostatic forces also play a role in stabilization.[68] Intercalation, first described in 1961 by Lerman,[69] is a noncovalent interaction in which the drug is held rigidly perpendicular to the helix axis. This causes the base pairs to separate vertically, thereby distorting the sugar-phosphate backbone and decreasing the pitch of the helix. (Figure 6.18 shows the intercalation of ethidium bromide into B-DNA.) Intercalation, apparently, is an energetically favorable process, because it occurs so readily. Presumably, the van der Waals forces that hold the intercalated molecules to the base pairs are stronger than those found between the stacked base pairs.

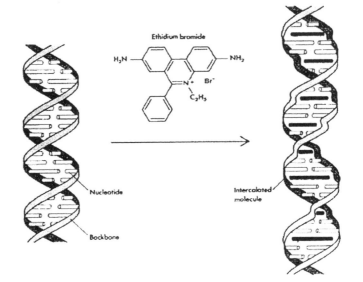

Figure 6.18 ▶ Intercalation of ethidium bromide into B-DNA [Adapted from Watson J. D., Hopkins N. H., Roberts J. W., Steitz J. A., Weiner A. M., *Molecular Biology of the Gene*, 4th ed., Vol. I, p. 255. Benjamin/Cummings Publishing: Menlo Park, CA; Copyright 1987 Benjamin/Cummings Publishing Company.]

Much of the binding energy is the result of the removal of the drug molecule from the aqueous medium and a hydrophobic effect. Intercalation occurs preferentially (by 7–13 kcal/mol) into pyrimidine-3′,5′-purine sequences rather than into purine-3′,5′-pyrimidine sequences.[70] Intercalators do not bind between every base pair. The *neighbor exclusion principle* states that intercalators can, at most, bind at alternate possible base pair sites on DNA, because saturation is reached at a maximum of one intercalator between every second site.[71] One explanation for this principle is that binding in one site causes a conformational change in the adjacent site, which prevents binding of the intercalator in that adjacent site (known as *negative cooperativity*).

In general, intercalation does not disrupt the Watson-Crick hydrogen bonding, but it does destroy the regular helical structure, unwinds the DNA at the site of binding, and, as a result of this, interferes with the action of DNA-binding enzymes such as DNA topoisomerases and DNA polymerases. Interference with topoisomerases alters the degree of supercoiling of DNA; interference with DNA polymerases inhibits the elongation of the DNA chain in the 5′ to 3′ direction and also prevents the correction of mistakes in the DNA by inhibiting the clipping out (via hydrolysis of the phosphodiester bond) of mismatched residues at the terminus. Most intercalators display either no sequence preferences in their binding or a slight G-C preference, which contrasts with the A-T binding preference of groove binders. Furthermore, groove binders, in general, exhibit significantly greater binding selectivity than intercalators. Groove binders interact with more base pairs than intercalators as they lie along the groove. Also, grooves are distinct in A-T and G-C regions, which adds to the potential for specificity in groove binding.

For simple cationic intercalators, intercalation involves two steps. First, the cation interacts with the negatively charged DNA sugar-phosphate backbone. Then the intercalator diffuses along the surface of the helix until it encounters gaps between base pairs that have separated because of thermal motion, thereby creating a cavity for intercalation.

Although the drugs in this section are categorized as being intercalators, it is now believed that intercalation of a drug into DNA is only the first step in the events that eventually lead to DNA damage by other mechanisms. For many classes of antitumor agents, there is an involvement of the DNA topoisomerases (see Section 6.2.C, p. 334) subsequent to intercalation.[72] DNA-topoisomerase I complex is the target for the antitumor agent topotecan hydrochloride (**6.12**, Hycamtin),[73] whereas DNA topoisomerase II is the target for a variety of classes of antitumor drugs, such as anthracyclines, anthracenediones, acridines, actinomycins, and ellipticines.[74] The quinolone antibacterial drugs, such as nalidixic acid (**6.13**, NegGram), act on bacterial DNA topoisomerase II. The evidence to date suggests that drugs that cause topoisomerase-induced DNA damage interfere with the breakage-rejoining reaction (see Scheme 6.1) by trapping the *cleavable complex*.[75] The cleavable complex may be stabilized by the formation of a reversible nonproductive (noncleavable) drug–DNA–topoisomerase ternary complex. It has been hypothesized that this ternary complex may collide with transcription and replication complexes; on collision, the ternary complex may lose its reversibility and generate lethal double-strand DNA breaks.[76] It is not clear if the drug binds to DNA first, then topoisomerase II forms the ternary complex, or if the drug binds to a topoisomerase II–DNA complex. An example of this phenomenon is TAS-103 (**6.14**), an antineoplastic agent that first intercalates into DNA, then kills cells by increasing the amount of DNA cleavage mediated by topoisomerase II as a result of its inhibition of the religation of the DNA from the cleavable complex.[77]

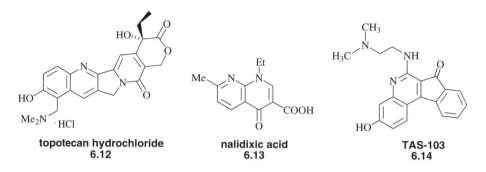

topotecan hydrochloride
6.12

nalidixic acid
6.13

TAS-103
6.14

A study of a series of anthracycline analogs showed that DNA intercalation of these compounds is required, but not sufficient for topoisomerase II-targeted activity.[78] There was a strong correlation between the potency of intercalation and cleavable complex formation. However, there does not appear to be a correlation between DNA intercalation and antitumor activity. Some strong intercalators do not induce cleavable complexes, possibly because of certain structural requirements for the binding of the intercalated drug to the topoisomerase. Also, epipodophyllotoxins, such as the anticancer drug etoposide (**6.15**, Etopophos), are nonintercalating DNA topoisomerase II poisons.

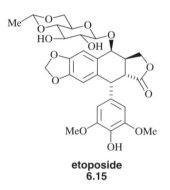

etoposide
6.15

Although the mechanism of topoisomerase II-induced DNA damage is not clear, the ternary complex appears to be lethal to proliferating cells. Selective sensitivity of proliferating tumor cells to the cytotoxic effects of DNA topoisomerase II poisons may be the result of the high levels of DNA topoisomerase II found in proliferating cells and the very low levels found in quiescent cells.

Intercalation may not be the direct cause for DNA damage, but it does produce a conformational change (unwinding) in the double helix. This, then, can result in the positioning of the drug in the DNA appropriately for binding with the topoisomerase in the ternary complex, or it can position the drug for subsequent reactions, as will be discussed in Sections 6.3.B and 6.3.C.

Three classes of drug molecules that have been well characterized as intercalators of DNA are the acridines (**6.16**), the actinomycins (**6.17**), and the anthracyclines (**6.18**).

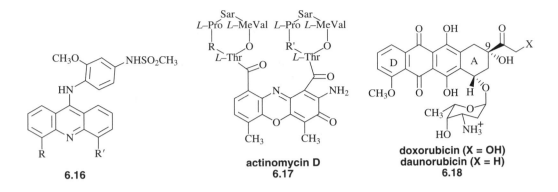

actinomycin D
6.17

doxorubicin (X = OH)
daunorubicin (X = H)
6.18

6.16

a. Amsacrine, an Acridine Analog

Acridine compounds, which were by-products of aniline dye manufacture, were first used in clinical medicine in the late 19th century against malaria.[79] By the First World War acridine derivatives such as proflavine (**6.19**) were in widespread use as local antibacterial agents. After the Second World War another acridine derivative, aminacrine (**6.20**, Monacrin), was the principal acridine antibacterial agent used.[80] In the 1960s and early 1970s, various anilino-substituted analogs of 9-anilinoacridine (**6.21**) were prepared and tested for antitumor activity[81] on the basis of the reported antitumor activity of **6.21** (R = H, R′ = Me₂N).[82] Although the 3,4-diamino analog had good antitumor activity, it was unstable to air oxidation. From structure–activity relationships, it was reasoned that an electron donor group was needed on the anilino ring; consequently, a sulfonamide group, which would be partially anionic at physiological pH, was selected. In fact, **6.21** (R = H, R′ = NHSO₂Me) was equally as potent as the 3,4-diamino analog.[83] This was used as a lead compound, and it was found that the most potent analogs had other electron-donating substituents in addition to the sulfonamide group. Amsacrine (**6.21**, R = OMe, R′ = NHSO₂Me; Amsidyl) was the most potent of those tested;[84] its main use is now in the treatment of leukemia.[85] Mutations in two residues of the human topoisomerase IIα isozyme leads to resistance to amsacrine by diminishing the binding of amsacrine to the cleavable complex;[86] amsacrine stabilizes the cleavable complex, which is detrimental to the cell.

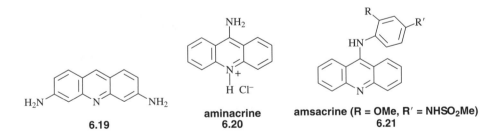

aminacrine
6.20

amsacrine (R = OMe, R′ = NHSO₂Me)
6.21

6.19

Paradoxically, although a very wide variation of structures of 9-anilinoacridines can be tolerated with retention of antitumor activity, among the active derivatives, large differences in potency are observed with small changes in structure.[87] The antitumor activity was parabolically related to drug lipophilicity as measured by log *P* values (see Chapter 2, Section 2.2.F.2.b, p. 55); compounds with log *P* values close to that of amsacrine were most potent. There also is a close correlation between the electronic properties (σ constant; see Chapter 2,

Section 2.2.F.1, p. 51) of groups at the *para*-position of the anilino ring and acridine pK_a values. Furthermore, when lipophilic and electronic effects of a series of bulky substituents at various positions on the 9-anilinoacridine framework are taken into account, the steric effects of the group play a dominant role. These results are consistent with the mode of action of 9-anilinoacridines as intercalators of double-stranded (duplex) DNA.[88] Earlier studies showed that these compounds unwound closed circular duplex DNA.[89] By analogy with the crystal structure of 9-aminoacridine bound to a dinucleotide,[90] Denny *et al.*[91] hypothesized that the anilino ring, which lies almost at right angles to the plane of the acridine chromophore, is lodged in the minor groove with the 1′-substituent (the sulfonamide group) oriented at a 90° angle to the helical axis. The sulfonamide may interact with a second macromolecule, such as a regulatory protein, and this ternary complex could mediate the biological effects of the 9-anilinoacridines. Other studies of the rates of dissociation of amsacrine from DNA suggest that the anilino group may bind in the major groove.[92]

Amsacrine lacks broad spectrum clinical activity and is difficult to formulate because of its low aqueous solubility. It was thought that the relatively high pK_a of the compound (8.02 for the acridine nitrogen) was important in limiting *in vivo* distribution; consequently, analogs with improved solubility and high DNA binding, but with lower pK_a, were sought. A compound was found (**6.16**, R = Me, R′ = CONHMe) that had all of the desirable physicochemical properties, showed superior antileukemic activity compared with amsacrine, and was broader in its spectrum of action.[93]

b. Dactinomycin, the Parent Actinomycin Analog

Actinomycin D (now called dactinomycin; **6.17**, R = R′ = *D*-Val; Cosmegen) was the first of a family of chromopeptide antibiotics isolated from a culture of a *Streptomyces* strain in 1940.[94] In 1952 these compounds were found to have antitumor activity and were used clinically.[95] Dactinomycin binds to double-stranded DNA and, depending on its concentration, inhibits DNA-directed RNA synthesis or DNA synthesis. RNA chain initiation is not prevented, but chain elongation is blocked.[96] The phenoxazone chromophore intercalates between bases in the DNA.[97] Binding depends on the presence of guanine; the 2-amino group of guanine is important for the formation of a stable drug–DNA complex.[98] X-ray crystal structures of a 1:2 complex of dactinomycin with deoxyguanosine,[99] deoxyguanylyl-3′,5′-deoxycytidine (Figure 6.19),[100] and a complex of dactinomycin with d(ATGCAT)[101] are models for the intercalation of dactinomycin into DNA. These structures suggest that the phenoxazone ring can intercalate between deoxyguanosines and that the cyclic peptide substituents can be involved in strong hydrogen bonding and hydrophobic interactions with DNA.[102] In particular, two crucial hydrogen bonds stabilize the DNA binding complex. One strong hydrogen bond exists between neighboring cyclic pentapeptide chains connecting the N–H of one *D*-valine residue with the C=O of the other *D*-valine residue. Another strong hydrogen bond connects the guanine 2-amino group with the carbonyl oxygen of the *L*-threonine residue. A weaker hydrogen bond connects the guanine N-3 ring nitrogen with the NH group on this same *L*-threonine residue. Stacking forces are primarily responsible for the recognition and preferential binding of a guanine base to dactinomycin.[103] The biological activity appears to depend on the very slow rate of DNA-dactinomycin dissociation, which reflects the intermolecular hydrogen bonds, the planar interactions between the purine rings and the chromophore, and the numerous van der Waals interactions between the polypeptide side chains and the DNA. The peptide substituents, which lie in the minor groove, may block the progression of the RNA polymerase along the DNA. It is clear, then, that intercalators

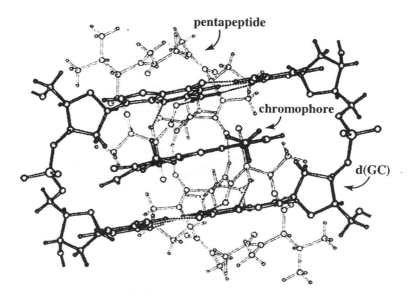

Figure 6.19 ► X-ray structure of a 1:2 complex of dactinomycin with d(GC) [Reprinted from *Journal of Molecular Biology*, Vol. 68, "Stereochemistry of actinomycin binding to DNA. II. Detailed molecular model of actinomycin DNA complex and its implications", pp. 26–34, Copyright 1972 Academic Press, with permission from Elsevier.]

can cause cytotoxicity by interfering with the normal cellular processing of DNA, such as replication and transcription.

Resistance to dactinomycin is associated with an overly active efflux pump mediated by overexpression of the P170 membrane glycoprotein and impaired drug uptake.[104] The cell membrane composition may affect the rate of drug diffusion.[105] There also appears to be a correlation between the ability of the cell to retain dactinomycin and the effectiveness of the drug.[106]

c. Doxorubicin (Adriamycin) and Daunorubicin (Daunomycin), Anthracycline Antitumor Antibiotics

The anthracycline class of antitumor antibiotics exemplified by doxorubicin (**6.18**, X = OH; previously called adriamycin; Doxil) and daunorubicin (**6.18**, X = H; also called daunomycin; Cerubidine) are isolated from different species of *Streptomyces*. Although these two compounds differ by only one hydroxyl group, there is a major difference in their antitumor activity. Whereas daunorubicin is active only against leukemia, doxorubicin is active against leukemia as well as a broad spectrum of solid tumors.

There is some controversy as to whether the mechanism of action of these compounds is related to their ability to intercalate into the DNA or to cause DNA strand breakage.[107] The vast majority of the intracellular drug is in the nucleus, where it intercalates into the DNA double helix (and forms the ternary complex with DNA topoisomerase II), with consequent inhibition of replication and transcription. X-ray[108] and NMR[109] studies of model daunorubicin-oligonucleotide complexes show that the oligonucleotides form a six base pair right-handed double helix with two daunorubicin molecules intercalated in the d(CpG) sequences. The tetracyclic chromophore is oriented orthogonal to the long dimension of the DNA base pairs, and the ring that has the amino sugar substituent (A ring) protrudes into the

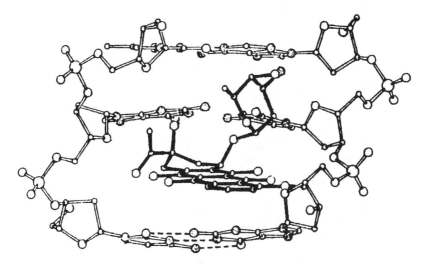

Figure 6.20 ▶ X-ray structure of daunorubicin intercalated into an oligonucleotide [Reprinted with permission from Myers, C. E., Jr.; Chabner, B. A. In *Cancer Chemotherapy: Principles and Practice* (Chabner, B. A.; Collins, J. M., Eds.), p. 356. Lippincott, Philadelphia, Pennsylvania. Copyright 1990 Lippincott.]

minor groove (Figure 6.20). Substituents on the A ring hydrogen bond to base pairs above and below the intercalation site. The amino sugar nearly covers the minor groove, but without bonding to the DNA. Ring D protrudes into the major groove. The complex is stabilized by stacking energies and by hydrogen bonding of the hydroxyl and carbonyl groups at C-9 of the A ring. X-ray diffraction analysis of the DNA–daunorubicin complex indicates that there is no interaction between the ionized amino group of the drug and any part of the double helix; it sits in the center of the minor groove. This is consistent with structure–activity studies[110] that indicate that modification of the amino group does not necessarily affect biological activity.

The same mechanisms of drug resistance that were found for dactinomycin apply to these drugs as well. In addition to intercalation, and topoisomerase II-induced DNA damage, another mechanism of action of the anthracycline antitumor antibiotics involves radical-induced DNA strand breakage; this mechanism is discussed in Section 6.3.C.1, p. 369.

d. Bisintercalating Agents

Once success with intercalating agents was realized, it was thought that *bifunctional intercalating agents* (also called *bisintercalating agents*), in which two potential intercalating molecules are tethered together so that each could intercalate into different DNA strands, would have enhanced affinity for DNA and slower dissociation rates. In general, the DNA affinity of bisintercalating agents is greater than that of their monointercalating counterparts, in some cases approaching values typical of those observed with natural repressor proteins. However, this affinity may result from the polycationic nature of the high-affinity bisintercalators and not necessarily because of the intercalating group. Quinoxaline antibiotics such as triostin A (**6.22**, R = CH_2S-SCH_2) and echinomycin (**6.22**, R = $CH(SCH_3)$-SCH_2), which have two potential intercalator molecules tethered to a cyclic depsipeptide,[111] fail to bind to DNA without the cyclic depsipeptide. This may be the result of the lack of rigidity to maintain the two quinoxaline groups in the correct orientation for bisintercalation with two base pairs between each intercalation site (consistent with the neighbor exclusion principle discussed

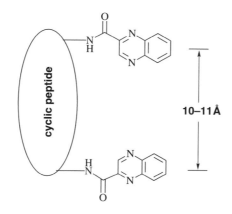

Figure 6.21 ▶ A general structure of bisquinoxaline intercalators

in this section). Both the rigidity and length of the linker chain between the two intercalator molecules are important to constrain the bisintercalator in the ideal configuration to form a sandwich with two base pairs.[112] A general structure of bisquinoxaline intercalators is shown in Figure 6.21. The synthetic diacridines with flexible linker chains are ineffective. For the most part, much of the expected enhancement in free energy of bisintercalation is not observed.[113] This is probably the result of the unfavorable entropy associated with loss of rotational freedom. In general, these agents do not show a remarkable improvement in specificity compared to that found for the monofunctional intercalators, although there are effective bisintercalators that utilize flexible linkers.[114]

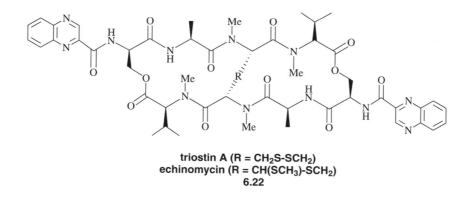

triostin A (R = CH$_2$S-SCH$_2$)
echinomycin (R = CH(SCH$_3$)-SCH$_2$)
6.22

6.3.B DNA Alkylators

The difference between the DNA alkylators and the DNA intercalators is akin to the difference between irreversible and reversible enzyme inhibitors (see Chapter 5). The intercalators (reversible enzyme inhibitors) bind to the DNA (enzyme) with noncovalent interactions. The DNA alkylators (irreversible inhibitors) react with the DNA (enzyme) to form covalent bonds. Several of the important classes of alkylating drugs that have different alkylation mechanisms are discussed next.

B.1   Nitrogen Mustards

a.   Lead Discovery

Sulfur mustard (**6.23**) is a highly toxic nerve gas that was used in World Wars I and II. Autopsies of soldiers killed in World War I by sulfur mustard revealed leukopenia (low white blood cell count), bone marrow aplasia (defective development), dissolution of lymphoid tissues, and ulceration of the gastrointestinal tract.[115] All of these lesions indicated that sulfur mustard has a profound effect on rapidly dividing cells and suggested that related compounds may be effective as antitumor agents. In fact, in 1931 sulfur mustard was injected directly into tumors in humans,[116] but this procedure turned out to be too toxic for systemic use. Because of the potential antitumor effects of sulfur mustards, a less toxic form was sought. Gilman and others examined the antitumor effects of the isosteric nitrogen mustards (**6.24**), less reactive alkylating agents, and in 1942 the first clinical trials of a nitrogen mustard were initiated. However, this research was classified during World War II, so the usefulness of nitrogen mustard in the treatment of cancer was not known until 1946, when these studies became declassified and Gilman published a review of his findings.[117] Soon thereafter several other summaries of clinical research carried out during the war appeared.[118] This work marks the beginning of modern cancer chemotherapy.

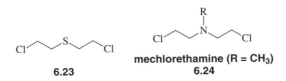

mechlorethamine (R = CH$_3$)

6.23   6.24

Prior to a discussion of lead modification and other classes of alkylating agents, let's take a brief excursion into the chemistry of alkylating agents in general.

b.   Chemistry of Alkylating Agents

According to Ross,[119] a *biological alkylating agent* is a compound that can replace a hydrogen atom with an alkyl group under physiological conditions (pH 7.4, 37°C, aqueous solution). These alkylation reactions are generally described in terms of substitution reactions by N, O, and S heteroatomic nucleophiles with the electrophilic alkylating agent, although Michael addition reactions also are important. The two most common types of nucleophilic substitution reactions are S_N1 (Scheme 6.2A), a stepwise reaction via an intermediate carbenium ion, and S_N2 (Scheme 6.2B), a concerted reaction. In general, the relative rates of nucleophilic substitution at physiological pH are in the order thiolate > amino > phosphate > carboxylate.[120] For DNA the most reactive nucleophilic sites are *N*-7 of guanine > *N*-3 of adenine > *N*-7 of adenine > *N*-3 of guanine > *N*-1 of adenine > *N*-1 of cytosine[121] [see **6.25** and **6.26** for the numbering system of purines (A and G) and pyrimidines (C and T), respectively]. The *N*-3 of cytosine, the *O*-6 of guanine, and the phosphate groups also can be alkylated. Quantum mechanical calculations[122] confirm that the *N*-7 position of guanine is the most nucleophilic site. The reactivity of various nucleophilic sites on DNA is strongly controlled by steric, electronic, and hydrogen bonding effects. For example, some of the nucleophilic sites are in the interior of the DNA double helix and are sterically blocked. Also, only nucleophilic centers in the major and minor grooves or in the walls of the double helix are readily accessible to alkylating agents. In addition to these steric effects, the nucleophilicity of various sites on the purine and pyrimidine bases of the DNA is diminished because of their involvement in Watson-Crick hydrogen bonding.

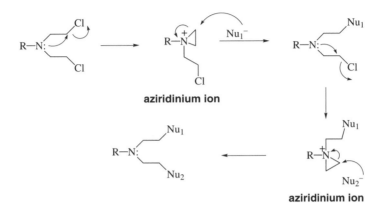

(A) Alkyl—X $\underset{}{\overset{S_N1}{\rightleftharpoons}}$ Alkyl$^+$ X$^-$ $\xrightarrow{\text{Nu}^-}$ Alkyl—Nu

(B) Alkyl—X $\xrightarrow[S_N2]{\text{Nu}^-}$ Alkyl—Nu + X$^-$

Scheme 6.2 ▶ Nucleophilic substitution mechanisms

aziridinium ion

aziridinium ion

Scheme 6.3 ▶ Alkylations by nitrogen mustards

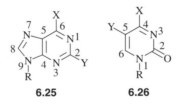

6.25   **6.26**

The reaction order for nucleophilic substitution depends on the chemical structure of the alkylating agent. Simple alkylating agents such as ethylenimines and methanesulfonates undergo S_N2 reactions. Alkylating agents such as nitrogen mustards, which have a nucleophile capable of anchimeric assistance (neighboring group participation), can undergo S_N1- or S_N2-type reactions, depending on the relative rates of the aziridinium ion formation and the nucleophilic attack on the aziridinium ion (Scheme 6.3).[123] When aziridinium ion formation is fast, the overall reaction rate is second order (S_N2), but when aziridinium ion formation is slower than nucleophilic attack, the overall reaction is first order (S_N1).

In the case of the nitrogen mustards, which are bifunctional alkylating agents (i.e., they have two electrophilic sites), the DNA undergoes intrastrand and interstrand cross-linking.[124] Although there does not appear to be a direct correlation between the chemical reactivity of the alkylating agent and the therapeutic or toxic effects,[125] the compounds that are able to cross-link DNA are much more effective than singly alkylating agents.[126] There is a rough correlation between antitumor efficacy and ability to induce mutations and inhibit DNA synthesis. Bardos *et al.*[125b] also found a relationship between the rate of solvolysis of the alkylating agent and its cytotoxic effect on tumor cells *in vitro*. The differences in the effectiveness of the various alkylating agents probably result from differences in pharmacokinetic factors, lipid solubility, ability to penetrate the central nervous system, membrane transport properties, detoxification reactions (see Chapter 7), and specific enzymatic reactions capable of repairing alkylated sites on the DNA.[127]

c. Lead Modification

The prototype of the nitrogen mustards is mechlorethamine (**6.24**, R = CH$_3$; Mustargen),[128] which is still used in the treatment of advanced Hodgkin's disease. Mechlorethamine is a bifunctional alkylating agent that reacts with the N-7 of two different guanines in DNA,[129] producing an interstrand cross-link (**6.27**) by the mechanism shown in Scheme 6.3. The formation of the N-7 ammonium ion makes the guanine more acidic and, therefore, appears to shift the equilibrium in favor of the enol tautomer.[130] Because guanine in this tautomeric form can make base pairs with thymine residues instead of cytosine (**6.28**), this leads to miscoding during replication. Furthermore, the N-7 alkylated guanosine is susceptible to hydrolysis, which results in the destruction of the purine nucleus and in DNA strand scission (Scheme 6.4). The deglycosylation (loss of the sugar) is a very slow reaction.[131] In addition to interstrand cross-links it also is possible for the second chloroethyl group of the nitrogen mustard to react with a thiol or amino group of a protein, resulting in a DNA–protein cross-link. Any of these reactions could explain both the mutagenic and cytotoxic effects of the nitrogen mustards.

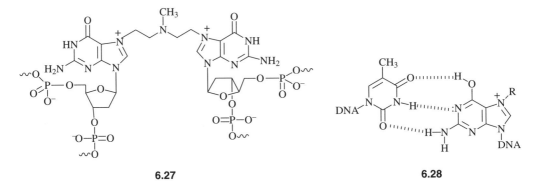

6.27   **6.28**

Mechlorethamine is quite unstable to hydrolysis. In fact, it is so reactive with water that it is marketed as a dry solid (HCl salt), and aqueous solutions are prepared immediately prior to injection; within minutes after administration, mechlorethamine reacts completely in the body. Because of this reactivity, a more stable analog was sought. Substitution of the methyl group of mechlorethamine with an electron withdrawing group, such as an aryl substituent (**6.29**), makes the nitrogen less nucleophilic, less able to participate in anchimeric assistance, and therefore slows down the rate of aziridinium ion formation (see Scheme 6.3); this decreases the reactivity of the nitrogen mustard.[132] As a result of this stabilization, some of these compounds could be administered orally, and they would be able to undergo absorption and distribution before extensive alkylation occurred. Simple aryl substituted nitrogen mustards are not water soluble enough for intravenous administration, but the solubility problem was solved with the use of carboxylate-containing aryl substituents. Direct substitution of the phenyl with a carboxylate (**6.29**, R = CO$_2$H), however, gave a compound that was too stable because the nonbonded electrons of the nitrogen could be delocalized into the carboxylate carbonyl; this compound was not very active. To increase the reactivity, the electron-withdrawing effect of the carboxylate was attenuated by the insertion of methylenes between the phenyl and carboxylate groups. The optimal number of methylenes was found to be three, giving the antitumor drug chlorambucil (**6.29**, R = (CH$_2$)$_3$CO$_2$H; Leukeran).[133] This approach

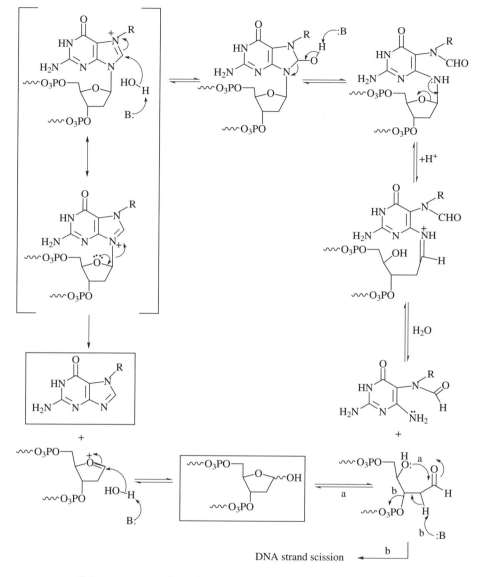

Scheme 6.4 ▶ Depurination of *N*-7 alkylated guanines in DNA

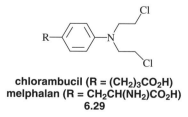

chlorambucil (R = (CH$_2$)$_3$CO$_2$H)
melphalan (R = CH$_2$CH(NH$_2$)CO$_2$H)
6.29

retained the water solubilizing effect of the carboxylate group, but the nitrogen lone pair could not be resonance stabilized, so its reactivity increased relative to **6.29**, R = CO$_2$H. Other nitrogen mustard analogs were prepared in an attempt to obtain an anticancer drug that would be targeted for a particular tissue. Because *L*-phenylalanine is a precursor to melanin, it was thought that *L*-phenylalanine nitrogen mustard (**6.29**, R = CH$_2$CH(NH$_2$)CO$_2$H; melphalan; Alkeran) might accumulate in melanomas. Although this analog is an effective, orally active anticancer drug (for multiple myelomas), it is not active against melanomas. The *L*-isomer (melphalan), the *D*-isomer (medphalan), and the racemic mixture (merphalan) have approximately equal potencies.[134] This lack of enantiospecificity suggests that there is no active transport of these compounds into cancer cells; however, a leucine carrier system appears to be involved in melphalan transport.[135]

d. Drug Resistance

A major problem that limits the effectiveness of alkylating agents in general is resistance in tumor cells. With the use of cells selected for resistance in culture and with repeatedly transplanted animal tumors, mechanisms of resistance have been found to include decreased drug entry into the cell, increased repair of the drug defect, increased nonprotein sulfhydryl content such as glutathione (to react with the alkylating agent), and increased levels of metabolic enzymes such as glutathione *S*-transferase. It is not clear, however, which of these mechanisms are clinically significant.

B.2   Ethylenimines

Because the reactive intermediate involved in DNA alkylation by nitrogen mustards (Scheme 6.3, p. 355) is an aziridinium ion, an obvious extension of the nitrogen mustards is aziridines (ethylenimines). Protonated ethylenimines are highly reactive (they are aziridinium ions) and would not be effective drugs. When electron-withdrawing groups are substituted on the aziridine nitrogen, however, the pK_a of the nitrogen is lowered to a point where the aziridine is not protonated at physiological pH. These aziridines are much less reactive. In general, two ethylenimine groups per molecule are required for antitumor activity; compounds with three or four aziridines are not significantly more potent.[136] Examples of antitumor ethylenimines include triethylenemelamine (**6.30**), carboquone (**6.31**), and diaziquone (**6.32**). By appropriate addition of lipophilic substituents to the benzoquinone ethylenimines, antitumor activity in the central nervous system can be achieved.[137]

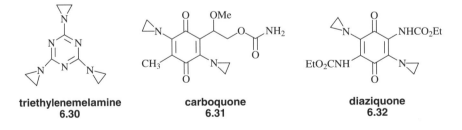

triethylenemelamine **6.30**	carboquone **6.31**	diaziquone **6.32**

B.3   Methanesulfonates

Methanesulfonate is an excellent leaving group. The most prominent example of this class of alkylating agents is the bifunctional anticancer drug busulfan[138] (**6.33**, *n* = 4; Myleran).

Compounds with one to eight methylene groups (**6.33**, $n = 1-8$) have antitumor activity, but maximum activity is obtained with four methylenes.[139] Alkylation of the N-7 position of guanine was demonstrated.[140] Unlike the nitrogen mustards (Section 6.3.B.1, p. 354), however, intrastrand, not interstrand, cross-links form.[141]

$$CH_3O_2SO-(CH_2)_n-OSO_2CH_3$$

busulfan (n = 4)
6.33

B.4 (+)-CC-1065 and Duocarmycins

(+)-CC-1065 (**6.34**),[142] (+)-duocarmycin A (**6.35**),[143] and (+)-duocarmycin SA (**6.36**),[144] natural products isolated from various strains of *Streptomyces*, are sequence-selective DNA alkylating agents.[145] Within each of these molecules is embedded a 4-spirocyclopropylcyclohexadienone (**6.37**), which is quite electrophilic because nucleophilic attack at the cyclopropane releases the electrons of the cyclopropane ring for delocalization into the dienone system of the cyclohexadienone (Scheme 6.5).[146] However, all of these natural products are stable toward nucleophiles at neutral pH because they also contain a nitrogen atom that is conjugated with the enone system, thereby sharply decreasing its electrophilicity (Scheme 6.6). Nonetheless, these molecules alkylate the N-3 atom of adenine bases selectively within the A-T-rich regions of the minor groove.[147] The reason for the selectivity of attachment appears to be because of a forced adoption of these molecules into helical conformations on binding at the narrower, deeper A-T-rich regions in the minor groove, which causes a greater degree of conformational change of the molecules. This conformational change twists the carbon–nitrogen bond of these molecules out of conjugation with the enone system, thereby decreasing the nitrogen stabilization of the enone; as a result, they become more like the spirocyclopropylcyclohexadienone system (**6.37**) that does not have a nitrogen in conjugation with the enone. Once the nitrogen is out of conjugation, nucleophilic attack at the cyclopropane is enhanced[148] (Scheme 6.7), and because this activation occurs most favorably in A-T-rich regions of the minor groove, that is where alkylation selectively occurs. The presence of the nitrogen atom increases the stability of the spirocyclopropyl group 10^3-10^4 times at pH 7.[149] The rate of DNA alkylation changes by less than a factor of 2 over a physiological relevant range spanning 2 pH units.[150] This indicates that the alkylation reaction is not acid catalyzed, supporting an S_N2-type reaction.

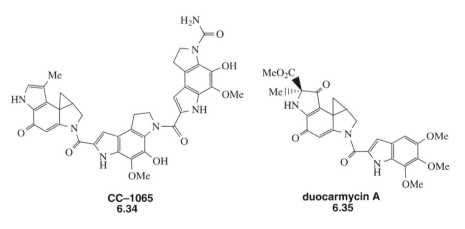

CC–1065
6.34

duocarmycin A
6.35

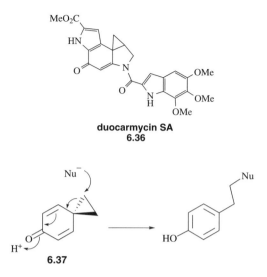

duocarmycin SA
6.36

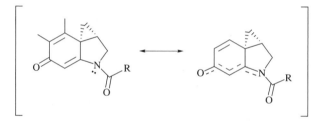

Scheme 6.5 ▶ Reaction of nucleophiles with 4-spirocyclopropylcyclohexadienone

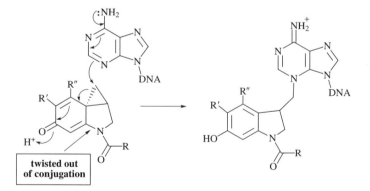

Scheme 6.6 ▶ Stabilization of the spirocyclopropylcyclohexadienone by nitrogen conjugation

Scheme 6.7 ▶ *N*-3 adenine alkylation by CC-1065 and related compounds

Therefore, these compounds start out as relatively unreactive minor groove binders, but as a result of binding at a particular region (A-T-rich regions), a conformational change activates these molecules for alkylation. This is the DNA equivalent of mechanism-based enzyme inactivation discussed in Chapter 5 (Section 5.5.C, p. 285).[151] This class of alkylating agents is activated by DNA, but there are many DNA alkylating agents that are stable until some biological species, such as one or more enzymes or a reducing agent, converts them into alkylating agents. Those compounds are called *metabolically activated alkylating agents.*

B.5 Metabolically Activated Alkylating Agents

a. Nitrosoureas

The nitrosoureas (**6.38**) were developed from the lead compound *N*-methylnitrosourea (**6.38**, R = CH$_3$, R' = H), which exhibited modest antitumor activity in animal tumor models.[152] Analogs with 2-chloroethyl substituents, such as carmustine (**6.38**, R = R' = CH$_2$CH$_2$Cl; BCNU; Gliadel) and lomustine (**6.38**, R = CH$_2$CH$_2$Cl, R' = cyclohexyl; CCNU; CeeNU), were found to possess much greater antitumor activity.[153] Because of their lipophilicity, the 2-chloroethyl analogs were able to cross the blood–brain barrier and, consequently, have been used in the treatment of brain tumors. Despite the potency of these antitumor drugs, they are less desirable than others because of severe problems of delayed and cumulative bone marrow toxicity.

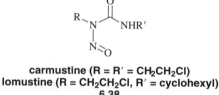

carmustine (R = R' = CH$_2$CH$_2$Cl)
lomustine (R = CH$_2$CH$_2$Cl, R' = cyclohexyl)
6.38

Extensive mechanistic studies have been carried out on nitrosoureas. Decomposition of the first active anticancer nitrosourea (**6.38**, R = CH$_3$, R' = H), produces methyl diazonium ion (**6.39**, Scheme 6.8), a potent methylating agent, and isocyanic acid (**6.40**), a carbamoylating agent.[154] Evidence that diazomethane is not the alkylating agent was provided by a model study[155] showing that under physiological conditions 1-trideuteriomethyl-3-nitro-1-nitrosoguanidine (**6.41**, Scheme 6.9) alkylates nucleophiles with the trideuteriomethyl group intact; if diazomethane were the alkylating agent, dideuteriomethyl groups would have resulted. It is now known that *N*-nitrosoamides (**6.42**) and *N*-nitrosourethanes (**6.43**), which produce alkylating, but not carbamoylating species, do have anticancer activity.[156] Furthermore, certain nitrosoureas that have little carbamoylating activity also are quite active, but nitrosoureas with no detectable alkylating activity are either very weakly active or inactive. Therefore, the alkylating, not carbamoylating, product appears to be the principal species responsible for the anticancer activity.

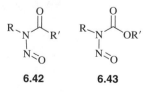

6.42   **6.43**

The carbamoylating isocyanate that is generated (e.g., **6.40** in Scheme 6.8) does not appear to be directly involved in the antitumor effects of nitrosoureas, but it does react

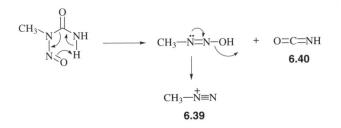

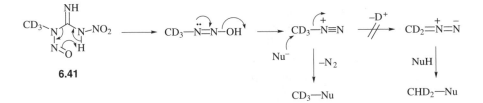

Scheme 6.8 ► Decomposition of *N*-methyl-*N*-nitrosourea

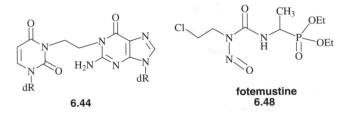

Scheme 6.9 ► Deuterium labeling experiment to determine mechanism of activation of nitrosoureas

with amines in proteins.[157] More importantly, it inhibits DNA polymerase[158] and other enzymes involved in the repair of DNA lesions,[159] such as O^6-alkylguanine-DNA alkyl-transferase, DNA nucleotidyl transferase, and DNA glycosylases. It also inhibits RNA synthesis and processing[160] and plays a role in the toxicity of the nitrosoureas.[161] The 2-chloroethyl-substituted analogs (**6.38**, R = CH$_2$CH$_2$Cl) react with DNA and produce an interstrand cross-link between a guanine on one strand and a cytosine residue on another.[162] 1-[N^3-Deoxycytidyl]-2-[N^1-deoxyguanosinyl]ethane (**6.44**) was isolated from the reaction of N,N'-bis(2-chloroethyl)-N-nitrosourea (carmustine; **6.38**, R = R′ = CH$_2$CH$_2$Cl). Because the same cross-link occurs with the mono-2-chloroethyl-substituted analog **6.38** (R = CH$_2$CH$_2$Cl, R′ = cyclohexyl), the mechanism shown in Scheme 6.10 was proposed. The same mechanism was proposed for fotemustine (**6.48**, Muphoran), and the reactive chloroethyldiazo hydroxide (**6.45**) was detected by electrospray ionization mass spectrometry.[163] To rationalize the regioselectivity of these alkylating agents, a kinetic analysis of the reaction was carried out, and an alternative reaction mechanism was suggested[164] (Scheme 6.11). The principal difference in these mechanisms is that in Scheme 6.11 a nucleoside on the DNA reacts with the intact drug to form a tetrahedral intermediate (**6.49**), which after cyclization to **6.50**, undergoes reaction with the O-6 of a guanine to give **6.51**, the precursor to the diazonium ion, equivalent to **6.46** in Scheme 6.10, which leads to interstrand cross-linking. Evidence for the cyclization of **6.46** to **6.47** (Scheme 6.10) and nucleophilic attack to give interstrand cross-linking comes from model chemistry for this reaction.[165]

6.44   **fotemustine**
                              6.48

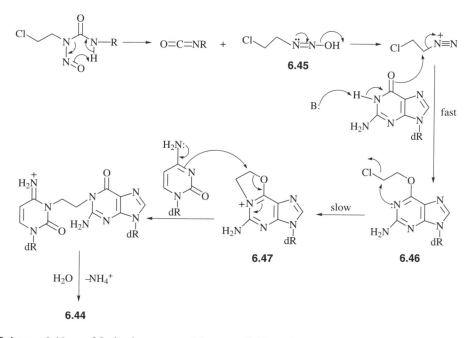

Scheme 6.10 ► Mechanism proposed for cross-linking of DNA by (2-chloroethyl)nitrosoureas

Evidence for the intermediacy of an O-6 guanine adduct such as **6.46** (Scheme 6.10) or **6.51** (Scheme 6.11) is based on the observation that cell lines capable of excising O-6 guanine adducts were resistant to cross-link formation[166] and that the addition of rat liver O^6-alkylguanine-DNA alkyltransferase, the enzyme that excises O-6 guanine adducts, prevents formation of the cross-links.[167] O^6-Alkylguanine adducts are one of the most important DNA modifications responsible for the induction of cancer, mutation, and cell death.[168] Repair of DNA adducts formed at the O-6 position of guanine by O^6-alkylguanine-DNA alkyltransferase is the predominant pathway for resistance.[169] The mechanism for repair by this enzyme is an S_N2 displacement of the O^6-alkyl group by an active site cysteine residue in the repair enzyme. Surprisingly, the alkyl group transferred to the active site cysteine is not removed; therefore, the enzyme can only turn over once and is inactivated by its substrate! By definition, then, O^6-alkylguanine-DNA alkyltransferase actually is not an enzyme (an enzyme is a catalyst, but in this case the original form of the catalyst is not regenerated). The alkylated form of the enzyme is rapidly degraded.[170]

b. Triazene Antitumor Drugs

Other antitumor drugs are metabolically converted into methyldiazonium ion, which methylates DNA. An example is the triazenoimidazoles, such as 5-(3,3-dimethyl-1-triazenyl)-1H-imidazole-4-carboxamide (**6.52**, dacarbazine; DTIC-Dome), which is active against a broad range of cancers but is used preferably for the treatment of melanotic melanoma.[171] Although dacarbazine is a structural analog of 5-aminoimidazole-4-carboxamide, an intermediate in purine biosynthesis, the cytotoxicity of **6.52** is a result of its conversion into an alkylating agent, not its structural similarity to the metabolic intermediate. With the use of [^{14}C-*methyl*]dacarbazine, it was shown that formaldehyde is generated and that the DNA becomes

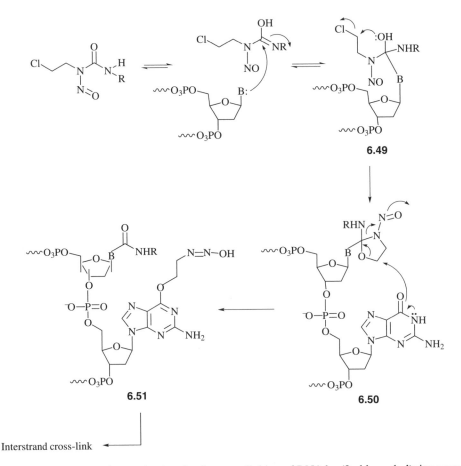

Scheme 6.11 ▶ Alternative mechanism for the cross-linking of DNA by (2-chloroethyl)nitrosoureas

methylated at the 7-position of guanine.[172] A mechanism that rationalizes these results is shown in Scheme 6.12.

c. Mitomycin C

Bioreductive alkylation is a process by which an inactive compound is metabolically reduced to an alkylating agent.[173] The prototype for antitumor antibiotics that act as bioreductive alkylating agents of DNA is mitomycin C (**6.53**, Scheme 6.13; Mutamycin), which contains three important carcinostatic functional groups, the quinone, the aziridine, and the carbamate groups.[174]

The mechanism proposed by Iyer and Szybalski[175] is shown in Scheme 6.13. Reduction of the quinone by one-electron reductants, such as cytochrome c reductase or cytochrome b_5 reductase, to the semiquinone (**6.54**, R = electron) or by the more common two-electron reductant, such as DT-diaphorase,[176] to the hydroquinone (**6.54**, R = H) converts the heterocyclic nitrogen from a vinylogous amide nitrogen (the nonbonded electrons of the nitrogen are in conjugation with the quinone carbonyl via the intermediate double bond), which is not nucleophilic, into an amine nitrogen, which can eliminate the β-methoxide ion (**6.54**). Tautomerization of the resultant immonium ion (**6.55**) gives **6.56**, which is set up for aziridine

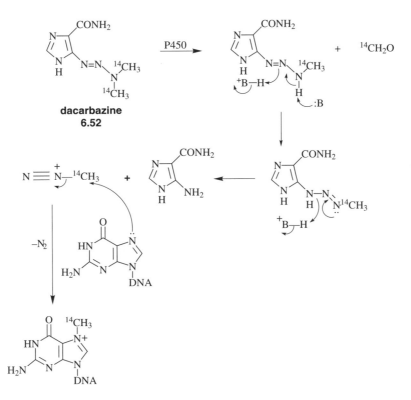

Scheme 6.12 ▶ Mechanism for the methylation of DNA by dacarbazine

ring opening. This activates the drug by unmasking the electrophilic site at C-1, which alkylates the DNA (**6.57**). A subsequent reaction of DNA at C-10 (**6.58**) results in the cross-linking of the DNA (**6.59**). Bean and Kohn[177] showed in chemical models that nucleophiles react most rapidly at C-1; the reaction at C-10 to displace the carbamate also occurs, but at a slower rate. Reduction of the quinone is necessary for the covalent reaction of **6.53** to DNA, but controversy exists as to whether the semiquinone (**6.54**, R = electron) or hydroquinone (**6.54**, R = H) is the viable intermediate.[178] Chemical model studies on the mechanism of action of mitomycin C indicate that the conversion of **6.53** to **6.58** can occur at the semiquinone stage[179] and the conversion of **6.58** to **6.59** occurs at the hydroquinone oxidation state.[180] Both monoalkylated[181] and bisalkylated DNA adducts have been identified; the extent of mono- and bisalkylation increases with increasing guanine base composition of the DNA.[182] Cross-links in the minor groove between two guanines at their C-2 amino groups form[183] with preferential interstrand cross-linking at 5′-CG rather than 5′-GC sequences.[184]

The bioreductive alkylation approach also was directed toward the design of new antineoplastic agents that may be selective for hypoxic (O_2-deficient) cells in solid tumors.[185] These cells are remote from blood vessels and are located at the center of the solid tumors. Hypoxia protects the tumor cells from radiation therapy, and because these cells are buried deep inside the tumor, appropriate concentrations of antitumor drugs may not reach them prior to drug metabolism. Because these cells might have a more efficient reducing environment, bioreductive alkylation seemed to be well suited. The bioreductive alkylation approach based on reduction of a quinone to the corresponding hydroquinone was utilized in the design of

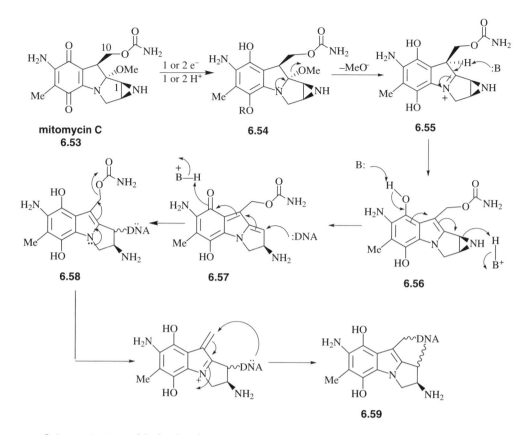

Scheme 6.13 ►   Mechanism for the bioactivation of mitomycin C and alkylation of DNA

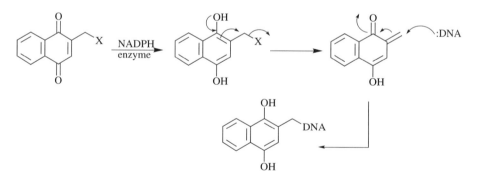

Scheme 6.14 ►   Bioreductive monoalkylating agents

both mono- (Scheme 6.14)[186] and bisalkylating agents (Scheme 6.15).[187] Electron-rich substituents lower the reduction potential of the quinones and make them more reactive.[188]

d.   Leinamycin

Leinamycin (**6.60**) is a potent antitumor agent that was isolated from a strain of *Streptomyces*, and its structure was elucidated by spectroscopy and X-ray crystallography.[189] The cytotoxic

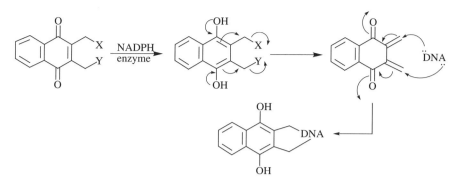

Scheme 6.15 ▶ Bioreductive bisalkylating agents

activity of this compound is initiated by a reaction with thiols.[190] Gates and coworkers,[191] using simple chemical model compounds containing 1,2-dithiolan-3-one 1-oxide groups (the five-membered ring heterocycle found in leinamycin), proposed that two of the intermediates generated by thiol addition to these model compounds are an oxathiolanone (**6.61**) and a hydrodisulfide (**6.62**) (Scheme 6.16). These intermediates would arise from thiol attack at the 1,2-dithiolan-3-one 1-oxide functional group to give a sulfenic acid intermediate, which cyclizes to the oxathiolanone.[192] Asai and coworkers[193] elegantly proved that the product of thiol addition to leinamycin in the presence of calf thymus DNA was an alkylated *N*-7 guanine adduct (**6.65**, Scheme 6.17), proposed to be derived from thiol attack on the 1,2-dithiolan-3-one 1-oxide to give oxathiolanone **6.63** (based on the above-cited model studies of Gates and coworkers), which rearranges to an episulfonium intermediate (**6.64**), a highly electrophilic species that alkylates the DNA. Although the episulfonium intermediate was too reactive to detect, epoxide **6.66** could be isolated and characterized and was shown to alkylate DNA; however, the product obtained is the one derived from attack on the episulfonium intermediate (**6.64**), indicating that the epoxide does not directly alkylate DNA, but is in equilibrium with **6.64**, which is the actual alkylating species.[194]

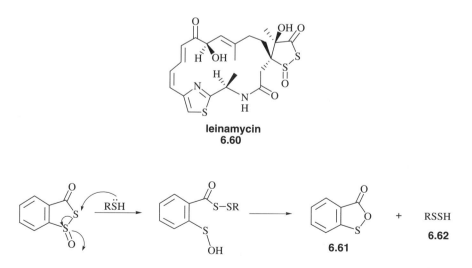

leinamycin
6.60

Scheme 6.16 ▶ Model reaction for the mechanism of activation of leinamycin

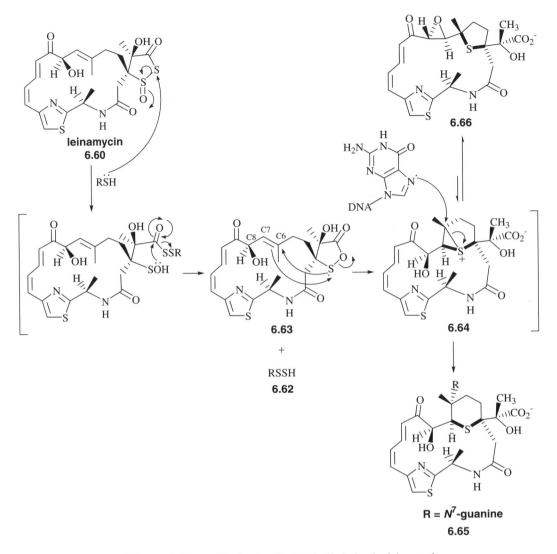

Scheme 6.17 ▶ Mechanism for DNA alkylation by leinamycin

This alkylation mechanism is only one way in which leinamycin modifies DNA. The other pathway involves further reactions of hydrodisulfide **6.62**, which can be oxidized by O_2 to give polysulfides, shown[195] to cause further thiol-dependent oxidative DNA damage, presumably via radical intermediates such as hydroxyl radicals (Scheme 6.18). This radical-induced DNA damage characterizes the third class of DNA-interactive agents, the DNA strand breakers (next section).

6.3.C   DNA Strand Breakers

Some DNA-interactive drugs initially intercalate into DNA, but then, under certain conditions, react in such a way as to generate radicals. These radicals typically abstract hydrogen atoms

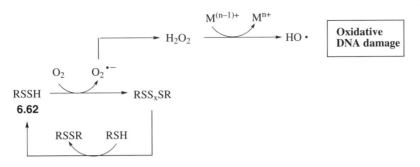

Scheme 6.18 ▶ Mechanism for hydrodisulfide activation of molecular oxygen to cause oxidative DNA damage

from the DNA sugar-phosphate backbone or from the DNA bases, leading to DNA strand scission. Therefore, these DNA-interactive compounds are metabolically activated radical generators. As examples of this mode of action of DNA-interactive drugs, we will consider the anthracycline antitumor antibiotics, bleomycin, tirapazamine (this one does not initially intercalate), and the enediyne antitumor antibiotics. Keep in mind that some of the compounds that lead to strand breakage act via the topoisomerase-induced mechanisms discussed in Section 6.3.A.3.

C.1 Anthracycline Antitumor Antibiotics

Doxorubicin (**6.18**, X = OH) and daunorubicin (**6.18**, X = H) are anthracyclines that were discussed in the section on DNA intercalators (Section 6.3.A.3, p. 346); however, these drugs also cause oxygen-dependent DNA damage.[196] Several mechanisms have been proposed to account for this destruction of DNA. Anthracyclines cause protein-associated breaks that correlate with their cytotoxicity,[197] and these breaks may be caused by the reaction of anthra-cyclines on topoisomerase II, an enzyme that promotes DNA strand cleavage and reannealing (see Section 6.2.C, p. 334).

Another mechanism for DNA damage is electron transfer chemistry. A one-electron reduc-tion of the anthracyclines, probably catalyzed by flavoenzymes such as NADPH cytochrome P450 reductase,[198] produces the anthracycline semiquinone radical (**6.67**, Scheme 6.19), which can transfer an electron to oxygen to regenerate the anthracycline and produce superoxide ($O_2^{-\cdot}$). Both the superoxide and anthracycline semiquinone radical anions can generate hydroxyl radical (HO·) (Scheme 6.20), which is known to cause DNA strand breaks.[199]

A third possibility for the mechanism of DNA damage by anthracyclines is the formation of a ferric complex, which binds to DNA by a mechanism different from intercalation and significantly tighter.[200] The binding constant for the doxorubicin-ferric complex (**6.68**) is 10^{33}. This ferric complex could react with superoxide to give oxygen and the corresponding ferrous complex (Scheme 6.20). The reaction of ferrous ions with hydrogen peroxide is known as the *Fenton reaction*,[201] which is used in the standard method for the generation of hydroxyl radicals when doing *DNA footprinting*, a tech-nique that indiscriminately cleaves the DNA to determine where protein–DNA interactions occur.[202]

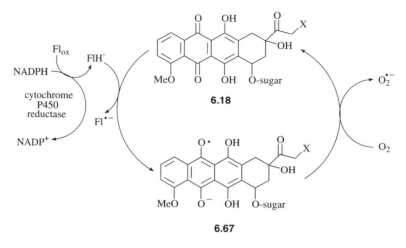

Scheme 6.19 ▶   Electron transfer mechanism for DNA damage by anthracyclines

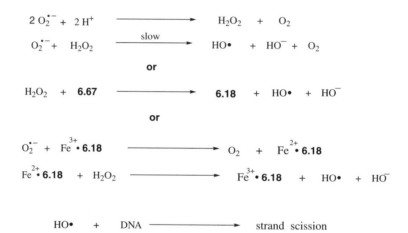

$$2\ O_2^{\bullet-} + 2\ H^+ \longrightarrow H_2O_2 + O_2$$

$$O_2^{\bullet-} + H_2O_2 \xrightarrow{\text{slow}} HO\bullet + HO^- + O_2$$

or

$$H_2O_2 + \textbf{6.67} \longrightarrow \textbf{6.18} + HO\bullet + HO^-$$

or

$$O_2^{\bullet-} + Fe^{3+}\bullet\textbf{6.18} \longrightarrow O_2 + Fe^{2+}\bullet\textbf{6.18}$$

$$Fe^{2+}\bullet\textbf{6.18} + H_2O_2 \longrightarrow Fe^{3+}\bullet\textbf{6.18} + HO\bullet + HO^-$$

$$HO\bullet + DNA \longrightarrow strand\ scission$$

Scheme 6.20 ▶   Anthracycline semiquinone generation of hydroxyl radicals

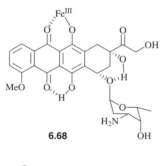

6.68

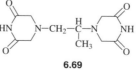

6.69

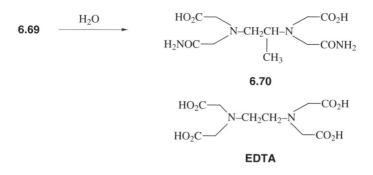

Scheme 6.21 ► Conversion of iron chelator prodrug **6.69** into iron chelator **6.70**

Because the generation of the hydroxyl radicals occurs adjacent to DNA, it is unlikely that radical scavengers would be an effective method of cell protection, as was shown with anthracycline antibiotic-induced cardiac toxicity. However, an iron chelator (**6.69**) does prevent doxorubicin-induced cardiac toxicity in humans.[203] This iron chelator actually is a prodrug (see Chapter 8) that, because of its nonpolar nature, enters the cell. Once inside the cell it is hydrolyzed to the active iron chelator **6.70** (Scheme 6.21), which is structurally related to the well-known iron chelator EDTA.

Gianni *et al.*[204] showed that the iron–doxorubicin (**6.18**, X = OH) complex is more reactive than the iron–daunorubicin (**6.18**, X = H) complex because the hydroxymethyl ketone side-chain of doxorubicin reacts spontaneously with iron to produce Fe^{2+}, HO·, and H_2O_2.

Drug resistance is attributed to elevated gene expression of the P170 glycoprotein that pumps drugs out of cells (efflux pump), to decreased activity of the repair enzyme topoisomerase II, and to increased glutathione production, which scavenges free radicals and peroxides.

C.2 Bleomycin

The anticancer drug bleomycin (**6.71**, BLM; Blenoxane) is actually a mixture of several glycopeptide antibiotics isolated from a strain of the fungus *Streptomyces verticellus*; the major component is bleomycin A_2 (**6.71**, R = $NH(CH_2)_3S^+(CH_3)_2$).[205] It cleaves double-stranded DNA selectively at 5′-GC and 5′-GT sites in the minor groove by a process that is both metal ion and oxygen dependent.[206] There are three principal domains in bleomycin.[207] The pyrimidine, the β-aminoalanine, and the β-hydroxyimidazole moieties make up the first domain, which is involved in the formation of a stable complex with iron (II). This complex interacts with O_2 to give a ternary complex (**6.72**, bleomycin A_2 complex), which is believed to be responsible for the DNA cleaving activity.[208] The second domain is comprised of the bithiazole moiety (the five-membered N and S heterocycles) and the attached sulfonium ion-containing side-chain. The bithiazole is important for sequence selectivity, presumably because of its intercalation properties with DNA;[209] possibly, the sulfonium ion is attracted electrostatically to a phosphate group.[210] The gulose and carbamoylated mannose disaccharide moiety, the third domain, may be responsible for selective accumulation of bleomycin in some cancer cells, but it does not appear to be involved in DNA cleavage.

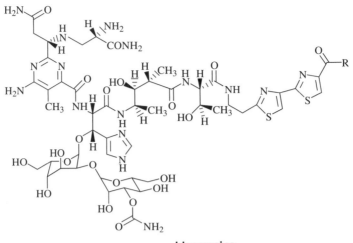

bleomycins
6.71

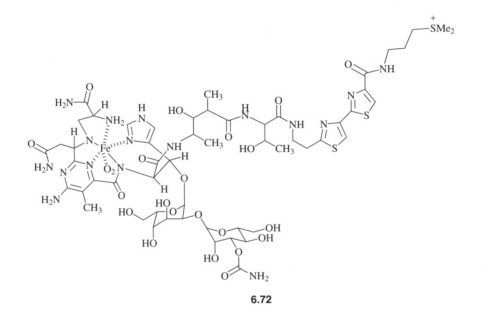

6.72

The primary mechanism of action of bleomycin is the generation of single- and double-strand breaks in DNA. This results from the production of radicals by a 1:1:1 ternary complex of bleomycin, Fe(II), and O_2 (Scheme 6.22). Activation of this ternary complex may be self-initiated by the transfer of an electron from a second unit of the ternary complex or activation may be initiated by a microsomal NAD(P)H-cytochrome P450 reductase-catalyzed reduction.[211] The activated bleomycin binds tightly to guanine bases in DNA, principally via the amino-terminal tripeptide (called tripeptide S) containing the bithiazole unit.[212] Binding by the bithiazole to G-T and G-C sequences is favored. Evidence for intercalation comes from the observation that when bleomycin, its tripeptide S moiety, or just bithiazole is mixed with DNA, a lengthening of the linear DNA and an unwinding of supercoiled circular DNA result.[213]

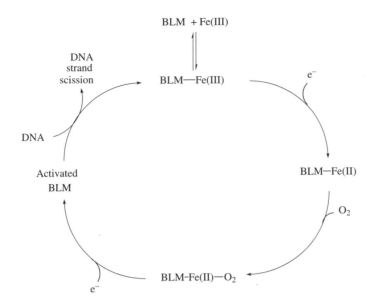

Scheme 6.22 ▶ Cycle of events involved in DNA cleavage by bleomycin (BLM)

Activation of the bound ternary complex is believed to occur by a reaction related to that for heme-dependent enzyme activation (see Section 4.3.D, p. 212) because it catalyzes the same reactions that are observed with these enzymes. As in the case of heme-dependent reactions, addition of an electron to the bleomycin–Fe(II)–O_2 complex would give a bleomycin–Fe(III)–OOH complex. This ferric hydroperoxo complex has been detected by UV-visible spectroscopy,[214] EPR spectrometry,[215] Mössbauer spectroscopy,[216] electrospray ionization mass spectrometry,[217] and XAS spectroscopy.[218] At least three mechanisms are possible for the formation of an activated bleomycin from the bleomycin–Fe(III)–OOH complex that leads to DNA strand scission: (1) Analogous to the case of the heme-dependent enzymes, addition of two protons would activate the O–O bond for heterolytic cleavage, giving a bleomycin–Fe(V)=O species, which could abstract a hydrogen atom from DNA; (2) the O–O bond could break homolytically to give a bleomycin–Fe(IV)=O species and hydroxyl radical, both of which could abstract a hydrogen atom from DNA; (3) Solomon and coworkers[219] favor a mechanism in which the bleomycin–Fe(III)–OOH complex undergoes a concerted reaction with DNA with concomitant O–O bond homolysis to give Fe(IV)=O, water, and a DNA radical. There is no experimental evidence that differentiates these, but density-functional theory calculations[220] show that heterolysis of the O–O bond of bleomycin–Fe(III)–OOH (mechanism 1) is predicted to be unfavorable by at least 40 kcal/mol, which is more than 150 kcal/mol less favorable than heterolysis of heme–OOH; direct reaction of bleomycin–Fe(III)–OOH with DNA (mechanism 3), however, is predicted to be close to thermoneutral. Mechanism 2 could be questioned because hydroxyl radicals are less selective than what is observed for bleomycin,[221] and homolytic cleavage of the O–O bond may have a large activation energy.[222]

The two major monomeric products formed when activated bleomycin reacts with DNA are nucleic base propenals (**6.76**, Scheme 6.23) and nucleic acid bases. Base propenal formation consumes an equivalent of O_2 in addition to that required for bleomycin activation and is accompanied by DNA strand scission with the production of 3′-phosphoglycolate (**6.77**) and

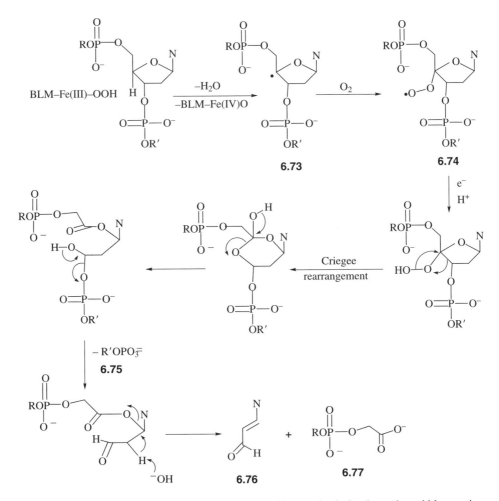

Scheme 6.23 ▶   Base propenal formation and DNA strand scission by activated bleomycin

5'-phosphate-modified DNA fragments (**6.75**, Scheme 6.23).[223] DNA base formation does not require additional O_2 and results in destabilization of the DNA sugar-phosphate backbone. Evidence for the C-4' radical (**6.73**) and the peroxy radical (**6.74**) comes from model studies of Giese and coworkers who used chemical methods to generate a C-4' radical in a single-stranded oligonucleotide.[224] They detected the C-4' radical and the peroxy radical and obtained similar (as well as additional) products to those observed from the reaction of bleomycin and DNA. Based on isotope studies with $^{18}O_2$ and $H_2{}^{18}O$ and other chemical precedence, two modified mechanisms are presented in Scheme 6.24. Mechanism A invokes a modified Criegee rearrangement to account for the ^{18}O labeling results, in which hydroxide is released rather than being added back to the adjacent carbon.[225] Also, the order of the steps is different from those in Scheme 6.23, but the same products are accounted for. Mechanism B involves a cleavage of the sugar-phosphate bond early in the mechanism based on other chemical precedence, and uses a Grob fragmentation step to break the hydroperoxide bond.[226]

Additional single-strand cleavage of the DNA occurs in the presence of alkali (the *alkali-labile lesion*).[227] The alkali-labile lesion was identified as the 4'-keto aldehyde (**6.78**),[228]

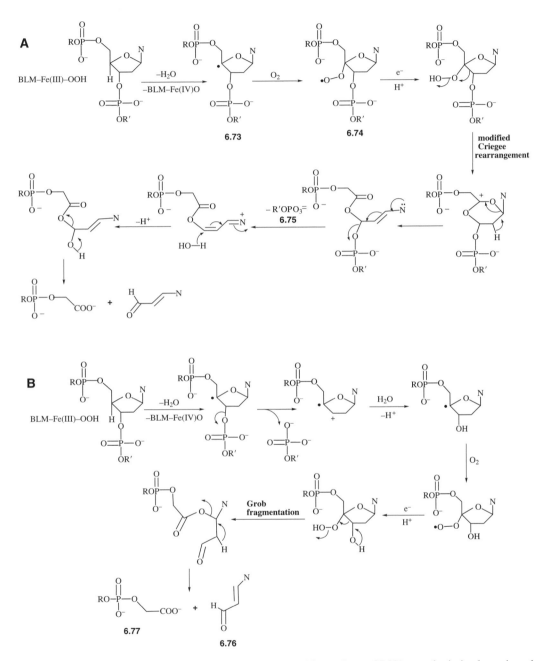

Scheme 6.24 ▶ Alternative mechanisms for base propenal formation and DNA strand scission by activated bleomycin: (A) Modified Criegee mechanism and (B) Grob fragmentation mechanism

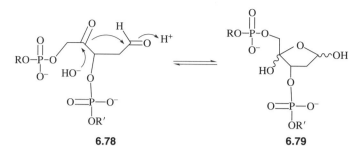

Scheme 6.25 ►   Alkali-labile lesion produced by bleomycin and hydroxide

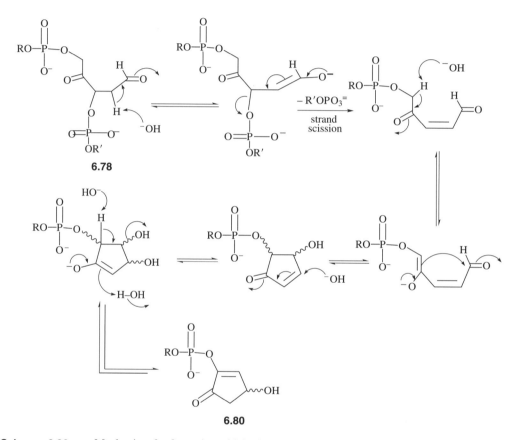

Scheme 6.26 ►   Mechanism for formation of 2-hydroxycyclopentenone analog and DNA strand scission
by bleomycin with hydroxide

which is probably in equilibrium with **6.79** (Scheme 6.25). When the alkali-labile product
(**6.78**) was heated with hydroxide, the 2-hydroxycyclopentenone analog **6.80** was isolated;
the mechanism shown in Scheme 6.26 accounts for the formation of that product. Under
more mild conditions Rabow *et al.*[229] were able to isolate the NaBH$_4$-reduced alkali-labile
product and show that it is formed in stoichiometric amounts relative to nucleic acid base
release, thereby establishing its relevance to the reaction with activated bleomycin. When
the reaction was carried out with $^{18}O_2$ or $H_2{}^{18}O$, it was found that the oxygen incorporated

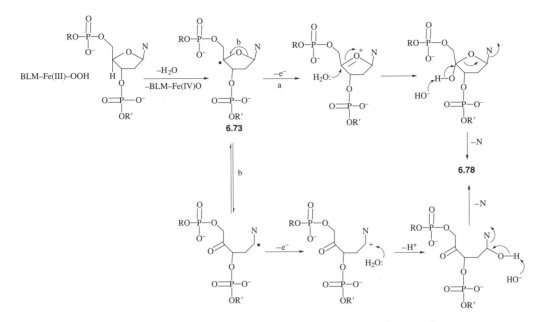

Scheme 6.27 ▶ Mechanism to account for incorporation of water into C-4′ and C-1′ of the sugar moiety during reaction of activated bleomycin with DNA

at C-4′ comes from H_2O, not O_2, and that some H_2O also is incorporated at C-1′.[230] The mechanism given in Scheme 6.27 rationalizes those observations.

DNA strand scission is sequence selective, occurring most frequently at 5′-GC-3′ and 5′-GT-3′ sequences.[231] The specificity for cleavage of DNA at a residue located at the 3′ side of G appears to be absolute. Preference for cleavage at 5′-GC and 5′-GT instead of the corresponding 5′-AC or 5′-AT sites can be attributed to reduced binding affinity of bleomycin at adenine because of one less hydrogen bond relative to that of guanine.[232] An important mechanism for resistance of cells to bleomycin is expression of bleomycin hydrolase, an aminopeptidase that hydrolyzes the carboxamide group of the L-aminoalaninecarboxamide substituent in the metal-free antibiotic.[233] The deamido bleomycin is less effective in the activation of oxygen by its Fe(II) complex.[234]

C.3 Tirapazamine

Tirapazamine (**6.81**) is a bioreductively activated antitumor agent that selectively kills oxygen-poor (hypoxic) cells in solid tumors.[235] One-electron reduction, possibly by enzymes such as NADPH-cytochrome P450 reductase or xanthine oxidase, produces the key radical intermediate (**6.82**), which can undergo homolytic cleavage to **6.83** and hydroxyl radicals, highly reactive radicals that readily degrade DNA (Scheme 6.28).[236] The DNA damage caused by reduction of tirapazamine is typical of hydroxyl radicals, namely, damage to both the DNA backbone and to the heterocyclic bases.[237] The activated form (**6.82**) is rapidly destroyed in normally oxygenated cells by reaction with O_2,[238] but under hypoxic conditions, this radical causes oxidative cleavage of the DNA backbone.[239] The reason tirapazamine can efficiently cause DNA strand cleavage under hypoxic conditions is because not only does it initiate the formation of deoxyribose radicals by hydroxyl

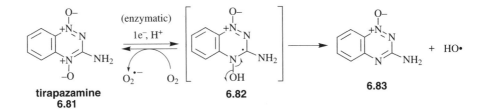

Scheme 6.28 ▶ Mechanism for formation of hydroxyl radicals by tirapazamine

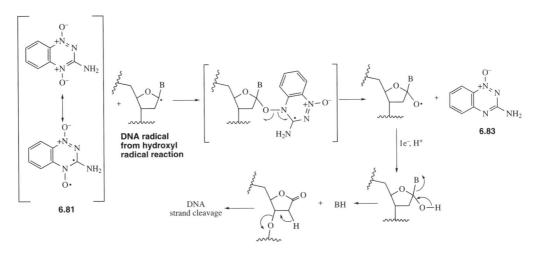

Scheme 6.29 ▶ Mechanism for DNA strand cleavage by tirapazamine

radical abstraction of a hydrogen atom, but it also reacts with these DNA radicals and converts them into strand breaks, thereby serving as a surrogate for molecular oxygen (Scheme 6.29).[240]

C.4 Enediyne Antitumor Antibiotics

Except for neocarzinostatin (**6.84**),[241] which was isolated in 1965, other members of the enediyne antibiotic class of antitumor agents, such as the esperamicins (**6.85**),[242] the calicheamicins (**6.86**),[243] dynemicin A (**6.87**),[244] kedarcidin (**6.88**),[245] C-1027 (**6.89**),[246] and N1999A2 (**6.90**),[247] were not isolated from various microorganisms until the mid-1980s. Because their common structural feature is a macrocyclic ring containing at least one double bond and two triple bonds, they are referred to as *enediyne antitumor antibiotics*.[248] Several of the enediyne antibiotics, such as neocarzinostatin, kedarcidin, and C-1027, are stabilized in nature as a noncovalent complex with a protein and are referred to as *chromoprotein enediyne antibiotics* because of the chromophoric properties of the compounds embedded in the protein. The proteins associated with kedarcidin and neocarzinostatin (and possibly all of the chromoprotein enediyne antibiotics) not only stabilize the enediyne chromophore, but also have been shown to proteolyze selectively basic peptides and proteins, such as histones, which, as we saw earlier (Section 6.2.C, p. 334), are essential for packing of the DNA into chromatin.[249] Therefore, these compounds are dual DNA cleaving agents and histone proteolytic agents.

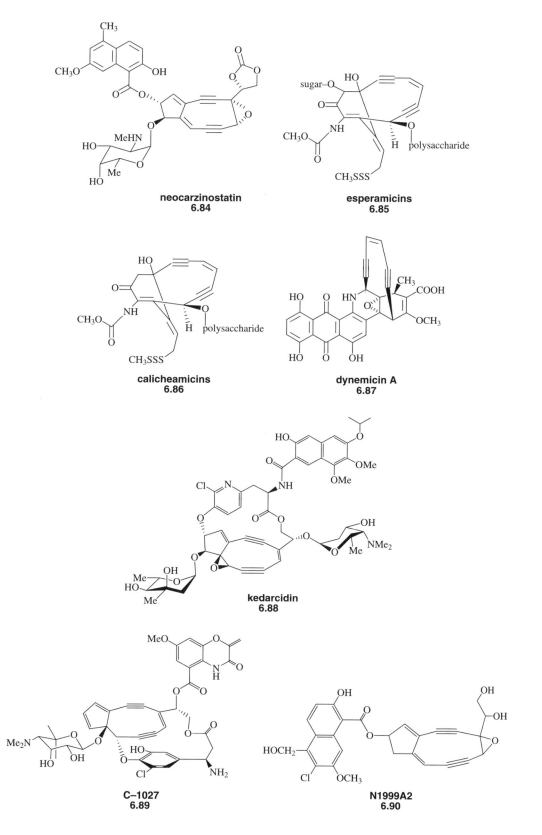

neocarzinostatin
6.84

esperamicins
6.85

calicheamicins
6.86

dynemicin A
6.87

kedarcidin
6.88

C-1027
6.89

N1999A2
6.90

All of these compounds appear to share two modes of action. Intercalation of part of the molecule into the minor groove of DNA[250] and reduction of the molecule by either thiol or NADPH, which triggers a reaction that leads to the generation of radicals, cleaving the DNA. It is not clear if DNA binding must precede the activation process.[251] Although knowledge of the chemistry involved in the latter process is only beginning to surface, we will be able to draw reasonable mechanisms for DNA damage by each of these classes of compounds.

Generally, two tests are used to demonstrate minor groove binding to B-DNA. One indication is an asymmetric cleavage pattern on the 3'-side of the opposite strand of the DNA helix.[252] The other test is inhibition of DNA cleavage by the known minor groove binders distamycin A and netropsin (similar to substrate protection of enzymes from inhibition; see Section 5.4.A, p. 241).

There are two phases to the mechanism of DNA degradation by the enediyne antibiotics after binding. First, there is the activation of the antitumor agent (except for C-1027,[253] activation requires addition of a thiol or reducing agent), then there is the action of the activated antitumor agent on DNA. This is much akin to the process of DNA degradation that we discussed for bleomycin (see Section 6.3.C.2, p. 371) and tirapazamine (see Section 6.3.C.3, p. 377). Most of what is now known about the mechanism of the reaction of activated enediynes with DNA comes from studies with neocarzinostatin, which will be discussed last. Because the chemistry of the activation of esperamicins and calicheamicins is virtually identical, we will first take a look at these compounds.

a. Esperamicins and Calicheamicins

The important structural features of these molecules (**6.85** and **6.86**) are a bicyclo[7.3.1] ring system, an allylic trisulfide attached to the bridgehead carbon, a 3-ene-1,5-diyne as part of the macrocycle, and an α,β-unsaturated ketone in which the double bond is at the bridgehead of the bicyclic system. It is believed that the enediyne moiety partially inserts into the minor groove, and then undergoes a reaction with either a thiol or NADPH, which reduces the trisulfide to the corresponding thiolate (**6.91**, Scheme 6.30).[254] Michael addition of this thiolate into the α,β-unsaturated ketone gives the dihydrothiophene (**6.92**) in which

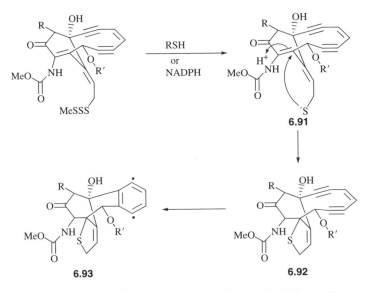

Scheme 6.30 ▶ Activation of esperamicins and calicheamicins

the bridgehead carbon hybridization has changed from sp^2 in **6.91** to sp^3 in **6.92**. This change in geometry at the bridgehead may be sufficient to allow the two triple bonds to interact with each other and to trigger a Bergman rearrangement,[255] giving the 1,4-dehydrobenzene biradical (**6.93**), which is the activated esperamicin or calicheamicin. The reaction of this biradical with DNA is, presumably, identical to that for the biradicals produced from all of the enediyne antibiotics.

b. Dynemicin A

Another member of the enediyne class of antitumor antibiotics is dynemicin A (**6.87**), which combines the structural features of both the anthracycline antitumor agents and the enediynes. The anthraquinone part of dynemicin A intercalates into the minor groove of DNA,[256] then, depending on whether activation is initiated by NADPH or a thiol, a reductive mechanism (Scheme 6.31) or nucleophilic mechanism (Scheme 6.32) of activation, respectively, is possible.[257] In the reductive activation mechanism (Scheme 6.31), either NADPH or a thiol reduces the quinone to a hydroquinone (**6.94**), which leads to opening of the epoxide. Following protonation to **6.95**, the geometry of the triple bonds becomes more favorable for the Bergman reaction to take place, leading to the generation of the 1,4-dehydrobenzene biradical (**6.96**). In the nucleophilic mechanism (Scheme 6.32), reaction of a thiol gives **6.97**, which is very similar in structure to the reduced intermediate **6.95** (Scheme 6.31). Bergman reaction of **6.97** gives the 1,4-dehydrobenzene biradical **6.98**. Either biradical (**6.96** or **6.98**) could be responsible for DNA degradation by the mechanism described in detail for neocarzinostatin in the next section.

c. Neocarzinostatin (Zinostatin)

The oldest known member of the enediyne family of antibiotics is neocarzinostatin (now called zinostatin; **6.84**; SMANCS), the only one of the enediynes (not in prodrug form) currently approved for clinical use, although not in the United States. More mechanistic studies have been performed with this compound than any of the more recent additions to the enediyne family. The naphthoate ester moiety is believed to intercalate into DNA, thereby positioning the epoxybicyclo[7.3.0]dodecadienediyne portion of the chromophore in the minor groove.[258] Activation by a thiol generates an intermediate (**6.99**, Scheme 6.33) capable of undergoing a Bergman rearrangement to biradical **6.100**.[259] This biradical differs from the biradicals generated by activation of esperamicin and calicheamicin (**6.93**, Scheme 6.30) and dynemicin A (**6.96**, Scheme 6.31, or **6.98**, Scheme 6.32) in that it is not a 1,4-dehydrobenzene biradical, but it is very similar.

Surprisingly, the reaction of small thiol reducing agents with zinostatin that is still complexed to its protein leads to deactivation of the chromophore rather than to activation.[260] Myers and coworkers[261] proposed that the mechanism for deactivation involves the same initial attack of the chromophore by the thiol, but they suggest that the epoxide is not disposed for opening because it is sequestered within a hydrophobic pocket of the protein. Therefore, the reaction proceeds as shown in Scheme 6.34, in which proposed biradical intermediate **6.101** is rapidly protonated to give **6.102**. σ,π-Biradicals, such as **6.101**, in contrast to σ,σ-biradicals, such as **6.100** (Scheme 6.33), are known to undergo polar addition reactions, for example, protonation, so this may account for the deactivation process.[262]

The highly reactive biradical **6.100** or **6.101** is responsible for DNA strand scission, which consumes one equivalent of O_2 per strand break. Both the C-4' and the C-5' hydrogens of DNA sugar-phosphate residues are accessible to the biradical within the minor groove, and either can be abstracted. In the presence of O_2 two different mechanisms of DNA cleavage can result

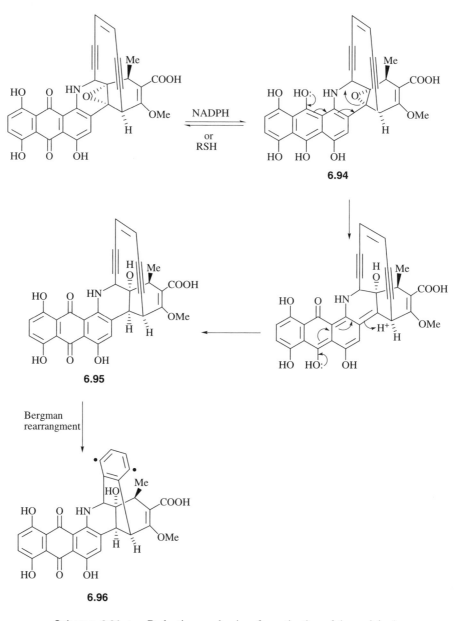

Scheme 6.31 ▶ Reductive mechanism for activation of dynemicin A

(Scheme 6.35).[263] About 80% of the cytotoxic lesions are the result of single-strand cleavages, mostly caused by C-5′ hydrogen atom abstraction from thymidine or deoxyadenosine residues (pathway a). This leads to the formation of the nucleoside 5′-aldehyde (**6.103**) and the 3′-phosphate (**6.104**). Abstraction of the C-4′ hydrogen atom (pathway b) gives a radical (**6.105**) that partitions between a modified basic carbohydrate terminus (**6.106**, pathway c) and a 3′-phosphoglycolate terminus (**6.107**, pathway d), as was observed with bleomycin (see Section 6.3.C.2, p. 371). Also, as in the case of bleomycin, double-strand breaks occur when the zinostatin-treated DNA is heated in alkali. It appears that single-strand breaks at AGT

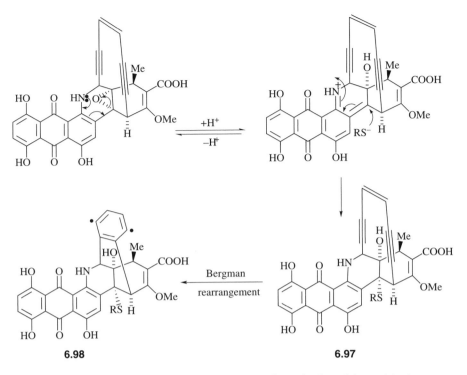

6.98   **6.97**

Scheme 6.32 ▶ Nucleophilic mechanism for activation of dynemicin A

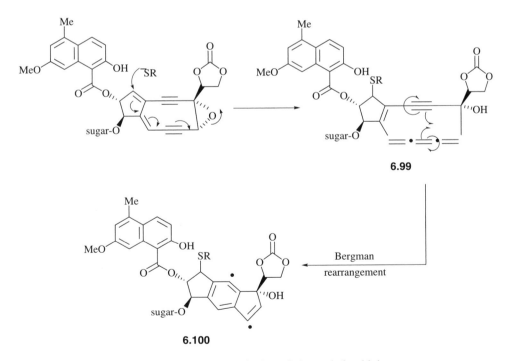

6.99

6.100

Scheme 6.33 ▶ Activation of zinostatin by thiols

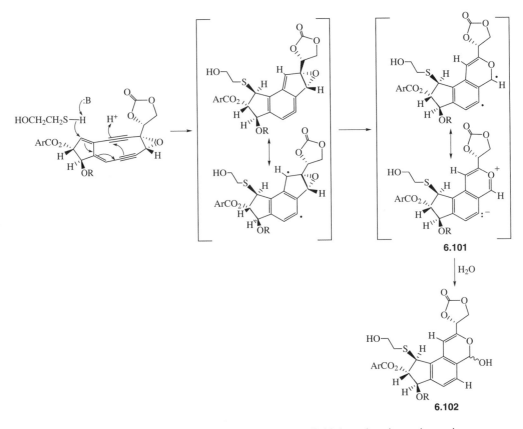

Scheme 6.34 ▶   Polar addition reaction by small thiols to deactivate zinostatin

sequences occur with both *C*-4′ and *C*-5′ oxidation, but ACT breaks occur only with *C*-5′ oxidation.

A catalytic antibody[264] was raised that catalyzes a Bergman cyclization of an enediyne (**6.108**), but instead of the expected tetralin (**6.109**), the biradical reacted with O_2, producing the corresponding quinone (**6.110**, Scheme 6.36).[265] Quinones also are known to be highly cytotoxic,[266] so this suggests that there may be more than one mechanism for cytotoxicity produced by enediyne antibiotics.

The enediynes, in general, have been too toxic for clinical use, but a conjugate of an antibody-calicheamicin derivative (gemtuzumab ozogamicin, Mylotarg) is now used for the treatment of acute myeloid leukemia (see Section 8.2.A.3.d, p. 519).

All of these highly cytotoxic antibiotics were isolated from various microorganisms, all of which also contain reducing agents to activate the cytotoxins and DNA that can be degraded. So why do these organisms produce these compounds, and why are the organisms not destroyed by them? Generally, these compounds are produced for protection against other microorganisms. The organism that produces the antibiotic protects itself in at least two ways. First, as was noted for the chromoprotein enediyne antibiotics, they produce proteins that sequester, stabilize, and transport the compounds out of the cell, thereby protecting the bacteria that produced them.[267] Another mechanism for protection is the production of enzymes that catalyze reactions to destroy the antibiotic in case it is accidentally released

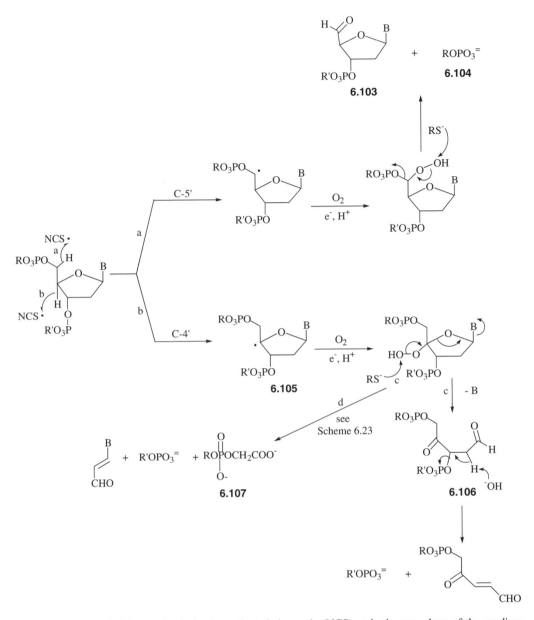

Scheme 6.35 ▶ DNA strand scission by activated zinostatin (NCS) and other members of the enediyne antibiotics. NCS is neocarzinostatin.

within the cell. A protein was identified that is responsible for self-resistance of the organism from its own production of calicheamicin.[268] As noted above, the neocarzinostatin chromoprotein complex is deactivated by small thiol reducing agents, but not by larger thiols. Many prokaryotes, such as bacteria, use small thiols (cysteine or hydrogen sulfide) for metabolism in contrast to eukaryotes, which utilize the tripeptide thiol, glutathione.[269] The susceptibility of the neocarzinostatin chromoprotein to small thiols may be a mechanism for protection by the organism if the neocarzinostatin complex is not transported quickly enough out of the cell.

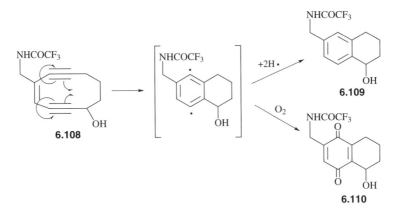

Scheme 6.36 ▶ Catalytic antibody-catalyzed conversion of an enediyne into a quinone via oxygenation of the corresponding benzene biradical

C.5   Sequence Specificity for DNA Strand Scission

Many studies have been carried out to determine cleavage site preferences in the minor groove for each of the strand breakers.[270] The preferences are rationalized as occurring because of selective binding at particular sites based on the structure of the compound. Evidence from *in vitro* and *in vivo* work suggests that adduct formation (alkylation)[271] and radical-mediated strand breakage[272] by small organic molecules is not markedly altered by chromatin structure, probably because of the accessibility of the outer surface of DNA in chromatin to small molecules. Therefore, *in vitro* studies of DNA do provide a reasonable model for the *in vivo* sequence preferences of DNA-damaging agents.

Although many of the DNA-interactive drugs react with the constituents of the minor groove, it is not yet clear if there is sufficient sequence specificity information in the minor groove of DNA to allow for the *design* of minor groove-selective agents. Nature has opted, for example, with protein–DNA interactions, to utilize major groove interactions.

6.4   Epilogue to Receptor-Interactive Agents

Chapters 3 through 6 have dealt with the structures and functions of various protein and nucleic acid receptors as well as with classes of drugs that interact with these receptors and how the medicinal chemist can begin to design molecules that selectively interact with these receptors. In Chapter 7 we turn our attention to the heroic efforts our bodies make to destroy and excrete xenobiotics, including drugs. Then in Chapter 8 we discuss the heroic efforts medicinal chemists make to outwit these metabolic processes by protecting the drugs until they reach their desired sites of action.

6.5   General References

DNA Structure and Function

Alberts, B.; Bray, D.; Lewis, J.; Raff, M.; Roberts, K.; Watson, J. D. *Molecular Biology of the Cell*, 2nd ed., Garland Publishing, New York, 1989.

Blackburn, G. M.; Gait, M. J. (Eds.) *Nucleic Acids in Chemistry and Biology*, 2nd ed., Oxford University Press, Oxford, 1996.

Saenger, W. *Principles of Nucleic Acid Structure*, Springer-Verlag, New York, 1984.

Sinden, R. R. *DNA Structure and Function*, Academic Press, San Diego, 1994.

Watson, J. D.; Baker, T. A.; Bell, S. P.; Gann, A.; Levine, M.; Losick, R. *Molecular Biology of the Gene*, 5th ed., Benjamin Cummings, San Francisco, CA, 2004.

Wolffe, A. P. *Chromatin: Structure and Function*, 3rd ed., Academic Press, San Diego, 1998.

Topoisomerases

Champoux, J. J. *Annu. Rev. Biochem.* **2001**, *70*, 369.

Wang, J. C. Q. *Rev. Biophys.* **1998**, *31*, 107.

Antitumor Drugs

Borders, D. B.; Doyle, T. W. (Eds.) *Enediyne Antibiotics as Antitumor Agents*, Marcel Dekker, New York, 1995.

Chabner, B. A.; Collins, J. M. *Cancer Chemotherapy: Principles & Practice*, J. B. Lippincott, Philadelphia, 1990.

Gates, K. S. In *Comprehensive Natural Products Chemistry*, Barton, D., Nakanishi, K., Meth-Cohn, O. (Eds.), Pergamon Press, Oxford, 1999, Vol. 7, pp. 491–552.

Neidle, S.; Waring, M. J. *Molecular Aspects of Anti-cancer Drug Action*, Verlag Chemie, Weinheim, 1983.

Pratt, W. B.; Ruddon, R. W. *The Anticancer Drugs*, Oxford University Press, New York, 1979.

Remers, W. A. *Antineoplastic Agents*, Wiley, New York, 1984.

Wilman, D. E. V. *The Chemistry of Antitumour Agents*, Blackie, Glasgow, 1990.

Drug Resistance

Andersson, B.; Murray, D. (Eds.) *Clinically Relevant Resistance in Cancer Chemotherapy*, Kluwer Academic Publishers, The Netherlands, 2002.

Brown, R.; Boger-Brown, U. (Eds.) *Cytotoxic Drug Resistance Mechanisms*, Humana Press, 1999.

Goldie, J. H.; Coldman, A. J. *Drug Resistance in Cancer: Mechanisms and Models*, Cambridge University Press, Cambridge, 1998.

Gupta, R. S. *Drug Resistance in Mammalian Cells*, CRC Press, Boca Raton, FL, 1989, Vols. I and II.

Woolley, P. V. III; Tew, K. D. *Mechanisms of Drug Resistance in Neoplastic Cells*, Academic Press, San Diego, 1988.

 # 6.6 PROBLEMS

(Answers can be found in Appendix at the end of the book.)

1. Draw a mechanism for the topoisomerase-catalyzed cleavage of DNA that gives the free 5′-end.

2. Show how isoguanosine (**1**) could form a Watson-Crick base pair with isocytidine (**2**).

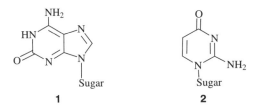

1   2

3. What properties would you incorporate into the design of DNA intercalating agents?

4. Alkylating agents are used in cancer chemotherapy to interfere with DNA biosynthesis or replication. Why are these useful considering that normal cells also need DNA?

5. Draw a mechanism for the intrastand cross-linkage of two guanines by busulfan (**6.33**).

6. Draw a mechanism from **6.51** (Scheme 6.11) to **6.44**.

7. Your DNA alkylating agent was found to lead to tumor cell resistance as a result of the induction of a new enzyme that repaired the modified DNA. What do you do next?

8. Amino-seco-CBI-TMI (**3**) is a minor groove alkylating agent. Show a mechanism for alkylation of DNA that does not involve direct S_N2 displacement of the chloride by DNA.

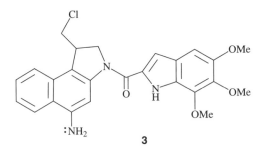

3

9. *N*,*N*-Diethylnitrosoamine (**4**) is carcinogenic, resulting in ethylation of DNA. Two other products formed are acetaldehyde and N_2. Under anaerobic conditions or in the absence of NADPH, no carcinogenicity is observed. Draw a mechanism for DNA ethylation by this compound.

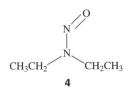

CH_3CH_2 CH_2CH_3

4

10. Compounds **6.31** and **6.32** are stable until some sort of activation occurs for antitumor activity. Show what is necessary for activation and explain why.

11. Why are enediyne antitumor agents stable until a thiol or NADPH reduction of some other part of the molecule occurs?

12. Draw a mechanism for the formation of a 1,4-benzene diradical from an activated enediyne (Bergman rearrangement).

13. Based on your knowledge of intercalating agents and strand breakers, design a new class of enediyne antitumor agents that utilizes both concepts.

14. A new antitumor agent was discovered from a random screen of soil microbes having the structure **5**. The compound is not active until it undergoes one electron reduction after which it was found to cause DNA strand breakage. Draw a possible mechanism for its activity.

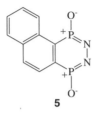

5

15. Draw a mechanism for how kedarcidin (**6**) might act as an antitumor agent (a reducing agent is required).

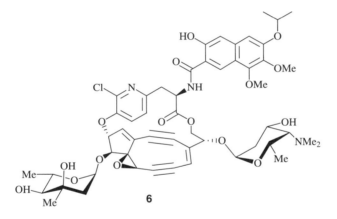

6

6.7 **References**

1. (a) Sancar, A.; Sancar, G. B. *Annu. Rev. Biochem.* **1988**, *57*, 29. (b) Friedberg, E. C. In *DNA Repair*, W. H. Freeman, San Francisco, 1984.

2. Stewart, Z. A.; Pietenpol, J. A. *Chem. Res. Toxicol.* **2001**, *14*, 243.

3. Blagosklonny, M. V.; Pardee, A. B. *Cancer Res.* **2001**, *61*, 4301.

4. Brown, J. M.; Wouters, B. G. *Cancer Res.* **1999**, *59*, 1391.

5. Waldman, T.; Zhang, Y; Dillehay, L.; Yu, J.; Kinzler, K.; Vogelstein, B.; Williams, J. *Nat. Med.* **1997**, *3*, 1034.

6. Doll, R.; Peto, R. *J. Natl. Cancer Inst.* **1981**, *66*, 1192.

7. (a) Tuteja, N.; Tuteja, R. *Crit. Rev. Biochem. Mol. Biol.* **2001**, *36*, 261. (b) Sancar, A. *Science* **1994**, *266*, 1954.

8. (a) Carell, T.; Burgdorf, L. T.; Kundu, L. M.; Cichon, M. *Curr. Opin. Chem. Biol.* **2001**, *5*, 491. (b) Yasui, A.; Eker, A. P. M. *Contemp. Cancer Res.* **1998**, 2 (DNA Damage and Repair, Vol. 2), 9. (c) Sancar, A. *Science* **1996**, *272*, 48. (d) Stege, H.; Roza, L.; Vink, A. A.; Grewe, M.; Ruzicka, T.; Grether-Beck, S.; Krutmann, J. *Proc. Natl. Acad. Sci. USA* **2000**, *97*, 1790.

9. (a) Setlow, R. B.; Carrier, W. *Proc. Natl. Acad. Sci. USA* **1964**, *51*, 226. (b) Boyce, R.; Howard-Flanders, P. *Proc. Natl. Acad. Sci. USA* **1964**, *51*, 293. (c) Pettijohn, D.; Hanawalt, P. C. *J. Mol. Biol.* **1964**, *7*, 395.

10. Paus, R.; Cotsarelis, G. *N. Engl. J. Med.* **1999**, *341*, 491.

11. Ohtsubo, M; Theodoras, A. M.; Schumacher, J.; Roberts, J. M.; Pagano, M. *Mol. Cell Biol.* **1995**, *15*, 2612.

12. (a) Davis, S. T.; Benson, B. G.; Bramson, H. N.; Chapman, D. E.; Dickerson, S. H.; Dold, K. M.; Eberwein, D. J.; Edelstein, M.; Frye, S. V.; Gampe, R. T., Jr.; Griffin, R. J.; Harris, P. A.; Hassell, A. M.; Holmes, W. D.; Hunter, R. N.; Knick, V. B.; Lackey, K.; Lovejoy, B.; Luzzio, M. J.; Murray, D.; Parker, P.; Rocque, W. J.; Shewchuk, L.; Veal, J. M.; Walker, D. H.; Kuyper, L. F. *Science* **2001**, *291*, 134.

(b) A retraction of reference 12(a) was published in *Science* **2002**, *298*, 2327. It was stated that the biological activity of the reported compound was not reproducible, but that they were pursuing other CDK inhibitors that block chemotherapy-induced alopecia in the neonatal rat model.

13. Sun, L.; Tran, N.; Tang, F.; App, H.; Hirth, P.; McMahon, G.; Tang, C. *J. Med. Chem.* **1998**, *41*, 2588.

14. Borison, H. L.; Brand, E. D.; Orkand, R. K. *Am. J. Physiol.* **1968**, *192*, 410.

15. Pittillo, R. F.; Schabel, F. M., Jr.; Wilcox, W. S.; Skipper, H. E. *Cancer Chemother. Rept.* **1965**, *47*, 1.

16. Collins, A. R. S.; Squires, S.; Johnson, R. T. *Nucleic Acids Res.* **1982**, *10*, 1203.

17. Cadman, E.; Heimer, R.; Davis, L. *Science* **1979**, *205*, 1135.

18. (a) Greenberger, L. M.; Williams, S. S.; Horwitz, S. B. *J. Biol. Chem.* **1987**, *262*, 13685. (b) Riordan, J. R.; Deuchars, K.; Kartner, N.; Alon, N.; Trent, J. Ling, V. *Nature* **1985**, *316*, 817.

19. Marchand, D. H.; Remmel, R. P.; Abdel-Monem, M. M. *Drug Metab. Dispos.* **1988**, *16*, 85.

20. (a) Watson, J. D.; Crick, F. H. C. *Nature* **1953**, *171*, 737. (b) Crick, F. H. C.; Watson, J. D. *Proc. Roy. Soc.* (*London*) *Ser. A* **1954**, *223*, 80.

21. Watson, J. D. *The Double Helix*, Weidenfeld and Nicholson, London, 1968.

22. Dekker, C. A.; Michelson, A. M.; Todd, A. R. *J. Chem. Soc.* **1953**, *947.*

23. Zamenhof, S.; Braverman, G.; Chargaff, E. *Biochim. Biophys. Acta* **1952**, *9*, 402.

24. Astbury, W. T. *Symp. Soc. Exp. Biol. I. Nucleic Acid* **1947**, *66.*

25. Gulland, J. M. *Cold Spring Harbor Symp. Quant. Biol.* **1947**, *12*, 95.

26. Wilkins, M. H. F. *Science,* **1963**, *140*, 941.

27. Maddox, B. *Rosalind Franklin, Dark Lady of DNA*, HarperCollins, London and New York, 2002.

28. Furberg, S. *Acta Crystallogr.* **1950**, *3*, 325.

29. Hanlon, S. *Biochem. Biophys. Res. Commun.* **1966**, *23*, 861.

30. Herskovits, T. T. *Arch. Biochem. Biophys.* **1962**, *97*, 474.

31. Beak. P. *Acc. Chem. Res.* **1977**, *10*, 186.

32. Wolfenden, R. V. *J. Mol. Biol.* **1969**, *40*, 307.

33. Chenon, M.-T.; Pugmire, R. J.; Grant, D. M.; Panzica, R. P.; Townsend, L. B. *J. Am. Chem. Soc.* **1975**, *97*, 4636.

34. (a) Kool, E. T. *Acc. Chem. Res.* **2002**, (b) Schweitzer, B. A.; Kool, E. T. *J. Org. Chem.* **1994**, *59*, 7238.

35. Guckian, K. M.; Kool, E. T. *Angew. Chem. Int. Ed.* **1998**, *36*, 2825.

36. Guckian, K. M.; Krugh, T. R.; Kool, E. T. *Nat. Struct. Biol.* **1998**, *5*, 954.

37. Guckian, K. M.; Krugh, T. R.; Kool, E. T. *J. Am. Chem. Soc.* **2000**, *122*, 6841.

38. Moran, S.; Ren, R. X.-F.; Rumney, S.; Kool, E. T. *J. Am. Chem. Soc.* **1997**, *119*, 2056.

39. (a) Kunkel, T. A.; Bebenek, K. *Annu. Rev. Biochem.* **2000**, *69*, 497. (b) Goodman, M. F. *Proc. Natl. Acad. Sci. USA* **1997**, *94*, 10493.

40. Moran, S.; Ren, R. X.-F.; Kool, E. T. *Proc. Natl. Acad. Sci. USA* **1997**, *94*, 10506.

41. Morales, J. C.; Kool, E. T. *Nat. Struct. Biol.* **1998**, *5*, 950.

42. (a) Kool, E. T. *Annu. Rev. Biophys. Biomol. Struct.* **2001**, *30*, 1. (b) Kool, E. T. *Annu. Rev. Biochem.* **2002**, *71*, 191.

43. (a) Richmond, T. J. *Chimia* **2001**, *55*, 487. (b) Widom, J. *Curr. Biol.* **1997**, 7, R653. (c) Ramakrishnan, V. *Annu. Rev. Biophys. Biomol. Struct.* **1997**, *26*, 83.

44. (a) Wang, J. C. *Annu. Rev. Biochem.* **1996**, *65*, 635. (b) Wang, J. C. *Nat. Rev. Mol. Cell Biol.,* **2002**, *3*, 430.

45. Liu, L. F. *Annu. Rev. Biochem.* **1989**, *58*, 351.

46. Wang, J. C. *Q. Rev. Biophys.* **1998**, *31*, 107.

47. Champoux, J. J. *Annu. Rev. Biochem.* **2001**, *70*, 369.

48. Halligan, B. D.; Edwards, K. A.; Liu, L. F. *J. Biol. Chem.* **1985**, *260*, 2475.

49. Hanai, R.; Caron, P. R.; Wang, J. C. *Proc. Natl. Acad. Sci. USA* **1996**, *93*, 3653.

50. (a) Khodursky, A. B.; Zechiedrich, E. L.; Cozzarelli, N. R. *Proc. Natl. Acad. Sci. USA* **1995**, *92*, 11801. (b) Adams, D. E.; Shekhtman, E. M.; Zechiedrich, E. L.; Schmid, M. B.; Cozzarelli, N. R. *Cell* **1992**, *71*, 277.

51. (a) Slesarev, A. I.; Belova. G. I.; Lake, J. A.; Kozyavkin, S. A. *Meth. Enzymol.* **2001**, *334*, 179. (b) Belova, G. I.; Prasad, R.; Nazimov, I. V.; Wilson, S. H.; Slesarev, A. I. *J. Biol. Chem.* **2002**, *277*, 4959.

52. (a) Buhler, C.; Lebbink, J. H. G.; Bocs, C.; Ladenstein, R.; Forterre, P. *J. Biol. Chem.* **2001**, *276*, 37215. (b) Bocs, C.; Buhler, C.; Forterre, P.; Bergerat, A. *Meth. Enzymol.* **2001**, *334*, 172.

53. (a) Wang, J. C. *J. Mol. Biol.* **1971**, *55*, 523. (b) Tse, Y.-C.; Kirkegaard, K.; Wang, J. C. *J. Biol. Chem.* **1980**, *255*, 5560. (c) Champoux, J. J. *J. Biol. Chem.* **1981**, *256*, 4805.

54. (a) Brown, P. O.; Cozzarelli, N. R. *Proc. Natl. Acad. Sci. USA* **1981**, *78*, 843. (b) Tse, Y.; Wang, J. C. *Cell* 1980, 22, 269.

55. Li, Z.; Mondragón, A.; DiGate, R. *J. Mol. Cell* **2001**, *7*, 301–7.

56. Lima, C. D.; Wang, J. C.; Mondragón, A. *Nature* **1994**, *367*, 138.

57. (a) Ono, K.; Ikegami, Y.; Nishizawa, M.; Andoh, T. *Jpn. J. Cancer Res.* **1992**, *83*, 1018. (b) Park, J.-S.; Park, S.; Lee, Y.; Kong, J.-Y.; Kim, W. J.; Koo, H.-S. *J. Biochem. Mol. Biol.* **1995**, *28*, 464.

58. Patel, S.; Keller, B. A.; Fisher, L. M. *Mol. Pharmacol.* **2000**, *57*, 784.

59. (a) Dervan, P. B. *Science* **1986**, *232*, 464. (b) Hurley, L. H.; Boyd, F. L. *Annu. Rep. Med. Chem.* **1987**, *22*, 259. (c) Hurley, L. H. *J. Med. Chem.* **1989**, *32*, 2027.

60. (a) Kielkopf, C. L.; White, S.; Szewczyk, J.; Turner, J. M.; Baird, E. E.; Dervan, P. B.; Rees, D. C. *Science* **1998**, *282*, 111. (b) von Hippel, P. H.; Berg, O. G. *Proc. Natl. Acad. Sci. USA* **1986**, *83*, 1608.

61. Branden, C; Tooze, J. *Introduction to Protein Structure*, Garland Press, New York, 1991, p. 83.

62. Luger, K.; Mäder, A. W.; Richmond, R. K.; Sargent, D. F.; Richmond, T. J. *Nature* **1997**, *389*, 251.

63. Polach, K.J.; Widom, J. *J. Mol. Biol.* **1995**, *254*, 130.

64. Haq, I. *Arch. Biochem. Biophys.* **2002**, *403*, 1.

65. (a) Burridge, J. M.; Quarendon, P.; Reynolds, C. A.; Goodford, P. J. *J. Mol. Graphics* **1987**, *5*, 165. (b) Zakrzewska, K.; Lavery, R.; Pullman, B. *Nucleic Acids Res.* **1983**, *11*, 8825.

66. (a) Goodsell, D. S.; Kopka, M. L.; Dickerson, R. E. *Biochemistry* **1995**, *34*, 4983. (b) Nunn, C. M.; Garman, E.; Neidle, S. *Biochemistry* **1997**, *36*, 4792.

67. Kopka, M. L.; Yoon, C.; Goodsell, D.; Pjura, P.; Dickerson, R. E. *Proc. Natl. Acad. Sci. USA* **1985**, *82*, 1376.

68. Neidle, S.; Abraham, Z. *CRC Crit. Rev. Biochem.* **1984**, *17*, 73.

69. Lerman, L. S. *J. Mol. Biol.* **1961**, *3*, 18.

70. (a) Krugh, T. R.; Reinhardt, C. G. *J. Mol. Biol.* **1975**, *97*, 133. (b) Nuss, M. E.; Marsh, F. J.; Kollman, P. A. *J. Am. Chem. Soc.* **1979**, *101*, 825.

71. (a) Kapur, A.; Beck, J. L.; Sheil, M. M. *Rapid Commun. Mass Spectrom.* **1999**, *13*, 2489. (b) Rao, S. N.; Kollman, P. A. *Proc. Natl. Acad. Sci. USA* **1987**, *84*, 5735.

72. (a) Topcu, Z. *J. Clin. Pharm. Therap.* **2001**, *26*, 405. (b) Toonen, T. R.; Hande, K. R. *Cancer Chemother. Biol. Response Mod.* **2001**, *19*, 129. (c) Stewart, C. F. *Cancer Chemother. Biol. Response Mod.* **2001**, *19*, 85. (d) Holden, J. A. *Curr. Med. Chem. Anti-Cancer Ag.* **2001**, *1*, 1. (e) Froelich-Ammon, S. J.; Osheroff, N. *J. Biol. Chem.* **1995**, *270*, 21429. (f) Entire issue of *Biochim. Biophys. Acta* **1998**, *1400*, 1.

73. Ulukan, H.; Swaan, P. W. *Drugs* **2002**, *62*, 2039.

74. Liu, L. F. *Annu. Rev. Biochem.* **1989**, *58*, 351.

75. Nelson, E. M.; Tewey, K. M.; Liu, L. F. *Proc. Natl. Acad. Sci. USA* **1984**, *81*, 1361.

76. Zhang, H.; D'Arpa, P.; Liu, L. F. *Cancer Cells* **1990**, *2*, 23.

77. (a) Byl, J. A. W.; Fortune, J. M.; Burden, D. A.; Nitiss, J. L.; Utsugi, T.; Yamada, Y.; Osheroff, N. *Biochemistry* **1999**, *38*, 15573. (b) Fortune, J. M.; Velea, L.; Graves, D. E.; Utsugi, T.; Yamada, Y.; Osheroff, N. *Biochemistry* **1999**, *38*, 15580.

78. (a) Bodley, A.; Liu, L. F.; Israel, M.; Seshadri, R.; Koseki, Y.; Giuliani, F. C.; Kirschenbaum, S.; Silber, R.; Potmesil, M. *Cancer Res.*, **1989**, *49*, 5969. (b) D'Arpa, P.; Liu, L. F. *Biochim. Biophys. Acta* **1989**, *989*, 163.

79. Mannaberg, J. *Arch. Klin. Med.* **1897**, *59*, 185.

80. Albert, A. *The Acridines*, 2nd ed., Edward Arnold, London, 1966.

81. Cain, B. F.; Atwell, G. J.; Seelye, R. N. *J. Med. Chem.* **1971**, *14*, 311.

82. Goldin, A.; Serpick, A. A.; Mantel, N. *Cancer Chemother. Rep.* **1966**, *50*, 173.

83. Atwell, G. J.; Cain, B. F.; Seelye, R. N. *J. Med. Chem.* **1972**, *15*, 611.

84. Denny, W. A.; Cain, B. F.; Atwell, G. J.; Hansch, C.; Panthananickal, A.; Leo, A. *J. Med. Chem.* **1982**, *25*, 276.

85. Zittoun, R. *Cancer Treat. Rep.* **1985**, *69*, 1447.

86. Patel, S.; Keller, B. A.; Fisher, L. M. *Mol. Pharmacol.* **2000**, *57*, 784.

87. Cain, B. F.; Atwell, G. J.; Denny, W. A. *J. Med. Chem.* **1975**, *18*, 1110.

88. Waring, M. J. *Eur. J. Cancer* **1976**, *12*, 995.

89. Braithwaite, A. W.; Baguley, B. C. *Biochemistry* **1980**, *19*, 1101.

90. Sakore, T. D.; Reddy, B. S.; Sobell, H. M. *J. Mol. Biol.* **1979**, *135*, 763.

91. Denny, W. A.; Baguley, B. C.; Cain, B. F.; Waring, M. J. In *Molecular Aspects of Anti-Cancer Drug Action*, Neidle, S.; Waring, M. J. (Eds.), Verlag Chemie, Weinheim, 1983, pp. 1–34.

92. Denny, W. A.; Wakelin, L. P. G. *Cancer Res.* **1986**, *46*, 1717.

93. Baguley, B. C.; Denny, W. A.; Atwell, G. J.; Finlay, G. J.; Rewcastle, G. W.; Twigden, S. J.; Wilson, W. R. *Cancer Res.* **1984**, *44*, 3245.

94. Waksman, S. A.; Woodruff, H. B. *Proc. Soc. Exp. Biol. Med.* **1940**, *45*, 609.

95. Schulte, G. *Z. Krebsforsch.* **1952**, *58*, 500.

96. Sobell, H. M. *Proc. Natl. Acad. Sci. USA* **1985**, 82, 5328.

97. Müller, W.; Crothers, D. M. *J. Mol. Biol.* **1968**, *35*, 251.

98. Cerami, A.; Reich, E.; Ward, D. C.; Goldberg, I. H. *Proc. Natl. Acad. Sci. USA* **1967**, 57, 1036.

99. Sobell, H. M.; Jain, S. C. *J. Mol. Biol.* **1972**, *68*, 21.

100. Takusagawa, F.; Dabrow, M.; Neidle, S.; Berman, H. M. *Nature* **1982**, *296*, 466.

101. Takusagawa, F.; Goldstein, B. M.; Youngster, S.; Jones, R. A.; Berman, H. M. *J. Biol. Chem.* **1984**, *259*, 4714.

102. Takusagawa, F. *J. Antibiot.* **1985**, *38*, 1596.

103. Chiao, Y.-C. C.; Krugh, T. R. *Biochemistry* **1977**, *16*, 747.

104. (a) Pastan, I.; Gottesman, M. *N. Engl. J. Med.* **1987**, *316*, 1388. (b) Polet, H. *J. Pharmacol. Exp. Ther.* **1975**, *192*, 270. (c) Bowen, D.; Goldman, I. D. *Cancer Res.* **1975**, *35*, 3054.

105. Bosmann, H. B. *Nature* **1971**, *233*, 566.

106. Schwarz, H. S. *Cancer Chemother.* **1974**, *58*, 55.

107. Gewirtz, D. A. *Biochem. Pharmacol.* **1999**, *57*, 727.

108. Quigley, G. J.; Wang, A. H.-J.; Ughetto, G.; van der Marel, G.; van Boom, J. H.; Rich, A. *Proc. Natl. Acad. Sci. USA* **1980**, *77*, 7204.

109. Patel, D. J.; Kozlowski, S. A.; Rice, J. A. *Proc. Natl. Acad. Sci. USA* **1981**, *78*, 3333.

110. Henry, D. W. *Cancer Treat. Rep.* **1979**, *63*, 845.

111. Ughetto, G.; Wang, A. H. J.; Quigley, G. J.; Van der Marel, G. A.; Van Boom, J. H.; Rich, A. *Nucleic Acids Res.* **1985**, *13*, 2305.

112. Wright, R. G. McR.; Wakelin, L. P. G.; Fieldes, A.; Acheson, R. M.; Waring, M. J. *Biochemistry* **1980**, *19*, 5825.

113. Wakelin, L. P. G. *Med. Res. Rev.* **1986**, *6*, 275.

114. (a) Guelev, V.; Lee, J.; Ward, J.; Sorey, S.; Hoffman, D. W.; Iverson, B. L., *Chem. Biol.* **2001**, *8*, 415. (b) Leng, F.; Priebe, W.; Chaires, J. B. *Biochemistry* **1998**, *37*, 1743.

115. Krumbhaar, E. B.; Krumbhaar, H. D. *J. Med. Res.* **1919**, *40*, 497.

116. Adair, F. E.; Bagg, H. *J. Ann. Surgery* **1931**, *93*, 190.

117. Gilman, A.; Philips, F. S. *Science* **1946**, *103*, 409.

118. (a) Rhoads, C. P. *J. Am. Med. Assoc.* **1946**, *131*, 656. (b) Goodman, L. S.; Wintrobe, M. M.; Dameshek, W.; Goodman, M. J.; Gilman, A.; McLennan, M. *J. Am. Med. Assoc.* **1946**, *132*, 126.

119. Ross, W. C. J. In *Biological Alkylating Agents*, Butterworths, London, 1962.

120. Montgomery, J. A.; Johnston, T. P.; Shealy, Y. F. In *Burger's Medicinal Chemistry*, 4th ed., Wolff, M. E. (Ed.), Wiley, New York, 1979, Part II, p. 595.

121. (a) Beranek, D. T. *Mutation Res.* **1990**, *231*, 11. (b) Lawley, P. D.; Phillips, D. H. *Mutation Res.* **1996**, *355*, 13. (c) Lawley, P. D.; Brookes, P. *Biochem. J.* **1963**, *89*, 127.

122. Pullman, A.; Pullman, B. *Int. J. Quant. Chem., Quant. Biol. Symp.* **1980**, *7*, 245.

123. Price, C. C. In *Handbook of Experimental Pharmacology*; Sartorelli, A. C.; Johns, D. J. (Eds.), Springer-Verlag, Berlin, 1974, Vol. 38, Part 2, p. 4.

124. (a) Kohn, K. W.; Spears, C. L.; Doty, P. *J. Mol. Biol.* **1966**, *19*, 266. (b) Lawley, P. D.; Brookes, P. *J. Mol. Biol.* **1967**, *25*, 143.

125. (a) Colvin, M.; Chabner, B. A. In *Cancer Chemotherapy: Principles and Practice*; Chabner, B. A.; Collins, S. M. (Eds.), Lippincott, Philadelphia,1990, p. 276. (b) Bardos, T. J.; Chmielewicz, Z. F.; Hebborn, P. *Ann. N.Y. Acad. Sci.* **1969**, *163*, 1006.

126. (a) Niculescu-Duvaz, I.; Baracu, I.; Balaban, A. T. In *The Chemistry of Antitumour Agents*, Wilman, D. E. V. (Ed.), Blackie, Glasgow, 1990, p. 63. (b) Kohn, K. W.; Erickson, L. C.; Laurent, G.; Ducore, J. M.; Sharkey, N. A.; Ewig, R. A. G. In *Nitrosoureas, Current Status and New Developments*, Prestayo, W.; Crooke, S. T.; Karter, S. K.; Schein, P. S. (Eds.), Academic Press, New York, 1981, p. 69.

127. Harris, A. L. *Cancer Surv.* **1985**, 4, 601.

128. Prelog, V.; Stepan, V. *Coll. Czech. Chem. Commun.* **1935**, *7*, 93.

129. Brookes, P.; Lawley, P. D. *Biochem. J.* **1961**, *80*, 496.

130. (a) Oida, T.; Humphreys, W. G.; Guengerich, F. P. *Biochemistry* **1991**, *30*, 10513. (b) Persmark, M.; Guengerich, F. P. *Biochemistry* **1994**, *33*, 8662.

131. Greenberg, M. M.; Hantosi, Z.; Wiederholt, C. J.; Rithner, C. D. *Biochemistry* **2001**, *40*, 15856.

132. Haddow, A.; Kon, G. A. R.; Ross, W. C. J. *Nature* **1948**, 162, 824.

133. Everett, J. L.; Roberts, J. J.; Ross, W. C. J. *J. Chem. Soc.* **1953**, 2386.

134. Schmidt, L. H.; Fradkin, R.; Sullivan, R.; Flowers, A. *Cancer Chemother. Rep.* **1965**, *Suppl. 2*, 1–1528.

135. Vistica, D. T. *Pharmacol. Ther.* **1983**, *22*, 379.

136. Goldin, A.; Wood, H. B., Jr. *Ann. N.Y. Acad. Sci.* **1969**, *163*, 954.

137. Khan, A. H.; Driscoll, J. S. *J. Med. Chem.* **1976**, *19*, 313.

138. Haddow, A.; Timmis, G. M. *Lancet* **1953**, *1*, 207.

139. Timmis, G. M.; Hudson, R. F. *Ann. N.Y. Acad. Sci.* **1958**, *68*, 727.

140. Brookes, P.; Lawley, P. D. *J. Chem. Soc.* **1961**, 3923.

141. Tong, W. P.; Ludlam, D. B. *Biochim. Biophys. Acta* **1980**, *608*, 174.

142. Hanka, L. J.; Dietz, A.; Gerpheide, S. A.; Kuentzel, S. L.; Martin, D. G. *J. Antibiot.* **1978**, *31*, 1211.

143. Takahashi, I.; Takahashi, K.; Ichimura, M.; Morimoto, M.; Asano, K.; Kawamoto, I.; Tomita, F.; Nakano, H. *J. Antibiot.* **1988**, *41*, 1915.

144. Ichimura, M.; Ogawa, T.; Takahashi, K.; Kobayashi, E.; Kawamoto, I.; Yasuzawa, T.; Takahashi, I.; Nakano, H. *J. Antibiot.* **1990**, *43*, 1037.

145. (a) Boger, D. L.; Garbaccio, R. M. *Acc. Chem. Res.* **1999**, *32*, 1043. (b) Boger, D. L.; Johnson, D. S. *Proc. Natl. Acad. Sci. USA* **1995**, *92*, 3642. (c) Searcey, M. *Curr. Pharm. Design* **2002**, *8*, 1375.

146. Baird, R.; Winstein, S. *J. Am. Chem. Soc.* **1963**, *85*, 567.

147. (a) Hurley, L. H.; Reynolds, V. L.; Swenson, D. H.; Petzold, G. L.; Scahill, T. A. *Science* **1984**, *226*, 843. (b) Hurley, L. H.; Lee, C.-S.; McGovren, J. O.; Mitchell, M.; Warpehoski, M. A.; Kelly, R. C.; Aristoff, P. A. *Biochemistry* **1988**, *27*, 3886.

148. Lin, C. H.; Beale, J. M.; Hurley, L. H. *Biochemistry* **1991**, *30*, 3597.

149. Boger, D. L.; Turnbull, P. *J. Org. Chem.* **1998**, *63*, 8004.

150. (a) Boger, D. L.; Boyce, C. W.; Johnson, D. S. *Bioorg. Med. Chem. Lett.* **1997**, *7*, 233. (b) Boger, D. L.; Garbaccio, R. M. *J. Org. Chem.* **1999**, *64*, 5666.

151. Warpehoski, M. A.; Harper, D. E. *J. Am. Chem. Soc.* **1995**, *117*, 2951.

152. Skinner, W. A.; Gram, H. F.; Greene, M. O. *J. Med. Pharm. Chem.* **1960**, *2*, 299.

153. Schabel, F. M., Jr.; Johnston, T. P.; McCaleb, G. S.; Montgomery, J. A.; Laster, W. R.; Skipper, H. E. *Cancer Res.* **1963**, *23*, 725.

154. Montgomery, J. A.; James, R.; McCaleb, G. S.; Johnston, T. P. *J. Med. Chem.* **1967**, *10*, 668.

155. Wheeler, G. P. In *Handbook of Experimental Pharmacology*, Sartorelli, A. C.; Johns, D. G. (Eds.), Springer-Verlag, Berlin, 1974, Vol. 38, Part 2, p. 7.

156. Johnston, T. P.; Montgomery, J. A. *Cancer Treat. Rep.* **1986**, *70*, 13.

157. Schmall, B.; Cheng, C. J.; Fujimura, S.; Gersten, N.; Grunberger, D.; Weinstein, I. B. *Cancer Res.* **1973**, *33*, 1921.

158. Baril, B. B.; Baril, E. F.; Laszlo, J.; Wheeler, G. P. *Cancer Res.* **1975**, *35*, 1.

159. (a) Kann, H. E., Jr.; Blumenstein, B. A.; Petkas, A.; Schott, M. A. *Cancer Res.* **1980**, *40*, 771. (b) Robins, P.; Harris, A. L.; Goldsmith, I.; Lindahl, T. *Nucleic Acids Res.* **1983**, *11*, 7743.

160. Kann, H. E., Jr.; Kohn, K. W.; Widerlite, L.; Gullion, D. *Cancer Res.* **1974**, *34*, 1982.

161. Panasci, L. C.; Green, D.; Nagourney, R.; Fox, P.; Schein, P. S. *Cancer Res.* **1977**, *37*, 2615.

162. (a) Tong, W. P.; Kirk, M. C.; Ludlum, D. B. *Cancer Res.* **1982**, *42*, 3102. (b) Tong, W. P.; Kirk, M. C.; Ludlum, D. B. *Biochem. Pharmacol.* **1983**, *32*, 2011. (c) Lown, J. W.; McLaughlin, L. W.; Chang, Y.-M. *Bioorg. Chem.* **1978**, *7*, 97.

163. Hayes, M. T.; Bartley, J.; Parsons, P. G.; Eaglesham, G.K.; Prakash, A. S. *Biochemistry* **1997**, *36*, 10646.

164. Buckley, N.; Brent, T. P. *J. Am. Chem. Soc.* **1988**, *110*, 7520.

165. Piper, J. R.; Laseter, A. G.; Johnston, T. P.; Montgomery, J. A. *J. Med. Chem.* **1980**, *23*, 1136.

166. (a) Erickson, L. C.; Sharkey, N. A.; Kohn, K. W. *Nature* **1980**, *288*, 727. (b) Brent, T. P.; Houghton, P. J.; Houghton, J. A. *Proc. Natl. Acad. Sci. USA* **1985**, *82*, 2985.

167. Ludlum, D. B.; Mehta, J. R.; Tong, W. P. *Cancer Res.* **1986**, *46*, 3353.

168. (a) Dolan, M. E.; Pegg, A. E. *Clin. Cancer Res.* **1997**, *3*, 837. (b) Karran, P. *Carcinogenesis* **2001**, *22*, 1931.

169. (a) Pegg, A. *Mutat. Res.* **2000**, *462*, 83. (b) Sekiguchi, M.; Nakabeppu, Y.; Sakumi, K.; Tuzuki, T. *J. Cancer Res. Clin. Oncol.* **1996**, *122*, 199.

170. Srivenugopal, K. S.; Yuan, X. H.; Bigner, D. D.; Friedman, H. S.; Ali-Osman, F. *Biochemistry* **1996**, *35*, 1328. (b) Xu-Welliver, M.; Pegg, A. E. *Carcinogenesis* **2002**, *23*, 823.

171. Comis, R. L. *Cancer Treat. Rep.* **1976**, *60*, 165.

172. (a) Skibba, J. L.; Ramirez, G.; Beal, D. D.; Bryan, G. T. *Biochem. Pharmacol.* **1970**, *19*, 2043. (b) Mizuno, N. S.; Humphrey, E. W. *Cancer Chemother. Rep.* (Part 1) **1972**, *56*, 465.

173. (a) Moore, H. W.; Czerniak, R. *Med. Res. Rev.* **1981**, *1*, 249. (b) Moore, H. W. *Science* **1977**, *197*, 527.

174. Tomasz, M.; Palom, Y. *Pharmacol. Therap.* **1997**, *76*, 73.

175. Iyer, V. N.; Szybalski, W. A. *Science* **1964**, *145*, 55.

176. (a) Fitzsimmons, S. A.; Workman, P.; Grever, M.; Paull, K.; Camalier, R.; Lewis, A. D. *J. Natl. Cancer Inst.* **1996**, *88*, 259. (b) Kumar, G. S.; Lipman, R.; Cummings, J.; Tomasz, M. *Biochemistry* **1997**, *36*, 14128.

177. Bean, M.; Kohn, H. *J. Org. Chem.* **1985**, *50*, 293.

178. (a) Franck, R. W.; Tomasz, M. In *The Chemistry of Antitumor Agents*; Wilman, D. E. V. (Ed.), Blackie and Son, Glasgow, 1990, p. 379. (b) Remers, W. A. *The Chemistry of Antitumor Antibiotics*, Wiley, New York, 1979, Vol. 1, p. 271.

179. (a) Kohn, H.; Zein, N.; Lin, X. Q.; Ding, J.-Q.; Kadish, K. M. *J. Am. Chem. Soc.* **1987**, *109*, 1833. (b) Danishefsky, S. J.; Egbertson, M. *J. Am. Chem. Soc.* **1986**, *108*, 4648. (c) Andrews, P. A.; Pan, S.-S.; Bachur, N. R. *J. Am. Chem. Soc.* **1986**, *108*, 4158.

180. Kohn, H.; Hong, Y. P. *J. Am. Chem. Soc.* **1990**, *112*, 4596.

181. Tomasz, M.; Chowdary, D.; Lipman, R.; Shimotakahara, S.; Veiro, D.; Walker, V.; Verdine, G. L. *Proc. Natl. Acad. Sci. USA* **1986**, *83*, 6702.

182. (a) Borowy-Borowski, H.; Lipman, R.; Chowdary, D.; Tomasz, M. *Biochemistry* **1990**, *29*, 2992. (b) Borowy-Borowski, H.; Lipman, R.; Tomasz, M. *Biochemistry* **1990**, *29*, 2999.

183. (a) Tomasz, M.; Lipman, R.; Chowdary, D.; Pawlak, J.; Verdine, G. L.; Nakanishi, K. *Science* **1987**, *235*, 1204. (b) Tomasz, M.; Lipman, R.; McGuinness, B. F.; Nakanishi, K. *J. Am. Chem. Soc.* **1988**, *110*, 5892.

184. Millard, J. T.; Weidner, M. F.; Raucher, S.; Hopkins, P. B. *J. Am. Chem. Soc.* **1990**, *112*, 3637.

185. Kennedy, K. A.; Teicher, B. A.; Rockwell, S.; Sartorelli, A. C. *Biochem. Pharmacol.* **1980**, *29*, 1.

186. Antonini, I.; Lin, T.-S.; Cosby, L. A.; Dai, Y.-R.; Sartorelli, A. C. *J. Med. Chem.* **1982**, *25*, 730.

187. Lin, A. J.; Lillis, B. J.; Sartorelli, A. C. *J. Med. Chem.* **1975**, *18*, 917.

188. (a) Lin, A. J.; Sartorelli, A. C. *Biochem. Pharmacol.* **1976**, *25*, 206. (b) Prakash, G.; Hodnett, E. M. *J. Med. Chem.* **1978**, *21*, 369.

189. Hara, M.; Takahashi, I.; Yoshida, M.; Kawamoto, I.; Morimoto, M.; Nakano, H. *J. Antibiot.* **1989**, *42*, 333.

190. Hara, M.; Saitoh, Y.; Nakano, H. *Biochemistry* **1990**, *29*, 5676.

191. Behroozi, S. J.; Kim, W.; Gates, K. S. *J. Org. Chem.* **1995**, *60*, 3964.

192. Gates, K. S. *Chem. Res. Toxicol.* **2000**, *13*, 953. (b) Mitra, K.; Gates, K. S. *Recent Res. Devel. Org. Chem.* **1999**, *3*, 311.

193. Asai, A.; Hara, M.; Kakita, S.; Kanda, Y.; Yoshida, M.; Saito, H.; Saitoh, Y. *J. Am. Chem. Soc.* **1996**, *118*, 6802.

194. Asai, A.; Saito, H.; Saitoh, Y. *Bioorg. Med. Chem.* **1997**, *5*, 723.

195. (a) Mitra, K.; Kim, W.; Daniels, J. S.; Gates, K. S. *J. Am. Chem. Soc.* **1997**, *119*, 11691. (b) Breydo, L.; Gates, K. S. *Bioorg. Med. Chem. Lett.* **2000**, *10*, 885. (c) Behroozi, S. B.; Kim, W.; Gates, K. S. *Biochemistry* **1996**, *35*, 1768.

196. (a) Lown, J. W.; Sim, S.-K.; Majumdar, K. C.; Chang, R.-Y. *Biochem. Biophys. Res. Commun.* **1977**, *76*, 705. (b) Bachur, N. R.; Gordon, S. L.; Gee, M. V. *Cancer Res.* **1977**, *38*, 1745.

197. Ross, W. A.; Glaubiger, D. L.; Kohn, K. W. *Biochim. Biophys. Acta* **1978**, *519*, 23.

198. Pan, S.-S.; Pedersen, L.; Bachur, N. R. *Mol. Pharmacol.* **1981**, *19*, 184.

199. Hertzberg, R. P.; Dervan, P. B. *Biochemistry* **1984**, *23*, 3934.

200. Garnier-Suillerot, A. In *Anthracycline and Anthracenedione-Based Anticancer Agents*, Lown, J. W. (Ed.), Elsevier, Amsterdam, **1988**, pp. 129–157.

201. Walling, C. *Acc. Chem. Res.* **1975**, *8*, 125.

202. Tullius, T. D.; Dombrowski, B. A.; Churchill, M. E. A.; Kam, L. *Methods Enzymol*, **1987**, *155*, 537.

203. Speyer, J. L.; Green, M. D.; Kramer, E.; Rey, M.; Sanger, J.; Ward, C.; Dubin, N.; Ferran, V.; Stecy, P.; Zeleniuch-Jaquotte, A.; Wernz, J.; Feit, F.; Slater, W.; Blum, R.; Mugia, F. *N. Engl. J. Med.* **1988**, *319*, 745.

204. Gianni, L.; Vigano, L.; Lanzi, C.; Niggeler, M.; Malatesta, V. *J. Natl. Cancer Inst.* **1988**, *80*, 1104.

205. Umezawa, H.; Suhara, Y.; Takita, T.; Maeda, K. *J. Antibiot. Ser. A.* **1966**, *19*, 210.

206. (a) Stubbe, J.; Kozarich, J. W.; Wu, W.; Vanderwall, D. E. *Acc. Chem. Res.* **1996**, *29*, 322. (b) Povirk, L. F.; Han, Y.-H.; Steighner, R. J. *Biochemistry* **1989**, *28*, 5808. (c) Steighner, R. J.; Povirk, L. F. *Proc. Natl. Acad. Sci. USA* **1990**, *87*, 8350.

207. Stubbe, J.; Kozarich, J. W. *Chem. Rev.* **1987**, *87*, 1107.

208. (a) Sausville, E. A.; Stein, R. W.; Peisach, J.; Horwitz, S. B. *Biochemistry* **1978**, *17*, 2746. (b) Hecht, S. M. *Acc. Chem. Res.* **1986**, *19*, 383.

209. Hamamichi, N.; Hecht, S. M. *J. Am. Chem. Soc.* **1992**, *114*, 6278.

210. Fisher, L. M.; Kuroda, R.; Sakai, T. T. *Biochemistry* **1985**, *24*, 3199.

211. (a) Ciriolo, M. R.; Magliozzo, R. S.; Peisach, J. *J. Biol. Chem.* **1987**, *262*, 6290. (b) Mahmutoglu, I.; Kappus, H. *Biochem. Pharmacol.* **1987**, *36*, 3677.

212. Umezawa, H.; Takita, T.; Sugiura, Y.; Otsuka, M.; Kobayashi, S.;Ohno, M. *Tetrahedron* **1984**, *40*, 501.

213. Povirk, L. F.; Hogan, M.; Dattagupta, N. *Biochemistry* **1979**, *18*, 96.

214. Burger, R. M.; Peisach, J.; Horwitz, S. B. *J. Biol. Chem.* **1981**, *256*, 11636.

215. Heimbrook, D. C.; Mulholland, R. L., Jr.; Hecht, S. M. *J. Am. Chem. Soc.* **1986**, *108*, 7839.

216. Burger, R. M.; Kent, T. A.; Horwitz, S. B.; Munck, E.; Peisach, J. *J. Biol. Chem.* **1983**, *258*, 1559.

217. Sam, J. W.; Tang, X.-J.; Peisach, J. *J. Am. Chem. Soc.* **1994**, *116*, 5250.

218. Westre, T. E.; Loeb, K. E.; Zaleski, J. M.; Hedman, B.; Hodgson, K. O.; Solomon, E. I. *J. Am. Chem. Soc.* **1995**, *117*, 1309.

219. Solomon, E. I.; Brunold, T. C.; Davis, M. I.; Kemsley, J. N.; Lee, S.-K.; Lehnert, N.; Neese, F.; Skulan, A. J.; Yang, Y.-S.; Zhou, J. *Chem. Rev.* **2000**, *100*, 235.

220. Neese, F.; Zaleski, J. M.; Loeb, K. E.; Solomon, E. I. *J. Am. Chem. Soc.* **2000**, *122*, 11703.

221. Burger, R. M. *Chem. Rev.* **1998**, *98*, 1153.

222. Solomon, E. I.; Sundaram, U. M.; Machonkin, T. E. *Chem. Rev.* **1996**, *96*, 2563.

223. (a) Giloni, L.; Takeshita, M.; Johnson, F.; Iden, C.; Grollman, A. P. *J. Biol. Chem.* **1981**, *256*, 8608. (b) Murugesan, N.; Xu, C.; Ehrenfeld, G. M.; Sugiyama, H.; Kilkuskie, R. E.; Rodriguez, L. O.; Chang, L.-H.; Hecht, S. M. *Biochemistry* **1985**, *24*, 5735.

224. (a) Giese, B.; Beyrich-Graf, X.; Erdmann, P.; Petretta, M.; Schwitter, U. *Chem. Biol.* **1995**, *2*, 367. (b) Giese, B.; Erdmann, P.; Giraud, L.; Göbel, T.; Petretta, M.; Schaefer, T. *Tetrahedron Lett.* **1994**, *35*, 2683.

225. McGall, G. H.; Rabow, L. E.; Ashley, G. W.; Wu, S. H.; Kozarich, J. W.; Stubbe, J. *J. Am. Chem. Soc.* **1992**, *114*, 4958.

226. Giese, B.; Beyrich-Graf, X.; Erdmann, P.; Giraud, L.; Imwinkelried, P.; Muller, S. N.; Schwitter, U. *J. Am. Chem. Soc.* **1995**, *117*, 6146.

227. Ross, S. L.; Moses, R. E. *Biochemistry* **1978**, *17*, 581.

228. Sugiyama, H.; Xu, C.; Murugesan, N.; Hecht, S. M.; van der Marel, G. A.; van Boom, J. H. *Biochemistry* **1988**, *27*, 58.

229. Rabow, L. E.; Stubbe, J.; Kozarich, J. W. *J. Am. Chem. Soc.* **1990**, *112*, 3196.

230. Rabow, L. E.; McGall, G. H.; Stubbe, J.; Kozarich, J. W. *J. Am. Chem. Soc.* **1990**, *112*, 3203.

231. (a) D'Andrea, A. D.; Haseltine, W. A. *Proc. Natl. Acad. Sci. USA* **1978**, *75*, 3608. (b) Takeshita, M.; Grollman, A. P.; Ohtsubo, E.; Ohtsubo, H. *Proc. Natl. Acad. Sci. USA* **1978**, *75*, 5983.

232. (a) Boger, D. L.; Ramsey, T. M.; Cai, H.; Hoehn, S. T.; Kozarich, J. W.; Stubbe, J. *J. Am. Chem. Soc.* **1998**, *120*, 53. (b) Bailly, C.; Waring, M. *J. Am. Chem. Soc.* **1995**, *117*, 7311.

233. Umezawa, H.; Takeuchi, S.; Hori, T.; Sawa, T.; Ishizuka, T.; Ichikawa, T.; Komai, T. *J. Antibiot.* **1972**, *25*, 409.

234. Sugiura, Y.; Muraoka, Y.; Fujii, A.; Takita, T.; Umezawa, H. *J. Antibiot.* **1979**, *32*, 756.

235. Brown, J. M. *Cancer Res.* **1999**, *59*, 5863.

236. (a) Daniels, J. S.; Gates, K. S. *J. Am. Chem. Soc.* **1996**, *118*, 3380. (b) Patterson, L. H.; Taiwo, F. A. *Biochem. Pharmacol.* **2000**, *60*, 1933.

237. Kotandeniya, D.; Ganley, B.; Gates, K. S. *Bioorg. Med. Chem. Lett.* **2002**, *12*, 2325.

238. (a) Wardman, P.; Priyadarsini, K. I.; Dennis, M. F.; Everett, S. A.; Naylor, M. A.; Patel, K. B.; Stratford, I. J.; Stratford, M. R. L.; Tracy, M. *Br. J. Cancer* **1996**, *74*, S70. (b) Lloyd, R. V.; Duling, D. R.; Rumyantseva, G. V.; Mason, R. P.; Bridson, P. K. *Mol. Pharmacol.* **1991**, *40*, 440.

239. (a) Laderoute, K. L.; Wardman, P.; Rauth, M. *Biochem. Pharmacol.* **1988**, *37*, 1487. (b) Fitzsimmons, S. A.; Lewis, A. D.; Riley, R. J.; Workman, P. *Carcinogenesis* **1994**, *15*, 1503.

240. (a) Hwang, J.-T.; Greenberg, M. M.; Fuchs, T.; Gates, K. S. *Biochemistry* **1999**, *38*, 14248. (b) Daniels, J. S.; Gates, K. S.; Tronche, C.; Greenberg, M. M. *Chem. Res. Toxicol.* **1998**, *11*, 1254. (c) Jones, G. D. D.; Weinfeld, M. *Cancer Res.* **1996**, *56*, 1584.

241. Ishida, N.; Miyazaki, K.; Kumagai, K.; Rikimaru, M. *J. Antibiot.* **1965**, *18*, 68.

242. Konishi, M.; Ohkuma, H.; Saitoh, K.; Kawaguchi, H.; Golik, J.; Dubay, G.; Groenewold, G.; Krishnan, B.; Doyle, T. W. *J. Antibiot.* **1985**, *38*, 1605.

243. (a) Lee, M. D.; Ellestad, G. A.; Borders, D. B. *Acc. Chem. Res.* **1991**, *24*, 235. (b) Lee, M. D.; Dunne, T. S.; Siegel, M. M.; Chang, C. C.; Morton, G. O.; Borders, D. B. *J. Am. Chem. Soc.* **1987**, *109*, 3464.

244. Konishi, M.; Ohkuma, H.; Matsumoto, K.; Tsuno, T.; Kamei, H.; Miyaki, T.; Oki, T.; Kawaguchi, H.; Van Duyne, G. D.; Clardy, J. *J. Antibiot.* **1989**, *42*, 1449.

245. (a) Lam, K. S.; Hesler, G. A.; Gustavson, D. R.; Crosswell, A. R.; Veitch, J. M.; Forenza, S.; Tomita, K. *J. Antibiot.* **1991**, *44*, 472. (b) Leet, J. E.; Schroeder, D. R.; Langley, D. R.; Colson, K. L.; Huang, S.; Klohr, S. E.; Lee, M. S.; Golik, J.; Hofstead, S. J.; Doyle, T. W.; Matson, J. A. *J. Am. Chem. Soc.* **1993**, *115*, 8432.

246. (a) Hu, J.; Xue, Y.-C.; Xie, M.-Y.; Zhang, R.; Otani, T.; Minami, Y.; Yamada, Y.; Marunaka, T. *J. Antibiot.* **1988**, *41*, 1575. (b) Minami, Y.; Yoshida, K.-i.; Azuma, R.; Saeki, M.; Otani, T. *Tetrahedron Lett.* **1993**, *34*, 2633. (c) Yoshida, K.-i.; Minami, Y.; Azuma, R.; Saeki, M.; Otani, T. *Tetrahedron Lett.* **1993**, *34*, 2637.

247. Ando, T.; Ishii, M.; Kajiura, T.; Kameyama, T.; Miwa, K.; Sugiura, Y. *Tetrahedron Lett.* **1998**, *39*, 6495.

248. Smith, A. L.; Nicolaou, K. C. *J. Med. Chem.* **1996**, *39*, 2103.

249. Zein, N.; Casazza, A. M.; Doyle, T. W.; Leet, J. E.; Schroeder, D. R.; Solomon, W.; Nadler, S. G. *Proc. Natl. Acad. Sci. USA* **1993**, *90*, 8009.

250. (a) Kumar, R. A.; Ikemoto, N.; Patel, D. J. *J. Mol. Biol.* **1997**, *265*, 173. (b) Kumar, R. A.; Ikemoto, N.; Patel, D. J. *J. Mol. Biol.* **1997**, *265*, 187. (c) Gao, X.; Stassinopoulos, A.; Gu, J.; Goldberg, I. H. *Bioorg. Med. Chem.* **1995**, *3*, 795. (d) Gao, X.; Stassinopoulos, A.; Rice, J. S.; Goldberg, I. H. *Biochemistry* **1995**, *34*, 40.

251. (a) Myers, A. G.; Cohen, S. B.; Tom, N. J.; Madar, D. J.; Fraley, M. E. *J. Am. Chem. Soc.* **1995**, *117*, 7574. (b) Myers, A. G.; Cohen, S. B.; Kwon, B. M. *J. Am. Chem. Soc.* **1994**, *116*, 1255.

252. Sluka, J. P.; Horvath, S. J.; Bruist, M. F.; Simon, M. I.; Dervan, P. B. *Science* **1987**, *238*, 1129.

253. Xu, Y.; Zhen, Y.; Goldberg, I. H. *Biochemistry* **1994**, *33*, 5947.

254. (a) Long, B. H.; Golik, J.; Forenza, S.; Ward, B.; Rehfuss, R.; Dabrowiak, J. C.; Catino, J. J.; Musial, S. T.; Brookshire, K. W.; Doyle, T. W. *Proc. Natl. Acad. Sci. USA* **1989**, *86*, 2. (b) Zein, N.; Sinha, A. M.; McGahren, W. J.; Ellestad, G. A. *Science* **1988**, *240*, 1198.

255. Lockhart, T. P.; Comita, P. B.; Bergman, R. G. *J. Am. Chem. Soc.* **1981**, *103*, 4082.

256. Sugiura, Y.; Shiraki, T.; Konishi, M.; Oki, T. *Proc. Natl. Acad. Sci. USA* **1990**, *87*, 3831.

257. Sugiura, Y.; Arawaka, T.; Uesugi, M.; Siraki, T.; Ohkuma, H.; Konishi, M. *Biochemistry* **1991**, *30*, 2989.

258. (a) Povirk, L. F.; Dattagupta, N.; Warf, B. C.; Goldberg, I. H. *Biochemistry* **1981**, *20*, 4007. (b) Lee, S. H.; Goldberg, I. H. *Biochemistry* **1989**, *28*, 1019.

259. Myers, A. G.; Cohen, S. B.; Kwon, B.-M. *J. Am. Chem. Soc.* **1994**, *116*, 1670.

260. Sugiyama, H.; Yamashita, K.; Fujiwara, T.; Saito, I. *Tetrahedron* **1994**, *50*, 1311.

261. Myers, A. G.; Arvedson, S. P.; Lee, R. W. *J. Am. Chem. Soc.* **1996**, *118*, 4725.

262. Myers, A. G.; Dragovich, P. S.; Kuo, E. Y. *J. Am. Chem. Soc.* **1992**, *114*, 9369.

263. (a) Frank, B. L.; Worth, L., Jr.; Christner, D. F.; Kozarich, J. W.; Stubbe, J.; Kappen, L. S.; Goldberg, I. H. *J. Am. Chem. Soc.* **1991**, *113*, 2271. (b) Kappen, L. S.; Goldberg, I. H. *Nucleic Acids Res.* **1985**, *13*, 1637. (c) Kappen, L. S.; Goldberg, I. H.; Frank, B. L.; Worth, L., Jr.; Christner, D. F.; Kozarich, J. W.; Stubbe, J. *Biochemistry* **1991**, *30*, 2034.

264. (a) Blackburn, G. M.; Datta, A.; Denham, H.; Wentworth, P., Jr. *Adv. Phys. Org. Chem.* **1998**, *31*, 249. (b) Wentworth, P., Jr.; Janda, K. D. *Curr. Opin. Chem. Biol.* **1998**, *2*, 138.

265. Jones, L. H.; Harwig, C. W.; Wentworth, P. Jr.; Simeonov, A.; Wentworth, A. D.; Py, S.; Ashley, J. A.; Lerner, R. A.; Janda, K. D. *J. Am. Chem. Soc.* **2001**, *123*, 3607.

266. Lin, A. J.; Pardini, R. S.; Cosby, L. A.; Lillis, B. J.; Shansky, C. W.; Sartorelli, A. C. *J. Med. Chem.* **1973**, *16*, 1268.

267. Thorson, J. S.; Shen, B.; Whitwam, R. E.; Liu, W.; Li, Y.; Ahlert, J. *Bioorg. Chem.* **1999**, *27*, 172.

268. Whitwam, R. E.; Ahlert, J.; Holman, T. R.; Ruppen, M.; Thorson, J. S. *J. Am. Chem. Soc.* **2000**, *122*, 1556.

269. Newton, G. L.; Fahey, R. C. In *Glutathione: Metabolism and Physiological Functions*, Via, J. (Ed.), CRC Press, Boca Raton, FL, 1990, pp. 69–77.

270. Reddy, B. S. P.; Sharma, S. K.; Lown, J. W. *Curr. Med. Chem.* **2001**, *8*, 475. (b) Goldberg, I. H.; Kappen, L. S.; Xu, Y. J.; Stassinopoulos, A.; Zeng, X.; Xi, Z.; Yang, C. F. *NATO ASI Series, Series C* **1996**, *479*, 1. (c) Stubbe, J.; Kozarich, J. W.; Wu, W.; Vanderwall, D. E. *Acc. Chem. Res.* **1996**, *29*, 322. (d) Fox, K.R.; Nightingale, K. P. *Nucleic Acids Mol. Biol.* **1994**, *8*, 167. (e) Smith, A. L.; Nicolaou, K. C. *J. Med. Chem.* **1996**, *39*, 2103. (f) Sugiura, Y.; Shiraki, T.; Konishi, M.; Oki, T. *Proc. Natl. Acad. Sci. USA* **1990**, *87*, 3831.

271. Cloutier, J.-F.; Drouin, R.; Castonguay, A. *Chem. Res. Toxicol.* **1999**, *12*, 840.

272. Wu, J.; Xu, J.; Dedon, P. C. *Biochemistry* **1999**, *38*, 15641.

Drug Metabolism

7.1 Introduction

When a foreign organism enters the body, the immune system produces antibodies to interact with and destroy it. Small molecules, however, do not stimulate antibody production. So how has the human body evolved to protect itself against low molecular weight environmental pollutants? The principal mechanism is the use of nonspecific enzymes that transform the foreign compounds (often highly nonpolar molecules) into polar molecules that are excreted by the normal bodily processes. Although this mechanism to rid the body of *xenobiotics* (molecules foreign to the organism) is highly desirable, especially when one considers all of the foreign materials to which we are exposed every day, it can cause problems when the foreign agent is a drug that we want to enter and be retained in the body sufficiently long to be effective. The enzymatic biotransformations of drugs is known as *drug metabolism*. Because many drugs have structures similar to those of endogenous compounds, drugs may get metabolized by specific enzymes for the related natural substrates as well as by nonspecific enzymes.

The principal site of drug metabolism is the liver, but the kidneys, lungs, and GI tract also are important metabolic sites. When a drug is taken orally (the most common route of administration), it is usually absorbed through the mucous membrane of the small intestine or from the stomach. Once out of the GI tract it is carried by the bloodstream to the liver where it is usually first metabolized. Metabolism by liver enzymes prior to the drug reaching the systemic circulation is called the *presystemic* or *first-pass effect*, which may result in complete deactivation of the drug. If a large fraction of the drug is metabolized, then larger or multiple doses of the drug will be required to get the desired effect. Another undesirable effect of drug metabolism is that occasionally the metabolites of a drug may be toxic, even though the drug is not.

The first-pass effect sometimes can be avoided by changing the route of drug administration. The *sublingual route* (the drug is placed under the tongue) bypasses the liver. After absorption through the buccal cavity, the drug enters the systemic circulation. This is the route employed with nitroglycerin (**7.1**, Nitrostat), a drug used for the treatment of angina pectoris that is converted by mitochondrial aldehyde dehydrogenase to nitrite ion, which is then reduced to nitric oxide,[1] a second messenger molecule that dilates blood vessels in the heart. The *rectal route*, in the form of a solid suppository or in solution as an enema, leads to absorption through the colon mucosa. Ergotamine (**7.2**, Ergomar), a drug for migraine headaches, is administered this way (who would have guessed?). *Intravenous* (i.v.) *injection* introduces the drug directly into the systemic circulation and is used when a rapid therapeutic response is desired. The effects are almost immediate when drugs are administered by this route, because the total blood circulation time in humans is 15–20 sec. *Intramuscular* (i.m.) *injection* is used when large volumes of drugs need to be administered, if slow absorption is desirable, or if the drug is unstable in the gastric acid of the stomach. A *subcutaneous* (s.c.) *injection* delivers the drug through the loose connective tissue of the subcutaneous layer of the skin. Another method of administration, particularly for gaseous or highly volatile drugs such as general anesthetics, is by *pulmonary absorption* through the respiratory tract. The asthma drug isoproterenol (**7.3**, Isuprel) is metabolized in the intestines and liver, but administration by aerosol inhalation is effective at getting the drug directly to the bronchi. *Topical application* of the drug to the skin or a mucous membrane is used for local effects; few drugs readily penetrate the intact skin. Not all drugs can be administered by these alternate routes, so their structures may have to be altered to minimize the first-pass effect or to permit them to be administered by one of these alternate routes. These structural modification approaches in drug design to avoid the first-pass effect are discussed in Chapter 8. Even if the first-pass effect is avoided, there are many enzymes in tissues other than the liver that are capable of catalyzing drug metabolism reactions. Once a drug has reached its site of action and elicited the desired response, it usually is desirable for the drug to be metabolized and eliminated at a reasonable rate. Otherwise, it may remain in the body and produce the effect longer than desired or it could accumulate and become toxic to the cells.

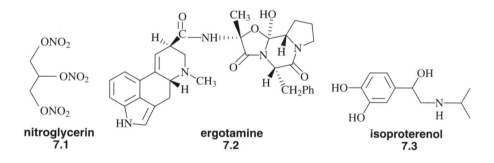

nitroglycerin
7.1

ergotamine
7.2

isoproterenol
7.3

Drug metabolism studies are essential for evaluating the potential safety and efficacy of drugs. Consequently, prior to approval of a drug for human use, an understanding of the metabolic pathways and disposition of the drug in humans and in preclinical animal species is required. The animal species used for metabolism studies are often those in which the toxicological evaluations are conducted. Additional toxicological studies have to be carried out on metabolites found in humans that were not observed in the animal metabolism studies. Metabolism studies also can be a useful lead modification approach. For example, after many years on the drug market, terfenadine (**7.4**, R = CH$_3$; Seldane) was removed because it was found to cause life-threatening cardiac arrhythmias when coadministered with inhibitors of hepatic cytochrome P450, such as erythromycin and ketoconazole.[2] The active metabolite of terfenadine, fexofenadine (**7.4**, R = COOH; Allegra), however, produces no arrhythmias, and it has replaced terfenadine on the market.[3]

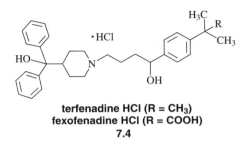

terfenadine HCl (R = CH₃)
fexofenadine HCl (R = COOH)
7.4

Once the metabolic products are known, it is possible to design a compound that is inactive when administered, but which utilizes the metabolic enzymes to convert it into the active form. These compounds are known as *prodrugs*, and are discussed in Chapter 8. In this chapter we consider the various reactions that are involved in the biotransformations of drugs. Because only very small quantities of drugs generally are required to elicit the appropriate response, it may be difficult to detect all of the metabolic products. To increase the sensitivity of the detection process, drug candidates are typically radioactively labeled. Radioactive compounds are useful for studying all aspects of absorption, distribution, metabolism, and excretion (ADME).[4] Metabolite studies often can be done directly by tandem mass spectrometry/mass spectrometry techniques.[5] In the next two sections we look briefly at how radiolabeling is carried out, how metabolites are detected, and how their structures are elucidated.

7.2 Synthesis of Radioactive Compounds

Because of the sensitivity of detection of particles of radioactive decay, a common approach used for detection, quantification, and profiling of metabolites in whole-animal studies is the incorporation of a radioactive label, typically a weak β-emitter such as ^{14}C[6] or ^{3}H,[7] into the drug molecule. When this approach is used, it does not matter how few metabolites are produced or how small the quantities of metabolites, even in the presence of a large number of endogenous compounds. Only the radioactively labeled compounds are isolated from the urine and the feces of the animals, and the structures of these metabolites are elucidated (see Section 7.3). If one of the carbon atoms of a drug is metabolized to carbon dioxide, as is the case of erythromycin (**7.5**, Erythrocin), ^{14}C labeling of the carbon atom that becomes CO_2 (the NMe₂ methyl groups of erythromycin are oxidized to CO_2), makes it possible to measure the rate of metabolism of the compound by measuring the rate of exhaled $^{14}CO_2$.[8] To incorporate a radioactive label into a compound, a synthesis must be designed so that a commercially available, radioactively labeled compound or reagent can be used in one of the steps. It is highly preferably to incorporate the radioactive moiety in a step at or near the end of the synthesis because once the radioactivity is introduced, the scale of the reaction is generally diminished and special precautions and procedures regarding radiation safety and disposal of radioactive waste must be followed. Often the radioactive synthesis is quite different from or longer than the synthesis of the unlabeled compound in order to use a commercially available radioactive material. It is preferable to prepare a [^{14}C]-labeled analog; when tritium is incorporated into the drug, the site of incorporation must be such that loss of the tritium by exchange with the medium does not occur even after an early metabolic step.

Generally, only one radioactive label is incorporated into a drug because drug metabolism typically leads to a modified structure with little fragmentation of the molecule. If you go back far enough in a synthesis, however, it is possible to synthesize a drug with several carbon atoms radioactively labeled. Radioactive labeling at multiple sites of a molecule would permit the identification of more fragments of the drug and consequently the elucidation of metabolite structures and the fate of the molecule *in vivo*.

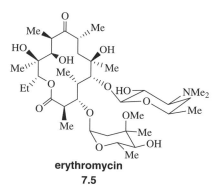

erythromycin
7.5

Industrially, the radioactive drug is synthesized with high *specific radioactivity* (a measure of the amount of radioactivity per mole of compound), often >57 mCi/mmol of ^{14}C (the theoretical maximum is 64 mCi/mmol). When needed, the specific radioactivity is diluted with non-[^{14}C]-labeled drug for use in metabolism studies. Typically, commercially available radioactive compounds have relatively low specific radioactivities. This means that may be only one in 10^6 or fewer molecules actually contains the radioactive tag; the remainder of the molecules are unlabeled and are carriers of the relatively few radioactive molecules. In the case of ^{14}C there will be no difference in the reactivity of the labeled and unlabeled molecules, so the statistical amount of radioactivity in the products formed is the same as that in the starting materials. The specific radioactivity of the metabolites formed during metabolism, then, should be identical to the specific radioactivity of the drug. In the case of tritiated drugs, however, if a carbon-hydrogen bond is broken, the radiolabel will be lost as tritiated water, and satisfactory recovery of total radioactivity in animal studies cannot occur. Also, a *kinetic isotope effect* will occur on those molecules that are tritiated. This will lead to metabolite formation with a lower specific radioactivity than that of the drug. As a result, quantitation of the various metabolic pathways, where some involve C–H bond cleavage and others do not, may require knowledge of the tritium isotope effect. This, then, is another reason why it is preferable to use [^{14}C]-labeling of a drug for metabolism studies rather than tritium labeling.

If the drug is a natural product or derivative of a natural product, the easiest procedure for incorporation of a radioactive label could be a biosynthetic approach, namely, to grow the organism that produces the natural product in the presence of a radioactive precursor, and let Nature incorporate the radioactivity into the molecule. Because of the volume of media generally involved and, therefore, the large amount of radioactive precursor required, this could be a very expensive approach; however, generally the expense is compensated by the ease of the method and the attractive yield of product obtained.

An example of a drug class that could use this approach is the penicillins, which are biosynthesized by *Penicillium* fungi from valine, cysteine, and various carboxylic acids (Scheme 7.1). Valine is commercially available with a ^{14}C label at the carboxylate carbon or it may be

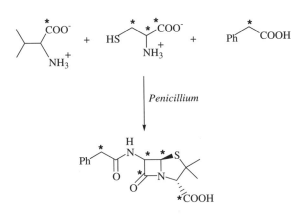

Scheme 7.1 ▶ Biosynthesis of penicillins

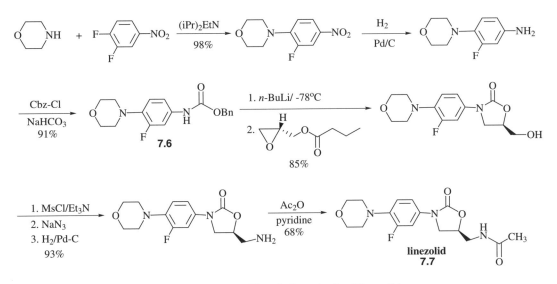

Scheme 7.2 ▶ Chemical synthesis of linezolid

obtained uniformly labeled; that is, all of the carbon atoms are labeled to some small extent with ^{14}C (albeit very few molecules would contain all of the carbon atoms labeled in the same molecule). It also can be purchased with a tritium label at the 2- and 3-positions or at the 3- and 4-positions. Cysteine is available uniformly labeled in ^{14}C or with a ^{35}S label. Penicillin G could be produced if phenylacetic acid (available with a ^{14}C label at either the 1- or 2-position) were inoculated into the *Penicillium* growth medium.

If the drug is not a natural product (the more common case) a chemical synthesis must be carried out. For example, the synthesis of the first in a new class of antibacterial drugs, linezolid (**7.7**, Zyvox), is shown in Scheme 7.2.[9] The last step in the synthesis, acetylation of the primary amine, can be carried out with [^{14}C]acetic anhydride to incorporate a radioactive label in the acetyl group of **7.7**. The oxazolidinone carbonyl carbon also could have been labeled using [*carbonyl*-^{14}C]Cbz-Cl to make **7.6**, but radioactive Cbz-Cl is not commercially available, so it would have had to be synthesized from [^{14}C]phosgene and benzyl alcohol.

Once the radioactive drug has been synthesized, it is used in metabolism studies in pre-clinical species usually first in rats, mice, or guinea pigs, then in dogs or monkeys. Typically, the urine and feces are collected from the animals, and the major radioactive compounds are isolated and their structures determined (see Section 7.3.C, p. 413). After demonstration of drug safety in animals following chronic dosing at elevated doses and satisfactory recovery of the radioactive dose (>95% in the urine and feces; some of the radioactivity may be lost as CO_2, detected in the breath), then the drug can be tested for safety and tolerability in phase I clinical trials with healthy human subjects. Once the safety is assured, the radioactive drug can be administered to humans during late phase I or early phase II clinical trials to obtain the human metabolic profile. In fact, [*acetyl*-^{14}C] **7.7** was used in a phase I human metabolism study.[10] The Food and Drug Administration approves a maximum absorbed dose of 3 rem of radioactivity to a specific organ in a healthy adult volunteer for drug metabolism studies.[11] These radioactive levels are estimated from a determination of the absorbed dose in animal models, then 10–100 times lower amounts are used in the human studies. On rare occasions, other fluids such as saliva, cerebrospinal fluid, eye fluids, perspiration, or breath may be examined as well as various organs and tissues. Generally the toxicological animal model species used are considered adequate if all of the major metabolites observed in humans are also observed in the animal models, even if more metabolites are observed in the animals. If a human metabolite is not formed in the animal toxicological model, a more relevant toxicological animal model has to be identified or additional toxicological studies need to be carried out with the metabolites unique to humans.[12]

If most of the radioactivity administered is not excreted from the animal, then it is dissected to determine the location of the radioactive compounds.[13] A newer methodology to determine tissue distribution of radiolabeled compounds in whole animals without dissection is quantitative whole-body autoradiography.[14]

From the above discussion it appears that drug metabolism studies are straightforward; however, until relatively recently these studies were difficult, at best, to carry out. The ready commercial availability of radioactively labeled precursors made the synthetic work much less tedious. The advent of *high-performance liquid chromatography* (HPLC) and the advancements in column packing materials permitted the separation of many metabolites very similar in structure. Metabolites that were previously overlooked can now be detected and identified. Structure elucidation by various types of mass spectrometry (MS) (see the next section) and by various techniques of nuclear magnetic resonance spectrometry has been relatively routine. As a result of these advances in instrumentation, more information can be gleaned from drug metabolism studies than ever before, and this can result in the discovery of new leads or in a basis for prodrug design (see Chapter 8). This, also, means that the Food and Drug Administration can demand that many more metabolites be identified and their pharmacological and toxicological properties be determined prior to drug approval (which is good news for the consumer, but bad news for the drug companies). The final step in the process to prove the identity of a metabolite is to synthesize it and demonstrate that its spectral and pharmacological properties are identical to those of the metabolite.

7.3 Analytical Methods in Drug Metabolism

The four principal steps in drug metabolism studies are isolation (extraction), separation (chromatography), identification (spectrometry), and quantification of the metabolites. Detection systems are sensitive enough to allow the isolation and identification of submicrogram quantities of metabolites. Often the isolation step can be omitted, and the urine sample or other biological sample injected directly into the HPLC or gas chromatograph for separation. For cleaner results, though, sample preparation is recommended.[15] Most pharmaceutical groups now rely most heavily on direct HPLC/electrospray (or atmospheric pressure chemical ionization) mass spectral analysis to identify drug metabolites, as described in Section 7.3.C, p. 413.

7.3.A Isolation

As discussed in Section 7.1, animals, including humans, usually convert drugs into more polar conjugates for excretion. Enzymatic hydrolysis (β-glucuronidase and arylsulfatase) of the conjugates releases the less polar drug metabolites for easier extraction and structure identification. A clean sample for analysis is preferred, especially with *in vivo* drug metabolism studies. Extensive older isolation methodologies, such as ion-pair extraction,[16] used to remove hydrophilic ionizable compounds from aqueous solution; salt-solvent pair extraction,[17] to separate metabolites into an ethyl acetate-soluble neutral and basic fraction, ethyl acetate-soluble acidic fraction, and a water-soluble fraction; and various ion-exchange resins such as the anion exchange resin DEAE-Sephadex,[18] the cation exchange resin Dowex 50,[19] and the nonionic resin Amberlite XAD-2,[20] used to separate acidic, basic, and neutral metabolites, respectively, from body fluids, have been replaced by high-throughput methodologies. With the advent of HPLC/MS analyses of metabolites described in Section 7.3.C, p. 413, often the isolation step can be eliminated using a fast-flow on-line extraction method.[21] Biological samples are injected directly into the liquid chromatography/mass spectrometer (LC/MS). A narrow-bore HPLC column packed with large particle size material extracts small molecule analytes but allows large molecules (such as proteins) to flow to the waste. The adsorbed analytes are then eluted through a column-switching valve onto an analytical column for LC/MS/MS analysis. For many assays simple protein precipitation or liquid extraction is sufficient.[22] Solid-phase extraction[23] and liquid/liquid extraction[24] have been automated to speed up the process. On-line solid-phase extraction[25] or direct plasma injection into the HPLC/MS[26] are other high-throughput methods of isolation.

7.3.B Separation

The three most important techniques for resolving mixtures of metabolites are *HPLC, capillary gas chromatography* (GC),[27] and *capillary electrophoresis* (CE).[28] HPLC is more versatile than GC because the metabolites can be charged or uncharged, they can be thermally unstable, and derivatization is unnecessary. Normal phase columns (silica gel) can be used for uncharged metabolites, and reversed phase columns (silica gel to which C4 to C18 alkyl chains are attached to give a hydrophobic environment) can be used for charged metabolites. For GC separation the metabolites must be volatilized. This often requires prior derivatization[29] in order for the metabolites to volatilize at lower temperatures. Carboxylic acids can be converted into the corresponding methyl esters with diazomethane; hydroxyl groups can be trimethylsilylated with bis-trimethylsilylacetamide or trimethylsilylimidazole in pyridine. Ketone carbonyls can be converted into *O*-substituted oximes. With radiolabeled compounds, the radioactivity can be monitored directly from the HPLC column using an in-line radioactivity detector.

7.3.C Identification

The two principal methods of metabolite structure identification are mass spectrometry and nuclear magnetic resonance spectrometry. It is preferable to link the separation and identification steps by running tandem LC-MS, tandem GC-MS, or tandem CE-MS. These methods are sufficiently sensitive to identify subnanogram amounts of material. The most popular methodology is *tandem LC-electrospray ionization mass spectrometry* by which a metabolite extract (or urine directly) can be injected into the HPLC, and each peak run directly into the mass spectrometer.[30] Similarly, *tandem CE-electrospray ionization mass spectrometry* has become a very valuable tool for separation of biomolecules and drug metabolites.[31] In liquid chromatography/tandem mass spectrometry/mass spectrometry (LC/MS/MS), the HPLC is connected to a mass spectrometer for parent ion data, and this is connected to a second mass spectrometer for fragmentation of the parent ion. This technique can provide both mass data and fragmentation data for each metabolite rapidly.[32] Ultra-fast gradient HPLC-tandem mass spectrometry can produce run times of less than 5 minutes.[33] In this way there is less chance for metabolite degradation or loss, and workup procedures for mass spectrometry sample preparation are eliminated. Mass spectrometric properties are determined using different ionization techniques. Common vacuum ionization sources include electron impact (EI), chemical ionization (CI), matrix-assisted laser desorption/ionization (MALDI), fast atom bombardment (FAB), and secondary-ion mass spectrometry (SIMS). The development of HPLC coupled to atmospheric pressure ionization sources, namely, electrospray ionization and atmospheric pressure chemical ionization mass spectrometry, have transformed the role of drug metabolism from its former minor role in drug discovery to its current important role in drug discovery and drug development. These latter LC/MS/MS methods are used not only for drug metabolism studies but also to investigate drug pharmacokinetics (absorption, bioavailability, and clearance). As indicated in Chapter 2, Section 2.2.F, p. 51, about three-quarters of drug candidates do not make it to clinical trials because of problems with pharmacokinetics in animals,[34] and about 40% of the molecules that fail in clinical trials do so because of pharmacokinetic problems, such as poor oral bioavailability or short plasma half-lives.[35] The trend in the pharmaceutical industry now is to initiate pharmacokinetic and metabolism studies as early as possible in the drug discovery process to aid in the selection of compounds that have the most drug-likeness and best chance for survival to avoid late attrition of drug candidates.[36] With these HPLC/atmospheric pressure ionization mass spectrometric techniques, assessment of *in vivo* plasma half-lives and metabolic degradation can be made rapidly on a large number of drug candidates.

A brief description of each of these mass spectrometric techniques follows. *Electron impact mass spectrometry* (EI-MS) involves the bombardment of the vaporized metabolite by high-energy electrons (0–100 eV), producing a molecular radical cation ($M^{+\cdot}$) having a mass equivalent to the molecular weight of the compound. The electron bombardment causes bond fission and the positively charged fragments produced are detected. The mass spectrum is a plot of the percentage of relative abundance of each ion produced versus the mass-to-charge ratio (m/z).

Chemical ionization mass spectrometry is important when compounds do not give spectra containing a molecular ion, generally because the molecular ion decomposes to give fragment ions. With CI-MS a reagent gas such as ammonia, isobutane, or methane is ionized in the mass spectrometer and then ion–molecule reactions such as protonation occur instead of electron–molecule reactions. This *soft ionization* process results in little fragmentation. Fragment ions in this case are almost always formed by loss of neutral molecules, and as a result, much less structural information can be gleaned relative to EI-MS.

A variety of mass spectral techniques for nonvolatile or higher mass compounds, including peptides and proteins, also are now available.[37] MALDI is a soft ionization technique that is important for analyzing biopolymers.[38] It has the ability to produce gas-phase ions with little or no molecular fragmentation. FAB ionization involves the bombardment of a liquid film containing the nonvolatile sample with a beam of energized atoms of xenon or argon. This method also is useful for thermally unstable compounds. SIMS is similar to FAB except that energetic ions (Xe^+ and Ar^+) instead of atoms are used in SIMS.[39]

Two important atmospheric pressure ionization techniques arose out of the need for an ionization source that provided even softer ionization (less fragmentation of the molecular ion) and as a convenient interface with a liquid chromatograph. With *electrospray ionization* (ESI), ions are generated in solution phase, then the carrier solvent is evaporated, and a gas-phase ion is produced.[40] In contrast to ESI, *atmospheric pressure chemical ionization* (APCI) is a gas-phase ionization process in which gas-phase molecules are isolated from the carrier solvent before ionization.[41] In general, ESI is more applicable to high molecular weight, more polar compounds because it requires less heat and can produce multiple charged ions, whereas APCI is more useful for less polar molecules. Nonetheless, for most compounds with some acidic or basic characteristics and with relatively low molecular weight, either technique is applicable.

In conjunction with HPLC profiling of metabolites, radioactively labeled drugs are useful to pinpoint retention times of metabolites for more focused mass spectrometric characterization. By incorporating a splitter into the HPLC sample stream that directs part of the effluent to a radioactivity detector and the rest to the mass spectrometer, simultaneous radioactivity and mass spectrometry monitoring can be carried out.[42]

In addition to tremendous advances in mass spectrometry, newer technologies in 2D and 3D nuclear magnetic resonance (NMR) spectrometry, particularly tandem LC-NMR (which became practical because of advances in solvent suppression techniques) have enhanced this analytical tool for studies in drug metabolism.[43] This continuous-flow method is particularly valuable, allowing ^1H and ^{19}F spectra to be obtained with only 5 ng or less of metabolite.[44] A mass spectrometer can be connected in tandem with the LC-NMR to give LC-NMR-MS spectrometry, which enables high-quality NMR and mass spectra to be obtained simultaneously from a single HPLC injection of biological fluid.[45]

7.3.D   Quantification

Quantification of drug metabolites is carried out by radioactive labeling, GC, HPLC, and mass spectrometry. The sensitivity and low volumes necessary for mass spectrometry reduce the assay development, sample preparation, and analysis time such that MS is ideally suited for the 96-well plate format of high-throughput metabolic screens.[46] In order for radioactive labeling techniques to be useful, the various radiolabeled metabolites are first separated by chromatography. Each is isolated and the rate of radioactive disintegration is determined by liquid scintillation counting methods. The amount of the metabolite isolated can be calculated from the specific radioactivity of the drug (see Section 7.2).

GC and HPLC both require the construction of a calibration curve of known quantities of a reference compound usually of similar structure to that of the metabolite. From the integration of the internal standard chromatography peak the amount of each metabolite formed can be determined.

Selected ion monitoring (SIM) is a highly selective method for detection and quantification of small quantities of metabolites. SIM uses a mass spectrometer as a selective detector of specific components in the effluent from a HPLC or gas chromatograph. By setting the spectrometer to detect characteristic fragment ions at a single m/z value, other compounds with the same retention times that do not produce those fragment ions will go undetected. When a full mass spectrum is recorded repetitively throughout a chromatogram, and a selected ion monitoring profile is reconstructed by computer, it is sometimes called *mass fragmentography*. Subpicogram quantities of metabolites in a mixture can be detected by the SIM method.

7.4 Pathways for Drug Deactivation and Elimination

7.4.A Introduction

The first mammalian drug metabolite that was isolated and characterized was hippuric acid (**7.8**) from benzoic acid in the early 19th century.[47] However, not until the late 1940s, when Mueller and Miller[48] demonstrated that the *in vivo* metabolism of 4-dimethyl-aminoazobenzene could be studied *in vitro* (see Section 7.4.B.1, p. 418), was the discipline of drug metabolism established. As a result of the ready commercial availability of radioisotopes and sophisticated separation, detection, and identification techniques that were developed in the latter half of the 20th century (see Section 7.3), drug metabolism studies have burgeoned.

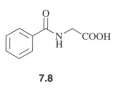

7.8

The function of drug metabolism is to convert a molecule that can cross biological membranes into one that is cleared, generally in the urine; each progressive metabolic step usually reduces the lipophilicity of the compound. The lipophilicity of the drug molecule will determine whether it undergoes direct renal clearance or is metabolically cleared. As the log $D_{7.4}$ (see Chapter 2, Section 2.2.F.2.b, p. 55) of the compound increases above zero, a marked decrease in direct renal clearance and a sharp increase in metabolic clearance occur (Figure 7.1 shows the results for a series of chromone-2-carboxylic acid derivatives),[49] indicating the contribution of lipophilicity to drug metabolism. Drug metabolism reactions have been divided into two general categories,[50] termed phase I and phase II reactions. *Phase I transformations* involve reactions that introduce or unmask a functional group, such as oxygenation or hydrolysis. *Phase II transformations* mostly generate highly polar derivatives (known as *conjugates*), such as glucuronides and sulfate esters, for excretion in the urine.

The rate and pathway of drug metabolism are affected by species, strain, sex, age, hormones, pregnancy, and liver diseases such as cirrhosis, hepatitis, porphyria, and hepatoma. Drug metabolism can have a variety of profound effects on drugs. It principally causes pharmacological deactivation of a drug by altering its structure so that it no longer interacts appropriately with the target receptor and becomes more susceptible to excretion. Drug metabolism, however, also can convert a pharmacologically inactive prodrug into an active

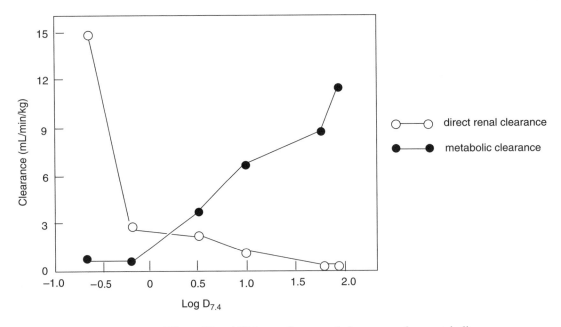

Figure 7.1 ► Effect of lipophilicity on direct renal clearance and on metabolism

drug (see Chapter 8). The pharmacological response of a drug may be altered if a metabolite has a new activity; in some cases, the metabolite has the same activity and a similar or different potency as the drug. A change in drug absorption and drug distribution (that is, the tissues or organs in which it is concentrated) also can result when it is converted into a much more polar species.

The majority of drug metabolizing enzymes also catalyze reactions on endogenous compounds. Consequently, the function of these enzymes may be metabolic disposition of endogenous cellular modulators, and it may be fortuitous that they also catalyze the metabolism of drugs and other xenobiotics. The greater affinity of the endogenous substrate over drugs in many cases seems to support this notion; however, many of these enzymes are very broad in specificity or are induced only on the addition of a particular chemical, so it is not clear that at least some of these enzymes have not evolved to protect the organism from undesirable substances.

As was discussed in Chapter 3 (Section 3.2.E.2, p. 143) and Chapter 4 (Section 4.1.B.1.a, p. 176), the interaction of a chiral molecule with a receptor or enzyme produces a diastereomeric complex. Therefore, it is not surprising that the processes in drug metabolism also are stereoselective, if not stereospecific.[51] *Stereoselectivity* can occur with enantiomers of drugs, in which case one enantiomer may be metabolized to a greater extent by one pathway and the other enantiomer predominantly by another pathway to give two different metabolites. Another type of stereoselectivity is the conversion of an achiral drug into a chiral metabolite. In both of these cases a difference in rates leads to unequal amounts of metabolites rather than exclusively to one metabolite. In many cases, however, a racemic drug is metabolized as if it were two different xenobiotics, each enantiomer displaying its own pharmacokinetic and pharmacodynamic profile. In fact, it was concluded by Hignite *et al.*[52] that "warfarin enantiomers should be treated as two drugs," Silber *et al.*[53] concluded that "*S*-(−)- and *R*-(+)-propranolol are essentially two distinct entities pharmacologically." These are just two of many examples of drugs whose enantiomers are metabolized by different routes. In some cases the inactive enantiomer can produce toxic metabolites which could be avoided by the administration of only the active enantiomer.[54] In other cases, the inactive isomer may inhibit the metabolism of the active isomer.

The metabolism of enantiomers may depend on the route of administration. For example, the racemic antiarrhythmic drug verapamil hydrochloride (**7.9**, Isoptin) is 16 times more potent when administered intravenously than when taken orally because of extensive hepatic presystemic elimination that occurs with oral administration (the first-pass effect; see Section 7.1).[55] The (−)-isomer, which is 10 times more potent than the (+)-isomer, is preferentially metabolized during hepatic metabolism and, therefore, there is much more of the less potent (+)-isomer available by the oral route than by the intravenous route.

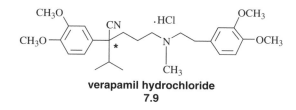

verapamil hydrochloride
7.9

In some cases one enantiomer of a drug can be metabolized to the other enantiomer. The therapeutically inactive *R*-isomer of the analgesic ibuprofen (**7.10**, Advil) is converted enzymatically[56] in the body to the active *S*-isomer.[57] The racemization occurs by initial conversion of the carboxylic acid group of ibuprofen to the corresponding *S*-CoA thioester, followed by racemization, then hydrolysis to the two enantiomers of ibuprofen.[58] If a racemic mixture is administered, a 70:30 mixture of *S*:*R* is excreted; if a 6:94 mixture of *S*:*R* is administered, an 80:20 mixture of *S*:*R* is excreted.

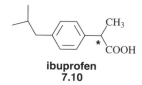

ibuprofen
7.10

In the next two sections the various types of metabolic biotransformations are described. I do not intend to imply that those reactions are the only ones that occur with those particular drugs; rather, they are just examples of the particular metabolic biotransformation under discussion. Some examples are included to show the effect of stereochemistry on drug metabolism. Keep in mind, however, that when a racemic mixture is administered, the metabolites observed may be different from those detected when a pure isomer is used. Many of the metabolites that are described are derived from *in vitro* studies (metabolites produced using either purified enzymes, subcellular fractions, whole cells, or tissue preparations). However, after analysis of a large amount of data on *in vitro* metabolism and a comparison with *in vivo* metabolism, it appears that it is valid to look at *in vitro* data as a reasonable guide for the prediction of *in vivo* metabolism.[59] It should be noted that some drugs are excreted without being metabolized at all.

7.4.B Phase I Transformations

B.1 Oxidative Reactions

Mueller and Miller[60] showed that the *in vivo* metabolism of 4-dimethylaminoazobenzene could be studied *in vitro* using rat liver homogenates. It was demonstrated that the *in vitro* system was functional only if nicotinamide adenine dinucleotide phosphate (NADP$^+$), molecular oxygen, and both the microsomal and soluble fractions from the liver homogenates were included. Later, Brodie and coworkers[61] found that the oxidative activity was in the microsomal fraction and that the soluble fraction could be replaced by either NADPH or a NADPH-generating system. This system was active toward a broad spectrum of structurally diverse compounds. Because it required both O_2 and a reducing system, it was classified as a *mixed function oxidase*;[62] that is, one atom of the O_2 is transferred to the substrate, and the other undergoes a two-electron reduction and is converted to water. This classification was confirmed when it was shown[63] that aromatic hydroxylation of acetanilide by liver microsomes in the presence of $^{18}O_2$ resulted in incorporation of one atom of ^{18}O into the product, and that a heme protein was an essential component for this reaction.[64] When this heme protein was reduced and exposed to carbon monoxide, a strong absorption in the visible spectrum at 450 nm resulted. Because of this observation, these microsomal oxidases were named *cytochrome P450*.

Cytochrome P450 now represents a superfamily of enzymes containing a heme cofactor but with structurally variable active sites that catalyze the same reaction on different substrates, namely, the oxidation of steroids, fatty acids, and xenobiotics.[65] These related cytochrome P450 enzymes are referred to as *isozymes*. Although about 500 genes encode different cytochrome P450 isozymes, only about 15 of these are very important in drug metabolism. The primary site for these enzymes is the liver, but they also are present in lung, kidney, gut, adrenal cortex, skin, brain, aorta, and other epithelial tissues. The heme is noncovalently bound to the apoprotein. Cytochrome P450 is associated with another enzyme, NADPH-cytochrome P450 reductase, a flavoenzyme that contains one molecule each of flavin adenine dinucleotide (FAD) and flavin mononucleotide (FMN).[66] Heme-dependent oxidation reactions were discussed in Chapter 4, Section 4.3.D, p. 212. As shown in Scheme 4.35, the NADPH-cytochrome P450 reductase reduces the flavin, which, in turn, transfers an electron to the heme–oxygen complex of cytochrome P450. Actually, the FAD accepts electrons from the NADPH, the FADH$^-$ then transfers electrons to the FMN, and the FMNH$^-$ donates the electron to the heme or heme–oxygen complex of cytochrome P450.[67]

In general, cytochrome P450 catalyzes either hydroxylation or epoxidation of various substrates (Table 7.1) and is believed to operate via radical intermediates (see Chapter 4, Scheme 4.35). When the concentrations of cytochrome P450 and other drug metabolizing enzymes are modified, drug metabolism becomes altered. Many drugs and environmental chemicals induce either their own metabolism or the metabolism of other drugs in humans as a result of their induction of cytochrome P450 and NADPH-cytochrome P450 reductase. These changes in pharmacokinetics and metabolism of the drugs when multiple drugs are taken together are known as *drug–drug interactions*.[68] One common mechanism leading to drug–drug interactions is the induction of cytochrome P450 isozymes by the administered drugs. For example, St. John's wort (*Hypericum perforatum*) is an herbal remedy for the treatment of depression,[69] which is sometimes referred to as Nature's Prozac. The constituent of this herb that contributed to the antidepressant activity, hyperforin (**7.11**),[70] was found to activate the pregnane X receptor, which is the key regulator of cytochrome P450 3A4 isozyme

TABLE 7.1 ▶ Classes of substrates for cytochrome P450

Functional Group	Product
R—⟨benzene ring⟩	R—⟨benzene ring⟩—OH
⟨alkene: R, R' / R, R'⟩	⟨epoxide: O, R', R, R'⟩
ArCH$_2$R	ArCHR, OH
R, R ⟨alkene⟩ CH$_2$R'	R, R ⟨alkene⟩ CHR', OH
⟨ketone: O=⟩ R—CH$_2$R'	⟨ketone: O=⟩ R—CHR', OH
RCH$_2$R'	RCHR', OH
RCH$_2$-X-R' (X = N, O, S, halogen)	(RCH-XR', OH) → RCHO + R'XH
R-X-R' (X = NR, S)	R—X⁺—R', O⁻

transcription.[71] Because of the increased expression of this P450 enzyme by hyperforin, drugs that are metabolized by this isozyme, which may be more than half of all drugs, are rapidly degraded when taken with St. John's wort.

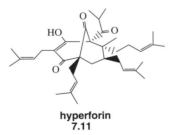

hyperforin
7.11

The other most common mechanism leading to drug–drug interactions arises when multiple drugs are administered and one of the drugs inhibits drug metabolism of the others as a result of its inhibition of cytochrome P450 isozymes or other enzymes. Adverse effects and toxicity of drugs from drug–drug interactions have been one of the major concerns in clinical practices and in new drug approval. A better characterization of the metabolic pathways and the enzymes involved in its metabolism is useful in understanding the underlying mechanisms of drug–drug interactions. On the bright side, as discussed in Chapter 5, Section 5.4.E.2, p. 269, a drug known to inhibit cytochrome P450 (such as ritonavir) could be used in combination with another drug (such as lopinavir) to block metabolism of the other drug intentionally.

Some properties of molecules lead to preferred metabolism by cytochrome P450. Lipophilicity is important for the binding of a molecule to cytochrome P450, which explains why increasing lipophilicity leads to an increase in metabolic clearance (see Figure 7.1, p. 416). The next most important physicochemical property of a molecule for cytochrome P450 metabolism is the presence of an ionizable group. This group may have a key role in binding to the active site and determine regioselectivity of the metabolic reaction.[72] If the ionizable group is a secondary or tertiary amine, it may direct the reaction to occur at that part of the molecule (see below). The reaction and site of reaction catalyzed by cytochrome P450s are determined by (1) the topography of the active site of the particular isozyme, (2) the degree of steric hindrance of the heme iron-oxo species to the site of reaction, and (3) the ease of hydrogen atom abstraction or electron transfer from the compound that is metabolized by the isozyme.

Another important family of enzymes involved in drug oxidation is the microsomal *flavin monooxygenases*[73] (see Chapter 4, Section 4.3.C, p. 205); mechanism of oxidation was described in Scheme 4.34, p. 212. According to this scheme, the flavin peroxide intermediate is an electrophilic species, indicating that the substrates for this enzyme are nucleophiles such as amines, thiols, and related compounds (Table 7.2). The enzyme contains one FAD molecule per subunit and, as in the case of cytochrome P450, requires NADPH to reduce the flavin. It has been found that nucleophilic compounds containing an anionic group are excluded from the active site of this enzyme. Because most endogenous nucleophiles contain negatively charged groups, this may be how Nature prevents normal cellular components from being oxidized by this enzyme. Oxygenation of some of these compounds leads to reactive intermediates,[74] for example, mercaptopyrimidines and thiocarbamides. Metabolism by flavin monooxygenases often can be differentiated from other oxygenases because of its unique stereoselectivity. For example, cytochrome P450s oxidize (*S*)-nicotine (**7.12**) to a mixture of *cis*- (**7.13**) and *trans*-*N*-1′-oxides (**7.14**), but flavin monooxygenase oxidizes it exclusively to the *trans*-*N*-1′-oxide.

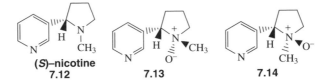

(*S*)–nicotine
7.12   **7.13**   **7.14**

Other enzymes involved in oxidative drug metabolism include prostaglandin H synthase (see Section 7.4.B.1.e, p. 430), alcohol dehydrogenase, aldehyde dehydrogenase (see Section 7.4.B.1.i, p. 447), xanthine oxidase, monoamine oxidase, and aromatase. These enzymes, however, are involved, for the most part, in the metabolism of endogenous compounds.

a. Aromatic Hydroxylation

In 1950 Boyland[75] hypothesized that aromatic compounds were metabolized initially to the corresponding epoxides (*arene oxides*). This postulate was confirmed in 1968 by a group at the National Institutes of Health (NIH) who isolated naphthalene 1,2-oxide from the microsomal oxidation of naphthalene[76] (Scheme 7.3). Kinetic isotope effect studies[77] now, however, indicate that a direct arene epoxidation is a highly unlikely process. Instead, an activated

TABLE 7.2 ▶ Classes of substrates for flavin monooxygenase

Functional Group	Product
R-NR'$_2$	R-$\overset{\text{O}^-}{\underset{+}{\text{N}}}R'_2$
R-NHR'	R-N$\overset{\text{OH}}{|}$R'
R-N$\overset{\text{OH}}{|}$R'	R=$\overset{\text{O}^-}{\underset{+}{\text{N}}}$R'
$\diagup\!\!=$NH	$\diagup\!\!-$NHOH
R-N-NHR' with R''	R-N$\overset{\text{O}^-}{\underset{|}{|+}}$-NHR' with R''
R-C$\overset{\text{S}}{\|}$NH$_2$	R-C$\overset{^+\text{S}\diagup\text{O}^-}{\|}NH_2$
RNH, R'N+H, -SH	RNH, R'N+H, -SO$_2^-$
2RSH	RSSR
RSSR	2 RSO$_2^-$

heme iron-oxo species may undergo electrophilic addition to the aromatic ring to give either a tetrahedral intermediate radical (**7.15**) or cation (**7.16**) (Scheme 7.4), similar to the mechanism described for the addition of the corresponding heme iron-oxo species to alkenes as was discussed in Chapter 4 (Section 4.3.D; Scheme 4.37). Because [1,2]-shifts of hydrogen and alkyl radicals are energetically unfavorable, an electron could be transferred to the Fe^{IV} of the heme, giving a carbocation species bound to Fe^{III} (**7.16**). A [1,2]-shift of hydride to a cation is energetically favorable, leading to the cyclohexadienone (**7.17**, pathway c), which would tautomerize to the phenol (**7.18**). Arene oxide (**7.19**) formation may arise from either **7.15** or **7.16**. Usually, arene oxides can undergo rearrangements to arenols, hydration (catalyzed by epoxide hydrolase) to the corresponding *trans*-diol, reaction with glutathione (catalyzed by glutathione *S*-transferase) to the β-hydroxy sulfide, and reactions with various macromolecular nucleophiles (Scheme 7.5).

The rearrangement of an arene oxide to an arenol is known as the *NIH shift* because a research group at the NIH proposed that mechanism based on studies with specifically deuterated substrates (Scheme 7.6).[78] Ring opening occurs in the direction that gives the more stable carbocation (*ortho*- or *para*-hydroxylation when R is electron donating). Also,

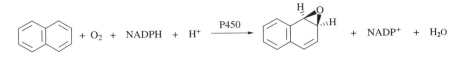

Scheme 7.3 ▶ Cytochrome P450 oxidation of naphthalene

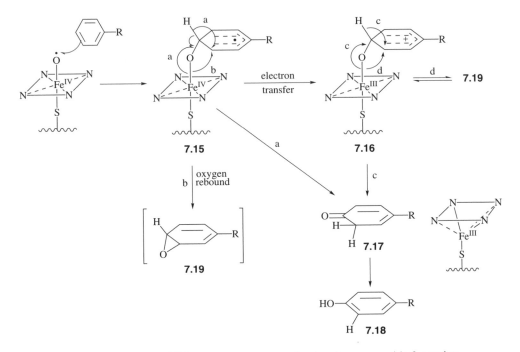

Scheme 7.4 ▶ Addition-rearrangement mechanism for arene oxide formation

because of an isotope effect on cleavage of the C–D bond, the proton is preferentially removed, leaving the migrated deuterium. Although there is an isotope effect on the cleavage of the C–D versus the C–H bond, this step is not the rate-determining step in the overall reaction; consequently, there is no overall isotope effect on this oxidation pathway when deuterium is incorporated into the substrate.

In competition with the [1,2]-hydride (deuteride) shift is deprotonation (dedeuteronation; Scheme 7.7). The percentage of each pathway depends on the degree of stabilization of the intermediate carbocation by R; the more stabilization (i.e., the greater electron donation of R), the less need for the higher energy hydride shift, and the more deprotonation (dedeuteronation) occurs.[79] For example, when R = NH_2, OH, $NHCOCF_3$ and $NHCOCH_3$, only 0–30% of the product phenols retain deuterium (NIH shift), but when R = Br, $CONH_2$, F, CN, and Cl, 40–54% deuterium retention is observed.

Whenever a deuterium is incorporated into a drug for metabolism studies, however, the problem of *metabolic switching* has to be considered. This is when deuteration at one site in the molecule changes the partition between two metabolic pathways because the deuterium isotope effect on C–H bond cleavage at the deuterated site leads to an increase in metabolism at a different site in the molecule.[80]

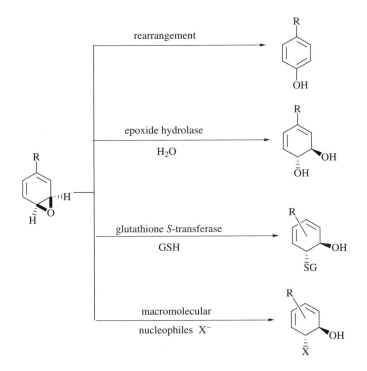

Scheme 7.5 ▶ Possible fates of arene oxides

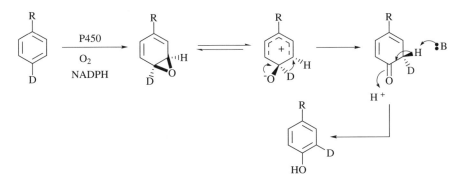

Scheme 7.6 ▶ Rearrangement of arene oxides to arenols (NIH shift)

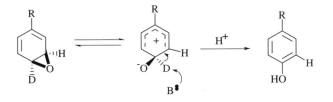

Scheme 7.7 ▶ Competing pathway for the NIH shift

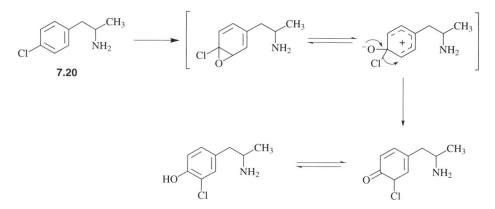

Scheme 7.8 ▶ NIH shift of a chloride ion

The NIH shift also occurs with substituents as well as with hydrogen. For example, rat liver metabolizes *p*-chloroamphetamine (**7.20**) to 3-chloro-4-hydroxyamphetamine[81] (Scheme 7.8). However, a common approach to slow down metabolism of aromatic compounds is to substitute the ring with a *para*-fluorine[82] or *para*-chlorine atom, which deactivates the ring and decreases the rate of oxidative metabolism.[83] For example, the anti-inflammatory drug diclofenac (**7.21**, Voltaren) is metabolized to 4-hydroxydiclofenac with a half-life of 1 hour. The related analog with a *para*-chloro substituent, fenclofenac (**7.22**, Flenac), has a half-life of more than 20 hours.[84]

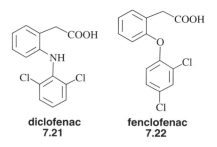

diclofenac   fenclofenac
7.21   **7.22**

A related substituent NIH shift was observed in the metabolic oxidation of the antiprotozoal agent tinidazole[85] (**7.23**, Scheme 7.9, Fasigyn).

As in the case of electrophilic aromatic substitution reactions, it appears that the more electron rich the aromatic ring (R is electron donating), the faster the microsomal hydroxylation will be.[86] Aniline (electron rich), for example, undergoes extensive *ortho*- and *para*-hydroxylation,[87] whereas the strongly electron-poor uricosuric drug probenecid (**7.24**, Benemid) undergoes no detectable aromatic hydroxylation.[88] In the case of drugs with two aromatic rings, the more electron-rich one (or less electron poor), generally, is hydroxylated. The antipsychotic drug chlorpromazine (**7.25**, R = H, Thorazine), for example, undergoes 7-hydroxylation (**7.25**, R = OH).[89] However, the binding orientation in the active site of the hydroxylases, as well as the activation of the aromatic rings, plays an important role in both the rate and position of hydroxylation on the aromatic ring.

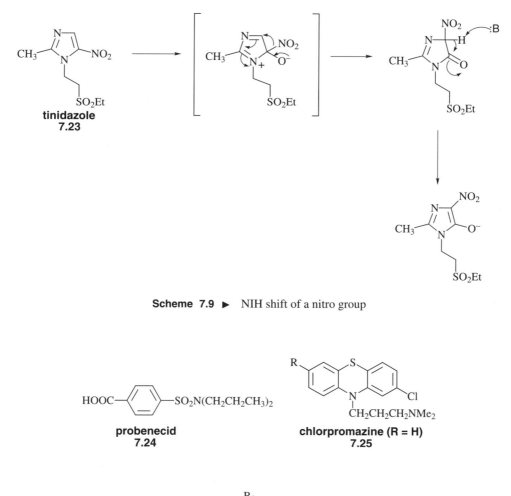

Scheme 7.9 ▶ NIH shift of a nitro group

probenecid
7.24

chlorpromazine (R = H)
7.25

phenytoin (R₁ = R₂ = R₃ = H)
7.26

Aromatic hydroxylation, as is the case for all metabolic reactions, is species specific. In humans, *para*-hydroxylation is a major route of metabolism for many phenyl-containing drugs. The site and stereoselectivity of hydroxylation depend on the animal studied. In humans, the antiepilepsy drug phenytoin (**7.26**; $R_1 = R_2 = R_3 = H$, Dilantin) is *para*-hydroxylated at the *pro-S* phenyl ring (**7.26**, $R_1 = OH$, $R_2 = R_3 = H$) 10 times more often than is the *pro-R* ring (**7.26**, $R_3 = OH$, $R_1 = R_2 = H$). In dogs, however, *meta*-hydroxylation of the *pro-R* phenyl ring (**7.26**, $R_2 = OH$, $R_1 = R_3 = H$) is the major pathway. The overall ratio of R_2:R_1:R_3 hydroxylation in dogs is 18:2:1.[90] *Meta*-hydroxylation may be catalyzed by an isozyme of cytochrome P450 that operates by a mechanism different from arene oxide formation.[91] For example, an important metabolite of chlorobenzene is 3-chlorophenol, but it was shown that neither 3- nor 4-chlorophenol oxide gave 3-chlorophenol in the presence of rat liver microsomes, suggesting that a direct oxygen insertion mechanism may be operative in this case. This mechanism also is consistent with the observation of a kinetic isotope effect $(k_H/k_D = 1.3-1.75)$ on the *in vivo* 3-hydroxylation.[92] An alternative explanation for *meta*-hydroxylation is that the *meta*-position is the most available site for electrophilic addition by the heme-oxo species.

Because of the reactivity of arene oxides, they can undergo rapid reactions with nucleophiles. If cellular nucleophiles react with these compounds, toxicity can result. Consequently, there are enzymes that catalyze deactivation reactions on these reactive species (see Scheme 7.5). Epoxide hydrolase (also referred to as epoxide hydratase or epoxide hydrase) catalyzes the hydration of highly electrophilic arene oxides to give *trans*-dihydrodiols.[93] The mechanism of this reaction is initial attack by an active site aspartate to give a covalent catalytic intermediate, which hydrolyzes to the *trans*-glycol (Scheme 7.10);[94] attack occurs predominantly at the less sterically hindered side.[95] The *trans*-dihydrodiol product can be oxidized further to give catechols; catechols also are generated by hydroxylation of arenols. Because of the instability of catechols to oxidation, they may be converted either to *ortho*-quinones or semiquinones.

Glutathione *S*-transferase is an important enzyme that protects the cell from the electrophilic arene oxide metabolites.[96] It catalyzes a nucleophilic substitution reaction of glutathione on various electrophiles.[97] Glutathione metabolites of naphthalene are shown in Scheme 7.11; only one of them was not detected.[98] Other reactions catalyzed by glutathione *S*-transferases are discussed in Section 7.4.C.5.

When the arene oxide escapes destruction by these enzymes, toxicity may result. An important example of this is the metabolism of benzo(a)pyrene (**7.27**, Scheme 7.12), a potent carcinogen found in soot and charcoal. The relationship between soot and cancer was noted by Sir Percival Scott, a British surgeon, in 1775 when he observed that chimney sweeps (people who cleaned out chimneys) frequently developed skin cancer. Metabolic activation of polyaromatic hydrocarbons can lead to the formation of covalent adducts with RNA, DNA, and proteins.[99] Covalent binding of these metabolites to DNA is the initial event that is responsible for malignant cellular transformation.[100] The key reactive intermediate responsible for alkylation of nucleic acids is $(+)$-7R,8S-dihydroxy-9R, 10R-oxy-7,8,9, 10-tetrahydrobenzo(a)pyrene (**7.28**, Scheme 7.12). This metabolite reacts with RNA to form a covalent adduct between the C-2 amino group of a guanosine and C-10 of the hydrocarbon[101]

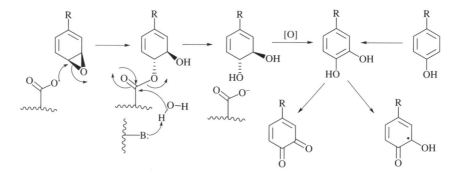

Scheme 7.10 ▶ Metabolic formation and oxidation of catechols

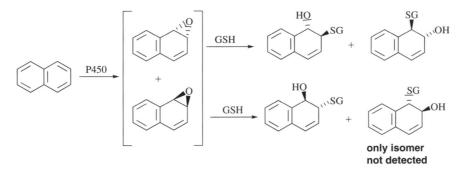

Scheme 7.11 ▶ Formation of glutathione adducts from naphthalene oxides

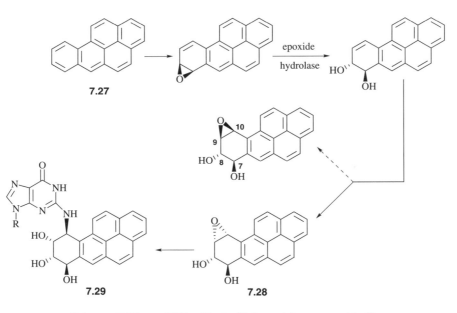

Scheme 7.12 ▶ RNA adduct with benzo(a)pyrene metabolite

(**7.29**). The reactive metabolite (**7.28**) also causes nicks in superhelical DNA of *E. coli*; an adduct between the *C*-2 amino group of deoxyguanosine and the *C*-10 position of the diol epoxide was isolated.[102]

b. Alkene Epoxidation

Because alkenes are more reactive than aromatic π-bonds, it is not surprising that alkenes also are metabolically epoxidized. An example of a drug that is metabolized by alkene epoxidation is the anticonvulsant agent carbamazepine (**7.30**, Scheme 7.13, Tegretol).[103] Carbamazepine-10, 11-epoxide (**7.31**) has been found to be an anticonvulsant agent as well, so the metabolite may be responsible for the anticonvulsant activity of carbamazepine.[104] The epoxide is converted stereoselectively into the corresponding (10S,11S)-diol (**7.32**) by epoxide hydrolase.

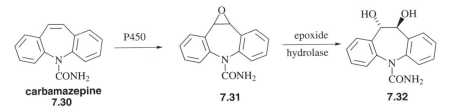

Scheme 7.13 ▶ Metabolism of carbamazepine

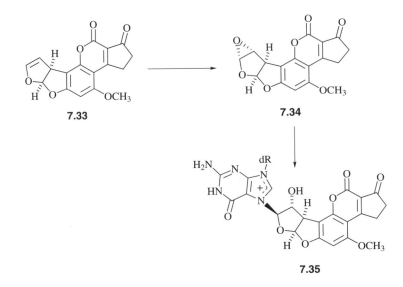

Scheme 7.14 ▶ Metabolic reactions of aflatoxin B₁

Metabolic epoxidation of an alkene also can lead to the formation of toxic products. The enol ether mycotoxin aflatoxin B₁ (**7.33**) is a mutagen in several strains of bacteria,[105] a hepatocarcinogen in animals,[106] and causes cancer in humans.[107] Aflatoxin B₁ is metabolized by a cytochrome P450 to the corresponding *exo*-8,9-epoxide (**7.34**),[108] which becomes covalently bound to cellular DNA (Scheme 7.14). With the use of radiolabeled **7.33**[109] and model studies,[110] it was shown that a covalent bond is formed between *C*-8 of aflatoxin B₁ and the *N*-7 of a guanine residue in DNA (**7.35**). The precursor epoxide **7.34**, has been synthesized and shown to undergo the expected reaction with a guanine-containing oligonucleotide. Mechanistic studies show that only the *exo* epoxide reacts with DNA, giving the *trans* adduct.[111]

c. Oxidations of Carbons Adjacent to sp² Centers

Carbons adjacent to aromatic, olefinic, and carbonyl or imine groups (all sp^2 hybridized) undergo metabolic oxidations. The oxidation mechanism is not clearly understood, but because the ease of oxidation parallels the C–H bond dissociation energies, it is likely that a typical cytochrome P450 oxidation is responsible.

Examples of benzylic oxidation are the metabolism of the antidepressent drug amitriptyline (**7.36**, R = R = H, Elavil), which is oxidized to **7.36** (R = H, R = OH and R = OH, R = H)[112] and the β_1-adrenoreceptor antagonist (β-blocker) antihypertensive drug metoprolol (**7.37**, Toprol-XL).[113] In the case of metoprolol both enantiomers (1R and 1S) of the hydroxylated drug are formed, but 1R-hydroxylation occurs to a greater extent. Furthermore, the ratio of the 1R- to 1S-isomers depends on the stereochemistry at the 2-position. 2R-Metoprolol gives a ratio of (1R, 2R)/(1S, 2R) metabolites of 9.4, whereas 2S-metoprolol gives a ratio of (1R, 2S)/(1S, 2S) of 26. Therefore, the stereochemistry in the methoxyethyl side chain is influenced by the stereochemistry in the *para* side chain. Hydroxylations generally are highly stereoselective, if not stereospecific. The stereochemistry at one part of the molecule may influence how the molecule binds in the active site of cytochrome P450, which will influence what part of the molecule is closest to the heme-oxo species for reaction, in this case, hydrogen atom abstraction.

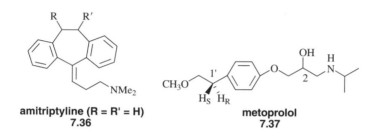

amitriptyline (R = R' = H)
7.36

metoprolol
7.37

The cytochrome P450-catalyzed metabolism of the antiarrhythmic drug quinidine (**7.38**, R = H, Quinidex) leads to allylic oxidation (**7.38**, R = OH).[114] The psychoactive constituent of marijuana, $\triangle^9$-tetrahydrocannabinol (**7.39**, R = R' = H) is extensively metabolized to all stereoisomers of **7.39** (R = H, R' = OH; R = OH, R' = H; and R = R' = OH).[115]

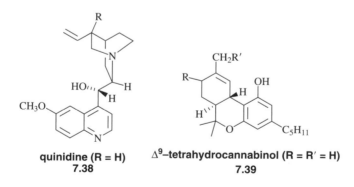

quinidine (R = H)
7.38

$\triangle^9$–**tetrahydrocannabinol (R = R' = H)**
7.39

The sedative-hypnotic, (+)-glutethimide (**7.40**, R = R' = H; Doriden) is converted to 5-hydroxyglutethimide (**7.40**, R = OH, R' = H).[116] This metabolite is pharmacologically active and may contribute to the comatose state of individuals who have taken toxic overdoses of the parent drug. The (−)-isomer is enantioselectively hydroxylated at the ethyl group to give **7.40** (R = H, R' = OH). The difference in hydroxylation sites for the two enantiomers may reflect the different orientations of binding of each enantiomer in one or more isozymes of cytochrome P450.

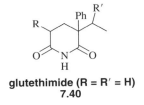

glutethimide (R = R′ = H)
7.40

d.   Oxidation at Aliphatic and Alicyclic Carbon Atoms

Metabolic oxidation at the terminal methyl group of an aliphatic side chain is referred to as *ω oxidation* and oxidation at the penultimate carbon is *ω−1 oxidation*. Because *ω* oxidation is a chemically unfavorable process (a primary radical or cation would be formed), the active site of the enzyme must be favorably disposed for this particular regiochemistry to proceed. In the case described above of (−)-glutethimide (**7.40**, R = R′ = H), *ω*−1 oxidation occurred. The anticonvulsant drug valproic acid (**7.41**, Depakene) undergoes both *ω* and *ω*−1 oxidations.[117]

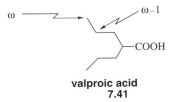

valproic acid
7.41

The coronary vasodilator perhexiline (**7.42**, R = H, Pexid) is metabolized to the alicyclic alcohol **7.42** (R = OH).[118]

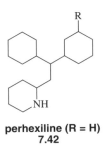

perhexiline (R = H)
7.42

If the carbon is beta to a carbonyl, hydroxylation may result in *C*-demethylation by a retro-aldol reaction.[119] For example, the *para*-amino metabolite of flutamide (**7.43**, Eulexin) undergoes a double *C*-demethylation, presumably via the *C*-hydroxylated intermediate (**7.44**) to give **7.45** (Scheme 7.15), which may occur with loss of formaldehyde. Further *C*-demethylation gives the isolated metabolite **7.46**.

e.   Oxidations of Carbon-Nitrogen Systems

The metabolism of organic nitrogen compounds is very complex. Two general classes of oxidation reactions will be considered for primary, secondary, and tertiary amines, respectively, namely, carbon- and nitrogen-oxidation reactions that lead to C−N bond cleavage and *N*-oxidation reactions that do not lead to C−N bond cleavage.

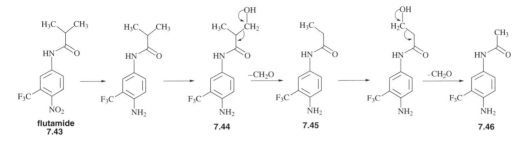

Scheme 7.15 ▶ *C*-demethylation of a flutamide metabolite

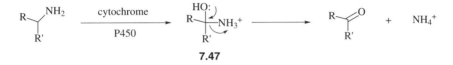

Scheme 7.16 ▶ Oxidative deamination of primary amines

TABLE 7.3 ▶ **Oxidative reactions of primary amines and amides**

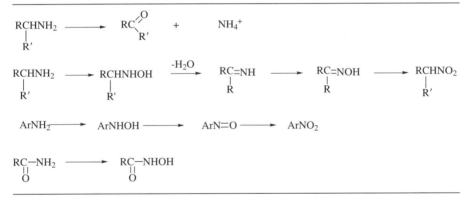

Primary amines and/or amides are metabolized by the oxidation reactions shown in Table 7.3. Primary aliphatic and arylalkyl amines having at least one α-carbon-hydrogen bond undergo cytochrome P450-catalyzed hydroxylation at the α-carbon (see Scheme 4.36, p. 213, in Chapter 4 for the general mechanism of carbon hydroxylation) to give the carbinolamine (**7.47**), which generally breaks down to the aldehyde or ketone and ammonia (Scheme 7.16). This process of oxidative cleavage of ammonia from the primary amine is known as *oxidative deamination*. As is predicted by this mechanism, primary aromatic amines and α,α-disubstituted aliphatic amines do not undergo oxidative deamination. A variety of endogenous arylalkyl amines, such as the neurotransmitters dopamine, norepinephrine, and serotonin, are oxidized by monoamine oxidase by an electron transfer mechanism (see Chapter 4, Scheme 4.32, p. 211, for the mechanism). This enzyme also may be involved in drug metabolism of arylalkyl amines with no α-substituents. An example of cytochrome P450-catalyzed primary amine oxidative deamination is the metabolism of amphetamine (**7.48**) to give 1-phenyl-2-propanone and ammonia.[120]

amphetamine
7.48

Another important oxidative pathway for primary amines is *N-oxidation* (hydroxylation of the nitrogen atom) to the corresponding hydroxylamine,[121] usually catalyzed by flavin monooxygenases (see Chapter 4, Scheme 4.34 for mechanism). It has been suggested[122] that the basic amines (pK_a 8–11) are oxidized by the flavoenzymes, nonbasic nitrogen-containing compounds such as amides are oxidized by cytochrome P450 enzymes, and compounds with intermediate basicity, such as aromatic amines, are oxidized by both enzymes. Cytochrome P450 enzymes, however, tend not to catalyze *N*-oxidation reactions when there are α-protons available. Amphetamine undergoes *N*-oxidation to the corresponding hydroxylamine (**7.49**, Scheme 7.17), the oxime (**7.50**), and the nitro compound (**7.51**). It could be argued that the 1-phenyl-2-propanone stated above as the product of oxidative deamination could be derived from hydrolysis of the oxime **7.50** (see Scheme 7.17). However, *N*-hydroxyamphetamine (**7.49**) metabolism to the oxime occurs only to a small extent, and the oxime is not hydrolyzed to the ketone *in vitro*.[123] Also, *in vitro* metabolism of amphetamine in $^{18}O_2$ leads to substantial enrichment of 1-phenyl-2-propanone with ^{18}O, indicating the relevance of a cytochrome P450-catalyzed oxidative deamination pathway to the ketone rather than a pathway involving hydrolysis of the oxime.

The amphetamine oxime is derived both from hydroxylation of **7.49** (pathway a) and from dehydration of **7.49** (pathway b) followed by hydroxylation of the imine (see Scheme 7.17).

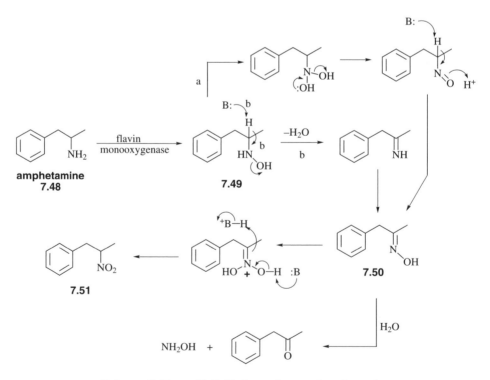

Scheme 7.17 ▶ *N*-Oxidation pathways of amphetamine

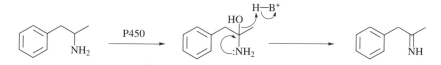

Scheme 7.18 ▶ Amphetamine imine formation via the carbinolamine

TABLE 7.4 ▶ **Oxidative reactions of secondary amines and amides**

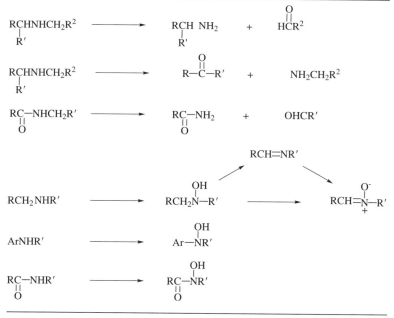

The imine also could be derived from dehydration of the carbinolamine (Scheme 7.18). Hydrolysis of the imine could lead to 1-phenyl-2-propanone.

Secondary aliphatic amines are metabolized by three different oxidative reactions: oxidative N-dealkylation, oxidative deamination, and N-oxidation (Table 7.4). When a small alkyl substituent of a secondary amine is cleaved from the parent amine to give a primary amine, the process is known as *oxidative N-dealkylation* (Scheme 7.19). On the basis of the reactions shown in Schemes 7.16 and 7.19, it is apparent that oxidative deamination and oxidative N-dealkylation are really the same reaction. The metabolic difference between the two is that in the former reaction a primary amine is oxidized to an aldehyde or ketone and ammonia; in the latter, secondary amines are converted to primary amines. Although mostly a matter of chemical semantics, oxidative N-dealkylation is a process in which a small alkyl group is cleaved from the main amine compound, whereas in oxidative deamination the main amine compound is oxidized with the loss of ammonia (in the case of a primary amine) or a small primary amine fragment (from a more complex secondary amine). It is not clear if the precise mechanisms of these two processes are the same, but from a reaction standpoint they are the same when viewed from the two sides of the nitrogen atom.

The secondary amine β-blocker propranolol (**7.52**, Inderal) is metabolized both by oxidative N-dealkylation (pathway a, Scheme 7.20) and by oxidative deamination (pathway b, Scheme 7.20).[124]

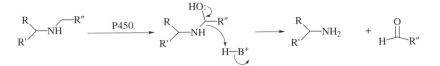

Scheme 7.19 ▶ Oxidative *N*-dealkylation of secondary amines

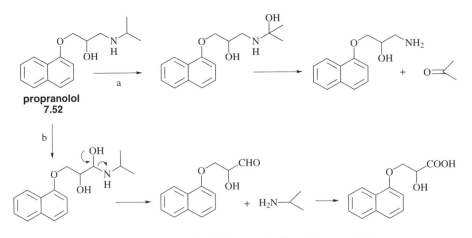

propranolol
7.52

Scheme 7.20 ▶ Oxidative metabolism of propranolol

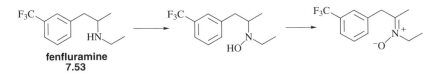

fenfluramine
7.53

Scheme 7.21 ▶ *N*-Oxidation of fenfluramine

Aldehydic metabolites often are oxidized by soluble aldehyde dehydrogenases to the corresponding carboxylic acids (see pathway b).

N-Oxidation of secondary amines leads to a variety of *N*-oxygenated products.[125] Secondary hydroxylamine formation is common, but these metabolites are susceptible to further oxidation to give nitrones. An example of this is in the metabolism of the anorectic drug fenfluramine (**7.53**, Scheme 7.21, Pondimin).[126]

Tertiary aliphatic amines generally undergo two metabolic reactions: oxidative *N*-dealkylation and *N*-oxidation reactions (Table 7.5). Oxidative *N*-dealkylation of tertiary amines (same mechanism as Scheme 7.19 for secondary amines) leads to the formation of secondary amines. Because of the basicity of tertiary amines relative to secondary amines, oxidative *N*-dealkylation of tertiary amines generally occurs more rapidly than that for secondary amines.[127] For example, the primary amine antihypertensive drug (calcium ion channel blocker) amlodipine (**7.54**, R = NH$_2$, Norvasc) is much more stable metabolically (11 mL/min/kg) than the corresponding secondary (12.5 mL/min/kg) or tertiary (25 mL/min/kg) analogs (**7.54**, R = NHCH$_3$ or N(CH$_3$)$_2$, respectively).[128] The low rate of metabolism may not solely be a result of the lower basicity of the primary amino functionality; it could also be because of the increased polarity.[129]

TABLE 7.5 ► Oxidative reactions of tertiary amines and amides

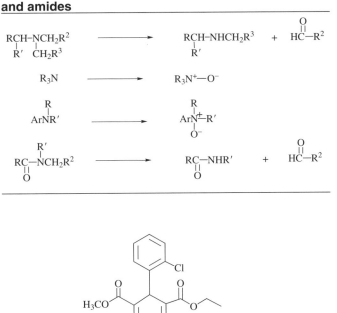

amlodipine (R = NH₂)
7.54

The tricyclic antidepressant drug imipramine (**7.55**, R = CH₃, Tofranil) is metabolized to the corresponding secondary amine, desmethylimipramine (desipramine, **7.55**, R = H, Norpramin) which also is an active antidepressant agent.[130] Very little oxidative *N*-demethylation of **7.55** (R = H) occurs. Enantioselective oxidative *N*-dealkylation can occur with chiral tertiary amines. The (*2S, 3R*)-(+)-enantiomer of propoxyphene (**7.56**, Darvon), an analgesic drug, is *N*-demethylated more slowly than the nonanalgesic (−)-enantiomer.[131]

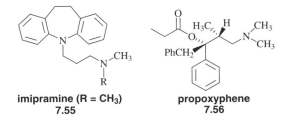

imipramine (R = CH₃)
7.55

propoxyphene
7.56

The antiparkinsonian drug selegiline (*R*-(−)-deprenyl, **7.57**, Scheme 7.22; Eldepryl) is a potent inactivator of MAO B (see Chapter 5, Section 5.5.C.3.d, p. 295). It is metabolized by various isozymes of cytochrome P450 to *R*-(−)-metamphetamine (**7.58**) and to *R*-(−)-amphetamine, which have only a weak CNS stimulant side effect.[132] The *S*-(+)-isomer is only a weak inhibitor of MAO B, but is metabolized by cytochrome P450 to *S*-(+)-metamphetamine, which is converted to *S*-(+)-amphetamine, both of which would produce an undesirable CNS stimulant side effect.[133] Because of this, only the (−)-enantiomer is used in the treatment of Parkinson's disease.

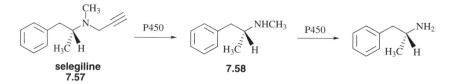

Scheme 7.22 ▶ Metabolism of selegiline (deprenyl)

When an alicyclic tertiary amine undergoes oxidation of the alicyclic carbon atoms attached to the nitrogen, a variety of oxidation products can result. Nicotine (**7.59**, Scheme 7.23) is α-hydroxylated both in the pyrrolidine ring (pathway a) and at the methyl carbon (pathway b). The carbinolamine **7.60** is oxidized further to the major metabolite, cotinine (**7.61**). Cotinine also can undergo hydroxylation, leading to a minor metabolite, γ-(3-pyridyl)-γ-oxo-N-methylbutyramide (**7.62**).[134] Carbinolamine **7.60** is in equilibrium with the iminium ion **7.63**, which has been trapped with cyanide ion *in vitro* to give **7.64**.[135] Oxidative N-demethylation of **7.59** to **7.65** also is observed as a metabolic pathway for nicotine.

Further evidence for metabolic iminium ion intermediates generated during tertiary amine oxidation is the isolation of the imidazolidinone **7.67** from the metabolism of the local anesthetic lidocaine[136] (**7.66**, Scheme 7.24; Xylocaine).

N-Oxidation of tertiary amines gives chemically stable tertiary amine N-oxides that do not undergo further oxidation unlike N-oxidation of primary and secondary amines. The antihypertensive drug guanethidine (**7.68**, Ismelin) is oxidized at the tertiary cyclic amine atom to give the N-oxide **7.69**.[137] The pyrrolidine nitrogen of nicotine (**7.59**) also is N-oxidized to 1R,2S- and 1S,2S-nicotine 1-N-oxide. One of the major metabolites of the antihistamine cyproheptadine (**7.70**, Periactin) in dogs is the α-N-oxide **7.71**; no β-N-oxide was detected.[138] Hydrogen peroxide oxidation of **7.70**, however, gives both α- and β-N-oxides.[139] N-Oxides are susceptible to bioreduction (see Section 7.4.B.2.e, p. 452), which regenerates the parent amine.

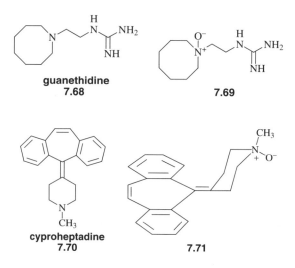

Aromatic amine oxidation is similar to that for aliphatic amines. N-Oxidation of aromatic amines appears to be an important process related to the carcinogenic and cytotoxic

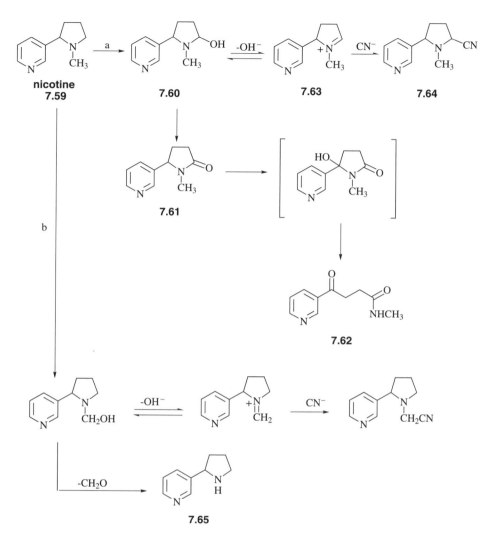

Scheme 7.23 ▶ Oxidative metabolism of nicotine leading to C–N bond cleavage

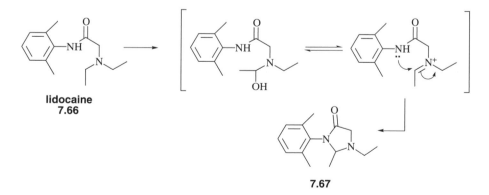

Scheme 7.24 ▶ Metabolism of lidocaine

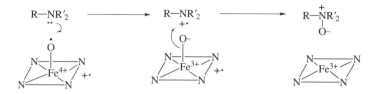

Scheme 7.25 ▶ Mechanism for cytochrome P450-catalyzed *N*-oxidation of tertiary aromatic amines

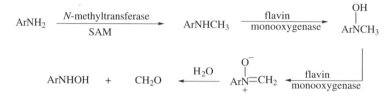

Scheme 7.26 ▶ Possible mechanism for *N*-oxidation of primary arylamines

properties of aromatic amines.[140] Two enzyme systems are responsible for *N*-oxidation of tertiary aromatic amines, a flavoprotein monooxygenase and cytochrome P450. The mechanism for the flavin-dependent reaction was discussed in Chapter 4, Section 4.3.C, p. 205 (see Scheme 4.34). Cytochrome P450-catalyzed *N*-oxidation occurs by an electron transfer reaction followed by oxygen rebound (Scheme 7.25). *N*-Oxidation by cytochrome P450 appears to occur only if there are no α-hydrogens available for abstraction, if the iminium radical is stabilized by electron donation, or if Bredt's rule prevents α-hydrogen abstraction.[141]

The mechanism for oxidation of primary arylamines may be different from that of secondary and tertiary arylamines. Most primary arylamines are not substrates of flavin monooxygenases; *N*-oxygenation of secondary and tertiary arylamines occurs readily. Ziegler *et al.*[142] showed that primary arylamines could be *N*-methylated by *S*-adenosylmethionine-dependent *amine N-methyltransferases*, and that the secondary arylamines produced were substrates for flavin monooxygenases (Scheme 7.26). The secondary hydroxylamine products of this *N*-oxygenation reaction are then further oxidized to nitrones, which, on hydrolysis, give the primary hydroxylamines and formaldehyde.

N-Demethylation of tertiary aromatic amines is believed to proceed by two mechanisms[143] (Scheme 7.27), one that involves intermediate carbinolamine formation by cytochrome P450 oxidation (pathway a) and one from flavin monooxygenation that produces the *N*-oxide, which rearranges to the carbinolamine (pathway b). Evidence to support carbinolamine formation is the isolation of the ^{18}O-labeled carbinolamine (**7.72**, R = OH) during cytochrome P450-catalyzed oxidation of *N*-methylcarbazole with $^{18}O_2$ (**7.72**, R = H).[144] Intrinsic isotope effects on the cytochrome P450-catalyzed oxidation of tertiary aromatic amines were found to be small.[145] Intrinsic isotope effects associated with deprotonation of aminium radicals are of low magnitude ($k_H/k_D < 3.6$), but those associated with direct hydrogen atom abstraction are large; this suggests that the mechanism for carbinolamine formation involves electron transfer and proton transfer (Scheme 7.28).

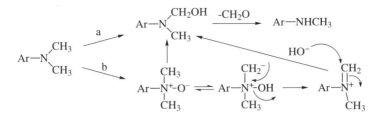

Scheme 7.27 ▶ Two pathways to *N*-demethylation of tertiary aromatic amines

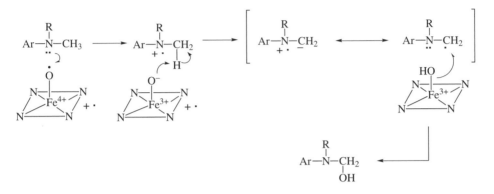

Scheme 7.28 ▶ Mechanism of carbinolamine formation during oxidation of tertiary aromatic amines

7.72

N-Oxidation of primary and secondary aromatic amines can lead to the generation of reactive electrophilic species that form covalent bonds to cellular macromolecules (Scheme 7.29).[146] The X^+ in Scheme 7.29 represents sulfation or acetylation of the hydroxylamine to give a good leaving group (OX^-) (see Sections 7.4.C.3 and 7.4.C.7). Attachment of aromatic amines to proteins, DNA, and RNA is known.[147]

Although C–N bond cleavage of heterocyclic aromatic nitrogen-containing compounds does not occur, *N*-oxidation to the aromatic *N*-oxide is prevalent. As noted in Chapter 5, Section 5.4.E.2, p. 269, for the discovery of the anti-HIV agent ritonavir, compounds with a pyridine substituent underwent extensive metabolic *N*-oxide formation.[148]

Amides also are metabolized both by oxidative *N*-dealkylation and *N*-oxidation (see Tables 7.3, 7.4, and 7.5). The sedative diazepam (**7.73**, R = CH₃, Valium) undergoes extensive oxidative *N*-demethylation to **7.73** (R = H).

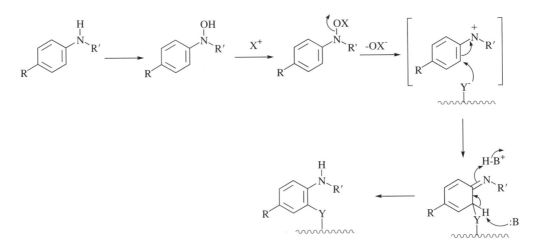

Scheme 7.29 ▶ Metabolic activation of primary and secondary aromatic amines

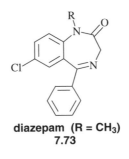

diazepam (R = CH₃)
7.73

As in the case of primary and secondary aromatic amines, which are activated to cytotoxic and carcinogenic metabolites, *N*-oxidation of primary and secondary aromatic amides also leads to electrophilic intermediates. For example, the carcinogenic agent 2-acetylaminofluorene (**7.74**, R = H) undergoes cytochrome P450-catalyzed oxidation to the *N*-hydroxy analog (**7.74**, R = OH). Activation of the hydroxyl group as in Scheme 7.29 leads to an electrophilic species capable of undergoing attack by cellular nucleophiles. Also, liver aryl-hydroxamic acid *N,O*-acyltransferase has been shown to catalyze the rearrangement of **7.74** (R = OH) to the corresponding *O*-acetyl hydroxylamine (**7.75**), which is activated for nucleophilic attack by the acyltransferase or by other cellular macromolecules (Scheme 7.30).[149] Covalent modification of arylhydroxamic acid *N,O*-acyltransferase prior to release of the activated species (**7.75**), is another example of mechanism-based enzyme inactivation (see Chapter 5, Section 5.5.C, p. 285). Release of **7.75** into the cell can lead to covalent bond formation with other cellular macromolecules, resulting in cytotoxicity and carcinogenicity.

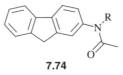

7.74

The analgesic agent acetaminophen (**7.76**, Scheme 7.31, Tylenol) is relatively nontoxic at therapeutic doses, but in larger doses, it causes severe liver necrosis.[150] The hepatotoxicity

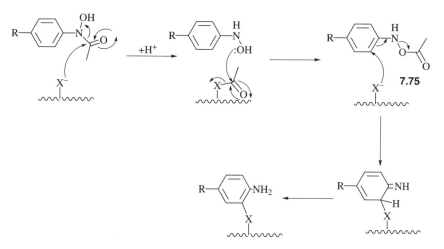

Scheme 7.30 ▶ Arylhydroxamic acid N,O-acyltransferase-catalyzed activation of N-hydroxy-2-acetylaminoarenes

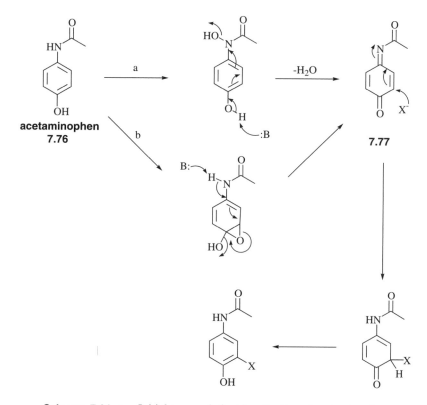

Scheme 7.31 ▶ Initial proposals for bioactivation of acetaminophen

arises from the depletion of liver glutathione levels as a result of the reaction of glutathione with an electrophilic metabolite of acetaminophen. When the glutathione levels drop by 80% or greater, then hepatic macromolecules react with the electrophilic metabolite leading to the observed liver necrosis. The original proposal for the mechanism of formation of the electrophilic metabolite (**7.77**) involves N-oxidation of the amide (Scheme 7.31, pathway a).[151] The X$^-$ in Scheme 7.31 represents either glutathione or a cellular macromolecular nucleophile (such as a protein amino acid residue). No evidence, however, for the formation of N-hydroxyacetaminophen could be found.[152] Another mechanism for the formation of the electrophilic metabolite **7.77** involves acetaminophen epoxidation (Scheme 7.31, pathway b). If this were correct, incubation of acetaminophen in the presence of $^{18}O_2$ should result in the incorporation of ^{18}O into metabolites; however, none was incorporated.[153] A third mechanistic possibility for the formation of **7.77** is a hydrogen atom abstraction mechanism via an acetaminophen radical (**7.78a** or **7.78b**, Scheme 7.32).[154] Isozymes of cytochrome P450 were shown to be responsible for the conversion of acetaminophen to electrophile **7.77**.[155] Because ethanol induces an isozyme of cytochrome P450 that could be responsible for the formation of **7.78**,[156] the hepatotoxicity of acetaminophen in ethanol-fed animals was compared with normal animals.[157] It was found that the alcoholic animals had increased hepatotoxicity, and that radical scavengers protected the animals. Furthermore, there is an increase in acetaminophen hepatotoxicity in alcoholics.[158] These results support a mechanism that involves cytochrome P450-dependent acetaminophen radical formation followed by second electron transfer (Scheme 7.32) as a mechanism for the generation of the electrophilic metabolite responsible for acetaminophen hepatotoxicity. Because of its reactivity, it was suggested that **7.78** may be responsible for the hepatotoxicity;[159] however, the benzoquinone imine (**7.77**) has been detected as a metabolite of the oxidation of acetaminophen by purified cytochrome P450[160] and by microsomes, NADPH, and O_2.[161] Furthermore, **7.77** reacts rapidly with glutathione *in vitro* to give the same conjugate as that found *in vivo*.

High doses of acetaminophen also cause renal damage in humans;[162] however, cytochrome P450 activity is low in the kidneys and, therefore, it may not be the major enzyme

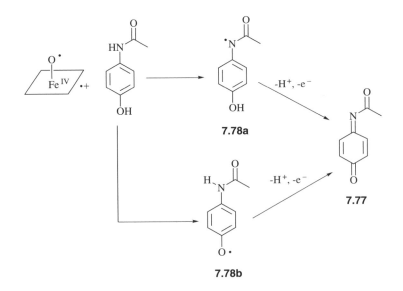

Scheme 7.32 ▶ Bioactivation of acetaminophen via a radical intermediate

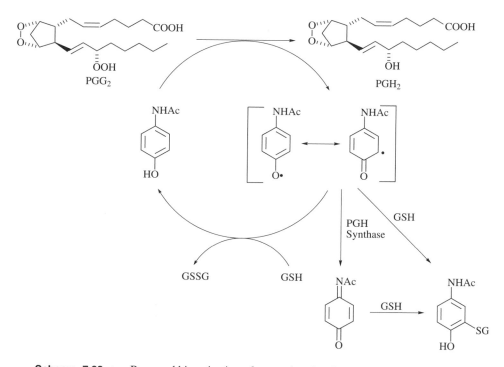

Scheme 7.33 ► Proposed bioactivation of acetaminophen by prostaglandin H synthase

responsible for acetaminophen toxicity there. Prostaglandin H synthase (also called cyclooxygenase), the enzyme that catalyzes the cyclooxygenation of arachidonic acid to prostaglandin G_2 (PGG_2) followed by the reduction of PGG_2 to PGH_2 (see Chapter 5, Scheme 5.16 in Section 5.5.B.2.b, p. 280), is present in high concentrations in the kidneys, and may be important in promoting acetaminophen toxicity in that organ. During the reduction of PGG_2 to PGH_2 prostaglandin H synthase, a heme-containing enzyme, can simultaneously co-oxidize a variety of substrates including certain drugs. Acetaminophen is metabolized to an intermediate that reacts with glutathione to form the glutathione conjugate.[163] By comparison of one- and two-electron reactions in the presence and absence of glutathione, it was found that prostaglandin H synthase also can catalyze the bioactivation of acetaminophen by one- and two-electron mechanisms[164] (Scheme 7.33). Prostaglandin H synthase may be another important drug-metabolizing enzyme in those tissues that are rich in this enzyme and low in the concentration of cytochrome P450.

f. Oxidations of Carbon-Oxygen Systems

Oxidative O-dealkylation is a common biotransformation which, as in the case of oxidative *N*-dealkylation (see Section 7.4.B.1.e), is catalyzed by microsomal mixed function oxidases. The mechanism appears to be the same as for oxidative *N*-dealkylation (see Scheme 7.19) and involves hydroxylation on the carbon attached to the oxygen followed by C–O bond cleavage to give the alcohol and the aldehyde or ketone. Although *O*-demethylation is rapid, dealkylation of longer chain *n*-alkyl substituents is generally slow; however, a study of alkoxy ethers of 7-hydroxycoumarin showed an increased rate of dealkylation up to propoxyl, then a decrease in rate.[165] Often $\omega - 1$ hydroxylation (see Section 7.4.B.1.d, p. 430) competes

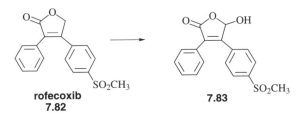

Scheme 7.34 ▶ Metabolic hydroxylation of rofecoxib

with *O*-dealkylation. Also, because of the greater steric bulk of a cyclopropyl group and its greater stability toward hydrogen atom abstraction, addition of this substituent provides compounds with longer plasma half-lives.[166] Lipophilic *N*-cyclopropyl amines, however, can be inactivators of heme and flavin oxidases (for example, see Chapter 5, Section 5.5.C.3.c, p. 292),[167] which may not be desirable.

A major metabolite of the anti-inflammatory drug indomethacin (**7.79**, R = CH₃, Indocin) is the *O*-demethylated compound (**7.79**, R = H).[168] Sometimes *O*-dealkylated metabolites also are pharmacologically active, as in the case of the narcotic analgesic codeine (**7.80**, R = CH₃), which is *O*-demethylated to morphine (**7.80**, R = H).[169] Nonequivalent methyl groups in drugs can be regioselectively *O*-demethylated; the blood pressure maintenance drug methoxamine (**7.81**, R = CH₃, Vasoxyl) gives exclusive 2-*O*-demethylation (**7.81**, R = H) in dogs.[170] *O*-Dealkylation can occur with a high degree of stereospecificity in the metabolism of chiral ethers.[171]

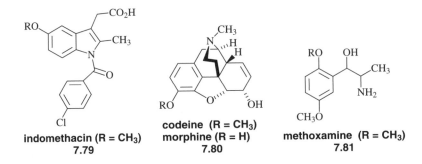

indomethacin (R = CH₃)
7.79

codeine (R = CH₃)
morphine (R = H)
7.80

methoxamine (R = CH₃)
7.81

Oxidation on the carbon next to the lactone oxygen leads to a stable α-hydroxy lactone (**7.83**) in the case of the antiarthritis (COX-2 selective inhibitor) drug rofecoxib (**7.82**, Scheme 7.34; Vioxx).[172]

g.   Oxidations of Carbon-Sulfur Systems

Fewer drugs contain sulfur than contain oxygen or nitrogen. The three principal types of biotransformations of carbon-sulfur systems are oxidative *S*-dealkylation, desulfuration, and *S*-oxidation. *Oxidative S-dealkylation* is not nearly as prevalent as the corresponding oxidative *N*- or *O*-dealkylations discussed above. The sedative methitural (**7.84**, R = CH₃) is metabolized by *S*-demethylation to **7.84** (R = H).[173]

methitural (R = CH₃)
7.84

Desulfuration is the conversion of a carbon-sulfur double bond (C=S) to a carbon-oxygen double bond (C=O). The anesthetic thiopental (**7.85**, X = S, Pentothal) undergoes desulfuration to pentobarbital (**7.85**, X = O, Nembutal [Sodium]).[174]

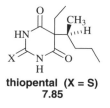

thiopental (X = S)
7.85

S-Oxidation of sulfur-containing drugs to the corresponding sulfoxide is a common metabolic transformation that is catalyzed both by flavin monooxygenase and by cytochrome P450.[175] Flavin monooxygenase produces exclusively the sulfoxide by a mechanism discussed in Chapter 4 (see Scheme 4.34, p. 212), but cytochrome P450 metabolizes sulfides to *S*-dealkylation products and to sulfoxides.[176] The common intermediate for these two metabolites is probably the sulfenium cation radical (**7.86**, Scheme 7.35); the mechanism for *S*-dealkylation may be related to that for cytochrome P450-catalyzed amine oxidations.[177] Both processes occur in the metabolism of the anthelmintic agent albendazole (**7.87**, Albenza).[178] Another example of *S*-oxidation is the oxidation of the antipsychotic drug thioridazine (**7.88**, R = SCH₃, Mellaril) in which case both sulfur atoms in the drug are metabolized to the corresponding sulfoxides.[179] The metabolite with only the methylthio substituent converted to the sulfoxide (**7.88**, R = S(O)CH₃) is twice as potent as thioridazine, and it also is used as an antipsychotic drug (mesoridazine, Serentil).

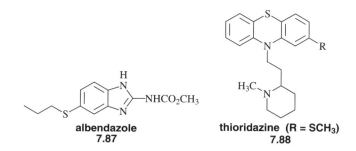

albendazole   **thioridazine (R = SCH₃)**
7.87   **7.88**

Some thiophene derivatives are known to form reactive metabolites.[180] The diuretic drug tienilic acid (**7.89**) is oxidized by cytochrome P450 to give electrophilic metabolites that covalently bind to liver proteins.[181] Incubation of **7.89** with liver microsomes in the presence of 2-mercaptoethanol to trap any electrophilic species generated (and to mimic a cysteine residue in a hepatic protein that might undergo reaction) gave **7.90** as an isolatable metabolite; the mechanism proposed for formation of the reactive metabolite (Scheme 7.36) involves initial formation of thiophene *S*-oxide.[182]

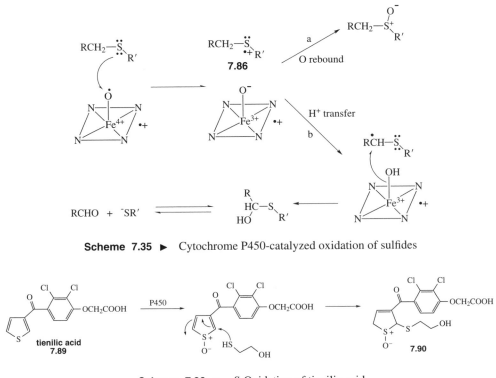

Scheme 7.35 ▶ Cytochrome P450-catalyzed oxidation of sulfides

Scheme 7.36 ▶ S-Oxidation of tienilic acid

Further oxidation of sulfoxides to sulfones also occurs; for example, the immunosuppressive drug oxisuran (**7.91**) is metabolized to the corresponding sulfone.[183] The two enantiomeric sulfoxide metabolites of the antiparasitic drug toltrazuril (**7.92**, Baycox) are oxidized further to the corresponding sulfones, but the rate of oxygenation of one enantiomer is seven times greater than that of the other.[184] It is not known if the same or different isozymes of cytochrome P450 are responsible for oxidation of the two enantiomers.

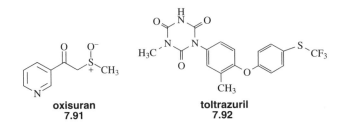

oxisuran
7.91

toltrazuril
7.92

h. Other Oxidative Reactions

Oxidative dehalogenation, oxidative aromatization, and oxidation of arenols to quinones are other important drug metabolism pathways that are catalyzed by cytochrome P450 enzymes. *Oxidative dehalogenation* occurs with the volatile anesthetic halothane (**7.93**, Scheme 7.37, Fluothane), which is metabolized to trifluoroacetic acid.[185] The acid chloride is responsible for the covalent binding of halothane to liver microsomes.[186]

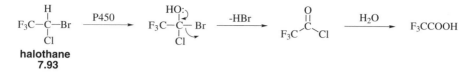

Scheme 7.37 ▶ Oxidative dehalogenation of halothane

Oxidative aromatization of the A ring in norgestrel (**7.94**, R = Et, Ovrette), a component of oral contraceptives, by a cytochrome P450 isozyme gives the corresponding phenol (**7.95**, R = Et).[187]

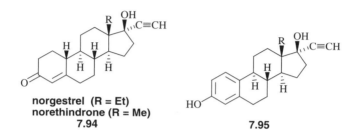

norgestrel (R = Et)
norethindrone (R = Me)
7.94

7.95

Catechols may be converted enzymatically to electrophilic *ortho*-quinones. Morphine, for example, is metabolized to a minor extent to its 2,3-catechol (**7.96**), which is oxidized to the *ortho*-quinone (**7.97**).[188]

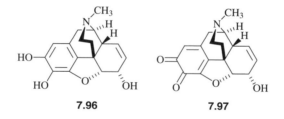

7.96

7.97

i. Alcohol and Aldehyde Oxidations

The oxidation of alcohols to aldehydes and of aldehydes to carboxylic acids, in general, is catalyzed by alcohol dehydrogenase and aldehyde dehydrogenase, respectively, soluble enzymes that require NAD^+ or $NADP^+$ as the cofactor (see Chapter 4, Section 4.3.B, p. 200).[189] They are found in highest concentration in the liver, but also in virtually every other organ. Cytochrome P450 isozymes also catalyze the oxidation of alcohols to aldehydes and aldehydes to carboxylic acids.[190] The reaction catalyzed by alcohol dehydrogenase is the oxidation of primary alcohols to the aldehyde and the reverse reaction (Scheme 7.38). As is apparent from this equation, the equilibrium in solution is pH dependent. The oxidation of alcohols is favored at higher pH (ca. pH 10) and the reduction of aldehydes is favored at lower pH (ca. pH 7). Based on this equilibrium, at physiological pH reduction should be favored. However, this is generally not observed because further oxidation of the aldehyde to the carboxylic acid (catalyzed by aldehyde dehydrogenase) is generally a more rapid process. Aldehydes are usually metabolized to the acid; relatively few examples of aldehyde reduction *in vivo* are known.[191] Actually, very few drugs contain an aldehyde group. The main exposure to

$$RCH_2OH + NAD^+ \rightleftharpoons RCHO + NADH + H^+$$
$$RCHO + NAD^+ + H_2O \rightleftharpoons RCOOH + NADH + H^+$$

Scheme 7.38 ▶ Reactions catalyzed by alcohol dehydrogenase and aldehyde dehydrogenase

aldehydes is through ingestion of the metabolic precursors such as primary alcohols or various amines (see Section 7.4.B.1.e, p. 430). The primary alcohol of the anti-HIV (reverse transcriptase inhibitor) drug abacavir sulfate (**7.98**, R = CH_2OH; Ziagen) is oxidized to the corresponding carboxylic acid (**7.98**, R = COOH),[192] presumably via the aldehyde, by alcohol and aldehyde dehydrogenases. The antihypertensive drug losartan potassium (**7.99**, R = CH_2OH; Cozaar) is metabolized by an isozyme of cytochrome P450 to the corresponding carboxylic acid (**7.99**, R = COOH).[193] The carboxylic acid metabolite is 10 times more potent an antagonist of the angiotensin II receptor than is losartan.[194]

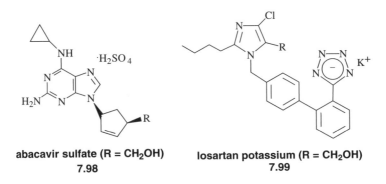

abacavir sulfate (R = CH_2OH)   losartan potassium (R = CH_2OH)
7.98   7.99

B.2 Reductive Reactions

Oxidative processes are, by far, the major pathways of drug metabolism, but reductive reactions are important for biotransformations of the functional groups listed in Table 7.6. Reductive reactions are important for the formation of hydroxyl and amino groups that render the drug more hydrophilic and set it up for phase II conjugation (see Section 7.4.C).

a. Carbonyl Reduction

Carbonyl reduction typically is catalyzed by aldo-keto reductases that require NADPH or NADH as the coenzyme. As described in the previous section, alcohol dehydrogenase catalyzes the reduction of aldehydes as well as the oxidation of alcohols. It is not common, however, to observe reduction of aldehydes to alcohols. A large variety of aliphatic and aromatic ketones, however, are reduced to alcohols by NADPH-dependent ketone reductases.[195] As discussed in Chapter 4, Section 4.3.B, p. 200, the NADPH carbonyl reductases are stereospecific with regard to hydride transfer from the pyridine nucleotide cofactor. In general, the aldehyde reductases exhibit A-(pro-4R)-hydrogen specificity and ketone reductases exhibit B-(pro-4S)-hydrogen specificity.[196] Stereoselectivity for enantiomer substrates and stereospecific reduction of the ketone carbonyl also are typical. The reduction of the anticoagulant drug warfarin (**7.100**, R = H, Coumadin) is selective for the R-(+)-enantiomer; reduction of the S-(−)-isomer occurs only at high substrate concentrations.[197] R-Warfarin is reduced in humans principally to the R,S-warfarin alcohol, whereas S-warfarin is metabolized mainly

TABLE 7.6 ▶ Classes of substrates for reductive reactions

Functional group	Product
(ketone, R–CO–R', C=O)	(alcohol, R–CH(OH)–R', OH)
RNO_2	RNHOH
RNO	RNHOH
RNHOH	RNH_2
RN=NR'	$RNH_2 + R'NH_2$
$R_3N—O^-$	R_3N
(quinone, R, R', R, R' with two C=O)	(semiquinone radical anion, R, R', R, R' with C=O and O^-)
R—X	$R• + X^-$

to 7-hydroxywarfarin (**7.100**, R = OH) but also is reduced to a 4:1 mixture of the S,S-alcohol and the S,R-alcohol. These studies were carried out by administration of the enantiomers separately to human volunteers. When racemic warfarin was administered to human volunteers a different picture emerged.[198] Of the 50% of the drug recovered as metabolites, 19% arose from the (R)-isomer and 31% from the (S)-isomer. The difference in the enantiomeric selectivity is a result of the difference in the rate of clearance from the body of the (S)-isomer relative to that of the (R)-isomer.[199] The major metabolite was (S)-7-hydroxywarfarin (22%); (S)-6-hydroxywarfarin (6%) also was obtained. In this study the main metabolite from (R)-warfarin was not the R,S-warfarin alcohol (6%), but rather (R)-6-hydroxywarfarin (9%); (R)-7-hydroxywarfarin (3%) also was obtained. Regioisomeric aromatic hydroxylation results from the selectivity of different isozymes of cytochrome P450.[200] By comparison of the results from the two warfarin studies it is apparent that one enantiomer can have an effect on the metabolism of the other. This is yet another reason to administer pure enantiomers, rather than racemic mixtures, as drugs. Enzymatic reduction of ketones in general produces the S-alcohol as the major metabolite,[201] even in the presence of chiral centers.

warfarin (R = H)
7.100

Species variation in the stereochemistry of the reduction of ketones is not uncommon. Naltrexone (**7.101**, ReVia), an opioid antagonist used to treat narcotics addicts who have been withdrawn from opiates in rehabilitation programs, is reduced to the 6α-alcohol (**7.102**, $R^1 =$ OH, $R^2 =$ H) in the chicken[202] and to the 6β-alcohol (**7.102**, $R^1 =$ H, $R^2 =$ OH) in rabbits and humans.[203]

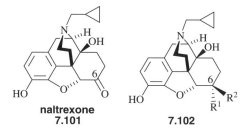

naltrexone
7.101   **7.102**

αβ-Unsaturated ketones can be metabolized to saturated alcohols (reduction of both the carbon–carbon double bond and the carbonyl group). Norgestrel (**7.94**, R = Et, Ovrette) and norethindrone (**7.94**, R = Me), used in oral contraceptives, are reduced in women to the 5β-*H*-3α,17β-diol (**7.103**; $R^1 =$ H, $R^2 =$ OH, $R^3 =$ Et) and 5β-H-3β,17β-diol (**7.103**, $R^1 =$ OH, $R^2 =$ H, $R^3 =$ Me) derivatives, respectively.[204] The Δ^4-double bond of both drugs is reduced to give the 5β-product, but the 3-keto group is reduced to the 3α-epimer in the case of norgestrel and to the 3β-epimer with norethindrone, even though the only difference in the two molecules is a 13β-ethyl group versus a 13β-methyl group, respectively.

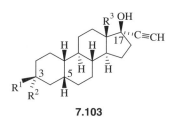

7.103

b. Nitro Reduction

Aromatic nitro reduction, catalyzed by cytochrome P450 in the presence of NADPH, but under anaerobic conditions (O_2 inhibits the reaction), and by the flavin-dependent NADPH-cytochrome P450 reductase (see Section 7.4.B.1), is a multistep process (Scheme 7.39); the reduction of the nitro group to the nitroso group (**7.105**) is the rate-determining step.[205] On the basis of electron paramagnetic resonance (EPR) spectra and the correlation of rates of radical formation with product formation, it has been proposed[206] that the nitro anion radical (**7.104**) is the first intermediate in the reduction of the nitro group. The reoxidation of this radical by oxygen to give the nitro compound back and superoxide radical anion may explain

$$R-NO_2 \xrightleftharpoons{e^-} R-NO_2^{\bullet-} \xrightarrow[2H^+]{e^-} R-NO \xrightarrow[H^+]{e^-} \left[R-\overset{H}{\underset{}{N}}-O^\bullet \right] \xrightarrow[H^+]{e^-} R-NHOH \xrightarrow[2H^+]{2e^-} R-NH_2$$

7.104   **7.105**

$O_2^{\bullet-}$ O_2

Scheme 7.39 ▶ Nitro group reduction

the inhibition of this metabolic pathway by oxygen.[207] Other enzymes that catalyze nitro group reduction are the bacterial nitro reductase in the gastrointestinal tract,[208] xanthine oxidase,[209] aldehyde oxidase,[210] and quinone reductase (DT-diaphorase).[211] Metabolic oxidation/reduction cycling of drugs may occur; the balance between the oxidative and reductive pathways is important in determining the pharmacological and toxicological profile of a drug.

An example of nitro reduction is the metabolism of the anticonvulsant drug clonazepam (**7.106**, R = NO_2, Klonopin) to its corresponding amine (**7.106**, R = NH_2).[212] In some *in vitro* experiments the reduced metabolite is not observed because it is easily air oxidized back to the parent compound. For example, the antiparasitic agent niridazole (**7.107**, R = NO_2) is reduced to the hydroxylamine metabolite (**7.107**, R = NHOH), which is reoxidized in air to **7.107** (R = NO_2).[213] The antibacterial drug nitrofurazone (**7.108**, Scheme 7.40, Furacin) is reduced both to the corresponding 5-hydroxylamino derivative (**7.109**) and the 5-amino derivative (**7.110**). The latter is unstable and tautomerizes to **7.111**.[214]

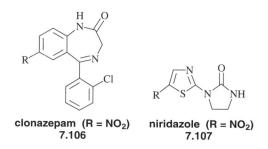

clonazepam (R = NO_2)
7.106

niridazole (R = NO_2)
7.107

c. Azo Reduction

Azo group (RN=NR) reduction is similar to nitro reduction in many ways. It, too, is mediated both by cytochrome P450 and by NADPH-cytochrome P450 reductase (see Section 7.4.B.1), and oxygen often inhibits the reaction. The initial reduction in the oxygen-sensitive

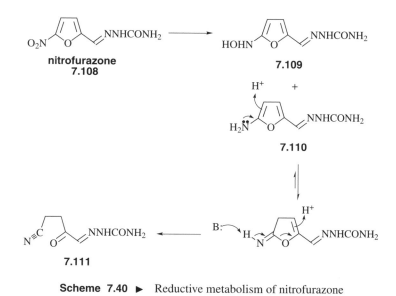

Scheme 7.40 ▶ Reductive metabolism of nitrofurazone

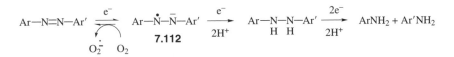

$$Ar-N=N-Ar' \underset{O_2^{\cdot-} \quad O_2}{\overset{e^-}{\rightleftarrows}} Ar-\overset{\cdot}{N}-\overset{-}{N}-Ar' \underset{2H^+}{\overset{e^-}{\longrightarrow}} \underset{H \quad H}{Ar-N-N-Ar'} \underset{2H^+}{\overset{2e^-}{\longrightarrow}} ArNH_2 + Ar'NH_2$$

$$\textbf{7.112}$$

Scheme 7.41 ▶ Azo group reduction

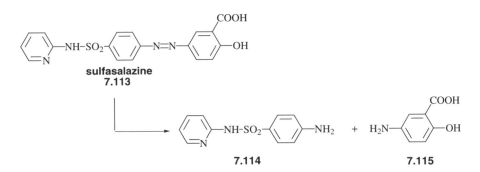

sulfasalazine
7.113

7.114   **7.115**

Scheme 7.42 ▶ Reductive metabolism of sulfasalazine

metabolism appears to proceed via the azo anion radical (**7.112**, Scheme 7.41);[215] the oxygen apparently reverses this process with concomitant conversion of the oxygen to the superoxide anion radical.[216] Oxygen-insensitive azoreductases presumably proceed by a two-electron reduction of the azo compound directly to the hydrazo intermediate.

Bacteria in the gastrointestinal tract also are important in azo reduction.[217] Reduction of sulfasalazine (**7.113**, Scheme 7.42, Azulfidine EN), used in the treatment of ulcerative colitis, to sulfapyridine (**7.114**) and 5-aminosalicylic acid (**7.115**) occurs primarily in the colon mediated by intestinal bacteria.[218]

d. Azido Reduction

The 3-azido group of the anti-AIDS drug zidovudine (**7.116**, X = N$_3$, Retrovir) is reduced by cytochrome P450 isozymes and NADPH-cytochrome P450 reductase to the corresponding amine (**7.116**, X = NH$_2$).[219]

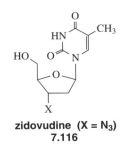

zidovudine (X = N$_3$)
7.116

e. Tertiary Amine Oxide Reduction

A wide variety of lipophilic aliphatic and aromatic tertiary amine oxides, such as imipramine *N*-oxide (**7.117**), are reduced to the corresponding amine by cytochrome P450 in the absence of oxygen.[220]

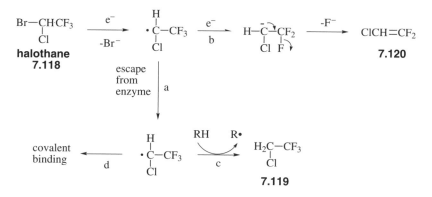

Scheme 7.43 ► Reductive dehalogenation of halothane

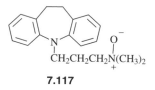

7.117

f. Reductive Dehalogenation

Under hypoxic or anaerobic conditions the volatile anesthetic halothane (**7.118**, Fluothane) is metabolized by a reductive dehalogenation mechanism by cytochrome P450[221] (Scheme 7.43), which differs from the normal oxidative P450 mechanism discussed in Section 7.4.B.1.h, p. 446). The first electron is transferred to halothane from cytochrome P450, which is reduced by NADPH cytochrome P450 reductase. This electron transfer ejects the bromide ion and produces the cytochrome P450-bound 1-chloro-2,2,2-trifluoroethyl radical. If this radical escapes from the active site (pathway a), it either can be reduced by hydrogen atom transfer (pathway c) to give 2-chloro-1,1,1-trifluoroethane (**7.119**) or form a covalent bond to cellular proteins (pathway d). A second electron reduction from **7.118** (pathway b) produces the carbanion; β-elimination of fluoride ion gives chlorodifluoroethylene (**7.120**). Pathway d, resulting in covalent attachment to proteins, has been proposed to mediate halothane hepatitis, a toxic reaction to halothane exposure in the liver. The second electron transfer (pathway b) is thought to be derived from cytochrome b_5; this leads to nonreactive products and it competes with pathway a, thereby leading to fewer reactive metabolites.

B.3 Carboxylation Reaction

Amino compounds may be susceptible to carboxylation by dissolved carbon dioxide, converting them to the corresponding carbamic acid derivatives, which may be further metabolized. For example, the serotonin reuptake inhibitor/antidepressant drug sertraline (**7.121**, R = H, Zoloft) is metabolized to the carbamic acid (**7.121**, R = COOH).[222]

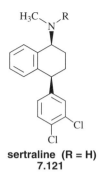

sertraline (R = H)
7.121

B.4 Hydrolytic Reactions

The hydrolytic metabolism of esters and amides leads to the formation of carboxylic acids, alcohols, and amines, all of which are susceptible to phase II conjugation reactions and excretion (see Section 7.4.C). As described in Chapter 4, Section 4.2.C, p. 182, enzyme-catalyzed hydrolysis can be acid and/or base catalyzed. Base-catalyzed hydrolysis is accelerated nonenzymatically when electron-withdrawing groups are substituted on either side of the ester or amide bond. When the carbonyl is in conjugation with a π-system, nonenzymatic base hydrolysis is decelerated relative to the aliphatic case. The effects on nonenzymatic hydrolysis rates also can be important in enzyme-catalyzed hydrolysis reactions.

A wide variety of nonspecific esterases and amidases involved in drug metabolism are found in plasma, liver, kidney, and intestine.[223] All mammalian tissues may contribute to the hydrolysis of a drug; however, the liver, the gastrointestinal tract, and the blood are sites of greatest hydrolytic capacity. Aspirin (**7.122**) is an example of a drug that is hydrolyzed in all human tissues.[224] The hydrolysis of xenobiotics is very similar in all mammals, but there are some exceptions, and large species differences can be observed. Some esterases catalyze hydrolysis of aliphatic esters and others aromatic esters. For example, only the benzoyl ester in cocaine (**7.123**, R = CH₃) is hydrolyzed by human liver *in vitro*, not the alicyclic ester;[225] however, *in vivo*, the major metabolite of cocaine is the alicyclic ester hydrolysis product, benzoylecgonine (**7.123**, R = H),[226] suggesting that the liver is not the primary site of metabolism *in vivo*.

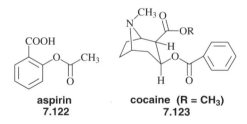

aspirin   **cocaine (R = CH₃)**
7.122   **7.123**

Generally, amides are more slowly hydrolyzed than esters. For example, the enzymatic hydrolysis of the antiarrhythmic drug procainamide (**7.124**, X = NH, Procanbid) is slow relative to that of the local aesthetic procaine (**7.124**, X = O, Novocaine).[227] The ester group, not the amide, is hydrolyzed in the anesthetic propanidid (**7.125**, Epontol).[228] However, the amide bond of the local anesthetic butanilicaine (**7.126**, Hostacain) is hydrolyzed by human liver at rates comparable to those of good ester substrates, possibly because of the electron-withdrawing groups attached.

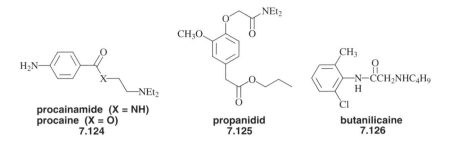

procainamide (X = NH)
procaine (X = O)
7.124

propanidid
7.125

butanilicaine
7.126

In some cases the hydrolysis of an ester or amide bond produces a toxic compound. Aromatic amines generated on hydrolysis of *N*-acylanilides become methemoglobin-forming agents after *N*-oxidation. Phenacetin (**7.127**, Acetophenetidin) causes methemoglobinemia in rats, which is reduced drastically when the carboxylesterase inhibitor *bis*(4-nitrophenyl)phosphate is coadministered.[229] The danger of using racemic mixtures as drugs is further exemplified by the observation that although both isomers of prilocaine (**7.128**, Citanest) have local anesthetic action, only the R-(−)-isomer is hydrolyzed to toluidine, which causes methemoglobinemia. The *S*-(+)-isomer, which is not hydrolyzed, does not cause this side effect.[230]

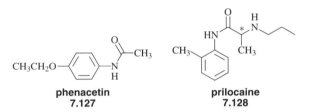

phenacetin
7.127

prilocaine
7.128

As shown in the above example, enzymatic hydrolysis often exhibits enantiomeric specificity. A racemic mixture of the anticonvulsant drug phensuximide (**7.129**, Milontin) is enzymatically *N*-demethylated and hydrolyzed stereospecifically to *R*-(−)-2-phenylsuccinamic acid (**7.130**).[231] Enantiomeric selectivity in the hydrolysis of the tranquilizer prodrug oxazepam acetate (**7.131**, Serax) may be organ selective. Preferential hydrolysis of the *R*-(−)-ester occurs in the liver, but the opposite is found in the brain.[232]

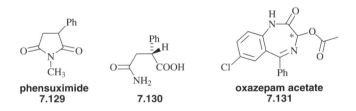

phensuximide
7.129

7.130

oxazepam acetate
7.131

Because of the stereoselectivity of various enzymes, one enantiomer may be a preferential substrate for one enzyme and the other enantiomer a substrate for a different enzyme. Both enantiomers of the hypnotic drug etomidate (**7.132**, Scheme 7.44, Amidate) are metabolized, but by different routes.[233] The active *R*-(+)-isomer is more rapidly hydrolyzed (pathway a) than is the *S*-(−)- isomer, but the S-(−)-isomer is more rapidly hydroxylated (pathway b) than is the *R*-(+)-isomer. *In vitro*, only the S-(−)-isomer produces acetophenone (**7.133**).

In addition to carboxylesterases and amidases, hydrolytic reactions also are carried out by various other mammalian enzymes such as phosphatases, *β*-glucuronidases, sulfatases,

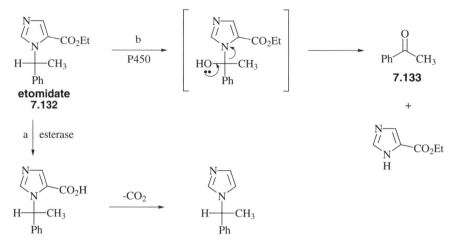

Scheme 7.44 ▶ Competitive metabolism of *R*- and *S*-etomidate

and deacetylases. We now turn our attention to the next phase of drug metabolism, that of transforming drugs and the phase I metabolites into conjugates for excretion.

7.4.C Phase II Transformations: Conjugation Reactions

C.1 Introduction

Phase II or *conjugating enzymes*, in general, catalyze the attachment of small polar endogenous molecules such as glucuronic acid, sulfate, and amino acids to drugs or, more often, to metabolites arising from phase I metabolic processes. This phase II modification further deactivates the drug, changes its physicochemical properties, and produces water-soluble metabolites that are readily excreted in the urine or bile. Phase II processes such as methylation and acetylation do not yield more polar metabolites, but serve primarily to terminate or attenuate biological activity. Metabolic reactions with the potent nucleophile glutathione serve to trap highly electrophilic metabolites before they covalently modify biologically important macromolecules such as proteins, RNA, and DNA. Many drugs are excreted without any modification at all.

Conjugation reactions take place primarily with hydroxyl, carboxyl, amino, heterocyclic nitrogen, and thiol groups. If these groups are not present in the drug, they are introduced or unmasked by the phase I reactions. For the most part, the conjugating moiety is an endogenous molecule that is first activated in a coenzyme form prior to its transfer to the acceptor group. The enzymes that catalyze these reactions are known as *transferases* (Table 7.7).

C.2 Glucuronic Acid Conjugation

Glucuronidation is the most common mammalian conjugation pathway; it occurs in all mammals except the cat. As shown in Table 7.7, the coenzyme form of glucuronic acid, namely, uridine 5-diphospho-α-D-glucuronic acid (UDP glucuronic acid, **7.136**, Scheme 7.45) is biosynthesized from α-D-glucose-1-phosphate (**7.134**) by phosphorylase-catalyzed conversion to the nucleotide sugar **7.135**, followed by UDP dehydrogenase-catalyzed oxidation. UDP glucuronic acid (**7.136**) contains D-glucuronic acid in the α-configuration at the anomeric carbon, but glucuronic acid conjugates (**7.137**) are β-glycosides. Therefore, the glucuronidation

TABLE 7.7 ▶ Mammalian phase II conjugating agents

Conjugate	Coenzyme form	Groups conjugated	Transferase enzyme
Glucuronide	Uridine-5′-diphospho-α-D-glucuronic acid (UDPGA)	-OH, -COOH, -NH₂, -NR₂, -SH, C-H	UDP-Glucuronosyl-transferase
Sulfate	3′-Phosphoadenosine-5′-phosphosulfate (PAPS)	-OH, -NH₂	Sulfotransferase
Glycine and glutamine	Activated acyl or aroyl coenzyme A cosubstrate	-COOH	Glycine N-acyltransferase / Glutamine N-acyltransferase
Glutathione	Glutathione (GSH)	Ar-X, arene oxide, epoxide, carbocation or related	Glutathione S-transferase
Acetyl	Acetyl coenzyme A	-OH, -NH₂	Acetyl-transferase
Methyl	S-Adenosyl methionine (SAM)	-OH, -NH₂, -SH, heterocyclic N	Methyl-transferase

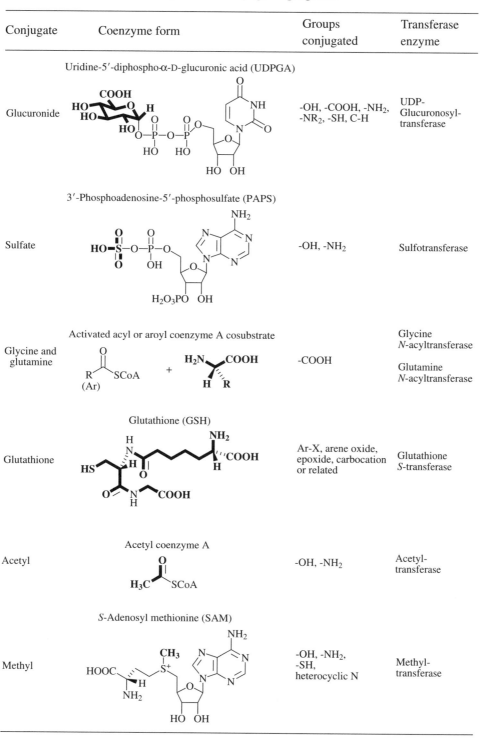

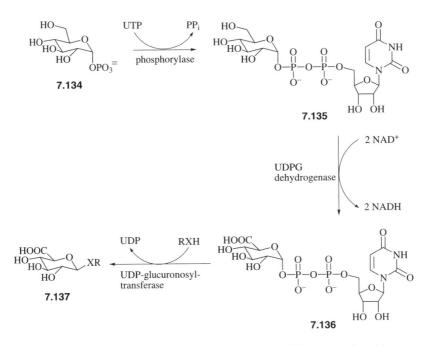

Scheme 7.45 ▶ Biosynthesis and reactions of UDP glucuronic acid

reaction involves an inversion of stereochemistry at the anomeric carbon. Because of the presence of the carboxylate and hydroxyl groups of the glucuronyl moiety, glucuronides are hydrophilic and, therefore, are set up for excretion. Glucuronides generally are excreted in the urine, but when the molecular weight of the conjugate exceeds 300, excretion in the bile becomes significant. There is some evidence that UDP glucuronosyltransferase is closely associated with cytochrome P450 so that as drugs become oxidized by the phase I cytochrome P450 reactions, the metabolites are efficiently conjugated.[234]

Four general classes of glucuronides have been established, the *O*-, *N*-, *S*-, and *C*-glucuronides; a small sampling of examples are given in Table 7.8; arrows point to the sites of glucuronidation.

Certain disease states (inborn errors of metabolism) are associated with defective glucuronide formation or attachment of the glucuronide to bilirubin, for example, Crigler-Najjar syndrome and Gilbert's disease; both are characterized by a deficiency of UDP-glucuronosyltransferase activity.[244] Neonates, which have undeveloped liver UDP-glucuronosyltransferase activity, may exhibit similar metabolic problems. In these cases there is a greater susceptibility to adverse effects caused by the accumulation of drugs that normally are glucuronidated. An example of this is the inability of neonates to conjugate the antibacterial drug chloramphenicol (**7.138**, Chloroptic), thereby leading to "gray baby syndrome" from the accumulation of toxic levels of the drug.[245]

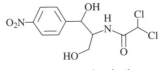

chloramphenicol
7.138

TABLE 7.8 ▶ Classes of compounds forming glucuronides with examples

Type	Example	Structure (arrow indicates site of glucuronidation)	Reference
O-Glucuronide			
Hydroxyl			
Phenol	**acetaminophen**	AcNH—⟨ ⟩—OH	235
alcohol	**chloramphenicol**	O_2N—⟨ ⟩ with OH, NHC(O)CHCl$_2$, OH	236
Carboxyl	**fenoprofen**	PhO—⟨ ⟩ CH(CH$_3$)—C(O)OH	237
N-Glucuronide			
Amine	**desipramine**	NHCH$_3$	238
Amide carbamate	**meprobamate**	OCONH$_2$ / O—C(O)—NH$_2$	239
Sulfonamide	**sulfadimethoxine**	H_2N—⟨ ⟩—SO_2NH—pyrimidine(OMe)(OMe)	240
S-Glucuronide			
Sulfhydryl	**methimazole**	imidazole—SH, N—CH$_3$	241
Carbodithioic acid	**disulfiram**	Et$_2$N—C(S)—SH (reduced metabolite)	242
C-Glucuronide	**phenylbutazone**	Ph, Ph pyrazolidine-dione with butyl	243

As in the case of phase I metabolism, phase II reactions also can be species specific, regioselective, and stereoselective. The antibacterial drug sulfadimethoxine (Albon, see Table 7.8) is glucuronidated in humans, but not in rats, guinea pigs, or rabbits. The bronchodilator fenoterol (**7.139**, Berotec) is conjugated as two different glucuronides, a *para*-glucuronide and a *meta*-glucuronide because there are both *para*- and *meta*-hydroxyl groups.[246] The *R*, *R*-(−)-isomer is conjugated with higher affinity but with lower velocity than is the *S*, *S*-(+)-isomer.

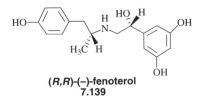

(*R,R*)-(−)-fenoterol
7.139

The antidepressant drug nortriptyline (**7.140**, R = H, Aventyl) is metabolized (cytochrome P450) predominantly to the *E*-(−)-hydroxy analog (**7.140**, R = OH; the absolute configurations of the (+)- and (−)-enantiomers are not known, so stereochemistry is not specified. This metabolite is converted stereospecifically into the corresponding *O*-glucuronide, but the stereospecificity is organ dependent.[247] The liver and kidney glucuronosyltransferases catalyze the conversion of only the *E*-(+)-isomer of **7.140** (R = OH) into the *O*-glucuronide, whereas the intestinal enzyme metabolizes only the *E*-(−)-isomer. The enantiomer that is not glucuronidated inhibits the glucuronidation of its antipode.

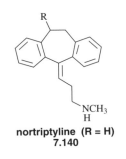

nortriptyline (R = H)
7.140

C.3 Sulfate Conjugation

Sulfate conjugation occurs less frequently than does glucuronidation presumably because of the limited availability of inorganic sulfate in mammals and the fewer number of functional groups (phenols, alcohols, arylamines, and *N*-hydroxy compounds) that undergo sulfate conjugation.[248]

There are three enzyme-catalyzed reactions involved in sulfate conjugation (Scheme 7.46). Inorganic sulfate is activated by the ATP sulfurylase-catalyzed reaction with ATP to give adenosine 5′-phosphosulfate (APS, **7.141**), which is phosphorylated in an APS phosphokinase-catalyzed reaction to 3-phosphoadenosine 5′-phosphosulfate (PAPS, **7.142**), the coenzyme form used for sulfation.[249] The acceptor molecule (RXH) undergoes sulfotransferase-catalyzed sulfation to **7.143** with release of 3-phosphoadenosine 5′-phosphate (PAP). There are a variety of sulfotransferases in the liver and other tissues. The main substrates for these enzymes are phenols, but aliphatic alcohols, amines, and, to a much lesser

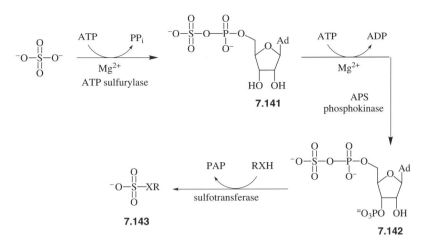

Scheme 7.46 ▶ Sulfate conjugation

extent, thiols also are sulfated. Often both glucuronidation and sulfation occur on the same substrates, but the K_m for sulfation is usually lower, so it predominates.[250] In addition to substrate binding differences, sulfotransferases are cytoplasmic (soluble) enzymes and glucuronosyltransferases are microsomal (membrane) enzymes. Sulfate conjugation tends to predominate at low doses, when there is less to diffuse into membranes[251] and with smaller, less lipid-soluble molecules.[252] However, lipophilicity does not necessarily mean that a compound will be glucuronidated rather than sulfated, because subcellular distances are small, diffusion out of membranes generally is rapid, and the K_m values with sulfotransferases are generally lower than those with glucuronosyltransferases. The bronchodilator albuterol (**7.144**, R = H, Albuterol) is metabolized to the corresponding sulfate ester (**7.144**), R = SO$_3^-$).[253] Note that although there are three hydroxyl groups in albuterol, phenolic sulfation predominates.

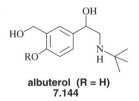

albuterol (R = H)
7.144

Phenolic O-glucuronidation often competes favorably with sulfation because of limited sulfate availability. In some cases the reverse situation occurs. Acetaminophen (Tylenol) is metabolized in adults mainly to the O-glucuronide, although some sulfate ester can be detected.[254] However, neonates and children 3–9 years old excrete primarily the acetaminophen sulfate conjugate because they have a limited capacity to conjugate with glucuronic acid.[255] Sulfation of aliphatic alcohols and arylamines occurs, but these are minor metabolic pathways. Sulfate conjugates can be hydrolyzed back to the parent compound by various sulfatases.

Sulfoconjugation plays an important role in the hepatotoxicity and carcinogenicity of N-hydroxyarylamides.[256] As described in Section 7.4.B.1.e, p. 430 (see Schemes 7.29 and 7.30), activated N-hydroxyarylamines are quite electrophilic and can react with protein and DNA nucleophiles. N-Hydroxy sulfation also activates these compounds as highly

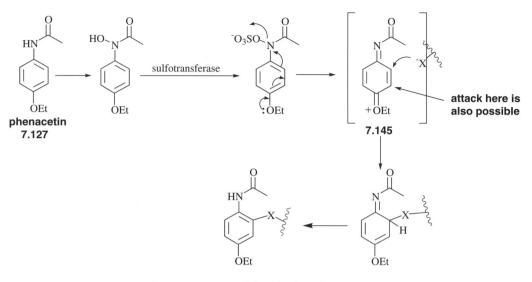

Scheme 7.47 ▶ Bioactivation of phenacetin

electrophilic nitrenium-like species. For example, sulfoconjugation of the *N*-hydroxylation metabolite of the analgesic phenacetin (**7.127**, Scheme 7.47) produces a metabolite (**7.145**) that may be responsible for hepatotoxicity and nephrotoxicity of that compound.[257]

C.4 Amino Acid Conjugation

The first mammalian drug metabolite isolated, hippuric acid (see **7.8**), was the product of glycine conjugation of benzoic acid. *Amino acid conjugation* of a variety of carboxylic acids, particularly aromatic, arylacetic, and heterocyclic carboxylic acids, leads to amide bond formation. The specific amino acid involved in conjugation within a class of animals usually depends on the bioavailability of that amino acid from endogenous and dietary sources. Glycine conjugates are the most common amino acid conjugates in animals; glycine conjugation in mammals follows the order herbivores > omnivores > carnivores. Conjugation with *L*-glutamine is most common in primate drug metabolism;[258] it does not occur to any significant extent in nonprimates. Taurine, arginine, asparagine, histidine, lysine, glutamate, aspartate, alanine, and serine conjugates also have been found in mammals.[259]

The mechanism of amino acid conjugation involves three steps (Scheme 7.48). The carboxylic acid is first activated by ATP to the AMP ester (**7.146**), which is converted to the corresponding coenzyme A thioester (**7.147**) with CoASH; these first two steps are catalyzed by acyl CoA synthetases (long-chain fatty acid-CoA ligases). The appropriate amino acid *N*-acyltransferase then catalyzes the condensation of the amino acid and coenzyme A thioester to give the amino acid conjugate (**7.148**). Conjugation does not take place with the AMP ester (**7.146**) directly because the AMP ester hydrolyzes readily; conversion to the CoA thioester produces a more hydrolytically stable product (**7.147**) that can be transported in the cell readily but is still quite reactive toward the appropriate amine nucleophiles.

As an example of how a drug can undergo both phase I and phase II transformations, Scheme 7.49 shows the metabolic pathway for the antihistamine brompheniramine (**7.149**, in cold remedies, such as Dimetapp).[260] This drug is converted by phase I metabolism in

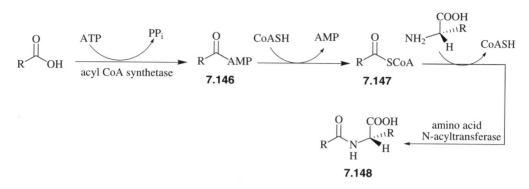

Scheme 7.48 ▶ Amino acid conjugation

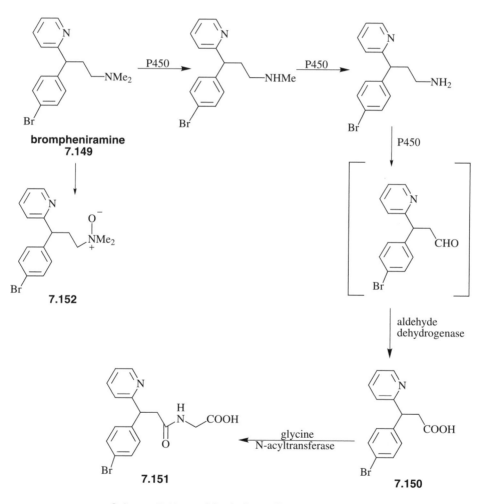

Scheme 7.49 ▶ Metabolism of brompheniramine

both dogs and humans (*N*-dealkylation, oxidative deamination, and aldehyde oxidation; see Section 7.4.B) to the carboxylic acid (**7.150**), which then is conjugated with glycine (phase II) to **7.151**. All of the metabolites shown in Scheme 7.49 (including **7.152**), except the aldehyde, were isolated from dog urine, and all except the aldehyde and the *N*-oxide (**7.152**) were isolated from human urine. The related antihistamine diphenhydramine (**7.153**, Benadryl) undergoes similar phase I oxidation, but is glutamine conjugated.[261]

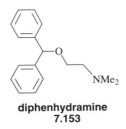

diphenhydramine
7.153

C.5   Glutathione Conjugation

The tripeptide glutathione (**7.154**) is found in virtually all mammalian tissues. It contains a reactive nucleophilic thiol group, and one of its functions appears to be as a scavenger of harmful electrophilic compounds ingested or produced by metabolism. Xenobiotics that are conjugated with glutathione are either highly electrophilic as such or are first metabolized to an electrophilic product prior to conjugation. Toxicity can result from the reaction of cellular nucleophiles with electrophilic metabolites (see Schemes 7.30–7.33) if glutathione does not first intercept these reactive compounds. Electrophilic species include any group capable of undergoing S_N2- or S_NAr-like reactions (e.g., alkyl halides, epoxides, and aryl halides), acylation reactions (e.g., anhydrides and sulfonate esters), conjugate additions (addition to a double or triple bond in conjugation with a carbonyl or related group), and reductions (e.g., disulfides and radicals). All of the reactions catalyzed by glutathione *S*-transferase also occur nonenzymatically, but at a slower rate.

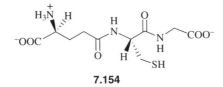

7.154

A few examples of glutathione conjugation are given in Scheme 7.50. Examples of S_N2 reactions are the glutathione conjugation of the leukemia drug busulfan (**7.155**, Myleran)[262] and of the coronary vasodilator nitroglycerin (**7.156**, Nitrostat).[263] The reaction of glutathione with the immunosuppressive drug azathioprine (**7.157**, Azathioprine)[264] is an example of an S_NAr reaction. These reactions are direct deactivations of the drugs. Morphine (**7.158**) has been reported to undergo oxidation by two different pathways, both of which lead to potent Michael acceptors that undergo subsequent glutathione conjugation. Pathway a, catalyzed by morphine 6-dehydrogenase, gives morphinone (**7.159**), which undergoes Michael addition with glutathione to **7.160**.[265] Pathway b is a cytochrome P450 catalyzed route that produces an electrophilic quinone methide (**7.161**). Glutathione addition occurs at the sterically less hindered 10α-position to give **7.162**.[266]

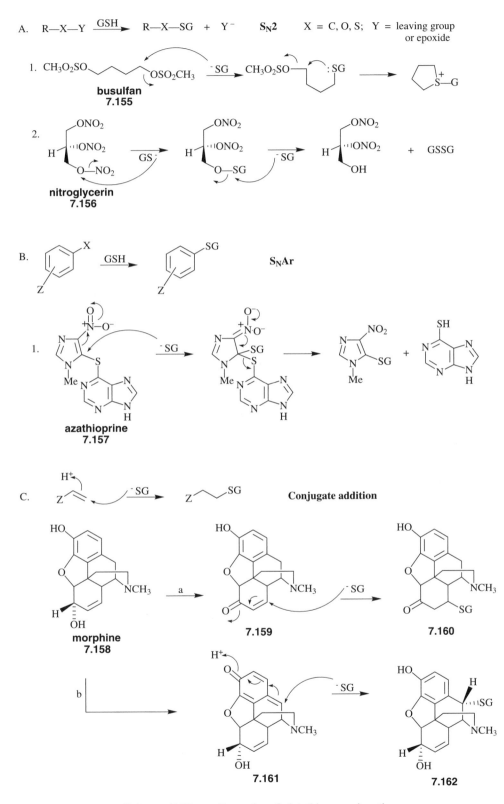

Scheme 7.50 ▶ Examples of glutathione conjugation

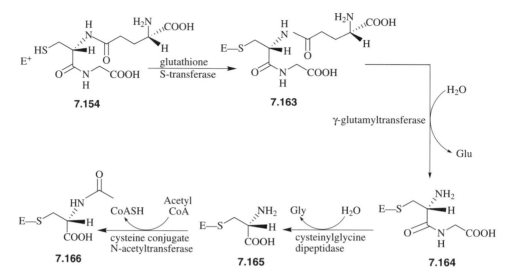

Scheme 7.51 ▶ Metabolism of glutathione conjugates to mercapturic acid conjugates

Glutathione conjugates are rarely excreted in the urine; because of their high molecular weight and amphiphilic character, when they are eliminated, it is in the bile. Most typically, however, glutathione conjugates are not excreted; instead they are metabolized further (sometimes referred to as phase III metabolism)[267] and are excreted ultimately as *N*-acetyl-*L*-cysteine (also known as mercapturic acid) conjugates (**7.166**, Scheme 7.51).[268] Formation of the mercapturic acid begins from the glutathione conjugate (**7.163**). The γ-glutamyl residue is hydrolyzed to glutamate and the cysteinylglycine conjugate **7.164** in a process catalyzed by γ-glutamyltranspeptidase. Cysteinylglycine dipeptidase-catalyzed hydrolysis of **7.164** leads to the release of glycine and the formation of the cysteine conjugate **7.165**, which is *N*-acetylated by acetyl CoA in a reaction catalyzed by cysteine conjugate *N*-acetyltransferase. Conjugation with glutathione occurs in the cytoplasm of most cells, especially in the liver and kidney where the glutathione concentration is 5–10 mM.

C.6 Water Conjugation

Epoxide hydrolase (also called epoxide hydratase or epoxide hydrase),[269] the enzyme principally involved in *water conjugation* (i.e., *hydration*), was discussed already in the context of hydration of arene oxides (see Section 7.4.B.1.a, p. 420). Because this enzyme catalyzes the hydration of endogenous epoxides, such as androstene oxide, at much faster rates than exogenous epoxides, it probably has an important role in endogenous metabolism (Procanbid).[270]

C.7 Acetyl Conjugation

Acetylation is an important route of metabolism for xenobiotics containing a primary amino group, including aliphatic and aromatic amines, amino acids, sulfonamides, hydrazines, and hydrazides. In all of the previously discussed conjugation reactions a *more* hydrophilic metabolite is formed. Acetylation, however, converts the ionized primary ammonium group (the aliphatic amine or hydrazine is protonated at physiological pH) to an uncharged amide which is less water soluble. The physicochemical consequences of N-acetylation, therefore, are different from those of the other conjugation reactions. The function of acetylation may be to deactivate the drug, although N-acetylprocainamide is as potent as the parent antiarrhythmic drug procainamide (Procanbid).[271]

Acetylation occurs widely in the animal kingdom; however, the dog and fox are unable to acetylate N-arylamines or the N^4-amino group of sulfonamides.[272] The extent of N-acetylation (and of other metabolic reactions) of a number of drugs in humans is a genetically determined individual characteristic. The genetic variability among individuals in the therapeutic response to the same dose of a drug may be the result of differences in absorption, distribution, metabolism, or elimination. A genetic alteration in an enzyme, known as *polymorphism*, can have serious consequences with regard to the safety of a drug as well as its therapeutic effectiveness.[273] A decrease in the rate of metabolism because of polymorphism of a drug metabolism enzyme would lead to an increase in blood levels of the drug, possibly leading to drug interactions or toxicity; an increase in drug metabolism as a result of an overly active drug metabolism enzyme may cause a reduced therapeutic response. Polymorphisms have been detected for many drug-metabolizing enzymes.[274] These polymorphisms give rise to subgroups in the population that differ in their ability to perform certain enzymatic biotransformations. The three phenotypes in *acetylation polymorphism* are homozygous fast, homozygous slow, and heterozygous (intermediate) acetylators.[275] The distribution of fast and slow acetylator phenotypes depends on the population studied, varying from mostly slow acetylators (Egyptians) to 50% fast/50% slow in the United States to 90% fast/10% slow in East Asians[276] to 100% fast acetylators in the Canadian Eskimo population.[277] Because of the differences in the rates of N-acetylation of certain drugs, there are significant individual variations in the therapeutic responses and toxicity to drugs exhibiting acetylation polymorphism. Slow acetylators in general often develop adverse reactions as a result of tissue exposure to higher concentrations of the drugs; however, this also may result in longer drug effectiveness. Fast acetylators are more likely to show an inadequate therapeutic response to standard doses of the drug. Examples of drugs exhibiting acetylation polymorphism are the antibacterial drug sulfamethazine (**7.167**, Sulmet), the antituberculosis drug isoniazid (**7.168**, Rifamate),[278] and dapsone (**7.169**, Dapsone),[279] used in the treatment of leprosy.

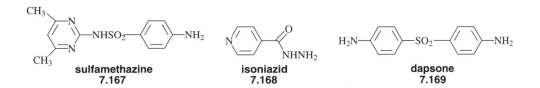

sulfamethazine
7.167

isoniazid
7.168

dapsone
7.169

Acetylation is a two-step covalent catalytic process (Scheme 7.52). First, acetyl CoA acetylates an active site amino acid residue of the soluble hepatic N-acetyltransferase (**7.170**), then the acetyl group is transferred to the substrate amino group.[280] Presumably, this two-step process allows the enzyme to have better control over the catalytic process.

One of the few examples of an aliphatic amine drug that is acetylated is cilastatin (**7.171**, R = H), which is metabolized to **7.171** (R = Ac).[281] Actually, cilastatin is not a drug, but is used in combination with the antibacterial drug imipenem (**7.172**) (the combination is called Primaxin). When imipenem is administered alone, it is rapidly hydrolyzed in the kidneys by dehydropeptidase I.[282] Cilastatin is a potent inhibitor of this enzyme, and it

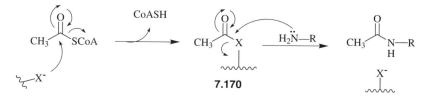

Scheme 7.52 ▶ *N*-Acetylation of amines

effectively prevents renal metabolism of imipenem when the two compounds are administered in combination.

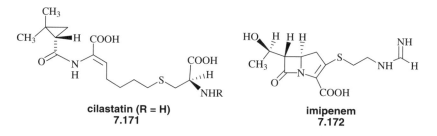

cilastatin (R = H)
7.171

imipenem
7.172

Aromatic amines resulting from phase I reduction of aromatic nitro compounds, such as the amine produced from the anticonvulsant drug clonazepam (see **7.106**, R = NO₂, Klonopin), also may be *N*-acetylated (**7.106**, R = NHAc).[283]

C.8 Fatty Acid and Cholesterol Conjugation

Compounds that contain a hydroxyl group can undergo conjugation reactions with a wide range of endogenous fatty acids, such as saturated acids from C_{10} to C_{16} and unsaturated acids such as oleic and linoleic acids.[284] The anti-inflammatory agent etofenamate (**7.173**, Rheumagel) is esterified by oleic, palmitic, linoleic, stearic, palmitoleic, myristic, and lauric acids[285] presumably as their coenzyme A thioesters.[286] These metabolites account for only 0.1% of the administered dose in rats and humans. Often these lipophilic fatty acid conjugates accumulate in the tissues rather than are excreted. For example, fatty acid ester metabolites of Δ^1-7-hydroxytetrahydrocannabinol (**7.174**) are deposited in the liver, spleen, adipose tissue, and bone marrow and account for the majority of tissue residues of this compound determined 15 days after administration.[287] "Flashbacks" by habitual cannabinoid users long after they stop using the drug are believed to be the result of the lipophilic conjugates retained in the tissues for extended periods of time which are released and converted back to the cannabinoid.[288]

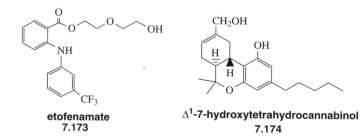

etofenamate
7.173

Δ^1-7-hydroxytetrahydrocannabinol
7.174

Cholesterol ester metabolites have been detected for drugs containing either an ester or a carboxylic acid. Transesterification of prednimustine (**7.175**), an ester of chlorambucil (Leukeran) and prednisolone, to cholesterol (see **2.77**) was catalyzed by the enzyme lecithin cholesterol acyltransferase.[289] The experimental hypolipidemic drug **7.176** was metabolized to the cholesterol ester, which deposited in the lipids in the liver; as a consequence, the development of this drug had to be discontinued.[290]

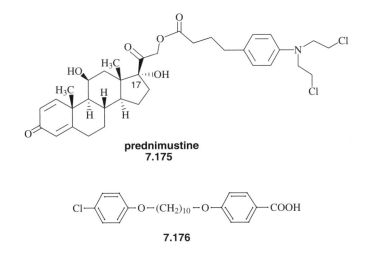

prednimustine
7.175

Cl—⟨⟩—O—(CH₂)₁₀—O—⟨⟩—COOH

$$Cl\text{---}\bigcirc\text{---}O\text{---}(CH_2)_{10}\text{---}O\text{---}\bigcirc\text{---}COOH$$

7.176

C.9 Methyl Conjugation

Methylation is a relatively minor conjugation pathway in drug metabolism, but it is very important in the biosynthesis of endogenous compounds such as epinephrine and melatonin, in the catabolism of biogenic amines such as norepinephrine, dopamine, serotonin, and histamine, and in modulating the activities of macromolecules such as proteins and nucleic acids.[291] Except when tertiary amines are converted to quaternary ammonium salts, methylation differs from almost all other conjugation reactions (excluding acetylation) in that it *reduces* the polarity and hydrophilicity of the substrates. In many cases this conjugation results in compounds with decreased biological activity. In general, xenobiotics that undergo methylation do share marked structural similarities to endogenous substrates that are methylated.[292]

Methylation is a two-step process (Scheme 7.53). First the methyl-transferring coenzyme, *S*-adenosylmethionine (SAM) (**7.178**), is biosynthesized mostly from methionine (**7.177**) in a reaction catalyzed by methionine adenosyltransferase. Some *S*-adenosylmethionine also is produced by the donation of the methyl group of N^5-methyltetrahydrofolate (see Section 4.3.B, p. 200) to *S*-adenosyl-*L*-homocysteine. Then the *S*-adenosylmethionine is utilized in the transfer of the activated methyl group (Nature's methyl iodide) to the acceptor molecules (RXH), which include catechols and phenols, amines, and thiols. A variety of methyltransferases, such as catechol-*O*-methyltransferase (COMT), phenol-*O*-methyltransferase, phenylethanolamine-*N*-methyltransferase, and nonspecific amine *N*-methyltransferases and thiol *S*-methyltransferases, are responsible for catalyzing the transfer of the methyl group from *S*-adenosylmethionine to RXH, which produces $RXCH_3$ and *S*-adenosylhomocysteine (**7.179**).[293]

COMT-catalyzed *O*-methylation of xenobiotic catechols leads to *O*-monomethylated catechol metabolites. Unlike the free catechols, these metabolites are not oxidized to reactive

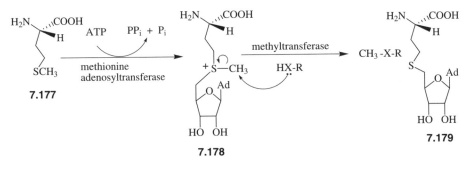

Scheme 7.53 ▶ Methylation of xenobiotics

ortho-quinonoid species that can produce toxic effects (see Section 7.4.B.1.h, p. 446). An example of this reaction is the metabolism of the β_1- and β_2-adrenergic bronchodilator isoproterenol (**7.180**, Isuprel), which is regioselectively methylated at the *C*-3 catechol OH group.[294] Compounds that are methylated by COMT must contain an aromatic 1,2-dihydroxy (catechol) group. Terbutaline (**7.181**, Brethine), another β_2-adrenergic bronchodilator related in structure to isoproterenol except that it contains a *meta*-dihydroxy arrangement of hydroxyl groups, does not undergo methylation.[295]

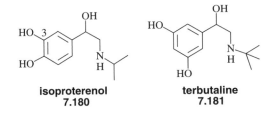

isoproterenol
7.180

terbutaline
7.181

Phenol hydroxyl groups also undergo methylation. Morphine (**7.182**, R = H), for example, is metabolized in humans to minor amounts of codeine (**7.182**, R = CH$_3$).[296]

morphine (R = H)
codeine (R = CH₃)
7.182

N-Methylation of xenobiotics is less common, but occurs occasionally. The *N*-dealkylated primary amine metabolite (**7.183**, R = H) of the coronary vasodilator (β-adrenergic blocker) oxprenolol (**7.183**, R = CH(CH$_3$)$_2$, Trasicor) is *N*-methylated to **7.183** (R = CH$_3$).[297]

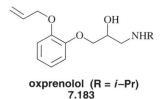

oxprenolol (R = *i*–Pr)
7.183

Heterocyclic nitrogen atoms also are susceptible to *N*-methylation, such as in the case of the sedative clomethiazole (**7.184**, Heminevrin).[298]

clomethiazole
7.184

Methylation of sulfhydryl groups in xenobiotics also is known. The thiol group of the antihypertensive drug captopril (**7.185**, Capoten)[299] and the antithyroid drug propylthiouracil (**7.186**, PTU)[300] undergo metabolic *S*-methylation.

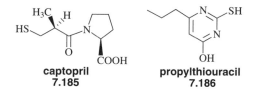

captopril   **propylthiouracil**
7.185   **7.186**

7.4.D Hard and Soft Drugs; Antedrugs

Rapid metabolism of a drug candidate can lead to the demise of the compound for further development because of poor bioavailability or short duration. However, sometimes the reverse situation also can be a problem, namely, a plasma half-life that is too long (the compound is not cleared at an adequate rate). Unless the drug can be eliminated, minor toxicities can accumulate and become serious problems. This phenomenon was observed during the discovery of the antiarthritis drug celecoxib (**7.187**, Celebrex; see Section 5.5.B.2.b, p. 280). One of the predecessors of **7.187** was **7.188** (in which a *para*-chloro group replaces the methyl group of celecoxib).[301] As we saw earlier halogens are sometimes incorporated into the *para*-position of aromatic compounds to block aromatic oxidation (see Section 7.4.B.1). This compound was more COX-2 selective than celecoxib, but had a plasma half-life in dogs of 680 h (about a month)! This may sound like a favorable property, expecially for a disease like arthritis in which the medication is taken every day anyway, but the compound accumulated in various tissues, and some liver toxicity resulted. To shorten the plasma half-life, the structure was modified to one that would more likely be metabolized and excreted, namely, by a bioisosteric replacement of the *para*-chloro group with a *para*-methyl group (which we saw is readily oxidized via the alcohol, then the aldehyde, to the carboxylic acid; see Sections 7.4.B.1.d, p. 430, and 7.4.B.1.i, p. 447). The resulting compound (celecoxib) has a plasma half-life in dogs of 9 h and in humans of 10–12 h (peak plasma levels at about 2 h).

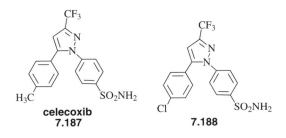

celecoxib
7.187

7.188

Bodor[302] has termed compounds like **7.188** hard drugs and those like celecoxib (**7.187**) soft drugs. *Hard drugs* are nonmetabolizable compounds, characterized either by high lipid solubility and accumulation in adipose tissues and organelles or high water solubility. They are poor substrates for the metabolizing enzymes; the potentially metabolically sensitive parts of these drugs are either sterically hindered or the hydrogen atoms are substituted with halogens to block oxidation.

Soft drugs (also referred to as *antedrugs*)[303] are biologically active drugs designed to have a predictable and controllable metabolism to nontoxic and inactive products after they have achieved their desired pharmacological effect. By building into the molecule the most desirable way in which the molecule could be deactivated and detoxified shortly after it has exerted its biological effect, the therapeutic index could be increased, providing a safer drug. The advantages of soft drugs can be significant: (1) elimination of toxic metabolites, thereby increasing the therapeutic index of the drug; (2) avoidance of pharmacologically active metabolites that can lead to long-term effects; (3) elimination of drug interactions resulting from metabolite inhibition of enzymes; and (4) simplification of pharmacokinetic problems caused by multiple active species. The general characteristics of a soft drug are[304] (1) that it has a close structural similarity to the lead; (2) that has a metabolically sensitive moiety built into the lead structure; (3) that the incorporated metabolically sensitive spot does not affect the overall physicochemical or steric properties of the lead compound; (4) that the built-in metabolism is the major, or preferably the only, metabolic route for drug deactivation; (5) that the rate of the predictable metabolism can be controlled; (6) that the products resulting from the metabolism are nontoxic and have no other biological activities; and (7) that the predicted metabolism does not lead to highly reactive intermediates.

Another example of this soft drug concept is the isosteric soft analog (**7.189**) of the hard antifungal cetylpyridinium chloride (**7.190**, Cepacol).[305] The soft analog has a metabolically soft spot (the ester group) built into the structure for detoxification by an esterase-catalyzed hydrolysis. Of course, incorporation of the ester group cannot interfere with binding of the compound to the appropriate receptor if this approach is to be effective.

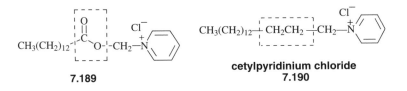

7.189

cetylpyridinium chloride
7.190

Another related approach to soft drug design is to utilize, as the basis for the design of the drug, a biologically inactive metabolite which is modified so that it becomes active, but which can be metabolically degraded to the inactive metabolite after its therapeutic action is completed. An example of this approach is the soft ophthalmic anti-inflammatory drug loteprednol etabonate (**7.191**, Alrex).[306] This glucocorticoid drug was designed based on the structure of an inactive metabolite of prednisolone (**7.192**), namely, **7.193**. The unfavorable responses, such as allergies and cataracts, that are observed in patients that undergo long-term glucocorticoid therapy have been proposed to occur because of the α-hydroxycarbonyl group at the 17β-position.[307] Bodor's group changed this to a carboxylic acid group (**7.193**), but the compound was inactive. The approach taken, then, involved making compounds that are active, but metabolically degraded to a biologically inactive metabolite, such as **7.193**, which does not bind to the glucocorticoid receptor. The active compound was found to be **7.191** (a potent binder of the glucocorticoid receptor). Compound **7.191** is metabolically degraded by nonspecific esterases to **7.193**, which is readily cleared. This approach is like doing a retro-synthesis in organic chemistry; you have to think backwards from the desired inactive metabolite to an active compound that can be metabolically degraded to the inactive metabolite.

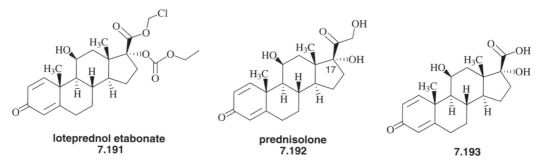

loteprednol etabonate
7.191

prednisolone
7.192

7.193

A soft drug, then, is an active drug that is degraded by metabolism *after* it carries out its therapeutic role. Another way of utilizing metabolism in drug design is to modify a drug so that it is not active, but before it reaches the site of action, is metabolically transformed into the active drug. These compounds are known as *prodrugs*, and they are the topic of the last chapter.

7.5 General References

Synthesis of Radioactive Compounds

Muccino, R. R. (Ed.) *Synthesis and Applications of Radioactive Compounds*, Elsevier, Amsterdam, 1986.

Analytical Methods in Drug Metabolism

Günther, H. *NMR Spectroscopy: Basic Principles, Concepts, and Applications in Chemistry*, 2nd ed., John Wiley & Sons, New York, 1995.
Lambert, J. B.; Mazzola, E. P. *Nuclear Magnetic Resonance Spectroscopy: An Introduction to Principles, Applications, and Experimental Methods*, Prentice-Hall, Upper Saddle River, NJ, 2003.
Lambert, J. B.; Shurvell, H. F.; Lightner, D. A.; Cooks, R. G. *Organic Structural Spectroscopy*, Prentice-Hall, Upper Saddle River, NJ, 1998.

Mass Spectrometry

Busch, K. L.; Glish, G. L.; McLuckey, S. A. *Mass Spectrometry/Mass Spectrometry: Techniques and Applications of Tandem Mass Spectrometry*, VCH, New York, 1988.

Cole, R. B. (Ed.) *Electrospray Ionization Mass Spectrometry*, John Wiley, New York, 1997.

McLafferty, F. W.; Turecek, F. *Interpretation of Mass Spectra*, 4th ed., University Science Books, 1996.

Rossi, D. T.; Sinz, M. W. (Eds.) *Mass Spectrometry in Drug Discovery*, Marcel Dekker, New York, 2002.

Watson, J. T. *Introduction to Mass Spectrometry*, 3rd ed., Lippincott Williams & Wilkins, Philadelphia, 1997.

HPLC

Cunico, R. L.; Gooding, K. M.; Wehr, T. *Basic HPLC and CE of Biomolecules*, Bay Bioanalytical Laboratory, 1998.

Snyder, L. R.; Kirkland, J.; Glajch, J. *Practical HPLC Method Development*, 2nd ed., Wiley-Interscience, New York, 1997.

Pathways for Drug Deactivation and Elimination: Books and Reviews

Armstrong, R. N. *CRC Crit. Rev. Biochem.* **1987**, *22*, 39.

Cheng, X.; Blumenthal, R. M. (Eds.) *S-Adenosylmethionine-Dependent Methyltransferases: Structures and Functions*, World Scientific, 2000.

Danielson, P. B. *Curr. Drug Metab.* **2002**, *3*, 561.

Dolphin, D.; Poulson, R.; Avramovic, O. *Glutathione*, Wiley, New York, 1989, Vols. 1 and 2.

Faber, K. (Ed.) *Biotransformations*, Springer Verlag, Berlin, 2000.

Gibson, G. G.; Skett, P. *Introduction to Drug Metabolism*, Chapman and Hall, London, 1986.

Guengerich, F. P. (Ed.) *Mammalian Cytochromes P450*, CRC Press, Boca Raton, FL, 1987, Vols. I and II.

Hawkins, D. R. (Ed.) *Biotransformations*, Royal Society of Chemistry, Cambridge, UK, annual series, beginning in 1989.

Ioannides, C. (Ed.) *Cytochromes P450: Metabolic and Toxicological Aspects*, CRC Press, Boca Raton, FL, 1996.

Ioannides, C. (Ed.) *Enzyme Systems That Metabolise Drugs and Other Xenobiotics (Current Toxicology Series)*, John Wiley & Sons, New York, 2002.

Kwon, Y. *Handbook of Essential Pharmacokinetics, Pharmacodynamics and Drug Metabolism for Industrial Scientists*, Plenum, New York, 2001.

Lewis, D. F. V. *Guide to Cytochromes P450. Structure and Function*, Taylor & Francis, New York, 2001.

Mulder, G. J. (Ed.) *Conjugation Reactions in Drug Metabolism*, Taylor & Francis, London, 1990.

Ortiz De Montellano, P. R. (Ed.) *Cytochrome P450: Structure, Mechanism, and Biochemistry*, 2nd ed., Plenum, New York, 1995.

Rodrigues, A. D.; Rushmore, T. H. *Curr. Drug Metab.* **2002**, *3*, 289.

Sies, H.; Ketterer, B. (Eds.) *Glutathione Conjugation: Mechanisms and Biological Significance*, Academic, London, 1988.

Smith, D. A.; Jones, B. C.; Walker, D. K. *Med. Res. Rev.* **1996**, *16*, 243.

Smith, D. A.; Van De Waterbeemd, H.; Walker, D. K. *Pharmacokinetics and Metabolism in Drug Design*, John Wiley, New York, 2000.

Weber, W. W. *The Acetylator Genes and Drug Response*, Oxford, New York, 1997.

Woolf, T. F., (Ed.) *Handbook of Drug Metabolism*, Marcel Dekker, New York, 1999.

Journals

Current Drug Metabolism
Drug Metabolism & Disposition Drug Metabolism & Drug Interactions
Drug Metabolism Reviews
Chemico-Biological Interactions

Software

Bio-Rad PredictIt™ Metabolism
Bio-Rad PredictIt® Toxicity

7.6 PROBLEMS

(Answers can be found in Appendix at the end of the book.)

1. How do you explain the observation that two enantiomers can be metabolized to completely different products?

2. Why do the two enantiomers of glutethimide (**7.40**) undergo P450-catalyzed hydroxylation at two different sites?

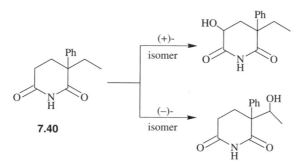

3. Draw a mechanism to account for the following metabolic reaction.

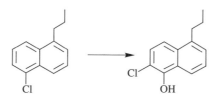

4. Compound **1** is metabolized to **2**; however, the metabolic enzyme that acts on **1** is slowly inactivated. Draw a mechanism for the conversion of **1** to **2** that also rationalizes the loss of enzyme activity (indicate at what point inactivation could occur).

5. Draw two mechanisms for oxidative aromatization (**7.94** ⟶ **7.95** in the text).

6. Show all of the steps (not the mechanisms) of this metabolic pathway, and note over each arrow used in your scheme the general types of enzymes involved.

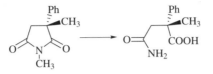

7. The antiestrogen drug toremifene (**3**) is metabolized into six major metabolites. Draw six reasonable metabolites for this drug, and note over each arrow what kind of reaction and enzyme are involved.

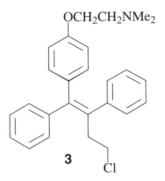

8. Compound **4** undergoes P450-catalyzed conversion to **5** and **6**. Draw a mechanism to account for this metabolic reaction.

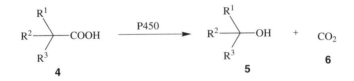

9. Compound **7** is metabolized by *O*-dealkylation to a dicarboxylic acid. Draw a mechanism.

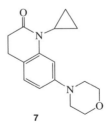

10. Why is glutathione in high concentrations (5–10 mM) in many tissues?

11. Which of the following hypothetical metabolic reactions are phase I and which are phase II? Name the reactions, and note the cofactors (no mechanisms needed).

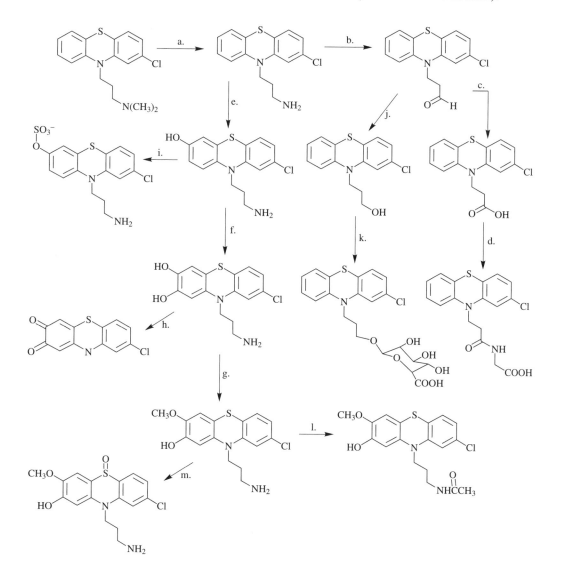

12. Compound **8** was metabolized to **9** (GSH is glutathione). Draw a reasonable mechanism to account for this metabolic reaction.

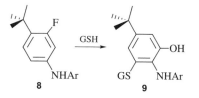

13. Predict the structures of the compounds that produce the following metabolites (you need to work backwards from the metabolite to the compound). Show the steps (not mechanisms), and name the enzymes.

A.

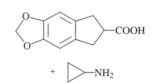

(2 steps)

B.

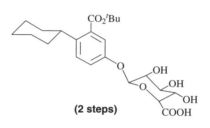

(2 steps)

C.

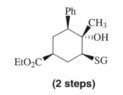

(2 steps)

14. Statistics show that about 50% of AIDS patients in the United States experience serious diarrhea (AIDS-related diarrhea or ARD). N^1, N^{14}-Diethylhomospermine (**10**) prevents diarrhea in rodents, but the metabolite, homospermine (**11**), accumulates and persists in tissues for a protracted period of time, which accounts for the chronic toxicity of **10**.

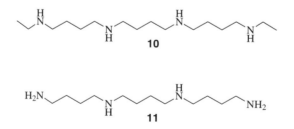

A. Suggest an approach you would take to get around the toxicity problem.

B. Draw a compound you might try to overcome the long plasma half-life, and note how it might work.

7.7 References

1. Chen, Z.; Zhang, J.; Stamler, J. S. *Proc. Natl. Acad. Sci. USA* **2002**, *99*, 8306.

2. (a) Honig, P. K.; Woosley, R. L.; Zamani, K.; Conner, D.P.; Cantilena, L. R., Jr. *Clin. Pharmacol. Ther.* **1992**, *52*, 231. (b) Honig, P. K.; Wortham, D. C.; Zamani, K.; Conner, D. P.; Mullin, J. C.; Cantilena, L.R. *J. Am. Med. Assoc.* **1993**, *269*, 1513.

3. Markham, A.; Wagstaff, A. *J. Drugs* **1998**, *55*, 269.

4. (a) Dalvie, D. *Curr. Pharm. Design* **2000**, *6*, 1009. (b) Dain, J. G.; Collins, J. M.; Robinson, W. T. *Pharm. Res.* **1994**, *11*, 925.

5. (a) Lim, C.-K.; Lord, G. *Biol. Pharm. Bull.* **2002**, *25*, 547. (b) Jackson, P. J.; Brownsill, R. D.; Taylor, A. R.; Walther, B. *J. Mass Spectrom.* **1995**, *30*, 446.

6. McCarthy, K. E. *Curr. Pharm. Design* **2000**, *6*, 1057.

7. Saljoughian, M.; Williams, P. G. *Curr. Pharm. Design* **2000**, *6*, 1029.

8. Hirth, J.; Watkins, P. B.; Strawderman, M.; Schott, A.; Bruno, R.; Baker, L. H. *Clin. Cancer Res.* **2000**, *6*, 1255.

9. Brickner, S. J.; Hutchinson, D. K.; Barbachyn, M. R.; Manninen, P. R.; Ulanowicz, D. A.; Garmon, S. A.; Grega, K. C.; Hendges, S. K.; Toops, D. S.; Ford, C. W.; Zurenko, G. E. *J. Med. Chem.* **1996**, *39*, 673.

10. Slatter, J. G.; Stalker, D. J.; Feenstra, K. L.; Welshman, I. R.; Bruss, J. B.; Sams, J. P.; Johnson, M. G.; Sanders, P. E.; Hauer, M. J.; Fagerness, P. E.; Stryd, R. P.; Peng, G. W.; Shobe, E. M. *Drug Metab. Dispos.* **2001**, *29*, 1136.

11. U.S. Code of Federal Regulations, Title 21, *Drugs Used in Research*, Part 361.1, U.S. Government Printing Office, Washington, DC, 1986, p. 160.

12. deSousa, G.; Florence, N.; Valles, B.; Coassolo, P.; Rahmani, R. *Cell Biol. Toxicol.* **1995**, *11*, 147.

13. Coe, R. A. *J. Regulatory Toxicol. Pharmacol.* **2000**, *31*, S1.

14. Solon, E. G.; Kraus, L. *J. Pharmacol. Toxicol. Meth.* **2002**, *46*, 73.

15. Henion, J.; Brewer, E.; Rule, G. *Anal. Chem.* **1998**, *70*, 650A.

16. (a) Schill, G.; Borg, K. O.; Modin, R.; Persson, B. A. In *Progress in Drug Metabolism*, Bridges, J. W.; Chasseaud, L. F. (Eds.) Wiley, New York, 1977, Vol. 2, p. 219. (b) Schill, G.; Modin, R.; Borg, K. O.; Persson, B. A. In *Drug Fate and Metabolism: Methods and Techniques*, Garrett, E. R.; Hirtz, J. L. (Eds.), Marcel Dekker, New York, 1977, Vol. 1, p. 135.

17. Horning, M. G.; Gregory, P.; Nowlin, J.; Stafford, M.; Letratanangkoon, K.; Butler, C.; Stillwell, W. G.; Hill, R. M. *Clin. Chem.* **1974**, *20*, 282.

18. Thompson, J. A.; Markey, S. P. *Anal. Chem.* **1975**, *47*, 1313.

19. Brodie, B. B.; Cho, A. K.; Gessa, G. L. In *Amphetamines and Related Compounds*; Costa, E.; Garattini, S. (Eds.), Raven, New York, 1970, p. 217.

20. Stolman, A.; Pranitis, P. A. F. *Clin. Toxicol.* **1977**, *10*, 49.

21. (a) Xia, Y. Q.; Whigan, D. B.; Powell, M. L.; Jemal, M. *Rapid Commun. Mass Spectrom.* **2000**, *14*, 105. (b) Wu, J. T.; Zeng, H.; Qian, M.; Brogdon, B. L.; Unger, S. E. *Anal. Chem.* **2000**, *72*, 61.

22. (a) Bennett, P.; Li, Y. T.; Edom, R.; Henion, J. *J. Mass Spectrom.* **1997**, *32*, 739. (b) Xia, Y.; Whigan, D.; Jemal, M. *Rapid Commun. Mass Spectrom.* **1999**, *13*, 1611.

23. Janiszewski, J.; Swyden, M.; Fouda, H. *J. Chromatogr. Sci.* **2000**, *38*, 255.

24. (a) Zhang, N.; Hoffman, K. L.; Li, W.; Rossi, D. T. *J. Pharm. Biomed. Anal.* **2000**, *22*, 131. (b) Steinborner, S.; Henion, J. *Anal. Chem.* **1999**, *71*, 2340.

25. Marchese, A.; McHugh, C.; Kehler, J.; Bi, H. *J. Mass. Spectrom.* **1998**, *33*, 1071.

26. (a) Needham, S. R.; Cole, M. J.; Fouda, H. G. *J. Chromatogr. B* **1998**, *718*, 87. (b) Jemal, M.; Huang, M.; Jiang, X.; Mao, Y.; Powell, M. *Rapid Commun. Mass Spectrom.* **1999**, *13*, 2125.

27. Maurer, H. H. *J. Chromatogr. B*: **1999**, *733* (1 + 2), 3.

28. (a) Hadley, M. R.; Camilleri, P.; Hutt, A. *J. Electrophoresis* **2000**, *21*, 1953. (b) Naylor, S.; Benson, L. M.; Tomlinson, A. J. *J. Chromatogr. A* **1996**, *735*(1 + 2), 415.

29. (a) Wells, R. J. *J. Chromatogr. A* **1999**, *843*, 1. (b) Little, J. L. *J. Chromatogr. A* **1999**, *844*, 1.

30. (a) Lim, C.-K.; Lord, G. *Biol. Pharm. Bull.* **2002**, *25*, 547. (b) Cole, M. J.; Janiszewski, J. S.; Fouda, H. G. *Practical Spectroscopy*; Pramanik, B. N.; Ganguly, A. K.; Gross, M. L. (Eds.), Marcel Dekker, New York, 2002, Vol. 32, pp. 211–249. (c) Gibbs, B.; Masse, R. *Innov. Pharm. Technol.* **2001**, *1*, 76, 78, 80, 82, 84.

31. (a) Von Brocke, A.; Nicholson, G.; Bayer, E. *Electrophoresis* **2001**, *22*, 1251. (b) Cherkaoui, S.; Rudaz, S.; Varesio, E.; Veuthey, J.-L. *Chimia* **1999**, *53*, 501.

32. (a) Deng, Y.; Zeng, H.; Wu, J.-T. *Recent Res. Develop. Anal. Chem.* **2001**, *1* 45. (b) Jemal, M. *Biomed. Chromatogr.* **2000**, *14*, 422.

33. Miller, J. D.; Chang, S. Y. *PharmaGenomics* **2001**, *1* (August), 46.

34. Edwards, R. A.; Zhang, K.; Ferth, L. *Drug Discovery World* **2002**, *3*, 67.

35. (a) Prentis, R. A.; Lis, Y.; Walker, S. R. *Brit. J. Clin. Pharmacol.* **1988**, *25*, 387. (b) Kennedy, T. *Drug Discovery Today* **1997**, *2*, 436.

36. Cole, M. J.; Janiszewski, J. S.; Fouda, H. G. In *Practical Spectroscopy*; Pramanik, B. N.; Ganguly, A. K.; Gross, M. L. (Eds.), Marcel Dekker, New York, 2002, Vol. 32, pp. 211–249.

37. Pramanik, B. N.; Bartner, P. L.; Chen, G. *Curr. Opin. Drug Dis. Develop.* **1999**, *2*, 401.

38. (a) Beavis, R. C.; Chait, B. T. *Meth. Enzymol.* **1996**, *270*, 519. (b) Hillenkamp, F.; Karas, M.; Beavis, R. C.; Chait, B. T. *Anal. Chem.* **1991**, *63*, 1193A.

39. Clerc, J.; Fourre, C.; Fragu, P. *Cell Biol. Internat.* **1997**, *21*, 619.

40. Cole, R. B. *Electrospray Ionization Mass Spectrometry*, Wiley, New York, 1997.

41. Willoughby, R.; Sheehan, E.; Mitrovich, S. *A Global View of LC/MS*, Global View, Pittsburgh, 1998.

42. (a) Schneider, R. P.; Fouda, H. G.; Inskeep, P. B. *Drug Metab. Dispos.* **1998**, *26*, 1149. (b) Prakash, C.; Kamel, A.; Anderson, W.; Howard, H. *Drug Metab. Dispos.* **1997**, *25*, 206.

43. (a) Pochapsky, S. S.; Pochapsky, T. C. *Curr. Top. Med. Chem.* **2001**, *1*, 427. (b) Lindon, J. C.; Nicholson, J. K.; Wilson, I. D. *J. Chromatogr. B* **2000**, *748*, 233. (c) Peng, S. X. *Biomed. Chromatogr.* **2000**, *14*, 430. (d) Wilson, I. D.; Nicholson, J. K.; Lindon, J. C. In *Handbook of Drug Metabolism*, Woolf, T. F. (Ed.), Marcel Dekker, New York, 1999, p. 523.

44. (a) Wu, N.; Webb, L.; Peck, T. L.; Sweedler, J. V. *Anal. Chem.* **1995**, *67*, 3101. (b) Behnke, B.; Schlotterbeck, G.; Tallarck, U.; Strohschein, S.; Tseng, L.-H.; Keller, T.; Albert, K.; Bayer, E. *Anal. Chem.* **1996**, *68*, 1110.

45. (a) Sandvoss, M.; Weltring, A.; Preiss, A.; Levsen, K.; Wuensch, G. *J. Chromatogr. A* **2001**, *917*, 75. (b) Dear, G. J.; Plumb, R. S.; Sweatman, B. C.; Ayrton, J.; Lindon, J. C.; Nicholson, J. K.; Ismail, I. M. *J. Chromatogr. B* **2000**, *748*, 281.

46. Hiller, D. L.; Zuzel, T. J.; Williams, J. A.; Cole, R. O. *Rapid Commun. Mass Spectrom.* **1997**, *11*, 593.

47. (a) Liberg, J. Poggendorff's *Ann. Phys. Chem.* **1829**, *17*, 389. (b) Lehmann, C. G. *J. Prakt. Chem.* **1835**, *6*, 113. (c) Ure, A. *Pharm. J. Trans.* **1841**, *1*, 24.

48. (a) Mueller, G. C.; Miller, J. A. *J. Biol. Chem.* **1948**, *176*, 535. (b) Mueller, G. C.; Miller, J. A. *J. Biol. Chem.* **1949**, *180*, 1125. (c) Mueller, G. C.; Miller, J. A. *J. Biol. Chem.* **1953**, *202*, 579.

49. Smith, D. A.; Brown, K.; Neale, M.G. *Drug Metab. Rev.* **1985**, *16*, 365.

50. (a) Woolf, T. F. (Ed.) *Handbook of Drug Metabolism*, Marcel Dekker, New York, 1999. (b) Williams, R. T. *Detoxification Mechanisms*, 2nd ed., Chapman & Hall, London, 1959.

51. Jamali, F.; Mehvar, R.; Pasutto, F. M. *J. Pharm. Sci.* **1989**, *78*, 695.

52. Hignite, C.; Utrecht, J.; Tschang, C.; Azarnoff, D. *Clin. Pharmacol. Ther.* **1980**, *28*, 99.

53. Silber, B.; Holford, N. H. G.; Riegelman, S. *J. Pharm. Sci.* **1982**, *71*, 699.

54. (a) Ariëns, E. J. *Med. Res. Rev.* **1986**, *6*, 451. (b) Simonyi, M. *Med. Res. Rev.* **1984**, *4*, 359.

55. Eichelbaum, M. *Biochem. Pharmacol.* **1988**, *37*, 93.

56. Adams, S. S.; Bresloff, P.; Mason, C. G. *J. Pharm. Pharmacol.* **1976**, *28*, 256.

57. (a) Kaiser, D. G.; Vangeissen, G. J.; Reischer, R. J.; Wechter, W. J. *J. Pharm. Sci.* **1976**, *65*, 269. (b) Lee, E.; Williams, K.; Day, R.; Graham, G.; Champion, D. Brit. *J. Clin. Pharmacol.* **1985**, *19*, 669.

58. Caldwell, J.; Hutt, A. J.; Fournel-Gigleux, S. *Biochem. Pharmacol.* **1988**, *37*, 105.

59. Houston, J. B. *Biochem. Pharmacol.* **1994**, *47*, 1469.

60. (a) Mueller, G. C.; Miller, J. A. *J. Biol. Chem.* **1948**, *176*, 535. (b) Mueller, G. C.; Miller, J. A. *J. Biol. Chem.* **1949**, *180*, 1125. (c) Mueller, G. C.; Miller, J. A. *J. Biol. Chem.* **1953**, *202*, 579.

61. Brodie, B. B.; Axelrod, J.; Cooper, J. R.; Gaudette, L.; LaDu, B. N.; Mitoma, C.; Udenfriend, S. *Science* **1953**, *121*, 603.

62. Mason, H. S. *Science* **1957**, *125*, 1185.

63. Posner, H. S.; Mitoma, C.; Rothberg, S.; Udenfriend, S. *Arch. Biochem. Biophys.* **1961**, *94*, 280.

64. (a) Klingenberg, M. *Arch. Biochem. Biophys.* **1958**, *75*, 376. (b) Garfinkel, D. *Arch. Biochem. Biophys.* **1958**, *77*, 493.

65. (a) Guengerich, F. P. *Chem. Res. Toxicol.* **2001**, *14*, 611. (b) Danielson, P. B. *Curr. Drug Metab.* **2002**, *3*, 561.

66. Kim, J.-J. P.; Roberts, D. L.; Djordjevic, S.; Wang, M.; Shea, T. M.; Masters, B. S. S. *Meth. Enzymol.* **1996**, *272*, 368.

67. (a) Vermilion, J. L.; Ballou, D. P.; Massey, V.; Coon, M. J. *J. Biol. Chem.* **1981**, *256*, 266. (b) Oprian, D. D.; Coon, M. J. *J. Biol. Chem.* **1982**, *257*, 8935. (c) Strobel, H. W.; Dignam, J. D.; Gum, J. R. *Int. Encycl. Pharmacol. Ther.* **1982**, *108*, 361.

68. Levy, R. H.; Thummel, K. E.; Trager, W. F. (Eds.) *Metabolic Drug Interactions*, Lippincott Williams & Wilkins, Philadelphia, 2000.

69. (a) Volz, H. P. *Pharmacopsychiatry* **1997**, *30* (Suppl. 2), 72. (b) Wheatley, D. *Pharmacopsychiatry* **1997**, *30* (Suppl. 2), 77.

70. Laakmann, G.; Schule, C.; Baghai, T.; Kieser, M. *Pharmacopsychiatry* **1998**, *31* (Suppl.), 54.

71. Moore, L. B.; Goodwin, B.; Jones, S. A.; Wisely, G. B.; Serabjit-Singh, C. J.; Willson, T. M.; Collins, J. L.; Kliewer, S. A. *Proc. Natl. Acad. Sci. USA* **2000**, *97*, 7500.

72. Smith, D. A.; Jones, B. C. *Biochem. Pharmacol.* **1992**, *44*, 2089.

73. (a) Ziegler, D. M. *Drug Metab. Rev.* **2002**, *34*, 503. (b) Ziegler, D. M. *Annu. Rev. Pharmacol. Toxicol.* **1993**, *33*, 179.

74. Hines, R. N.; Cashman, J. R.; Philpot, R. M.; Williams, D. E.; Ziegler, D. M. *Toxicol. Appl. Pharmacol.* **1994**, *125*, 1.

75. Boyland, E. *Biochem. Soc. Symp.* **1950**, *5*, 40.

76. Jerina, D. M.; Daly, J. W.; Witkop, B.; Zaltzman-Nirenberg, P.; Udenfriend, S. *Biochemistry* **1970**, *9*, 147.

77. Korzekwa, K. R.; Swinney, D. C.; Trager, W. F. *Biochemistry* **1989**, *28*, 9019.

78. (a) Guroff, G.; Daly, J. W.; Jerina, D. M.; Renson, J.; Witkop, B.; Udenfriend, S. *Science* **1967**, *157*, 1524. (b) Jerina, D. *Chem. Technol.* **1973**, *4*, 120.

79. (a) Daly, J.; Jerina, D.; Witkop, B. *Arch. Biochem. Biophys.* **1968**, *128*, 517. (b) Daly, J. W.; Jerina, D. M.; Witkop, B. *Experientia* **1972**, *28*, 1129.

80. Atkinson, J. K.; Hollenberg, P. F.; Ingold, K. U.; Johnson, C. C.; Le Tadic, M.-H.; Newcomb, M.; Putt, D. A. *Biochemistry* **1994**, *33*, 10630.

81. Parli, C. J.; Schmidt, B. *Res. Commun. Chem. Pathol. Pharmacol.* **1975**, *10*, 601.

82. Park, B. K.; Kitteringham, N. R.; O'Neill, P. M. *Annu. Rev. Pharmacol. Toxicol.* **2001**, *41*, 443.

83. Marchetti, P.; Navalesci, R. *Clin. Pharmacokinet.* **1989**, *16*, 100.

84. Verbeck, R. K.; Blackburn, J. L.; Loewen, G. R. *Clin. Pharmacokinet.* **1983**, *8*, 297.

85. Wood, S. G.; Scott, P. W.; Chasseaud, L. F.; Faulkner, J. K.; Matthews, R. W.; Henrick, K. *Xenobiotica* **1985**, *15*, 107.

86. Daly, J. In *Concepts in Biochemical Pharmacology*, Brodie, B. B.; Gillette, J. R. (Eds.), Springer-Verlag, Berlin, 1971, Part 2, p. 285.

87. Parke, D. V. *Biochem. J.* **1960**, *77*, 493.

88. Dayton, P. G.; Perel, J. M.; Cummingham, R. F.; Israeli, Z. H.; Weiner, I. M. *Drug Metab. Dispos.* **1973**, *1*, 742.

89. Perry, T. L.; Culling, C. F. A.; Berry, K.; Hansen, S. *Science* **1964**, *146*, 81.

90. Butler, T. C.; Dudley, K. H.; Johnson, D.; Roberts, S. B. *J. Pharmacol. Exp. Ther.* **1976**, *199*, 82.

91. (a) Selander, H. G.; Jerina, D. M.; Piccolo, D. E.; Berchtold, G. A. *J. Am. Chem. Soc.* **1975**, *97*, 4428. (b) Billings, R. E.; McMahon, R. E. *Mol. Pharmacol.* **1978**, *14*, 145.

92. Tomaszewski, J. E.; Jerina, D. M.; Daly, J. W. *Biochemistry* **1975**, *14*, 2024.

93. Fretland, A. J.; Omiecinski, C. J. *Chem.-Biol. Interact.* **2000**, *129*, 41.

94. (a) Armstrong, R. N.; Cassidy, C. S. *Drug Metab. Rev.* **2000**, *32*, 327. (b) Lacourciere, G. M.; Armstrong, R. N. *J. Am. Chem. Soc.* **1993**, *115*, 10466. (c) Borham, B.; Jones, A. D.; Pinot, F.; Grant, D. F.; Kurth, M. J.; Hammock, B. D. *J. Biol. Chem.* **1995**, *270*, 26923. (d) Hammock, B. D.; Pinot, F.; Beetham, J. K.; Grant, D. F.; Arand, M. E.; Oesch, F. *Biochem. Biophys. Res. Commun.* **1994**, *198*, 850.

95. (a) Hanzlik, R. P.; Edelman, M.; Michaely, W. J.; Scott, G. *J. Am. Chem. Soc.* **1976**, *98*, 1952. (b) Hanzlik, R. P.; Hiedeman, S.; Smith, D. *Biochem. Biophys. Res. Commun.* **1978**, *82*, 310.

96. (a) Rushmore, T. H.; Pickett, C. B. *J. Biol. Chem.* **1993**, **268**, 11475. (b) Eaton, D. L.; Bammler, T. K. *Toxicol. Sci.* **1999**, *49*, 156. (c) Schipper, D. L.; Wagenmans, M. J. H.; Wagener, D. J. T.; Peters, W. H. M. *Int. J. Oncol.* **1997**, *10*, 1261.

97. (a) Liu, S.; Zhang, P.; Ji, X.; Johnson, W. W.; Gilliland, G. L.; Armstrong, R. N. *J. Biol. Chem.* **1992**, *267*, 4296. (b) Thorson, J. S.; Shin, I.; Chapman, E.; Stenberg, G.; Mannervik, B.; Schultz, P. G. *J. Am. Chem. Soc.* **1998**, *120*, 451.

98. Buckpitt, A. R.; Castagnoli, N., Jr.; Nelson, S. D.; Jones, A. D.; Bahnson, L. S. *Drug Metab. Dispos.* **1987**, *15*, 491.

99. (a) Nebert, D. W.; Boobis, A. R.; Yagi, H.; Jerina, D. M; Khouri, R. E. In *Biological Reactive Intermediates*; Jollow, D. J.; Kocsis, J. J.; Snyder, R.; Vainio, H. (Eds.), Plenum, New York, 1977, p. 125. (b) Yamamoto, J.; Subramaniam, R.; Wolfe, A. R.; Meehan, T. *Biochemistry* **1990**, *29*, 3966.

100. Heidelberger, C. *Annu. Rev. Biochem.* **1975**, *44*, 79.

101. (a) Weinstein, I. B.; Jeffrey, A. M.; Jennette, K. W.; Blobstein, S. H.; Harvey, R. G.; Harris, C.; Autrup, H.; Kasai, H.; Nakanishi, K. *Science* **1976**, *193*, 592. (b) Koreeda, M.; Moore, P. D.; Yagi, H.; Yeh, H. J. C.; Jerina, D. M. *J. Am. Chem. Soc.* **1976**, *98*, 6720.

102. Straub, K. M.; Meehan, T.; Burlingame, A. L.; Calvin, M. *Proc. Natl. Acad. Sci. USA* **1977**, *74*, 5285.

103. Bellucci, G.; Berti, G.; Chiappe, C.; Lippi, A.; Marioni, F. *J. Med. Chem.* **1987**, *30*, 768.

104. Johannessen, S. I.; Gerna, N. M.; Bakke, J.; Strandjord, R. E.; Morselli, P. L. *Brit. J. Clin. Pharmacol.* **1976**, *3*, 575.

105. McCann, J.; Spingtain, N. E.; Ikobori, J.; Ames, B. N. *Proc. Natl. Acad. Sci. USA* **1975**, *72*, 979.

106. McMahon, G.; Davis, E. F.; Huber, L. J.; Kim, Y; Wogan, G. N. *Proc. Natl. Acad. Sci. USA* **1990**, *67*, 1104.

107. Wogan, G. N. *Prog. Clin. Biol Res.* **1992**, *374*, 123.

108. Iyer, R.; Harris, T. M. *Chem. Res. Toxicol.* **1993**, *6*, 313.

109. (a) Essigmann, J. M.; Croy, R. G.; Nadzan, A. M.; Busby, W. F., Jr.; Reinhold, V. N.; Bchi, G.; Wogan, G. N. *Proc. Natl. Acad. Sci. USA* **1977**, *74*, 1870. (b) Croy, R. G.; Essigmann, J. M.; Reinhold, V. N.; Wogan, G. N. *Proc. Natl. Acad. Sci. USA* **1978**, *75*, 1745.

110. Gopalakrishnan, S.; Stone, M. P.; Harris, T. M. *J. Am. Chem. Soc.* **1989**, *111*, 7232.

111. Iyer, R. S.; Coles, B. F.; Raney, K. D.; Thier, R.; Guengerich, F. P.; Harris, T. M. *J. Am. Chem. Soc.* **1994**, *116*, 1603.

112. Hucker, H. B. *Pharmacologist* **1962**, *4*, 171.

113. Shetty, H. U.; Nelson, W. L. *J. Med. Chem.* **1988**, *31*, 55.

114. Guengerich, F. P.; Müller-Enoch, D.; Blair, I. A. *Mol. Pharmacol.* **1986**, *30*, 287.

115. Nakahara, Y.; Cook, C. E. *J. Chromatogr.* **1988**, *434*, 247.

116. Keberle, H.; Reiss, W.; Hoffman, K. *Arch. Int. Pharmacodyn.* **1963**, *142*, 117.

117. Ponchaut, S.; van Hoof, F.; Veitch, K. *Biochem. Pharmacol.* **1992**, *43*, 2435.

118. Cooper, R. G.; Evans, D. A. P.; Whibley, E. J. *J. Med. Genet.* **1984**, *21*, 27.

119. Katchen, B.; Buxbaum, S. *J. Clin. Endocrinol. Metabol.* **1975**, *41*, 373.

120. Wright, J.; Cho, A. K.; Gal, J. *Life Sci.* **1977**, *20*, 467.

121. Coutts, R. T.; Beckett, A. H. *Drug Metab. Rev.* **1977**, *6*, 51.

122. Gorrod, J. W. *Chem.-Biol. Interact.* **1973**, *7*, 289.

123. Parli, C. H.; McMahon, R. E. *Drug Metab. Dispos.* **1973**, *1*, 337.

124. Bakke, O. M.; Davies, D. S.; Davies, L.; Dollery, C. T. *Life Sci.* **1973**, *13*, 1665.

125. Beckett, A. H.; Al-Sarraj, S. *J. Pharm. Pharmacol.* **1972**, *24*, 916.

126. Beckett, A. H.; Coutts, R. T.; Ogunbona, F. A. *J. Pharm. Pharmacol.* **1974**, *26*, 312.

127. Floyd, D. M.; Dimball, S. D.; Krapcho, J.; Das, J.; Turk, C.; Moquin, R. V.; Lago, M. W.; Duff, K. J.; Lee, V. G.; White, R. E.; Ridgewell, R. E.; Moreland, S.; Brittain, R. J.; Normandin, D. E.; Hedberg, S. A.; Cucinotta, G. G. *J. Med. Chem.* **1992**, *35*, 756.

128. Smith, D. A.; *Drug Metab. Rev.* **1991**, *23*, 355.

129. Atwal, K. S.; Swanson, B. N.; Unger, S. E.; Floyd, D. M.; Moreland, S.; Hedberg, A.; O'Reilly, B. C. *J. Med. Chem.* **1991**, *34*, 806.

130. Gram, T. E.; Wilson, J. T.; Fouts, J. R. *J. Pharmacol. Exp. Ther.* **1968**, *159*, 172.

131. Anders, M. W.; Cooper, M. J.; Takemori, A. E. *Drug Metab. Dispos.* **1973**, *1*, 642.

132. (a) Dragoni, S.; Bellik, L.; Frosini, M.; Sgaragli, G.; Marini, S.; Gervasi, P. G.; Valoti, M. *Xenobiotica* **2003**, *33*, 181. (b) Haberle, D.; Szoko, E.; Magyar, K. *Curr. Med. Chem.* **2002**, *9*, 47. (c) Shin, H.-S. *Drug Metab. Dispos.* **1997**, *25*, 657. (d) Bach, M. V.; Coutts, R. T.; Baker, G. B. *Xenobiotica* **2000**, *30*, 297.

133. (a) Tarjanyi, Zs.; Kalasz, H.; Szebeni, G.; Hollosi, I.; Bathori, M.; Furst, S. *J. Pharm. Biomed. Anal.* **1998**, *17*, 725. (b) Baker, G. B.; Urichuk, L. J.; McKenna, K. F.; Kennedy, S. H.; Scheinin, H. Anttila, M.; Dahl, M. L.; Karnani, H.; Nyman, L. *Cell. Mol. Neurobiol.* **1999**, *19*, 411. (c) Bach, M. V.; Coutts, R. T.; Baker, G. B. *Xenobiotica* **2000**, *30*, 297.

134. (a) Langone, J. J.; Van Vunakis, H. *Methods Enzymol.* **1982**, *84*, 628. (b) Schievelbein, H. *Int. Encyl. Pharmacol. Ther.* **1984**, *114*, 1.

135. Nguyen, T.-L.; Gruenke, L. D.; Castagnoli, N., Jr. *J. Med. Chem.* **1976**, *19*, 1168.

136. Nelson, S. D.; Garland, W. A.; Breck, G. D.; Trager, W. F. *J. Pharm. Sci.* **1977**, *66*, 1180.

137. McMartin, C.; Simpson, P. *Clin. Pharmacol. Ther.* **1971**, *12*, 73.

138. Hucker, H. B.; Balletto, A. J.; Stauffer, S. C.; Zacchei, A. G.; Arison, B. H. *Drug Metab. Dispos.* **1974**, *2*, 406.

139. Christy, M. E.; Anderson, P. S.; Arison, B. H.; Cochran, D. W.; Engelhardt, E. L. *J. Org. Chem.* **1977**, *42*, 378.

140. (a) Weisburger, J. H.; Weisburger, E. K. *Pharmacol. Rev.* **1973**, *25*, 1. (b) Miller, J. A. *Cancer Res.* **1970**, *30*, 559.

141. Bondon, A.; Macdonald, T. L.; Harris, T. M.; Guengerich, F. P. *J. Biol. Chem.* **1989**, *264*, 1988.

142. Ziegler, D. M.; Ansher, S. S.; Nagata, T.; Kadlubar, F. F.; Jakoby, W. B. *Proc. Natl. Acad. Sci. USA* **1988**, *85*, 2514.

143. (a) Barker, E. A.; Smuckler, E. A. *Mol. Pharmacol.* **1972**, *8*, 318. (b) Willi, P.; Bickel, M. H. *Arch. Biochem. Biophys.* **1973**, *156*, 772.

144. Gorrod, J. W.; Temple, D. *J. Xenobiotica* **1976**, *6*, 265.

145. Miwa, G. T.; Walsh, J. S.; Kedderis, G. L.; Hollenberg, P. F. *J. Biol. Chem.* **1983**, *258*, 14445.

146. Weisburger, E. K. *Ann. Rev. Pharmacol. Toxicol.* **1978**, *18*, 395.

147. (a) Lin, J.-K.; Miller, J. A.; Miller, E. C. *Biochemistry* **1969**, *8*, 1573. (b) Lin, J.-K.; Miller, J. A.; Miller, E. C. *Cancer Res.* **1975**, *35*, 844.

148. Kempf, D. J.; Marsh, K. C.; Fino, L. C.; Bryant, P.; Craig-Kennard, A.; Sham, H. L.; Zhao, C.; Vasavanonda, S.; Kohlbrenner, W. E. *Bioorg. Med. Chem.* **1994**, *2*, 847.

149. Wick, M. J.; Jantan, I.; Hanna, P. E. *Biochem. Pharmacol.* **1988**, *37*, 1225.

150. (a) Mitchell, J. R.; Jollow, D. J.; Potter, W. Z.; Gillette, J. R.; Brodie, B. B. *J. Pharmacol. Exp. Ther.* **1973**, *187*, 211. (b) Potter, W. Z.; Davis, D. C.; Mitchell, J. R.; Jollow, D. J.; Gillette, J. R.; Brodie, B. B. *J. Pharmacol. Exp. Ther.* **1973**, *187*, 203.

151. Jollow, D. J.; Thorgeirsson, S. S.; Potter, W. Z.; Hashimoto, M.; Mitchell, J. R. *Pharmacol.* **1974**, *12*, 251.

152. (a) Hinson, J. A.; Pohl, L. R.; Gillette, J. R. *Life Sci.* **1979**, *24*, 2133. (b) Nelson, S. D.; Forte, A. J.; Dahlin, D. C. *Biochem. Pharmacol.* **1980**, *29*, 1617.

153. (a) Nelson, S. D.; McMurty, R. J.; Mitchell, J. R. In *Biological Oxidation of Nitrogen*; Gorrod, J. W. (Ed.), Elsevier/North-Holland, Amsterdam, 1978, p. 319. (b) Hinson, J. A.; Pohl, L. R.; Monks, T. J.; Gillette, J. R.; Guengerich, F. P. *Drug Metab. Dispos.* **1980**, *8*, 289.

154. Nelson, S. D. *Drug Metab. Rev.* **1995**, *27*, 147.

155. (a) Thummel, K. E.; Lee, C. A.; Kunze, K. L.; Nelson, S. D.; Slattery, J. T. *Biochem. Pharmacol.* **1993**, *45*, 1563. (b) Patten, C. J.; Thomas, P. E.; Guy, R. L.; Lee, M.; Gonzalez, F. J.; Guengerich, F. P.; Yang, C. S. *Chem. Res. Toxicol.* **1993**, *6*, 511. (c) Raucy, J. L.; Lasker, J. M.; Lieber, C. S.; Black, M. *Arch. Biochem. Biophys.* **1989**, *271*, 270.

156. Ryan, D. E.; Koop, D. R.; Thomas, P. E.; Coon, M. J.; Levin, W. *Arch. Biochem. Biophys.* **1986**, *246*, 633.

157. Rosen, G. M.; Singletary, W. V. Jr; Rauckman, E. J.; Killenberg, P. G. *Biochem. Pharmacol.* **1983**, *32*, 2053. (b) Vendemiale, G.; Altomare, E.; Lieber, C. S. *Drug Metab. Dispos.* **1984**, *12*, 20.

158. (a) McClain, C. J.; Kromhout, J. P.; Peterson, F. J.; Holtzman, J. L. *J. Am. Med. Assoc.* **1980**, *244*, 251. (b) Hall, A. H.; Kulig, K. W.; Rumack, B. H. *Ann. Intern. Med.* **1986**, *105*, 624.

159. West, P. R.; Harman, L. S.; Josephy, P. D.; Mason, R. P. *Biochem. Pharmacol.* **1984**, *33*, 2933.

160. Dahlin, D. C.; Miwa, G. T.; Lu, A. Y. H.; Nelson, S. D. *Proc. Natl. Acad. Sci. USA* **1984**, *81*, 1327.

161. Potter, D. W.; Hinson, J. A. *J. Biol. Chem.* **1987**, *262*, 966.

162. (a) Mitchell, J. R.; Jollow, D. J.; Potter, W. Z.; Gillette, J. R.; Brodie, B. B. *J. Pharmacol. Exp. Ther.* **1973**, *187*, 211. (b) Potter, W. Z.; Davis, D. C.; Mitchell, J. R.; Jollow, D. J.; Gillette, J. R.; Brodie, B. B. *J. Pharmacol. Exp. Ther.* **1973**, *187*, 203.

163. Moldus, P.; Andersson, B.; Rahimtula, A.; Berggren, M. *Biochem. Pharmacol.* **1982**, *31*, 1363.

164. Potter, D. W.; Hinson, J. A. *J. Biol. Chem.* **1987**, *262*, 974.

165. Reen, R. K.; Ramakanth, S.; Wiebel, F. J.; Jain, M. P.; Singh, *J. Anal. Biochem.* **1991**, *194*, 243.

166. Manoury, P. M.; Binet, J. L.; Rousseau, J.; Leferre-Borg, F. M.; Cavero, I. G. *J. Med. Chem.* **1987**, *30*, 1003.

167. (a) Tullman, R. H.; Hanzlik, R. P. *Drug Metab. Rev.* **1984**, *15*, 1163. (b) Silverman, R. B. *Acc. Chem. Res.* **1995**, *28*, 335.

168. Duggan, D. E.; Hogans, A. F.; Kwan, K. C.; McMahon, F. G. *J. Pharmacol. Exp. Ther.* **1972**, *181*, 563.

169. Adler, T. K.; Fujimoto, J. M.; Way, E. L.; Baker, E. M. *J. Pharmacol. Exp. Ther.* **1955**, *114*, 251.

170. Klutch, A.; Bordun, M. *J. Med. Chem.* **1967**, *10*, 860.

171. Davis, P. J.; Abdel-Maksoud, Hamdy; Trainor, T. M.; Vouros, Paul; Neumeyer, J. L. *Biochem. Biophys. Res. Commun.* **1985**, *127*, 407.

172. (a) Baillie, T. A.; Halpin, R. A.; Matuszewski, B. K.; Geer, L. A.; Chavez-Eng, C. M.; Dean, D.; Braun, M.; Doss, G.; Jones, A.; Marks, T.; Melillo, D.; Vyas, K. P. *Drug Metab. Dispos.* **2001**, *29*, 1614. (b) Nicoll-Griffith, D. A.; Yergey, J. A.; Trimble, L. A.; Silva, J. M.; Li, C.; Chauret, N.; Gauthier, J. Y.; Grimm, E.; Leger, S.; Roy, P.; Therien, M.; Wang, Z.; Prasit, P.; Zamboni, R.; Young, R. N.; Brideau, C.; Chan, C.-C.; Mancini, J.; Riendeau, D. *Bioorg. Med. Chem. Lett.* **2000**, *10*, 2683.

173. Mazel, P.; Henderson, J. F.; Axelrod, J. *J. Pharmacol. Exp. Ther.* **1964**, *143*, 1.

174. Spector, E.; Shideman, F. E. *Biochem. Pharmacol.* **1959**, *2*, 182.

175. Mitchell, S. C.; Waring, R. H. *Drug Metab. Rev.* **1986**, *16*, 255.

176. Oae, S.; Mikami, A.; Matsuura, T.; Ogawa-Asada, K.; Watanabe, Y.; Fujimori, K.; Iyanagi, T. *Biochem. Biophys. Res. Commun.* **1985**, *131*, 567.

177. Goto, Y.; Matsui, T.; Ozaki, S.-i.; Watanabe, Y.; Fukuzumi, S. *J. Am. Chem. Soc.* **1999**, *121*, 9497.

178. Souhaili el Amri, H.; Fargetton, X.; Delatour, P.; Batt, A. M. *Xenobiotica* **1987**, *17*, 1159.

179. Gruenke, L. D.; Craig, J. C.; Dinovo, E. C.; Gottschalk, L. A.; Noble, E. P.; Biener, R. *Res. Commun. Chem. Pathol. Pharmacol.* **1975**, *10*, 221.

180. (a) Neau, E.; Dansette, P. M.; Andronik, V.; Mansuy, D. *Biochem. Pharmacol.* **1990**, *39*, 1101. (b) Dansette, P. M.; Amar, C.; Smith, C.; Pons, C.; Mansuy, D. *Biochem. Pharmacol.* **1990**, *39*, 911.

181. Dansette, P. M.; Amar, C.; Valadon, P.; Pons, C.; Beaune, P. H.; Mansuy, D. *Biochem. Pharmacol.* **1991**, *40*, 553.

182. Mansuy, D.; Valadon, P.; Erdelmeier, I.; Lopez-Garcia, P.; Amar, C.; Girault, J.-P.; Dansette, P. M. *J. Am. Chem. Soc.* **1991**, *113*, 7825.

183. Crew, M. C.; Melgar, M. D.; Haynes, L. J.; Gala, R. L.; DiCarlo, F. *J. Xenobiotica* **1972**, *2*, 431.

184. Benoit, E.; Buronfosse, T.; Moroni, P.; Delatour, P.; Riviere-J.-L. *Biochem. Pharmacol.* **1993**, *46*, 2337.

185. (a) Pohl, L. R.; Pumford, N. R.; Martin, J. L. *Eur. J. Haematol.* **1996**, *Suppl. 60*, 98. (b) Bourdi, M.; Amouzadeh, H. R.; Rushmore, T. H.; Martin, J. L.; Pohl, L. R. *Chem. Res. Toxicol.* **2001**, *14*, 362.

186. Martin, J. L.; Meinwald, J.; Radford, P.; Liu, Z.; Graf, M. L. M.; Pohl, L. R. *Drug Metab. Rev.* **1995**, *27*, 179.

187. Sisenwine, S. F.; Kimmel, H. B.; Lin, A. L.; Ruelius, H. W. *Drug Metab. Dispos.* **1975**, *3*, 180.

188. Misra, A. L.; Vadlamani, N. L.; Pontani, R. B.; Mul, S. J. *Biochem. Pharmacol.* **1973**, *22*, 2129.

189. Dolphin, D.; Poulson, R.; Avramović, O. (Eds.) *Pyridine Nucleotide Coenzymes*, Wiley, New York, 1987, Parts A and B.

190. Yun, C.-H.; Lee, H. S.; Lee, H.; Rho, J. K.; Jeong, H. G.; Guengerich, F. P. *Drug Metab. Dispos.* **1995**, *23*, 285.

191. (a) Kanamori, T.; Inoue, H.; Iwata, Y.; Ohmae, Y.; Kishi, T. *J. Anal. Toxicol.* **2002**, *26*, 61. (b) Lien, E. A.; Solheim, E.; Lea, O. A.; Lundgren, S.; Kvinnsland, S.; Uelandm, P. M. *Cancer Res.* **1989**, *49*, 2175.

192. Ravitch, J. R.; Moseley, C. G. *J. Chromatogr. B*. **2001**, *762*, 165.

193. (a) Stearns, R. A.; Chakravarty, P. K.; Chen, R.; Chiu, S. H. L. *Drug Metab. Dispos.* **1995**, *23*, 207. (b) Tamaki, T.; Nishiyama, A.; Kimura, S.; Aki, Y.; Yoshizumi, M.; Houchi, H.; Morita, K.; Abe, Y. *Cardiovasc. Drug Rev.* **1997**, *15*, 122. (c) Yun, C.-H.; Lee, H. S.; Lee, H.; Rho, J. K.; Jeong, H. G.; Guengerich, F. P. *Drug Metab. Dispos.* **1995**, *23*, 285.

194. Sachinidis, A.; Ko, Y.; Weisser, P.; Meyer zu Brickwedde, M. K.; Dusing, R.; Christian, R.; Wieczorek, A. J.; Vetter, H. *J. Hypertension* **1993**, *11*, 155.

195. (a) Bachur, N. R. *Science* **1976**, *193*, 595. (b) Wermuth, B. *Prog. Clin. Biol. Res.* **1985**, *174*, 209.

196. Felsted, R. L.; Richter, D. R.; Jones, D. M.; Bachur, N. R. *Biochem. Pharmacol.* **1980**, *29*, 1503.

197. (a) Chan, K. K.; Lewis, R. J.; Trager, W. F. *J. Med. Chem.* **1972**, *15*, 1265. (b) Sutcliffe, F. A.; MacNicoll, A. D.; Gibson, G. G. *Rev. Drug Metab. Drug Interact.* **1987**, *5*, 225. (c) Park, B. K. *Biochem. Pharmacol.* **1988**, *37*, 19. (d) Hermans, J. J. R.; Thijssen, H. H. W. *Adv. Exp. Med. Biol.* **1993**, *328*, 351.

198. Toon, S.; Lon, L. K.; Gibaldi, M.; Trager, W. F.; O'Reilly, R. A.; Mottey, C. H.; Goulart, D. A. *Clin. Pharmacol. Ther.* **1986**, *39*, 15.

199. Holford, N. H. G. *Clin. Pharmacokin.* **1986**, *11*, 483.

200. (a) Kaminsky, L. S.; Zhang, Z.-Y. *Pharmacol. Ther.* **1997**, *73*, 67. (b) Yamazaki, H.; Shimada, T. *Biochem. Pharmacol.* **1997**, *54*, 1195.

201. (a) Prelog, V. *Pure Appl. Chem.* **1964**, *9*, 119. (b) Horjales, E.; Bränden, C. I. *J. Biol. Chem.* **1985**, *260*, 15445.

202. Roerig, S.; Fujimoto, J. M.; Wang, R. I. H.; Lange, D. *Drug Metab. Dispos.* **1976**, *4*, 53.

203. Dayton, H. E.; Inturrisi, C. E. *Drug Metab. Dispos.* **1976**, *4*, 474.

204. Gerhards, E.; Hecker, W.; Hitze, H.; Nieuweboer, B.; Bellmann, O. *Acta Endocrinol.* **1971**, *68*, 219.

205. (a) Uehleke, H. *Naturwissenschaften* **1963**, *50*, 335. (b) Gillette, J. R. *Forschr. Arzneim-Forsch.* **1963**, *6*, 11.

206. (a) Mason, R. P.; Josephy, P. D. In *Toxicity of Nitroaromatic Compounds*; Rickert, D. (Ed.), Hemisphere, New York, 1985, p. 121. (b) Mason, R. P.; Holtzman, J. L. *Biochemistry* **1975**, *14*, 1626. (c) Moreno, S. N. *J. Comp. Biochem. Physiol.* **1988**, *91C*, 321.

207. Mason, R. P.; Holtzman, J. L. *Biochem. Biophys. Res. Commun.* **1975**, *67*, 1267.

208. (a) Scheline, R. R. *Pharmacol. Rev.* **1973**, *25*, 451. (b) Wheeler, L. A.; Soderberg, F. B.; Goldman, P. *J. Pharmacol. Exp. Ther.* **1975**, *194*, 135.

209. Morita, M.; Feller, D. R.; Gillette, J. R. *Biochem. Pharmacol.* **1971**, *20*, 217.

210. Wolpert, M. K.; Althaus, J. R.; Johns, D. G. *J. Pharmacol. Exp. Ther.* **1973**, *185*, 202.

211. Poirier, L. A.; Weisburger, J. H. *Biochem. Pharmacol.* **1974**, *23*, 661.

212. Garattini, S.; Marcucci, F.; Mussini, E. In *Psychotherapeutic Drugs*, Usdin, E.; Forrest, I. S. (Eds.), Marcel Dekker, New York, 1977, p. 1039, Part 2.

213. Feller, D. R.; Morita, M.; Gillette, J. R. *Biochem. Pharmacol.* **1971**, *20*, 203.

214. Tatsumi, K.; Kitamura, S.; Yoshimura, H. *Arch. Biochem. Biophys.* **1976**, *175*, 131.

215. (a) Mason, R. P.; Peterson, F. J.; Holtzman, J. L. *Biochem. Biophys. Res. Commun.* **1977**, *75*, 532. (b) Peterson, F. J.; Holtzman, J. L.; Crankshaw, D.; Mason, R. P. *Mol. Pharmacol.* **1988**, *34*, 597.

216. Mason, R. P.; Peterson, F. J.; Holtzman, J. L. *Mol. Pharmacol.* **1978**, *14*, 665.

217. (a) Scheline, R. R. *Pharmacol. Rev.* **1973**, *25*, 451. (b) Wheeler, L. A.; Soderberg, F. B.; Goldman, P. *J. Pharmacol. Exp. Ther.* **1975**, *194*, 135.

218. (a) Peppercorn, M. A.; Goldman, P. *J. Pharmacol. Exp. Ther.* **1972**, *181*, 555. (b) Schröder, H.; Gustafsson, B. E. *Xenobiotica* **1973**, *3*, 225.

219. (a) Eagling, V. A.; Howe, J. L.; Barry, M. J.; Back, D. *J. Biochem. Pharmacol.* **1994**, *48*, 267. (b) Stagg, M. P.; Cretten, E. M.; Kidd, L. *Clin. Pharmacol. Ther.* **1992**, *51*, 668.

220. Kato, R.; Iwasaki, K.; Noguchi, H. *Mol. Pharmacol.* **1978**, *14*, 654.

221. Tamura, S.; Kawata, S.; Sugiyama, T.; Tarui, S. *Biochim. Biophys. Acta* **1987**, *926*, 231.

222. Tremaine, L. M.; Welch, W. M.; Ronfeld, R. A. *Drug Metab. Dispos.* **1989**, *17*, 542.

223. (a) Williams, F. M. *Pharmacol. Ther.* **1987**, *34*, 99. (b) Heymann, E. In *Enzymatic Basis of Detoxification*, Jakoby, W. B. (Ed.) Academic Press, New York, 1980, Vol. II, p. 291.

224. Puetter, J. *Eur. J. Drug Metab. Pharmacokinet.* **1979**, *4*, 1.

225. Steward, D. J.; Inaba, T.; Lucassen, M.; Kalow, W. *Clin. Pharmacol. Ther.* **1979**, *25*, 464.

226. Kogan, M. J.; Verebey, K. G.; DePace, A. C.; Resnick, R. B.; Mul, S. J. *Anal. Chem.* **1977**, *49*, 1965.

227. Mark, L. C.; Kayden, H. J.; Steele, J. M.; Cooper, J. R.; Berlin, I.; Rovenstein, E. A.; Brodie, B. B. *J. Pharmacol. Exp. Ther.* **1951**, *102*, 5.

228. Junge, W.; Krisch, K. *CRC Crit. Rev. Toxicol.* **1975**, *3*, 371.

229. Heymann, E.; Krisch, K.; Buch, H.; Buzello, W. *Biochem. Pharmacol.* **1969**, *18*, 801.

230. Akerman, B.; Ross, S. *Acta Pharmacol. Toxicol.* **1970**, *28*, 445.

231. Dudley, K. H.; Roberts, S. B. *Drug Metab. Dispos.* **1978**, *6*, 133.

232. Maksay, G.; Tegyey, Z.; Ötvös, L. *J. Pharm. Sci.* **1978**, *67*, 1208.

233. (a) Heykants, J. J. P.; Meuldermans, W. E. G.; Michiels, L. J. M.; Lewi, P. J.; Janssen, P. A. *J. Arch. Int. Pharmacodyn. Ther.* **1975**, *216*, 113. (b) Meuldermans, W. E. G.; Lauwers, W. F. J.; Heykants, J. J. P. *Arch. Int. Pharmacodyn. Ther.* **1976**, *221*, 140.

234. Vainio, H. In *Mechanisms of Toxicity and Metabolism, Proceedings of the 6th International Congress of Pharmacology*, Karki, N. T. (Ed.), Pergamon, Oxford, 1976, Vol. 6, p. 53.

235. Cummings, A. J.; King, M. L.; Martin, B. K. *Brit. J. Pharmacol. Chemother.* **1967**, *29*, 150.

236. Nakagawa, T.; Masada, M.; Uno, T. *J. Chromatogr.* **1975**, *111*, 355.

237. Rubin, A.; Warrick, P.; Wolen, R. L.; Chernish, S. M.; Ridolfo, A. S.; Gruber, C. M., Jr. *J. Pharmacol. Exp. Ther.* **1972**, *183*, 449.

238. Bickel, M. H.; Minder, R.; diFrancesco, C. *Experientia* **1973**, *29*, 960.

239. Tsukamoto, H.; Yoshimura, H.; Tatsumi, K. *Chem. Pharm. Bull.* **1963**, *11*, 421.

240. Adamson, R. H.; Bridges, J. W.; Kibby, M. R.; Walker, S. R.; Williams, R. T. *Biochem. J.* **1970**, *118*, 41.

241. Sitar, D. S.; Thornhill, D. P. *J. Pharmacol. Exp. Ther.* **1973**, *184*, 242.

242. Dutton, G. J.; Illing, H. P. A. *Biochem. J.* **1972**, *129*, 539.

243. Dieterle, W.; Faigle, J. W.; Frueh, F.; Mory, H.; Theobald, W.; Alt, K. O.; Richter, W. J. *Arzneim-Forsch.* **1976**, *26*, 572.

244. (a) Kadakol, A.; Ghosh, S. S.; Sappal, B. S.; Sharma, G.; Chowdhury, J. R.; Chowdhury, N. R. *Human Mutation* **2000**, *16*, 297. (b) Tukey, R. H.; Strassburg, C. P. *Annu. Rev. Pharmacol. Toxicol.* **2000**, *40*, 581.

245. (a) Kasten, M. J. *Mayo Clinic Proc.* **1999**, *74*, 825. (b) Knight, M. *J. Clin. Pharmacol.* **1994**, *34*, 128.

246. Koster, A. Sj.; Frankhuijzen-Sierevogel, A. C.; Mentrup, A. *Biochem. Pharmacol.* **1986**, *35*, 1981.

247. Dahl-Puustinen, M.-L.; Dumont, E.; Bertilsson, L. *Drug Metab. Dispos.* **1989**, *17*, 433.

248. (a) Mulder, G. J.; Jakoby, W. B. In *Conjugation Reactions in Drug Metabolism*, Mulder, G. J. (Ed.), Taylor & Francis, London, 1990, p. 107. (b) Levy, G. *Fed. Proc.* **1986**, *45*, 2235. (c) Mulder, G. J. (Ed.) *Sulfation of Drugs and Related Compounds*, CRC Press, Boca Raton, FL, 1981.

249. (a) Jakoby, W. B.; Ziegler, D. M. *J. Biol. Chem.* **1990**, *265*, 20715. (b) Falany, C. N. *FASEB J.* **1997**, *11*, 206.

250. Pang, K. S. In *Conjugation Reactions in Drug Metabolism*, Mulder, G. J. (Ed.), Taylor & Francis, London, 1990, p. 5.

251. Capel, I. D.; French, M. R.; Milburn, P.; Smith, R. I.; Williams, R. T. *Xenobiotica* **1972**, *2*, 25.

252. (a) Mulder, G. J. In *Conjugation Reactions in Drug Metabolism*; Mulder, G. J. (Ed.), Taylor & Francis, London, 1990, p. 41. (b) Whitmer, D. I.; Ziurys, J. C.; Gollan, J. L. *J. Biol. Chem.* **1984**, *259*, 11969.

253. (a) Lin, C.; Li, Y.; McGlotten, J.; Morton, J. B.; Symchowicz, S. *Drug Metab. Dispos.* **1977**, *5*, 234. (b) Walle, T.; Walle, U. K.; Thornburg, K. R.; Schey, K. L. *Drug Metab. Dispos.* **1993**, *21*, 76.

254. Albert, K. S.; Sedman, A. J.; Wagner, J. G. *J. Pharmacokinet Biopharm.* **1974**, *2*, 381.

255. (a) Miller, R. P.; Roberts, R. J.; Fischer, L. *J. Clin. Pharmacol. Ther.* **1976**, *19*, 284. (b) Levy, G.; Khana, N. N.; Soda, D. M.; Tsuzuki, O.; Stern, L. *Pediatrics* **1975**, *55*, 818.

256. Mulder, G. J.; Meerman, J. H. H.; van den Goorbergh, A. M. In *Xenobiotic Conjugation Chemistry*, Paulson, G. D.; Caldwell, J.; Hutson, D. H.; Menn, J. J. (Eds.), American Chemical Society, Washington, DC, 1986, p. 282.

257. Mulder, G. J.; Hinson, J. A.; Gillette, J. R. *Biochem. Pharmacol.* **1977**, *26*, 189.

258. Smith, R. L.; Caldwell, J. In *Drug Metabolism: From Microbe to Man*, Parke, D. V.; Smith, R. L. (Eds.), Taylor & Francis, London, 1977, p. 331.

259. Killenberg, P. G.; Webster, L. T., Jr. In *Enzymatic Basis of Detoxification*, Jakoby, W. B. (Ed.), Academic, New York, 1980, Vol. II, p. 141.

260. Bruce, R. B.; Turnbull, L. B.; Newman, J. H.; Pitts, J. E. *J. Med. Chem.* **1968**, *11*, 1031.

261. Drach, J. C.; Howell, J. P.; Borondy, P. E.; Glazko, A. J. *Proc. Soc. Exp. Biol. Med.* **1970**, *135*, 849.

262. Marchand, D. H.; Remmel, R. P.; Abdel-Monem, M. M. *Drug Metab. Dispos.* **1988**, *16*, 85.

263. Needleman, P. In *Organic Nitrates*, Needleman, P. (Ed.), Springer-Verlag, Berlin, 1975, p. 57.

264. de Miranda, P.; Beacham, L. M., III; Creagh, T. H.; Elion, G. B. *J. Pharmacol. Exp. Ther.* **1973**, *187*, 588.

265. Ishida, T.; Kumagai, Y.; Ikeda, Y.; Ito, K.; Yano, M.; Toki, S.; Mihashi, K.; Fujioka, T.; Iwase, Y.; Hachiyama, S. *Drug Metab. Dispos.* **1989**, *17*, 77.

266. Correia, M. A.; Krowech, G.; Caldera-Munoz, P.; Yee, S. L.; Straub, K.; Castagnoli, N., Jr. *Chem. Biol. Interact.* **1984**, *51*, 13.

267. Suzuki, T.; Nishio, K.; Tanabe, S. *Curr. Drug Metab.* **2001**, *2*, 367.

268. Stevens, J. L.; Jones, D. P. In *Glutathione*, Dolphin, D.; Poulson, R.; Avramović, O. (Eds.), Wiley, New York, 1989, Part B, p. 45.

269. (a) Guenthner, T. M. In *Conjugation Reactions in Drug Metabolism*, Mulder, G. J. (Ed.), Taylor & Francis, London, 1990, p. 365. (b) Fretland, A. J.; Omiecinski, C. *J. Chem.-Biol. Interact.* **2000**, *129*, 41.

270. (a) Vogel-Bindel, U.; Bentley, P.; Oesch, F. *Eur. J. Biochem.* **1982**, *126*, 425. (b) Faendrich, F.; Degiuli, B.; Vogel-Bindel, U.; Arand, M.; Oesch, F. *Xenobiotica* **1995**, *25*, 239.

271. Elson, J.; Strong, J. M.; Atkinson, A. J., Jr. *Clin. Pharmacol. Ther.* **1975**, *17*, 134.

272. Williams, R. T. In *Biogenesis of Natural Compounds*, 2nd ed., Bernfeld, P. (Ed.), Pergamon, Oxford, 1967, p. 589.

273. Meyer, U. A. *Lancet* **2000**, *356*, 1667.

274. (a) Daly, A. K.; Cholerton, S.; Gregory, W.; Idle, J. R. *Pharmacol. Ther.* **1993**, *57*, 129. (b) Meyer, U. A.; Zanger, U. M. *Annu. Rev. Pharmacol. Toxicol.* **1997**, *37*, 269. (c) Lee, C. R.; Goldstein, J. A.; Pieper, J. A. *Pharmacogenetics* **2002**, *12*, 251. (d) Nagata, K.; Yamazoe, Y. *Drug Metab. Pharmacokin.* **2002**, *17*, 167.

275. Drayer, D. E.; Reidenberg, M. M. *Clin. Pharmacol. Ther.* **1977**, *22*, 251.

276. (a) Kalow, W. *Pharmacogenetics, Heredity and the Response to Drugs*, W. B. Saunders, Philadelphia, 1962. (b) Weber, W. W. In *Metabolic Conjugation and Metabolic Hydrolysis*, Fishman, W. H. (Ed.), Academic, New York, 1973, Vol. 3, p. 250.

277. (a) Kalow, W. *Pharmacogenetics, Heredity and the Response to Drugs*, W. B. Saunders, Philadelphia, 1962. (b) Weber, W. W. In *Metabolic Conjugation and Metabolic Hydrolysis*, Fishman, W. H. (Ed.), Academic Press, New York, 1973, Vol. 3, p. 250. (c) Lunde, P. K. M.; Frislid, K.; Hansteen, V. *Clin. Pharmacokinet.* **1977**, *2*, 182.

278. (a) Lunde, P. K. M.; Frislid, K.; Hansteen, V. *Clin. Pharmacokinet.* **1977**, *2*, 182. (b) Weber, W. W. In *Therapeutic Drugs*, Dollery, C. T. (Ed.), Churchill Livingstone, New York, 1986.

279. Patterson, E.; Radtke, H. E.; Weber, W. W. *Mol. Pharmacol.* **1980**, *17*, 367.

280. Dyda, F.; Klein, D. C.; Hickman, A. B. *Annu. Rev. Biophys Biomol. Struct.* **2000**, *29*, 81.

281. Lin, J. H.; Chen, I.-W.; Ulm, E. H. *Drug Metab. Dispos.* **1989**, *17*, 426.

282. Kropp, H.; Sundelof, J. G.; Hajdu, R.; Kahan, F. M. *Antimicrob. Agents Chemother.* **1982**, *22*, 62.

283. Eschenhof, E. *Arzneim.-Forsch.* **1973**, *23*, 390.

284. Caldwell, J.; Marsh, M. V. *Biochem. Pharmacol.* **1983**, *32*, 1667.

285. Dell, H. D.; Fiedler, J.; Kamp, R.; Gau, W.; Kurz, J.; Weber, B.; Wuensche, C. *Drug Metab. Dispos.* **1982**, *10*, 55.

286. Caldwell, J. *Biochem. Soc. Trans.* **1984**, *12*, 9.

287. Leighty, E. G.; Fentiman, A. F., Jr.; Foltz, R. L. *Res. Commun. Chem. Path. Pharmacol.* **1976**, *14*, 13.

288. (a) Leighty, E. G. *Biochem. Pharmacol.* **1973**, *22*, 1613. (b) Caldwell, J.; Parkash, M. K. In *Perspectives in Medicinal Chemistry*, Testa, B.; Kyburz, E.; Fuhrer, W.; Giger, R. (Eds.), VCH, Weinheim, 1993, p. 595.

289. Gunnarsson, P. O.; Johansson, S.-A.; Svensson, L *Xenobiotica* **1984**, *14*, 569.

290. Fears, R.; Baggaley, K. H.; Walker, P.; Hindley, R. M. *Xenobiotica* **1982**, *12*, 427.

291. (a) Thakker, D. R.; Creveling, C. R. In *Conjugation Reactions in Drug Metabolism*, Taylor & Francis, London, 1990, p. 193. (b) Ansher, S. S.; Jakoby, W. B. In *Conjugation Reactions in Drug Metabolism*, Taylor & Francis, London, 1990, p. 233. (c) Stevens, J. L.; Bakke, J. E. In *Conjugation Reactions in Drug Metabolism*, Taylor & Francis, London, 1990, p. 251.

292. Bonifacio, M. J.; Archer, M.; Rodrigues, M. L.; Matias, P. M.; Learmonth, D. A.; Carrondo, M. A.; Soares-Da-Silva, P. *Mol. Pharmacol.* **2002**, *62*, 795.

293. (a) Clarke, S.; Banfield, K. In *Homocysteine in Health and Disease*; Carmel, R.; Jacobsen, D. W. (Eds.), Cambridge University Press, Cambridge, UK, 2001, pp. 63–78. (b) Martin, J. L.; McMillan, F. M. *Curr. Opin. Struct. Biol.* **2002**, *12*, 783. (c) Usdin, E.; Borchardt, R. T.; Creveling, C. R. (Eds.) *The Biochemistry of S-Adenosylmethionine and Related Compounds*, Macmillan Press, London, 1982.

294. Morgan, C. D.; Sandler, M.; Davies, D. S.; Connolly, M.; Paterson, J. W.; Dollery, C. T. *Biochem. J.* **1969**, *114*, 8P.

295. Persson, K.; Persson, K. *Xenobiotica* **1972**, *2*, 375.

296. Brner, U.; Abbott, S. *Experientia* **1973**, *29*, 180.

297. Leeson, G. A.; Garteiz, D. A.; Knapp, W. C.; Wright, G. J. *Drug Metab. Dispos.* **1973**, *1*, 565.

298. Herbertz, G.; Metz, T.; Reinauer, H.; Staib, W. *Biochem. Pharmacol.* **1973**, *22*, 1541.

299. Drummer, O. H.; Miach, P.; Jarrott, B. *Biochem. Pharmacol.* **1983**, *32*, 1557.

300. Lindsay, R. H.; Hulsey, B. S.; Aboul-Enein, H. Y. *Biochem. Pharmacol.* **1975**, *24*, 463.

301. Penning, T.D.; Talley, J. J.; Bertenshaw, S. R.; Carter, J. S.; Collins, P. W.; Docter, S.; Graneto, M. J.; Lee, L. F.; Malecha, J. W.; Miyashiro, J. M.; Rogers, R. S.; Rogier, D. J.; Yu, S. S.; Anderson, G. D.; Burton, E. G.; Cogburn, J. N.; Gregory, S. A.; Koboldt, C. M.; Perkins, W. E.; Seibert, K.; Veenhuizen, A. W.; Zhang, Y. Y.; Isakson, P. C. *J. Med. Chem.* **1997**, *40*, 1347.

302. (a) Bodor, N. In *Design of Biopharmaceutical Properties of Prodrugs and Analogs*, Roche, E. B. (Ed.), Academy of Pharmaceutical Sciences, Washington, DC, 1977, Chap. 7. (b) Bodor, N. *Adv. Drug Res.* **1984**, *13*, 255. (c) Bodor, N. *Med. Res. Rev.* **1984**, *4*, 449.

303. Lee, H. J.; Cooperwood, J. S.; You, Z.; Ko, D.-H. *Arch. Pharmacal Res.* **2002**, *25*, 111.

304. Bodor, N.; Buchwald, P. *Med. Res. Rev.* **2000**, *20*, 58.

305. Bodor, N.; Kaminski, J. J.; Selk, S. *J. Med. Chem.* **1980**, *23*, 469.

306. Druzgala, P.; Hochhaus, G.; Bodor, N. *J. Steroid Biochem. Mol. Biol.* **1991**, *38*, 149.

307. Bucala, R.; Fishman, J.; Cerami, A. *Proc. Natl. Acad. Sci. USA* **1982**, *79*, 3320.

Chapter 8

Prodrugs and Drug Delivery Systems

8.1 Enzyme Activation of Drugs

The term *prodrug*, which was used initially by Albert,[1] is a pharmacologically inactive compound that is converted into an active drug by a metabolic biotransformation. A prodrug also can be activated by a nonenzymatic process such as hydrolysis, but in this case the compounds generally are inherently unstable and may cause stability problems. The prodrug to drug conversion can occur before absorption, during absorption, after absorption, or at a specific site in the body. In the ideal case a prodrug is converted to the drug as soon as the desired goal for designing the prodrug has been achieved.

As noted in Chapter 7, Section 7.4.D, p. 471, the concepts of prodrugs and soft drugs (antedrugs) are opposite. Whereas prodrugs are inactive compounds that require a metabolic conversion to the active form, a soft drug is pharmacologically active and uses metabolism as a means of promoting excretion. However, it is possible to design a *pro-soft drug*, a modified soft drug that requires metabolic activation for conversion to the active soft drug.

8.1.A Utility of Prodrugs

Prodrug design is a lead modification approach that is used to correct a flaw in a drug candidate. Below are numerous reasons why you might want to utilize a prodrug strategy in drug design.

A.1 Aqueous Solubility

Consider an injectable drug that is so insoluble in water that it would need to be taken up in more than a liter of saline to administer the appropriate dose! Or what if each dose of your opthalmic drug required a liter of saline for dissolution, but it was to be administered as eye drops! These drugs could be safe, effective, and potent, but they would not be viable for their applications. In these cases, a water-solubilizing group could be attached to the drugs, which is metabolically released after drug administration.

A.2 Absorption and Distribution

If the desired drug is not absorbed and transported to the target site in sufficient concentration, it can be made more water soluble or lipid soluble depending on the desired site of action. Once absorption has occurred or when the drug is at the appropriate site of action, the water- or lipid-soluble group is removed enzymatically.

A.3 Site Specificity

Specificity for a particular organ or tissue can be made if there are high concentrations of or uniqueness of enzymes present at that site that can cleave the appropriate appendages from the prodrug and unmask the drug. Alternatively, something that directs the drug to a particular type of tissue could be attached to the drug, which is released after the drug reaches the target tissue.

A.4 Instability

A drug may be rapidly metabolized and rendered inactive prior to when it reaches the site of action. The structure may be modified to block that metabolism until the drug is at the desired site.

A.5 Prolonged Release

It may be desirable to have a steady low concentration of a drug released over a long period of time. The drug may be altered so that it is metabolically converted to the active form slowly.

A.6 Toxicity

A drug may be toxic in its active form and would have a greater therapeutic index if it were administered in a nontoxic inactive form that was converted into the active form only at the site of action.

A.7 Poor Patient Acceptability

An active drug may have an unpleasant taste or odor, produce gastric irritation, or cause pain when administered (for example, when injected). The structure of the drug could be modified to alleviate these problems, but once administered, the prodrug would be metabolized to the active drug.

A.8   Formulation Problems

If the drug is a volatile liquid, it would be more desirable to have it in a solid form so that it could be formulated as a tablet. An inactive solid derivative could be prepared that would be converted in the body to the active drug.

8.1.B   Types of Prodrugs

There are several classifications of prodrugs. Some prodrugs are not designed as such; the biotransformations are fortuitous, and it is discovered only after isolation and testing of the metabolites that activation of the drug had occurred. In most cases a specific modification in a drug has been made on the basis of known metabolic transformations. It is expected that after administration, the prodrug will be appropriately metabolized to the active form. This has been termed *drug latentiation* to signify the rational design approach rather than *serendipity*.[2] The term *drug latentiation* has been refined even further by Wermuth[3] into two classes which he called carrier-linked prodrugs and bioprecursors.

A *carrier-linked prodrug* is a compound that contains an active drug linked to a carrier group that can be removed enzymatically, such as an ester which is hydrolyzed to an active carboxylic acid-containing drug. The bond to the carrier group must be labile enough to allow the active drug to be released efficiently *in vivo*, and the carrier group must be nontoxic and biologically inactive when detached from the drug. Carrier-linked prodrugs can be subdivided even further into bipartate, tripartate, and mutual prodrugs. A *bipartate prodrug* is a prodrug comprised of one carrier attached to the drug. When a carrier is connected to a linker that is connected to the drug, it is called a *tripartate prodrug*. A *mutual prodrug* consists of two, usually synergistic, drugs attached to each other (one drug is the carrier for the other and vice versa).

A *bioprecursor prodrug* is a compound that is metabolized by molecular modification into a new compound which is the active principle or which can be metabolized further to the active drug. For example, if the drug contains a carboxylic acid group, the bioprecursor may be a primary amine that is metabolized by oxidation to the aldehyde, which is further metabolized to the carboxylic acid drug (see Chapter 7, Section 7.4.B.1.e, p. 430). Unlike the carrier-linked prodrug, which is the active drug linked to a carrier, a bioprecursor prodrug contains a different structure that cannot be converted into the active drug by simple cleavage of a group from the prodrug.

The concept of prodrugs can be analogized to the use of protecting groups in organic synthesis.[4] If, for example, you wanted to carry out a reaction on a compound that contained a carboxylic acid group, it may be necessary first to protect the carboxylic acid as, say, an ester, so that the acidic proton of the carboxylic acid does not interfere with the desired reaction. After the desired synthetic transformation was completed, the carboxylic acid analog could be unmasked by deprotection, i.e., hydrolysis of the ester (Scheme 8.1A). This is analogous to a carrier-linked prodrug; an ester functionality can be used to make the properties of the drug more desirable until it reaches the appropriate biological site where it is "deprotected." Another type of protecting group in organic synthesis is one that has no resemblance to the desired functional group. For example, a terminal alkene can be oxidized with ozone to an aldehyde,[5] and the aldehyde can be oxidized to a carboxylic acid with hydrogen peroxide (Scheme 8.1B). As in the case of a bioprecursor prodrug, a drastic structural change is required to unmask the desired group. Oxidation is a common metabolic biotransformation for bioprecursor prodrugs.

When designing a prodrug, you should keep in mind that a particular metabolic transformation may be species specific (see Chapter 7). Therefore, a prodrug whose design was based on rat metabolism studies may not necessarily be effective in humans.

A. $RCO_2H \xrightarrow[\substack{HCl \\ \Delta}]{EtOH} RCO_2Et \xrightarrow[\text{on R}]{\text{reaction}} R'CO_2Et \xrightarrow[\Delta]{H_3O^+} R'CO_2H$

B. $RCH=CH_2 \xrightarrow[\text{on R}]{\text{reaction}} R'CH=CH_2 \xrightarrow[\substack{1.\ O_3 \\ 2.\ H_2O_2}]{} R'CO_2H$

Scheme 8.1 ▶ Protecting group analogy for a prodrug

8.2 Mechanisms of Drug Activation

8.2.A Carrier-Linked Prodrugs

An ideal drug carrier must (1) protect the drug until it is at the site of action; (2) localize the drug at the site of action; (3) allow for release of the drug chemically or enzymatically; (4) minimize host toxicity; (5) be biodegradable, biochemically inert, and nonimmunogenic; (6) be easily prepared inexpensively; and (7) be chemically and biochemically stable in its dosage form.

The most common reaction for activation of carrier-linked prodrugs is hydrolysis. First, let's consider the general functional groups involved, then look at specific examples for different types of prodrugs.

A.1 Carrier Linkages for Various Functional Groups

a. Alcohols, Carboxylic Acids, and Related Groups

There are several reasons why the most common prodrug form for drugs containing alcohol or carboxylic acid functional groups is an ester. First, esterases are ubiquitous, so metabolic regeneration of the drug is a facile process. Also, it is possible to prepare ester derivatives with virtually any degree of hydrophilicity or lipophilicity. Finally, a variety of stabilities of esters can be obtained by appropriate manipulation of electronic and steric factors. Therefore, a multitude of ester prodrugs can be prepared to accommodate a wide variety of problems that require the prodrug approach.

Alcohol-containing drugs can be acylated with aliphatic or aromatic carboxylic acids to decrease water solubility (increase lipophilicity) or with carboxylic acids containing amino or additional carboxylate groups to increase water solubility (Table 8.1).[6] Conversion to phosphate or sulfate esters also increases water solubility. By using these approaches a wide range of solubilities can be achieved that will affect the absorption and distribution properties of the drug. These derivatives also can have an important effect on the dosage form, that is, whether used in tablet form or in aqueous solution. One problem with the use of this prodrug approach is that in some cases certain esters are not very good substrates for the endogenous esterases, sulfatases, or phosphatases, and may not be hydrolyzed at a rapid enough rate. When that occurs, however, a different ester can be tried. Another approach to accelerate the hydrolysis rate could be to attach electron-withdrawing groups (if a base hydrolysis mechanism is relevant) or electron-donating groups (if an acid hydrolysis mechanism is important)[7] to the carboxylate side of the ester. Succinate esters can be used to accelerate the rate of hydrolysis by intramolecular catalysis (Scheme 8.2). If the ester is too reactive, substituents can be appended that cause steric hindrance to hydrolysis or esters of long-chain fatty acids can be

TABLE 8.1 ▶ Ester Analogs of Alcohols as Prodrugs

$$\text{Drug—OH} \longrightarrow \text{Drug—OX}$$

X	Effect on water solubility
$\overset{\displaystyle O}{\underset{\displaystyle \parallel}{C}}$—R	(R = aliphatic or aromatic) decreases
$\overset{\displaystyle O}{\underset{\displaystyle \parallel}{C}}$—CH$_2$NHMe$_2^+$	increases (pK_a ~ 8)
$\overset{\displaystyle O}{\underset{\displaystyle \parallel}{C}}$—CH$_2CH_2COO^-$	increases (pK_a ~ 5)
$\overset{\displaystyle O}{\underset{\displaystyle \parallel}{C}}$— (pyridine, —NH$^+$)	increases (pK_a ~ 4)
PO$_3$=	increases (pK_a ~ 2 and ~ 6)
$\overset{\displaystyle O}{\underset{\displaystyle \parallel}{C}}CH_2SO_3^-$	increases (pK_a ~ 1)

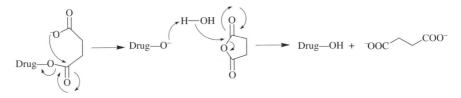

Scheme 8.2 ▶ Intramolecular catalysis of succinate esters

employed. Alcohol-containing drugs also can be converted into the corresponding acetals or ketals for rapid hydrolysis in the acidic medium of the gastrointestinal tract.

Enolic hydroxyl groups can be esterified to prodrug forms as well. A series of enol esters of the antirheumatic oxindole **8.1** were prepared, and it was found that the hemifumarate derivative (**8.2**) was more stable than the corresponding nonionizable esters at neutral pH.[8]

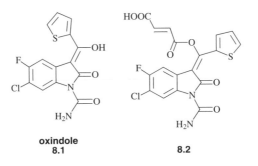

**oxindole
8.1**

8.2

Carboxylic acid-containing drugs can be esterified with various alcohols. The reactivity of the derivatized drug can be adjusted by appropriate structural manipulations as discussed above for ester prodrugs of alcohol-containing drugs. The pK_a of a carboxylic acid can be raised by conversion to a choline ester (**8.3**, R = R′ = Me; $pK_a \sim 7$) or an amino ester (**8.3**, R = H, R′ = H or Me; $pK_a \sim 9$). Likewise, phosphate- or phosphonate-containing drugs can be converted into ester prodrug forms, such as 3-phthalidyl esters (**8.4**).[9]

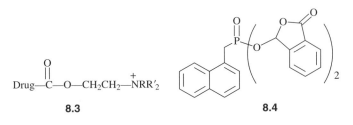

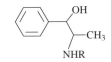

8.3   **8.4**

b. Amines

N-Acylation of amines to give amide prodrugs is not commonly used, in general, because of the stability of amides toward metabolic hydrolysis. Activated amides, generally of low basicity amines, or amides of amino acids are more susceptible to enzymatic cleavage (Table 8.2). Although carbamates in general are too stable, phenyl carbamates ($RNHCO_2Ph$) are rapidly cleaved by plasma enzymes,[10] and, therefore, they can be used as prodrugs.

The pK_a of amines can be lowered by approximately three units by conversion to their *N*-Mannich bases (Table 8.2, X = $CH_2NHCOAr$). This lowers the basicity of the amine so that at physiological pH few of the prodrug molecules are protonated, thereby increasing its lipophilicity. For example, the partition coefficient (see Chapter 2, Section 2.2.F.2.b, p. 55) between octanol and phosphate buffer, pH 7.4 for the *N*-Mannich base **8.5** (R = $CH_2NHCOPh$), derived from benzamide and the decongestant phenylpropanolamine hydrochloride (**8.5**, R = H · HCl; in several cold remedies, such as Entex), is almost 100 times greater than for the parent amine.[11] However, the rate of hydrolysis of *N*-Mannich bases depends on the amide carrier group; salicylamide and succinimide are more susceptible to hydrolysis than is benzamide.[12]

phenylpropanolamine hydrochloride (R = H · HCl)
8.5

TABLE 8.2 ▶ Prodrug Analogs of Amines

$$Drug-NH_2 \longrightarrow Drug-NHX$$

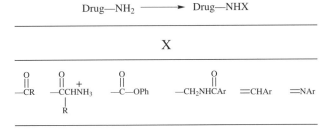

Another approach for lowering the pK_a of amines and, thereby, making them more lipophilic, is to convert them to imines (Schiff bases); however, imines often are too labile in aqueous solution. The anticonvulsant agent progabide (**8.6**; Gabrene)[13] is a prodrug form of γ-aminobutyric acid, an important inhibitory neurotransmitter (see Chapter 5, Section 5.5.C.3.a, p. 287); once inside the brain it is hydrolyzed to γ-aminobutyric acid.[14]

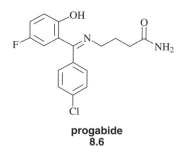

progabide
8.6

Because most solid tumors are *hypoxic* (have a low oxygen concentration), these cells are resistant to radiation therapy and to many chemotherapeutic approaches.[15] To take advantage of the reductive milieu, prodrugs that require reductive mechanisms are beneficial. Although most carrier-linked prodrugs require hydrolytic activation mechanisms, a reductive activation mechanism also is feasible. A general approach is the reduction of a nitroaromatic prodrug (**8.7**, X = NH or O; Y = O, S, NCH_3; Z = CH, N), converting the electron-withdrawing nitro group to an electron-donating hydroxylamino group (or amino group, depending on the rate of elimination versus the *in vivo* rate of reduction of the hydroxylamino to an amino group), which initiates release of the drug from the carrier. This can be used for release of amines (X = NH) or alcohols (X = O) from the carrier (Scheme 8.3).[16] The drawback to this approach is that **8.8** is quite electrophilic and will react with whatever nucleophiles may be present, including other proteins, unless it is trapped possibly by the reducing agent or by glutathione (see Chapter 7, Section 7.4.C.5).

c. Sulfonamides

Just like amines, sulfonamides can be acylated, but this generates an acidic proton, which makes these compounds amenable to conversion to water-soluble sodium salts. For example, the second-generation anti-inflammatory drug and COX-2 inhibitor valdecoxib (**8.9**, Bextra) (see Chapter 5, Section 5.5.B.2.b, p. 280) has been converted into parecoxib sodium (**8.10**,

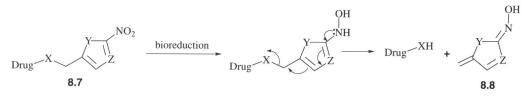

Scheme 8.3 ▶ Bioreductively activated carrier-linked prodrug

Dynastat), an injectable analgesic drug.[17] The plasma half-life in humans for conversion back to valdecoxib is about 5 minutes.

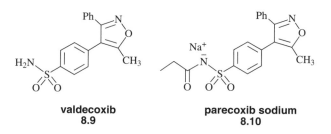

valdecoxib
8.9

parecoxib sodium
8.10

d. Carbonyl Compounds

The most important prodrug forms of aldehydes and ketones are Schiff bases, oximes, acetals (ketals), enol esters, oxazolidines, and thiazolidines (Table 8.3).

A more complete review of bioreversible derivatives of the functional groups was written by Bundgaard.[18]

A.2 Examples of Carrier-Linked Bipartate Prodrugs

a. Prodrugs for Increased Water Solubility

Prednisolone (**8.11**, R = R' = H; Prelone) and methylprednisolone (**8.11**, R = CH_3, R' = H; Depo-Medrol) are poorly water-soluble corticosteroid drugs. To permit aqueous injection or opthalmic delivery of these drugs, they must be converted into water-soluble forms, such as one of the ionic esters described in Section 8.2.A.1.a. However, there are two considerations in the choice of a solubilizing group: The ester must be stable enough in aqueous solution so that a ready-to-inject solution has a reasonably long shelf life (greater than 2 years; half-life about 13 years), but it must be hydrolyzed *in vivo* with a reasonably short half-life after administration (less than 10 minutes). For this optimal situation to occur, the *in vivo/in vitro*

TABLE 8.3 ▶ Prodrug Analogs of Carbonyl Compounds

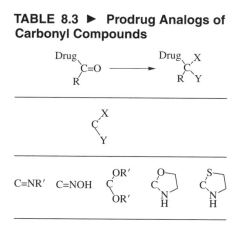

lability ratio would have to be on the order of 10^6. This is possible when the biotransformation is enzyme catalyzed.

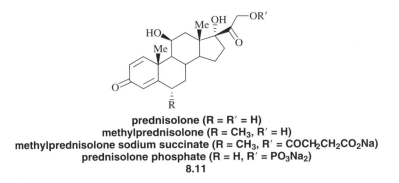

prednisolone (R = R' = H)
methylprednisolone (R = CH₃, R' = H)
methylprednisolone sodium succinate (R = CH₃, R' = COCH₂CH₂CO₂Na)
prednisolone phosphate (R = H, R' = PO₃Na₂)
8.11

The water-soluble prodrug form of methylprednisolone that is in medical use is methyl-prednisolone sodium succinate (**8.11**, R $=$ CH$_3$, R$'$ $=$ COCH$_2$CH$_2$CO$_2$Na; Solu-Medrol). However, the *in vitro* stability is low, probably because of intramolecular catalysis; consequently, it is distributed as a lyophilized (freeze-dried) powder that must be reconstituted with water and then used within 48 h. The lyophilization process adds to the cost of the drug and makes its use less convenient. On the basis of physical-organic chemical rationalizations, a series of more stable water-soluble methylprednisolone esters was synthesized, and several of the analogs were shown to have shelf lives in solution of greater than 2 years at room temperature.[19] Ester hydrolysis studies of these compounds in human and monkey serum indicated that derivatives having an anionic solubilizing moiety such as carboxylate or sulfonate are poorly or not hydrolyzed, but compounds with a cationic (tertiary amino) solubilizing moiety are hydrolyzed rapidly by serum esterases.[20] Prednisolone phosphate (**8.11**, R $=$ H, R$'$ $=$ PO$_3$Na$_2$; Pediapred) is prescribed as a water-soluble prodrug for prednisolone that is activated *in vivo* by phosphatases.

Because of the poor water solubility of the antitumor drug etoposide (**8.12**, R $=$ H; Vepesid), it has to be formulated with the detergent Tween 80, polyethylene glycol, and ethanol, all of which have been shown to be toxic.[21] Conversion to the corresponding phosphate ester, etoposide phosphate (**8.12**, R $=$ PO$_3$H$_2$; Etopophos), allows the drug to be delivered in a more concentrated form over a much shorter period of time without the detrimental vehicle.[22]

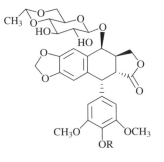

etoposide (R = H)
etoposide phosphate (R = PO₃H₂)
8.12

The local anesthetic benzocaine (**8.13**, R = H; one trade name is Americaine) has been converted into water-soluble amide prodrug forms with various amino acids (**8.13**, R = $^+NH_3CHR'CO$); amidase-catalyzed hydrolysis in human serum occurs rapidly.[23]

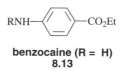

benzocaine (R = H)
8.13

b. Prodrugs for Improved Absorption and Distribution

The skin is designed to maintain the body fluids and prevent absorption of xenobiotics into the general circulation. Consequently, drugs applied to the skin are poorly absorbed.[24] Even steroids have low dermal permeability, particularly if they contain hydroxyl groups that can interact with the skin or binding sites in the keratin. Corticosteroids for the topical treatment of inflammatory, allergic, and pruritic skin conditions can be made more suitable for topical absorption by esterification or acetonidation. For example, both fluocinolone acetonide (**8.14**, R = H; one trade name is Synalar) and fluocinonide (**8.14**, R = $COCH_3$; Lidex) are prodrugs used for inflammatory and pruritic manifestations. Once absorbed through the skin, an esterase releases the drug.

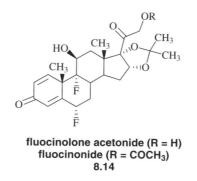

fluocinolone acetonide (R = H)
fluocinonide (R = COCH₃)
8.14

Dipivaloylepinephrine (dipivefrin, **8.15**, R = Me_3CCO; Propine), a prodrug for the antiglaucoma drug epinephrine (**8.15**, R = H; one trade name is Epifrin), is able to penetrate the cornea better than epinephrine. The cornea and aqueous humor have significant esterase activity.[25]

dipivefrin (R = Me₃CCO)
epinephrine (R = H)
8.15

c. Prodrugs for Site Specificity

The targeting of drugs for a specific site in the body by conversion to a prodrug is plausible when the physicochemical properties of the parent drug and prodrug are optimal for the target site. Keep in mind, however, that when the lipophilicity of a drug is increased, it will improve passive transport of the drug nonspecifically to all tissues.

Oxyphenisatin (**8.16**, R = H; Lavema) is a bowel sterilant that is active only when administered rectally. However, when the hydroxyl groups are acetylated (**8.16**, R = Ac, one trade name is Noloc), the prodrug, oxyphenisatin acetate, can be administered orally, and it is hydrolyzed at the site of action in the intestines to oxyphenisatin.

oxyphenisatin (R = H)
8.16

One important membrane that must be traversed for drug delivery into the brain is the *blood–brain barrier*,[26] a unique lipid-like protective barrier that prevents hydrophilic compounds from entering the brain unless they are actively transported. The blood–brain barrier also contains active enzyme systems to protect the central nervous system even further. Consequently, molecular size and lipophilicity are often necessary, not sufficient, criteria for gaining entry into the brain.[27]

As was discussed in Chapter 5, Section 5.5.C.3.a, p. 287, increasing the brain concentration of the inhibitory neurotransmitter γ-aminobutyric acid (GABA) results in anticonvulsant activity. However, GABA is too polar to cross the blood–brain barrier, so it is not an effective anticonvulsant drug. As mentioned above, progabide (**8.6**) is an effective lipophilic analog of GABA that crosses the blood–brain barrier, releases GABA inside the brain, and shows anticonvulsant activity.[28] Another related example of anticonvulsant drug design is a glyceryl lipid (**8.17**, R = linolenoyl) containing one GABA molecule and one vigabatrin molecule, a mechanism-based inactivator of GABA aminotransferase and anticonvulsant drug (see Chapter 5, Section 5.5.C.3.a, p. 287).[29] This compound inactivates GABA aminotransferase *in vitro* only if brain esterases are added to cleave the vigabatrin from the glyceryl lipid. It also is 300 times more potent than vigabatrin *in vivo* presumably because of its increased ability to enter the brain.

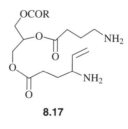

8.17

In the above examples, the lipophilicity of the drugs was increased so that they could diffuse through various membranes. Another approach for site-specific drug delivery is to design a prodrug that requires activation by an enzyme found predominantly at the desired site of action. For example, tumor cells contain a higher concentration of phosphatases and amidases than do normal cells. Consequently, a prodrug of a cytotoxic agent could be directed to tumor cells if either of these enzymes was important to the prodrug activation process. Diethylstilbestrol diphosphate (**8.18**, R $= PO_3^=$) was designed for site-specific delivery of diethylstilbestrol (**8.18**, R $=$ H; one trade name is Stilbestrol) to prostatic carcinoma tissue.[30] In general, though, this tumor-selective approach has not been very successful for several reasons: The appropriate prodrugs are too polar to reach the enzyme site, the relative enzymatic selectivity is insufficient, and the tumor cell perfusion rate is too poor.

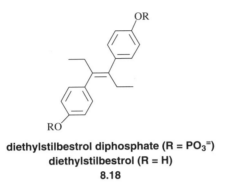

diethylstilbestrol diphosphate (R = PO$_3^=$)
diethylstilbestrol (R = H)
8.18

Several strategies, under the rubric of *enzyme-prodrug therapies*, have been developed to achieve selective activation of prodrugs at a desired site, typically in tumor cells.[31] All of these approaches involve two steps: In the first step a prodrug-activating enzyme is incorporated into the target tumor cells, and in the second step a nontoxic prodrug, which is a substrate of the exogenous enzyme that was incorporated into the tumors, is administered systemically. The prodrug is selectively converted into the active anticancer drug in a high local concentration inside the tumor cell. Certain criteria are important for this general approach to be effective:[32] (1) The prodrug-activating enzyme should be either of nonhuman origin or a human protein that is absent or expressed only at low concentrations in normal tissues; (2) the prodrug-activating enzyme must achieve adequate expression in the targeted tumor cells and have high catalytic activity; (3) the prodrug should be a good substrate for the enzyme incorporated in the tumors but not be activated by endogenous enzymes outside of the tumors; (4) the prodrug must be able to cross the tumor cell membrane for intracellular activation; (5) the cytotoxicity difference between the prodrug and its corresponding active drug should be high; (6) the activated drug should be highly diffusible or be actively taken up by adjacent nonexpressing cancer cells for what is known as a *bystander killing effect*, the ability of the drug to kill neighboring nonexpressing cells; and (7) the half-life of the active drug should be long enough to induce a bystander killing effect but short enough to avoid the drug leaking out of the tumor cells and causing damage elsewhere.

One strategy of this type is called *antibody-directed enzyme prodrug therapy* (abbreviated ADEPT). In the first step an antibody raised against a particular tumor cell line is conjugated with (attached to) the enzyme that is needed to activate an antitumor prodrug. After the antibody–enzyme conjugate is administered and has accumulated on the tumor cell, the excess conjugate not bound to the tumor cell is given enough time to clear from the blood and normal tissues. The prodrug is then administered. The enzyme conjugated with the antibody at the tumor cell surface catalyzes the conversion of the prodrug to the drug when it reaches the tumor cell. The advantage of this method, relative to direct administration of the prodrug, is the increased selectivity for release of high concentrations of the drug at the targeted cells. This advantage is only evident if enough time is allowed for clearance of the antibody–enzyme

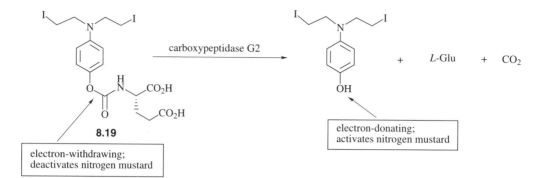

electron-withdrawing;
deactivates nitrogen mustard

8.19

electron-donating;
activates nitrogen mustard

Scheme 8.4 ▶   Nitrogen mustard activation by carboxypeptidase G2 for use with ADEPT

conjugate that is not bound to the tumor cells. An increase in the clearance rate of unbound antibody–enzyme conjugate would permit the administration of a higher concentration of the prodrug for a longer period of time. Galactosylation of the antibody leads to more rapid and efficient clearance of the unbound antibody–enzyme conjugate by galactose receptors in the liver.[33]

The drawbacks to ADEPT are the potential for immunogenicity and rejection of the antibody-enzyme conjugate, the potential for leakback of the active drug formed at the tumor, and the requirement of i.v. administration as well as the complexity of the treatment. Nonetheless, this approach is in the clinical trial stage.

An example of ADEPT is the delivery of a nitrogen mustard as a glutamic acid conjugate (**8.19**) after administration of a humanized monoclonal antibody[34] conjugated to the bacterial enzyme carboxypeptidase G2 (Scheme 8.4).[35] Humanization of antibodies raised from external sources minimizes immunogenicity. Note that the enzyme selected for prodrug activation is a bacterial enzyme so that there may be selectivity for prodrug activation by this enzyme in preference to that by human carboxypeptidase at sites other than at the tumor cells. The drawback to this approach is the potential for increased incidence of immunogenicity and rejection as a result of using a bacterial enzyme. This problem could be alleviated with the use of a humanized catalytic antibody[36] in place of the bacterial enzyme for activation of the prodrug. The reaction shown in Scheme 8.4 was also effected using a humanized catalytic antibody in a process termed *antibody-directed abzyme*[37] *prodrug therapy* (ADAPT).[38]

An even more effective approach for attaining selectivity for prodrug activation at the tumor cell would be to use a humanized catalytic antibody that not only does not exist in humans, but catalyzes a reaction not known to occur in humans. That way, the only site where the prodrug could be activated would be where the catalytic antibody resides, presumably directed with a monoclonal antibody to the desired tumor cells. Antibody 38C2[39] is a broad specificity catalytic antibody that catalyzes sequential retro-aldol and retro-Michael reactions, a combination of reactions not catalyzed by any known human enzyme. The abzyme was found to be long lived *in vivo*, to activate prodrugs selectively, and to potentiate the killing of colon and prostate cancer cell lines.[40] An example of the activation reactions catalyzed by this abzyme is shown in Scheme 8.5 for a doxorubicin prodrug (**8.20**).

A related strategy for improving the selectivity of cancer chemotherapy is called *gene-directed enzyme prodrug therapy* (GDEPT; also called *suicide gene therapy*).[41] In this approach a gene encoding the prodrug-activating enzyme is integrated into the genome of the target cancer cells under the control of tumor-selective promotors or by viral transfection.

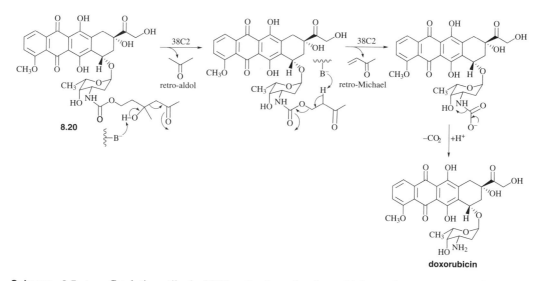

Scheme 8.5 ▶ Catalytic antibody 38C2 activation of a doxorubicin prodrug by tandem retro-aldol/retro-Michael reactions for use with ADAPT

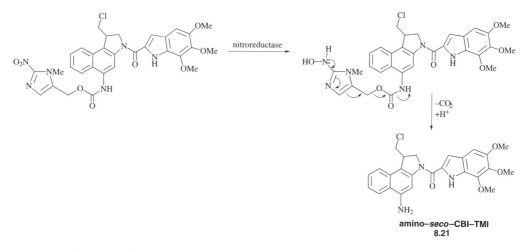

Scheme 8.6 ▶ Nitroreductase activation of a prodrug for use with GDEPT

These cells, then, express the enzyme that activates the prodrug added in the second step, as described above for ADEPT.

A common enzyme used for activation of a prodrug by GDEPT is an aerobic flavin-dependent nitroreductase from *E. coli* B, that catalyzes the reduction of aromatic nitro groups to the corresponding hydroxylamino group.[42] For example, the (2-nitroimidazol-5-yl)methyl carbamate prodrug of the minor groove alkylating agent amino-*seco*-CBI-TMI (**8.21**) is stable until nitroreductase reduces the aromatic nitro group to the corresponding hydroxylamino group. This initiates the elimination of the carbamate to give CO_2 and the free alkylating agent (Scheme 8.6).[43]

Virus-directed enzyme prodrug therapy (VDEPT) is another gene therapy strategy that uses viral vectors to deliver a gene that encodes a prodrug-activating enzyme.[44] Despite extensive use of retroviral and adenoviral vectors to deliver prodrug-activating enzyme genes, both vectors have some disadvantages. The principal disadvantage of a retroviral vector is that recombinant retroviruses only target dividing cells,[45] but most human tumor cells divide slowly. Even in a rapidly growing tumor nodule, only 6–20% of the cells are in a proliferating state. Therefore, the majority of the tumor would not be sensitive to killing by retroviral VDEPT. However, this drawback could be beneficial in some cases, such as for brain tumors, because in the brain only the tumor cells would be proliferating. This would allow for a high tumor:normal transfection differential for retroviral delivery. Although most viral vectors are engineered to be replication deficient, there is a slight risk of reversion to the wild-type virus (yikes!). Furthermore, retrovirus vectors are inserted into the host-cell DNA, which may cause mutagenesis of the host's genome (double yikes!).[46]

GDEPT and VDEPT have an advantage over ADEPT in that many enzymes need cofactors that are present only inside the cells. Therefore, enzymes delivered by ADEPT may need to gain access to the inside of tumor cells before they can optimally activate prodrugs. This is a problem because of the poor cell penetration of antibody-enzyme conjugates. In GDEPT gene-encoding enzymes can be specifically delivered to target tissues, which allows for the expression of the enzyme within the target cells.[47] Problems associated with GDEPT include insertional mutagenesis, anti-DNA antibody formation, local infection, and tumor nodule ulceration,[48] as well as difficulties with the selective delivery and expression of the genes.[49]

d. Prodrugs for Stability

Some prodrugs protect the drug from the *first-pass effect* (see Chapter 7, Section 7.1, p. 406). Propranolol (**8.22**, R = R′ = H; Inderal) is a widely used antihypertensive drug, but because of first-pass elimination, an oral dose has a much lower bioavailability than does intravenous injection. The major metabolites (see Chapter 7) are propranolol *O*-glucuronide (**8.22**, R = H, OR′ = glucuronide), *p*-hydroxypropranolol (**8.22**, R = OH, R′ = H) and its *O*-glucuronide (**8.22**, R = OH, OR′ = glucuronide). The hemisuccinate ester of propranolol (**8.22**, R = H, R′ = $COCH_2CH_2COOH$) was prepared to block glucuronide formation; following oral administration of propranolol hemisuccinate, the plasma levels of propranolol were eight times greater than when propranolol was used.[50]

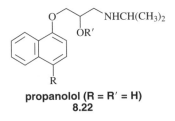

propanolol (R = R′ = H)
8.22

Naltrexone (**8.23**, R = H; Trexan), used in the treatment of opioid addiction, is nonaddicting and is well absorbed from the gastrointestinal tract. However, it undergoes extensive first-pass metabolism when given orally. Ester prodrugs, the anthranilate (**8.23**, R = CO-*o*-NO₂Ph) and the acetylsalicylate (**8.23**, R = CO-*o*-AcO-Ph), enhanced the bioavailability 45- and 28-fold, respectively, relative to **8.23** (R = H).[51]

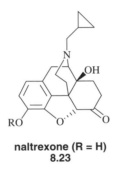

naltrexone (R = H)
8.23

e. Prodrugs for Slow and Prolonged Release

The utility of slow and prolonged release of drugs is several-fold: (1) It reduces the number and frequency of doses required. (2) It eliminates nighttime administration of drugs. (3) Because the drug is taken less frequently, it minimizes patient noncompliance. (4) When a fast-release drug is taken, there is a rapid surge of the drug throughout the body. As metabolism of the drug proceeds, the concentration of the drug diminishes. A slow-release drug would eliminate these peaks and valleys of fast-release drugs, which place a strain on cells. (5) Because a constant lower concentration of the drug is being released, it reduces the possibility of toxic levels of drugs. (6) It reduces gastrointestinal side effects. A common strategy in the design of slow-release prodrugs is to make a long-chain aliphatic ester, because these esters hydrolyze slowly, and to inject them intramuscularly.

Prolonged release drugs are quite important in the treatment of psychoses because these patients require medication for extended periods of time and often show high patient non-compliance rates. Haloperidol (**8.24**, R = H; Haldol) is a potent, orally active central nervous system depressant, sedative, and tranquilizer. However, peak plasma levels are observed between 2 and 6 h after administration. The ester prodrug haloperidol decanoate (**8.24**, R = $CO(CH_2)_8CH_3$; Haldol Decanoate) is injected intramuscularly as a solution in sesame oil, and its antipsychotic activity lasts for about 1 month.[52] The antipsychotic fluphenazine (**8.25**, R = H; one trade name is Prolixin) also has a short duration of activity (6–8 h). Fluphenazine enanthate (**8.25**, R = $CO(CH_2)_5CH_3$; Prolixin Enan) and fluphenazine decanoate (**8.25**, R = $CO(CH_2)_8CH_3$; Prolixin Dec), however, have durations of activity of about 1 month.[53]

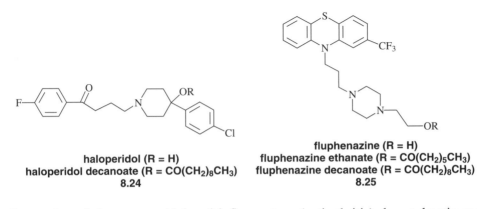

haloperidol (R = H)
haloperidol decanoate (R = $CO(CH_2)_8CH_3$)
8.24

fluphenazine (R = H)
fluphenazine ethanate (R = $CO(CH_2)_5CH_3$)
fluphenazine decanoate (R = $CO(CH_2)_8CH_3$)
8.25

Conversion of the nonsteroidal anti-inflammatory (antiarthritis) drug tolmetin sodium (**8.26**, R = O^-Na^+; Tolectin) to the corresponding glycine conjugate (**8.26**, R =

$NHCH_2COOH$) increases the potency and extends the peak concentration of tolmetin from 1 h to about 9 h because of the slow hydrolysis of the prodrug amide linkage.[54]

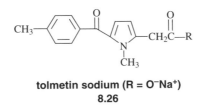

tolmetin sodium (R = O⁻Na⁺)
8.26

f.   Prodrugs to Minimize Toxicity

The prodrugs that were designed for improved absorption (Section 8.2.A.2.b, p. 507), for site specificity (Section 8.2.A.2.c, p. 507), for stability (Section 8.2.A.2.d, p. 512), and for slow release (Section 8.2.A.2.e, p. 513) also lowered the toxicity of the drug. For example, epinephrine (**8.15**, R = H) (see Section 8.2.A.2.b, p. 507), used in the treatment of glaucoma, has a number of ocular and systemic side effects associated with its use. The prodrug dipivaloylepinephrine (**8.15**, R = Me_3CCO) has been shown to be more potent than epinephrine in dogs and rabbits and nearly as effective in humans[55] with a significantly improved toxicological profile compared with epinephrine. Another example of the utility of the prodrug approach to lower toxicity of a drug can be found in the design of aspirin (**8.27**, R = H) analogs.[56] Side effects associated with the use of aspirin are gastric irritation and bleeding. The gastric irritation and ulcerogenicity associated with aspirin use may result from an accumulation of the acid in the gastric mucosal cells. Esterification of aspirin (**8.27**, R = alkyl) and other nonsteroidal anti-inflammatory agents greatly suppresses gastric ulcerogenic activity. However, esterification also renders the acetyl ester of aspirin extremely susceptible to enzymatic hydrolysis (the $t_{1/2}$ for deacetylation of aspirin in human plasma is about 2 h, but for deacylation of aspirin esters is 1–3 minutes). Esters of certain N, N-disubstituted 2-hydroxyacetamides (**8.27**, R = $CH_2CONR_1R_2$), were found to be chemically highly stable, but were hydrolyzed very rapidly by pseudocholinesterase in plasma and therefore are well suited as aspirin prodrugs to lower the gastric irritation effects of aspirin.

aspirin (R = H)
8.27

g.   Prodrugs to Encourage Patient Acceptance

A fundamental tenet in medicine is that in order for a drug to be effective, the patient has to take it! Painful injections and unpleasant taste or odor are the most common reasons for the lack of patient acceptance of a drug. An excellent example of how a prodrug can increase the potential for patient acceptance is related to the antibacterial drug clindamycin (**8.28**, R = H; one trade name is Cleocin). Whereas clindamycin causes pain on injection, the prodrug clindamycin phosphate (**8.28**, R = PO_3H_2; one trade name is Dalacin) is well tolerated; hydrolysis of the prodrug *in vivo* occurs with a $t_{1/2}$ of approximately 10 minutes.[57] Also, clindamycin has a bitter taste, so it is not well accepted orally by children. However, it was found that by increasing the chain length of 2-acyl esters of clindamycin the taste improved from bitter (acetate ester) to no bitter taste (palmitate ester).[58] Of course, when dealing with young children, it is not sufficient for a drug to be just tasteless; consequently, clindamycin palmitate (**8.28**, R = $CO(CH_2)_{14}CH_3$; Cleocin Pediatric) is sold for pediatric use in a cherry-flavored syrup. Bitter taste results from a compound dissolving in the saliva and interacting with a bitter taste receptor in the mouth. Esterification with long-chain fatty acids makes the drug less water soluble and unable to dissolve in the saliva. It also may alter the interaction of the compound with the taste receptor.

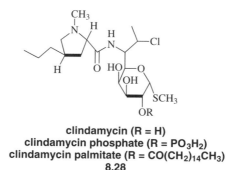

clindamycin (R = H)
clindamycin phosphate (R = PO₃H₂)
clindamycin palmitate (R = CO(CH₂)₁₄CH₃)
8.28

The antibacterial sulfa drug sulfisoxazole (**8.29**, R = H) also is bitter tasting, but sulfisoxazole acetyl (**8.29**, R = $COCH_3$; Gantrisin) is tasteless. For pediatric use this prodrug is combined with the tasteless prodrug form of erythromycin (i.e., erythromycin ethylsuccinate), in a strawberry-banana-flavored suspension (Pediazole).

sulfisoxazole (R = H)
sulfisoxazole acetyl (R = COCH₃)
8.29

h. Prodrugs to Eliminate Formulation Problems

Formaldehyde (CH_2O) is a flammable, colorless gas with a pungent odor that is used as a disinfectant. Solutions of high concentrations of formaldehyde are toxic. Consequently, it cannot be used directly in medicine. However, the reaction of formaldehyde with ammonia produces a stable adamantane-like solid compound, methenamine (**8.30**, one trade name is Hiprex). In acidic pH media, methenamine hydrolyzes to formaldehyde and ammonium ions. Because the pH of urine in the bladder is mildly acidic, methenamine is used as a urinary tract antiseptic.[59] To prevent hydrolysis of this prodrug in the acidic environment of the stomach, the tablets are enteric coated.

methenamine
8.30

The topical fungistatic prodrug triacetin (**8.31**, Captex 500) owes its activity to acetic acid, the product of skin esterase hydrolysis of triacetin.

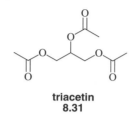

triacetin
8.31

A.3   Macromolecular Drug Carrier Systems

a.   General Strategy

Although the prodrug approach has been very fruitful in general, three areas need improvement: site specificity, protection of the drug from biodegradation, and minimization of side effects. Another carrier-linked bipartate prodrug approach that has been utilized to address these shortcomings is *macromolecular drug delivery*. This is a drug carrier system in which the drug is covalently attached to a macromolecule, such as a synthetic polymer, a glycoprotein, a lipoprotein, a lectin, a hormone, albumin, a liposome, DNA, dextran, an antibody, or a cell. The pharmacokinetic characteristics of these drugs change dramatically because the absorption and distribution of the drug depends on the physicochemical properties of the macromolecular carrier, not of the drug. These parameters can be altered by manipulation of the properties of the carrier. This approach has the potential advantage of targeting drugs for a specific site and improving the therapeutic index by minimizing interactions with nontarget tissues (i.e., lowering the toxicity) as well as reducing premature drug metabolism and excretion. However, it has the disadvantages that the macromolecules may not be well absorbed after oral administration, that alternative means of administration are required, and that they may be immunogenic. Although polymer conjugates generally cannot pass through membranes, they can gain access to the interior of a cell by *pinocytosis*, the process by which the cell membrane invaginates the particle and then pinches itself off to form an intracellular vesicle that moves into the cell and eventually fuses with lysosomes. Because the breakdown of proteins and other macromolecules is believed to occur in the lysosomes,[60] and because this breakdown then liberates the drug, the design of a macromolecular drug carrier system should be a fruitful approach to deliver the drug inside a cell. This approach has already been taken in Nature, although in this case the drug is not covalently bound to the macromolecular carrier. The antitumor antibiotic C-1027 consists of an enediyne (see Chapter 6, Section 6.3.C.4, p. 378) bound to a carrier protein; the protein protects the labile enediyne from destroying the host.[61]

Some of the macromolecular drug carrier systems exert their effects while the drug is still attached to the carrier, but these are not prodrugs. Several examples of macromolecular drug carrier systems follow.

b. Synthetic Polymers

Aspirin linked to poly(vinyl alcohol) (**8.32**) was shown to have the same potency as aspirin, but was less toxic. Another anti-inflammatory agent, ibuprofen (the carboxylic acid of **8.33**, Advil), was attached as a poly(oxyethylene) diester (**8.33**).[62] This macromolecular carrier system resulted in a sustained release of ibuprofen, giving prolonged anti-inflammatory activity and a higher plasma half-life relative to the free drug.

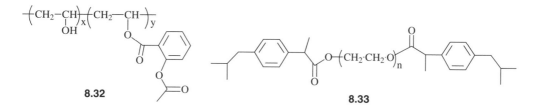

8.32   **8.33**

Because it is necessary for the drug to be released from the polymer backbone, steric hindrance by the polymer to chemical or enzymatic hydrolysis may cause problems. We saw a problem related to this in Chapter 2, Section 2.2.E.5.d (p. 41) during parallel synthesis on a solid support. To avoid steric hindrance by the polymer in the first synthetic step, a spacer was incorporated between the polymer and the first building block. When the steroid hormone testosterone was linked to poly(methacrylate) (**8.34**), no androgenic effect was observed, but when a spacer arm was inserted between the polymer and the testosterone (**8.35**), this macromolecular drug carrier was as effective as testosterone. The 3-thiabutyl oxide chain was attached to the polymer to enhance water solubility.

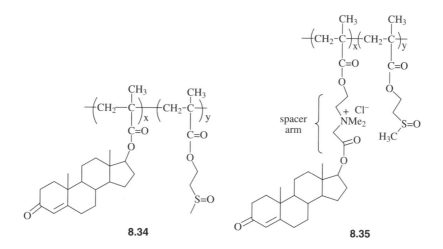

8.34   **8.35**

c. Poly(α-Amino Acids)

The disadvantage of using synthetic polymers is that they are generally not biodegradable, and they can take 5–12 months to be eliminated from the body. Poly(α-amino acids) are biodegradable (at least the *L*-isomers are); the rate of biodegradability depends on the choice of amino acid.

Conjugation of the antitumor drug methotrexate (Rheumatrex) to poly(*L*-lysine) (**8.36**; attachment of the polymer also may be to the α-carboxyl group) markedly increased the cellular uptake of the drug and provided a new way to overcome drug resistance related to deficient drug transport.[63] Because the activity of methotrexate is a function of its inhibitory properties of dihydrofolate reductase (see Chapter 4, Section 4.3.B, p. 200), and **8.36** is a poor inhibitor of this enzyme *in vitro*, the methotrexate must become detached from the polymer backbone inside the cell. Furthermore, attachment of methotrexate to poly(*D*-lysine), which, unlike poly(*L*-lysine) does not undergo proteolytic digestion inside the cell, gave a conjugate devoid of activity with resistant or normal cell lines. Methotrexate attached to poly(*L*-lysine) also is more inhibitory to the growth of human solid tumor cell lines than to the growth of human lymphocytes; free methotrexate is equally toxic to both kinds of cells.[64]

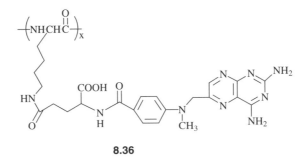

8.36

Research directed at a sustained release contraceptive resulted in the macromolecular drug delivery system **8.37**.[65] The contraceptive norethindrone (Nor-QD) was attached via a 17-carbonate linkage to poly-N^5-(3-hydroxypropyl)-*L*-glutamine. In rats the contraceptive agent was slowly released over a 9-month period.

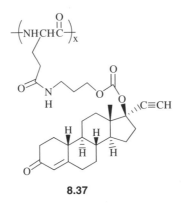

8.37

A general scheme for the design of a site-specific macromolecular drug delivery system was described by Ringsdorf[66] (**8.38**). A drug is attached to the polymer backbone usually through a spacer, so that it can be cleaved hydrolytically or enzymatically without steric hindrance. The desired solubility of the drug-polymer conjugate can be adjusted by attachment of an appropriate hydrophilic or hydrophobic ligand. Finally, site specificity, for example, to a particular cancer cell line, can be manipulated by attachment of a "homing device" such as an antibody raised against that cell line.

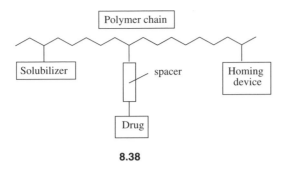

8.38

An elegant example of this approach in which a nitrogen mustard was delivered to tumor cells is shown in **8.39**.[67] Poly(*L*-glutamate) was used as the polymeric backbone so that the side-chain carboxylic acid groups could be functionalized appropriately. The water solubilizing groups are the unsubstituted glutamate side-chain carboxylate groups; the antitumor alkylating agent (the *p*-phenylenediamine mustard) is attached to the built-in spacer arm, that is, the glutamate side chain; and the homing device is an immunoglobulin (Ig) derived from a rabbit antiserum against mouse lymphoma cells. This macromolecular drug delivery system was much more effective than the individual components or a mixture of the components. Whereas none of the five control mice was alive and tumor free after 60 days, all five of the polymer prodrug-treated mice were. Also, the therapeutic index of *p*-phenylenediamine mustard is greatly enhanced (40-fold) when it is attached to the polymer system, because it is less toxic to normal proliferating cells. Similar results were obtained when the neutral and water-soluble polymer dextran was used.[68]

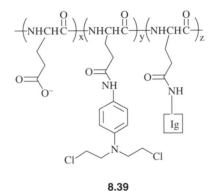

8.39

d. Other Macromolecular Supports

Because inhibitors of DNA synthesis generally are toxic to normal rapidly proliferating cells as well as to cancer cells, a targeted macromolecular approach to the delivery of the antitumor agents floxuridine (**8.40**, R = H; Sterile FUDR) and cytosine arabinoside (cytarabine, **8.41**, R = H; Cytosar-U) was taken to decrease their toxicity.[69] These drugs were conjugated to albumin because once proteins enter cells, they are rapidly broken down by lysosomal enzymes, and this would release the drugs from the albumin inside the cells. Because certain neoplastic proliferating cells are highly endocytic (high protein uptake) and normal cells with high protein uptake do not proliferate, selective toxicity to neoplastic or to DNA viruses that replicate in cells with high protein uptake could be accomplished. Both conjugates (**8.40** and

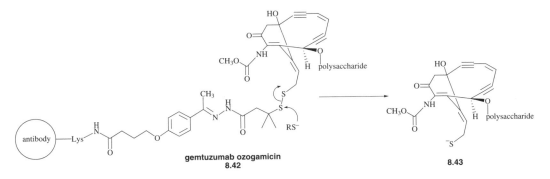

Scheme 8.7 ▶   Antibody-targeted chemotherapy, a prodrug for calicheamicin

8.41, R = albumin-CO) were shown to inhibit the growth of *Ectromelia* virus in mouse liver, whereas the free inhibitors were ineffective. The conjugates exert their antiviral activity in liver macrophages (cells with high protein uptake), suggesting that the drugs are concentrated in these cells.

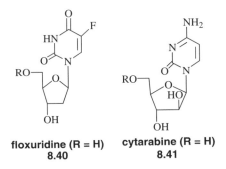

floxuridine (R = H)   cytarabine (R = H)
8.40   **8.41**

As mentioned in Chapter 6, Section 6.3.C.4.a, p. 380, calicheamicin is too toxic for use in cancer chemotherapy. To minimize its toxicity, *antibody-targeted chemotherapy* was undertaken using a slightly modified version of calicheamicin (a disulfide instead of a trisulfide), which was attached to a humanized antibody through a spacer to give the drug called gemtuzumab ozogamicin (**8.42**, Scheme 8.7; Mylotarg). Used for the treatment of CD33-positive acute myeloid leukemia (the antibody is specific for the CD33 antigen, a protein commonly expressed by myeloid leukemic cells),[70] this novel antibody conjugate does not release calicheamicin nonenzymatically to any significant extent and exhibits no immune response. Reduction of the disulfide bond releases the calicheamicin intermediate (**8.43**), which leads to DNA strand breakage as shown in Chapter 6, Section 6.3.C.4.a, Scheme 6.30, p. 380.

A.4 Tripartate Prodrugs

Bipartate prodrugs may be ineffective because the prodrug linkage is too labile (e.g., certain esters) or too stable (because of steric hindrance to hydrolysis). Katzenellenbogen and coworkers[71] designed a *tripartate* (also known as a *self-immolative*) *prodrug* to remedy this problem. In a tripartate prodrug the carrier is not connected directly to the drug, but rather to a linker that is attached to the drug (Scheme 8.8). This allows for different kinds of functional groups to be incorporated for varying stabilities, and it also displaces the drug farther from

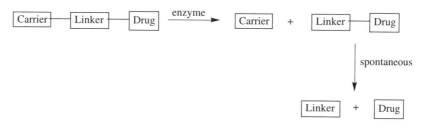

Scheme 8.8 ▶ Tripartate prodrugs

$$\text{Drug—X—CH}_2\text{—O—}\overset{\overset{\displaystyle O}{\|}}{\text{C}}\text{R} \xrightarrow{\text{esterase}} \text{Drug—X} \overset{\frown}{} \text{CH}_2 \overset{\frown}{} \text{O}^- + \text{RCOOH}$$

$$\downarrow \text{fast}$$

$$\text{Drug—X}^- + \text{CH}_2\text{O}$$

Scheme 8.9 ▶ Double prodrug concept

the hydrolysis site, which decreases the steric interference by the carrier. The drug-linker connection, however, must be designed so that it cleaves spontaneously (i.e., is self immolative) *after* the carrier has been detached. One approach to accomplish this has been termed the *double prodrug* or, in the case where X = COO, the *double ester* concept, generalized in Scheme 8.9[72] (X = COO, O, NH).

This strategy was employed in the design of prodrugs of ampicillin (**8.44**, one trade name is Omnipen), a β-lactam antibiotic that is poorly absorbed when administered orally. Because only 40% of the drug is absorbed, 2.5 times more drug must be administered orally than by injection. Having to take extra antibiotic can lead to a more rapid onset of resistance, and the nonabsorbed antibiotic may destroy important intestinal bacteria used in digestion and for the biosynthesis of cofactors.[73] A lipid-soluble prodrug of ampicillin would be a useful approach to increase absorption of this drug. However, although various simple alkyl and aryl esters of the thiazolidine carboxyl group are hydrolyzed rapidly to ampicillin in rodents, they are too stable in humans to be therapeutically useful. This suggests that the esterases in rodents and man are different and that, most likely, steric hindrance of the ester carbonyl by the thiazolidine ring is important in the human esterase. A solution to the problem was the construction of a "double ester," an acyloxymethyl ester[74] such as **8.45** (bacampicillin, R = CH_3, R′ = OEt; Penglobe)[75] or **8.45** (pivampicillin, R = H, R′ = t-Bu; Pondocillin)[76] (Scheme 8.10), which would extend the terminal ester carbonyl away from the thiazolidine ring and eliminate the inherent steric hindrance with the enzyme. Hydrolysis of the terminal ester (or carbonate, in the case of bacampicillin) gives an unstable hydroxymethyl ester (**8.46**) that spontaneously decomposes to ampicillin and either acetaldehyde (bacampacillin) or formaldehyde (pivampicillin). Bacampicillin is a nontoxic prodrug because it decomposes to ampicillin and all natural metabolites in the body: CO_2, acetaldehyde, and ethanol. (The usual recommended dose of bacampicillin is 400 mg twice a day; therefore only about 50 μl of ethanol would be released with each dose, so don't expect to get high.) Unlike ampicillin, bacampicillin is absorbed to the extent of 98–99%, and ampicillin is liberated into the bloodstream in less than 15 minutes.

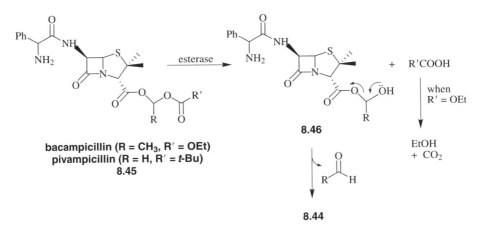

bacampicillin (R = CH$_3$, R' = OEt)
pivampicillin (R = H, R' = t-Bu)
8.45

8.46

8.44

+ R'COOH

when
R' = OEt

EtOH
+ CO$_2$

Scheme 8.10 ▶   Tripartate prodrugs of ampicillin

Because of the excellent absorption properties of bacampicillin, only one-half to one-third of the ampicillin dose is required orally.

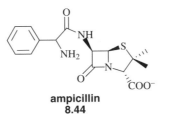

ampicillin
8.44

As mentioned in Section 8.2.A.2.c, p. 507, the blood–brain barrier is an important membrane for protection of the brain from polar, hydrophilic molecules that do not belong there. Bodor and coworkers have devised a reversible redox drug delivery system for getting drugs into the central nervous system, then once in, from preventing their efflux.[77] The approach is based on the attachment of a hydrophilic drug to a lipophilic carrier, a dihydropyridine (**8.47**), thereby making the prodrug overall sufficiently lipophilic to cross the blood–brain barrier passively (Scheme 8.11). Furthermore, the nitrogen atom in the dihydropyridine ring is conjugated through the double bond into the carbonyl (a vinylogous amide), thereby making the carbonyl less reactive toward nucleophiles, such as water, and therefore more stable to hydrolysis. Once inside the brain, the lipophilic carrier is converted enzymatically into a highly hydrophilic species (**8.48**) in which the pyridinium nitrogen atom is no longer conjugated into the carbonyl; in fact, the pyridinium group is electron withdrawing, so it now activates the carbonyl for nucleophilic attack. The drug is readily released by enzymatic hydrolysis, and the N-methylnicotinic acid (**8.49**) is relatively nontoxic and actively transported out of the brain. The XH group on the drug can be an amino, hydroxyl, or carboxyl group. A tripartate redox prodrug also could be prepared (**8.50**), which would decompose by the self-immolative reaction shown in Scheme 8.12. The oxidation of the dihydropyridine (**8.47**) to the pyridinium ion (**8.48**) (half-life generally 20–50 minutes) prevents the drug from escaping out of the brain because it becomes charged. This drives the equilibrium of the lipophilic precursor (**8.47**) throughout all of the tissues of the body to favor the brain. Any oxidation occurring outside

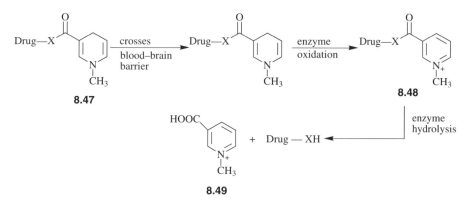

8.47   **8.48**

8.49

Scheme 8.11 ▶ Redox drug delivery system to cross the blood-brain barrier

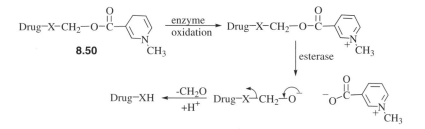

8.50

Scheme 8.12 ▶ Redox tripartate drug delivery system

of the brain produces a hydrophilic species that can be rapidly eliminated from the body (see Chapter 7). Although this is a carrier-linked prodrug, it requires enzymatic oxidation to target the drug to the brain. The oxidation reaction is a bioprecursor reaction (see Section 8.2.B.5.c, p. 532).

A tripartate example of this approach is the brain delivery of β-lactam antibiotics for the possible treatment of bacterial meningitis. The difficulty in purging the central nervous system of infections is that the cerebrospinal fluid contains less than 0.1% of the number of immunocompetent leukocytes found in the blood and almost no immunoglobins; consequently, antibody generation to these foreign organisms is not significant. Because β-lactam antibiotics are hydrophilic, they enter the brain very slowly, and they are actively transported out of the brain back into the bloodstream. Therefore, they are not as effective in the treatment of brain infections as elsewhere. Bodor and coworkers[78] prepared a variety of penicillin prodrugs in which the drug is attached to the dihydropyridine carrier through various linkers (**8.51**) and showed that β-lactam antibiotics could be delivered in high concentrations into the brain, presumably by the mechanism in Scheme 8.13.

The antitumor agent 5-fluorouracil (**8.52**, R = H; one trade name is Adrucil) also has been used in the treatment of certain skin diseases. However, because of its low lipophilicity, it does not produce optimal topical bioavailability. N-1-Acyloxymethyl derivatives (**8.52**, R = CH$_2$OCOR) were prepared for increased lipophilicity. These prodrugs were shown to penetrate the skin about five times faster than **8.52** (R = H) and to be metabolized to **8.52** (R = H) rapidly.[79] The mechanism for conversion of **8.52** (R = CH$_2$OCOR) to **8.52** (R = H) is the same as that shown in Scheme 8.10 for ampicillin derivatives.

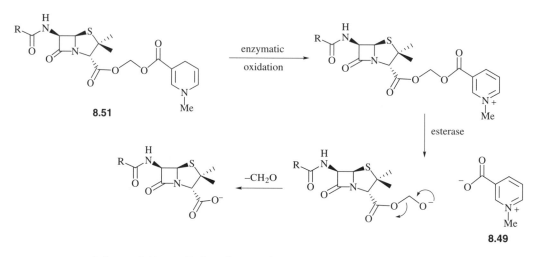

Scheme 8.13 ▶ Redox tripartate drug delivery of β-lactam antibiotics

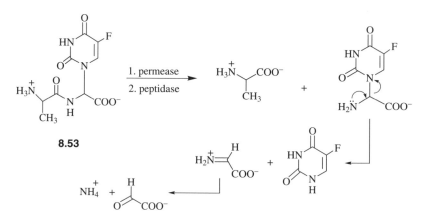

Scheme 8.14 ▶ Activation of peptidyl derivatives of 5-fluorouracil

5–fluorouracil (R = H)
8.52

Microorganisms have specialized transport systems for the uptake of peptides (permeases), and these transport systems generally have little side-chain specificity. Consequently, peptidyl derivatives of 5-fluorouracil (**8.53**) were designed as potential antifungal and antibacterial agents that would be substrates for both microbial permeases and peptidases.[80] In accord with the known stereochemical selectivity of peptide permeases, only the peptidyl prodrug with the *L,L* configuration was active. The mechanism for release of 5-fluorouracil after peptidase action is shown in Scheme 8.14.

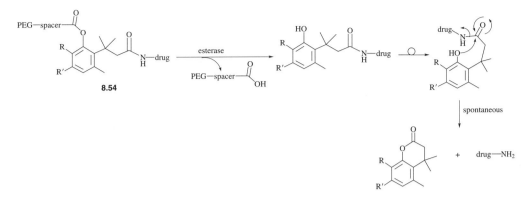

Scheme 8.15 ► Tripartate macromolecular drug delivery system

Amine-containing drugs can be solubilized as acid salts, but their rate of renal excretion is often high. If these drugs are converted into small amide prodrugs, they are no longer able to make salts, so the aqueous solubility decreases. A tripartate macromolecular drug delivery system was designed to retain water solubility of amine-containing antitumor agents without the high renal excretion (**8.54**, Scheme 8.15).[81] In this system 40-kDa polyethylene glycol (PEG) is incorporated in the carrier to retain water solubility, and the linker is an *o*-hydroxyphenyl-3,3-dimethylpropionic acid, which, after carrier hydrolysis, undergoes a rapid intramolecular lactonization[82] to release the amine-containing drug. The rate of hydrolysis of the carrier-linker bond can be controlled by varying the spacer and the substituents around the linker aromatic ring. This approach was taken for the drugs daunorubicin (**8.55**, Cerubidine) and for cytarabine (**8.41**, R = H). In an *in vivo* solid tumor panel, one of the PEG-daunorubicin prodrugs was more efficacious against ovarian tumors than daunorubicin.

daunorubicin
8.55

A.5 Mutual Prodrugs

When it is necessary for two synergistic drugs to be at the same site at the same time, a mutual prodrug approach should be considered. A *mutual prodrug* is a bipartate or tripartate prodrug in which the carrier is a synergistic drug with the drug to which it is linked. In Chapter 5, Section 5.5.B.2.a, p. 277, a form of resistance to β-lactam antibacterial drugs was discussed in Chapter 5, which these bacteria excrete a high concentration of the enzyme β-lactamase, which deactivates these antibiotics by hydrolysis of the β-lactam ring. For resistant bacteria, compounds that inhibit β-lactamase are given in combination with a β-lactam antibacterial drug. For example, the combination of the penicillin derivative amoxicillin (**8.56**, Amoxil) and the β-lactamase inactivator potassium clavulanate (**8.57**) (Augmentin) is used for oral treatment of infections caused by β-lactamase-producing bacteria. Another combination used is the ampicillin prodrug, pivampicillin (**8.45**, R = H; R′ = t-Bu, Pondocillin) plus the double ester (**8.58**, R = CH$_2$OCOCMe$_3$) of the β-lactamase inactivator penicillanic acid sulfone (**8.58**, R = H; Zosyn). However, if the two prodrugs are given separately, it is not clear that they are absorbed and transported to the site of action at the same time and in equivalent amounts. An example of a tripartate mutual prodrug is sultamicillin (**8.59**, Unasyn Oral), which on hydrolysis by an esterase produces ampicillin, penicillanic acid sulfone, and formaldehyde in a reaction like that shown in Scheme 8.10.[83] A mutual prodrug would have a high probability of success provided it is well absorbed, both components are released concomitantly and quantitatively after absorption, the maximal effect of the combination of the two drugs occurs at a 1:1 ratio, and the distribution and elimination of the two components are similar.

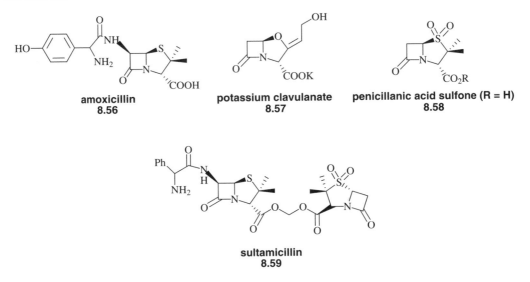

amoxicillin
8.56

potassium clavulanate
8.57

penicillanic acid sulfone (R = H)
8.58

sultamicillin
8.59

8.2.B   Bioprecursor Prodrugs

B.1   Origins

The birth of bioprecursor prodrugs occurred when it was demonstrated that the antibacterial agent prontosil was active only *in vivo* because it was metabolized to the actual drug sulfanilamide (see Chapter 5, Section 5.4.B.2, p. 254). In this case the azo prodrug prontosil was reduced to the amine sulfa drug. This exemplifies the bioprecursor strategy. The compounds discussed in Chapter 6, Section 6.3.B.5, p. 359, metabolically activated alkylating agents, are also examples of bioprecursor prodrugs, but because of their eventual alkylation of DNA, they were placed in that part of the book instead of here. Some of the examples here lead to DNA modification and could have equally been placed in Chapter 6 as well.

Whereas carrier-linked prodrugs rely largely on hydrolysis reactions for their effectiveness, bioprecursor prodrugs mostly utilize either oxidative or reductive activation reactions. The examples given below are arranged according to the type of metabolic activation reaction involved. The first example is the simplest of metabolic transformations, namely, protonation as a mechanism for prodrug activation.

B.2 Proton Activation: An Abbreviated Case History of the Discovery of Omeprazole

In Chapter 3 we discussed the development of the antiulcer drugs cimetidine and ranitidine (Section 3.2.G, p. 159). These compounds lowered gastric acid secretion by antagonizing the H_2 histamine receptor. Another approach for lowering gastric acid secretion is by inhibition of the enzyme H^+,K^+-ATPase (also known as the *proton pump*), which is responsible for acid secretion by the *parietal cell*, the cell in the gastric mucosa responsible for acidification of the stomach. This enzyme catalyzes a one-to-one exchange of proton and potassium ions.[84] In 1972 the Swedish pharmaceutical company Hässle was searching for a compound that could block gastric acid secretion and discovered a lead compound (**8.60**) in a random screen.[85] The liver toxicity caused by this compound was attributed to the thioamide group, so other sulfur-containing analogs were made, and **8.61** emerged with good antisecretory activity. A series of analogs of **8.61** with different heterocycles led to **8.62** with high activity. A metabolism study in dogs demonstrated that the corresponding sulfoxide (**8.63**, timoprazole) was more potent, but it also blocked the uptake of iodine into the thyroid gland, so it could not be used in humans. A variety of analogs of timoprazole were synthesized, and **8.64** (picoprazole) was found to have antisecretory activity without the iodine blockage activity. In 1977 it was found that picoprazole inhibited the enzyme H^+,K^+-ATPase. A SAR of analogs of picoprazole showed that electron-donating groups on the pyridine ring, which increased the pK_a of the pyridine ring, also increased the potency as an inhibitor of H^+,K^+-ATPase. The best analog was omeprazole (**8.65**, Prilosec).[86]

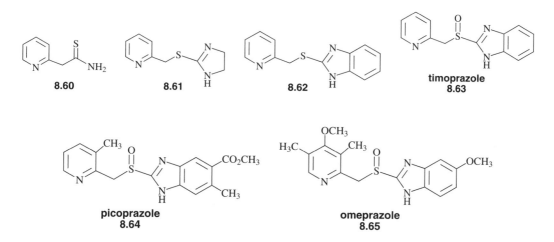

Studies with ^3H-labeled omeprazole showed that the compound concentrated in the gastric mucosa.[87] Later it was found to be bound to the enzyme H^+,K^+-ATPase in parietal cells.[88] Omeprazole is a relatively weak base, having a pK_a of only about 4. Therefore, the pyridine ring is not protonated at physiological pH, so it is lipid permeable and able to diffuse into the secretory canaliculus of the parietal cell. However, the pH in the parietal cell is below 1, so omeprazole becomes protonated *inside* the canaliculus of the cell, where it becomes trapped, then undergoes a proton-initiated transformation to **8.66**, which reacts covalently with a cysteine residue of H^+,K^+-ATPase (Scheme 8.16).[89] Omeprazole also inhibits human carbonic anhydrase isozymes I and II in erythrocytes and isozyme IV selectively in gastric mucosa.[90] Inhibition of carbonic anhydrase has been shown to be another mechanism for lowering gastric acid secretion.[91] This indicates that omeprazole may have

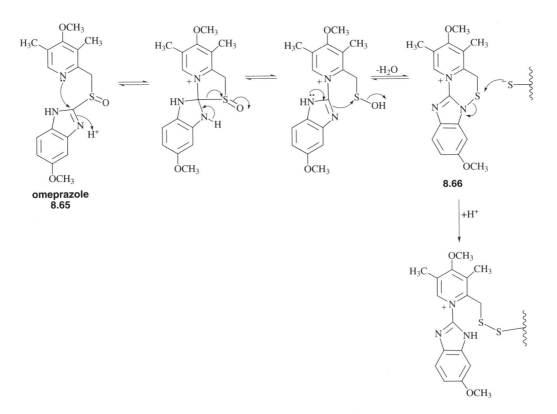

Scheme 8.16 ▶   Mechanism of inactivation of H^+, K^+-ATPase by omeprazole

a twofold mechanism of action, which may explain the greater effectiveness of the substituted benzimidazole class of antiulcer drugs compared to other classes of antiulcer drugs. Related analogs that are similar to omeprazole include lansoprazole (**8.67**, Prevacid),[92] rabeprazole (**8.68**, Aciphex),[93] and pantoprazole sodium (**8.69**, Protonix).[94]

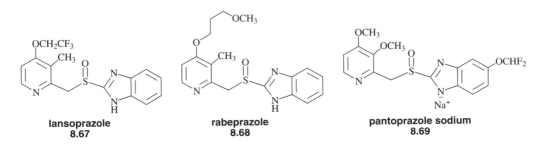

B.3   Hydrolytic Activation

Hydrolysis can be a mechanism for bioprecursor prodrug activation if the product of hydrolysis requires additional activation to become the active drug. Leinamycin (**8.70**, Scheme 8.17) is a potent antitumor agent described in Chapter 6, Section 6.3.B.5.d, p. 366, which is unstable and toxic. The half-life of aqueous stability was increased by a factor of up to fivefold by

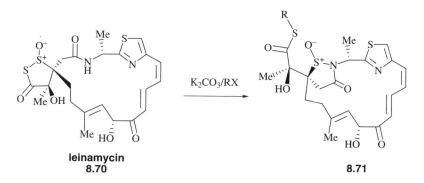

Scheme 8.17 ► Conversion of leinamycin into a series of prodrugs

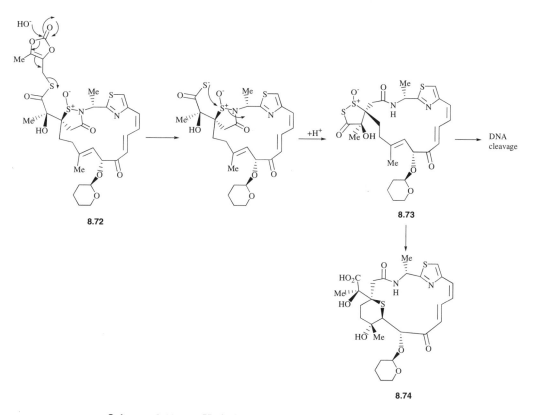

Scheme 8.18 ► Hydrolytic activation of leinamycin prodrugs

conversion into a series of prodrugs (**8.71**).[95] Incubation of analog **8.72** [Scheme 8.18; the most potent analog, in which the macrocyclic ring hydroxyl group also is protected as an (*R*)-2-tetrahydropyranyl prodrug] with fetal calf serum gave **8.74** as the major metabolite, the same metabolite produced from the (*R*)-2-tetrahydropyranyl ether of leinamycin (**8.73**), the proposed intermediate in the hydrolytic activation of **8.72**. The prodrug showed increased antitumor activity relative to leinamycin, presumably as a result of its increased metabolic stability.

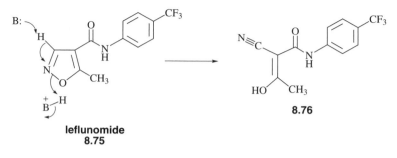

leflunomide
8.75

Scheme 8.19 ▶ Activation of leflunomide to the active drug

B.4 Elimination Activation

Another relatively simple prodrug activation mechanism is elimination. The rheumatoid arthritis drug leflunomide (**8.75**, Arava) is an immunomodulatory agent shown to inhibit pyrimidine biosynthesis in human T lymphocytes by blocking the enzyme dihydroorotate dehydrogenase.[96] Whereas leflunomide shows no inhibitory effect on dihydroorotate dehydrogenase at 1 μM concentration, its metabolite, **8.76**, is a potent inhibitor (K_i 179 nM).[97] Isoxazoles are known to undergo facile elimination to nitriles (Scheme 8.19).[98]

B.5 Oxidative Activation

a. N- and O-Dealkylations

Open-ring analogs of benzodiazepines, such as the anxiolytic drug alprazolam (**8.77**, X = H; Xanax) and the sedative triazolam (**8.77**, X = Cl; Halcion), undergo metabolic N-dealkylation and spontaneous cyclization (Scheme 8.20).[99]

An example of a bioprecursor prodrug that is activated by O-dealkylation is the analgesic and antipyretic agent phenacetin (**8.78**, R = CH$_2$CH$_3$, Acetophenetidin), which owes its activity to its conversion by O-dealkylative metabolism to acetaminophen (**8.78**, R = H; Tylenol).[100]

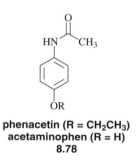

phenacetin (R = CH₂CH₃)
acetaminophen (R = H)
8.78

b. Oxidative Deamination

Because of the high concentration of phosphoramidases in neoplastic cells, hundreds of phosphamide analogs of nitrogen mustards were synthesized and tested as carrier-linked antitumor prodrugs. Cyclophosphamide (**8.79**, Scheme **8.21**; Cytoxan) emerged as an important drug for the treatment of a wide variety of malignant diseases; however, it was later found that it was inactive in tissue culture, suggesting that simple hydrolysis was not involved. Preincubation

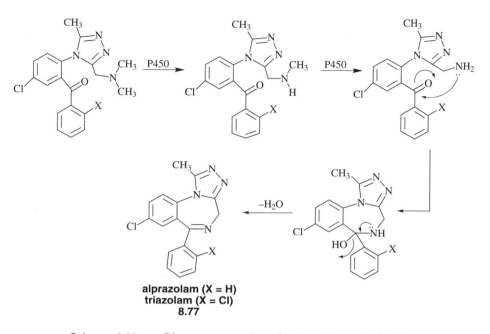

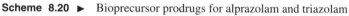

Scheme 8.20 ▶ Bioprecursor prodrugs for alprazolam and triazolam

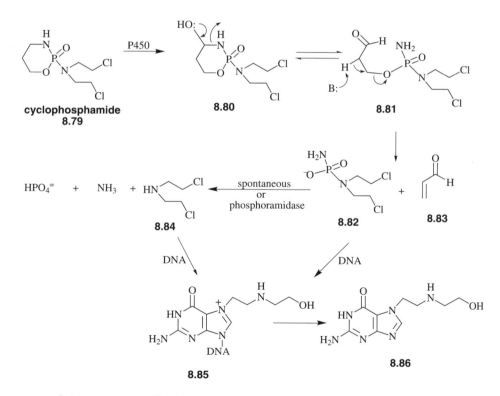

Scheme 8.21 ▶ Cytochrome P450-catalyzed activation of cyclophosphamide

of the compound with liver homogenates, however, activated it, suggesting that cyclophosphamide is a prodrug requiring an oxidative mechanism (see Chapter 7, Section 7.4.B.1.e, p. 430).[101] The activation mechanism is believed to be that shown in Scheme 8.21. (Other metabolites that are not shown in this scheme are derived from each of the intermediates.) It is not clear which of the toxic metabolites, the phosphoramide mustard (**8.82**), or the parent nitrogen mustard (**8.84**), is responsible for the therapeutic action; the major adduct isolated by HPLC from *in vitro* and *in vivo* studies in rat is *N*-(2-hydroxyethyl)-*N*-[2-(7-guaninyl)ethyl]amine (**8.86**),[102] the decomposition product of **8.85**. The reaction of nitrogen mustards with DNA was discussed in Chapter 6, Section 6.3.B.1, p. 354. Acrolein (**8.83**) is a potent Michael acceptor that may be responsible for the hemorrhagic cystitis side effect; administration of sulfhydryl compounds, which react readily with acrolein, can prevent this side effect. Aldehyde dehydrogenase catalyzes the oxidation of **8.80** to the corresponding cyclic amide and of **8.81** to the corresponding carboxylic acid; however, both of these metabolites are inactive. It has been suggested that these detoxification reactions occur to a greater extent in normal cells than with cancer cells, and may account for the selective toxicity of cyclophosphamide.[103]

c. *N*-Oxidation

The antitumor drug used against advanced Hodgkin's disease, procarbazine (**8.87**, Matulane), is believed to be activated by *N*-oxidation (Scheme 8.22); it is inert unless treated with liver homogenates or oxidized in neutral solution.[104] Those of you who have studied Chapter 7 are probably wondering why this circuitous mechanism to **8.89** and methylhydrazine, starting with an *N*-oxidation reaction, was written instead of a direct conversion of **8.87** to these same metabolites by an oxidative deamination mechanism. The reason is that azoprocarbazine (**8.88**) was identified as the initial metabolic product.[105] 7-Methylguanine (**8.90**) was identified in the urine of mice given procarbazine,[106] which suggests that an activated methylating agent such as methyl diazonium or methyl radical[107] is the reactive intermediate.

Another *N*-oxidation prodrug activation reaction is based on the reversible redox drug delivery strategy of Bodor and coworkers for getting drugs into the brain (see Section 8.2.A.4, p. 520). In the case of pralidoxime chloride (**8.91**, Protopam), an antidote for poisoning by organophosphorus pesticides and nerve toxins, the oxidation reaction converts prodrug **8.92**

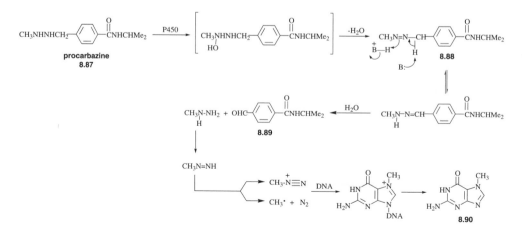

Scheme 8.22 ▶ Cytochrome P450-catalyzed activation of procarbazine

into the drug and also prevents efflux of the drug from the brain. The neurotoxic organophosphorus compounds exert their effects by reacting with acetylcholinesterase, the enzyme found in nervous tissue of all species of animals that catalyzes the hydrolysis of the excitatory neurotransmitter acetylcholine after this neurotransmitter has served its neurohumoral transmission function. Inhibition of the enzyme results in an accumulation of acetylcholine, which continues to act on the receptors in various muscles. Muscle cells in the airways contract and secrete mucous, both of which cause choking and difficulty in breathing, and eventually the muscles become paralyzed. Secretion of epinephrine increases the blood pressure and heart rate. Overactivation of nerve cells in the brain causes severe confusion, dizziness, and convulsions.

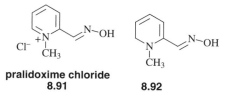

pralidoxime chloride
8.91   **8.92**

The active site of the enzyme contains two important binding sites, the site that binds the quaternary ammonium cation of acetylcholine and the ester site where the catalytic hydrolysis of the acetyl group occurs (Scheme 8.23).[108] An X-ray crystal structure of acetylcholinesterase revealed that the ammonium cation does not interact with an anionic residue, but with the π-electrons of a group of aromatic residues.[109] Stabilization of cations by aromatic π-electrons is well documented.[110]

Organophosphorus compounds, such as the nerve poison diisopropyl phosphorofluoridate (**8.93**), phosphorylate acetylcholinesterase at the ester site[111] (Scheme 8.24). It was thought that a nucleophilic agent may be capable of dephosphorylating the ester site, and this would reactivate the enzyme. Hydroxylamine appeared to be effective, but also was quite toxic. Because acetylcholinesterase has a cation binding site, quaternary amine analogs were designed, and 2-formyl-1-methylpyridinium chloride oxime (pralidoxime chloride, **8.91**) was found to be an effective reactivator of the enzyme (Scheme 8.25). However, **8.91** is very poorly soluble in lipids, so its generation is most likely restricted to the peripheral nervous system; little reactivation of brain acetylcholinesterase was observed *in vivo*.[112] Apparently, the effectiveness of **8.91** as an antidote for organophosphorus nerve poisons results from the fact that the primary damage done by these poisons is to the peripheral nervous system. To improve the permeability of **8.91** into the central nervous system, Bodor and coworkers[113] prepared

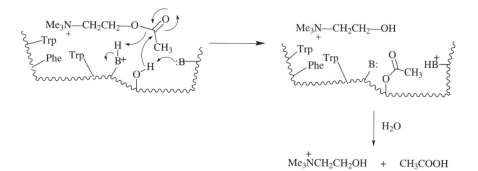

Scheme 8.23 ▶ Acetylcholinesterase-catalyzed hydrolysis of acetylcholine

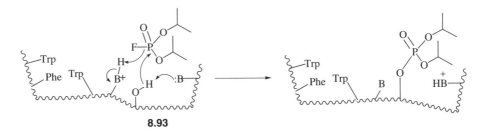

8.93

Scheme 8.24 ▶ Phosphorylation of acetylcholinesterase by diisopropyl phosphorofluoridate

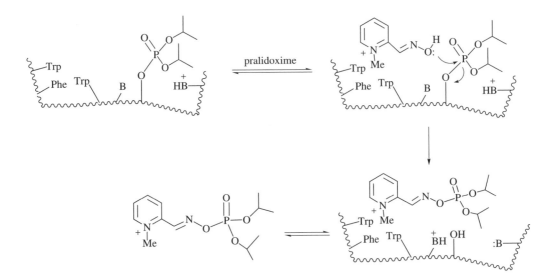

Scheme 8.25 ▶ Reactivation of phosphorylated acetylcholinesterase by pralidoxime chloride

the 5,6-dihydropyridine analog **8.92**. Because it is uncharged, its permeability through the blood–brain barrier is quite good. Once inside the brain it is oxidized to **8.91**.

It is interesting to note that whereas irreversible inactivators of acetylcholinesterase, such as organophosphorus nerve gases, are highly toxic, compounds that form weakly stable covalent bonds to the serine residue in the ester binding site are useful therapeutic agents. This inhibition of acetylcholinesterase results in the enhancement of cholinergic action by facilitating the transmission of impulses across neuromuscular junctions, which has a cholinomimetic effect on skeletal muscle. An example of this is neostigmine (**8.94**, Prostigmin), a drug used in the treatment of the neuromuscular disease myasthenia gravis. This drug carbamylates the active site serine residue of acetylcholinesterase; the carbamate, however, hydrolyzes slowly so that, in effect, **8.94** acts as a reversible inhibitor of the enzyme (Scheme 8.26). Therefore, the difference in effects of the acetylcholinesterase substrates and inhibitors is derived from the stabilities of the covalent adducts. The acetylated serine formed from acetylcholine (a substrate) hydrolyzes readily, the carbamoylated serine produced from neostigmine (an inhibitor) hydrolyzes slowly, and the phosphorylated serine from organophosphorus compounds (inactivators) is stable to hydrolysis.

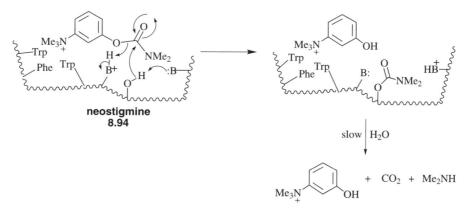

Scheme 8.26 ► Carbamoylation of acetylcholinesterase by neostigmine

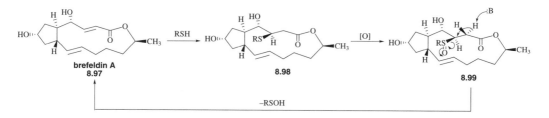

Scheme 8.27 ► Synthesis of brefeldin A prodrugs and their conversion back to brefeldin A

Reversible inhibitors of acetylcholinesterase, such as donepezil hydrochloride (**8.95**, Aricept)[114] and tacrine hydrochloride (**8.96**, Cognex)[115] are used in the treatment of Alzheimer's disease. The increase in acetylcholine as a result acetylcholinesterase inhibition enhances cholinergic neurotransmission involved in the memory circuit.

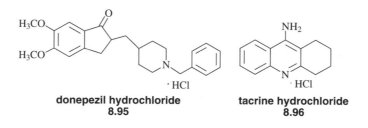

d. S-Oxidation

Brefeldin A (**8.97**, Scheme 8.27), an antitumor and antiviral antibiotic, has poor oral bioavailability and is rapidly cleared. A series of Michael addition sulfide prodrugs (**8.98**) was prepared having greater aqueous solubilities than brefeldin A.[116] The prodrugs were converted back to brefeldin A by S-oxidation to the sulfoxide (**8.99**) followed by a retro-Michael reaction.

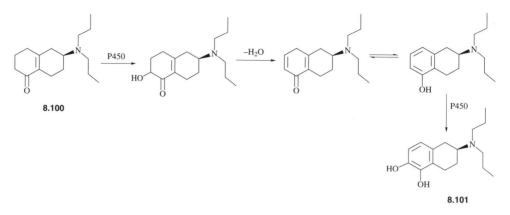

Scheme 8.28 ▶ Cytochrome P450-catalyzed conversion of a cyclohexenone to a catechol

e.   Aromatic Hydroxylation

Cyclohexenone derivative **8.100** was activated *in vivo* by hydroxylation to the potent dopamin-ergic catecholamine **8.101**. There is no evidence for the reactions shown in Scheme 8.28, but it depicts one possible pathway. The conversion of **8.100** to **8.101** suggests that cyclohexenones can be prodrugs for catechols with increased bioavailability.[117]

f.   Other Oxidations

Carbamazepine (**8.102**, one trade name is Cabatrol) is an anticonvulsant drug that is the metabolic precursor of the active agent, carbamazepine-10,11-oxide (**8.103**).[118]

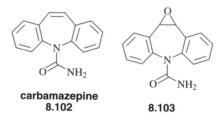

carbamazepine
8.102

8.103

Stimulation of pyruvate dehydrogenase results in a change of myocardial metabolism from fatty acid to glucose utilization. Because the latter requires less oxygen consumption, glucose utilization is beneficial to patients with ischemic heart disease in which arterial blood flow is blocked and therefore less oxygen is available.[119] Arylglyoxylic acids (**8.104**) are important stimulators of pyruvate dehydrogenase, but they have short durations of action. A series of phenylglycine analogs was synthesized (**8.105**), and it was found that *L*-(+)-2-(4-hydroxyphenyl)glycine (oxfenicine, **8.105**, R = OH) is a stable amino acid that is actively transported across lipid membranes and is rapidly transaminated (see Chapter 4, Section 4.3.A.3, p. 195) to 4-hydroxyphenylglyoxylic acid (**8.104**, R = OH).[120] This active transport system and rapid conversion of the prodrug to the drug allow a higher concentration of the active drug to remain at the desired site of action longer.

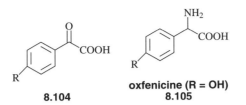

8.104

oxfenicine (R = OH)
8.105

B.6 Reductive Activation

a. Azo Reduction

As described in Section 8.2.B.1, p. 526, the paradigm for bioprecursor prodrugs, prontosil, is activated by reduction of its azo linkage to the true bacteriostatic agent, sulfanilamide (AVC).

Sulfasalazine (**8.106**, Azulfidine), which is used in the treatment of inflammatory bowel disease (ulcerative colitis), is reductively cleaved by anaerobic bacteria in the lower bowel to 5-aminosalicylic acid (**8.107**) and sulfapyridine (**8.108**); **8.107** is the therapeutic agent and **8.108** produces adverse side effects (Scheme 8.29).[121] A macromolecular drug delivery system was developed to improve the therapeutic index of this drug. The drug (**8.107**) was azo linked at the 5-position through a spacer to poly(vinyl amine) (**8.109**).[122] The advantages of this polymeric drug delivery system are that it is not absorbed or metabolized in the small intestine, **8.107** can be released by reduction at the disease site, and **8.108** is not released. The water-soluble polymer-linked drug (**8.109**) was more potent than **8.106** or **8.107** in the guinea pig ulcerative colitis model.

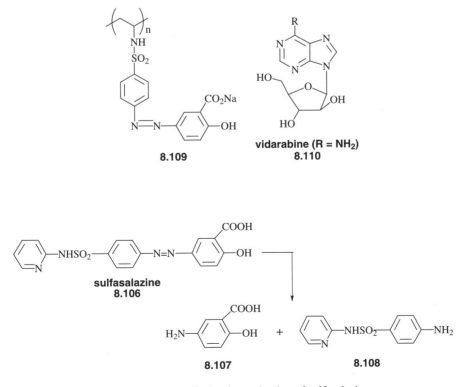

Scheme 8.29 ▶ Reductive activation of sulfasalazine

b. Azido Reduction

Vidarabine (**8.110**, R = NH_2; Vira-A) was originally discovered as an antitumor agent,[123] then later it was shown to be active against herpes simplex virus types 1 and 2.[124] However, the clinical use of vidarabine is limited because of its rapid deamination by adenosine deaminase[125] and its poor aqueous solubility. 9-(β-D-Arabinofuranosyl)-6-azidopurine, the 6-azido analog of vidarabine (**8.110**, R = N_3), however, is not a substrate for adenosine deaminase, and it is considerably more stable *in vivo*.[126] The 6-azido prodrug is activated by cytochrome P450-catalyzed reduction to vidarabine; *in vivo*, the half-life for vidarabine produced from the corresponding azide is 7–14 times higher than for vidarabine administered directly. Furthermore, whereas vidarabine was not found in the brain after its direct intravenous administration, significant levels of vidarabine were found in the brain after either oral or intravenous administration of the azido prodrug, which is useful for the treatment of brain infections.

c. Sulfoxide Reduction

The antiarthritis drug sulindac (**8.111**, Clinoril) is an indene isostere (see Chapter 2, Section 2.2.E.4, p. 29) of the nonsteroidal anti-inflammatory (antiarthritis) drug indomethacin (**8.112**, Indocin), which originally was designed as a serotonin (see **2.19**, p. 17) analog. Sulindac is less irritating to the gastrointestinal tract and produces many fewer and more mild central nervous system effects than does indomethacin.[127] The 5-fluoro group was substituted for the methoxyl group to improve the analgesic properties, and the *p*-methylsulfinyl group was substituted for the chlorine atom to increase the solubility. Sulindac is inactive *in vitro*, but is highly active *in vivo*. The corresponding sulfide, however, is active *in vitro* and *in vivo*. Therefore, sulindac is a prodrug for the sulfide, the metabolic reduction product.

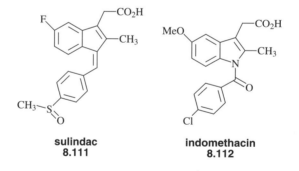

sulindac
8.111

indomethacin
8.112

d. Disulfide Reduction

Because thiamin (vitamin B_1, **8.114**) is a quaternary ammonium salt, it is poorly absorbed into the central nervous system and from the gastrointestinal tract. To increase its lipophilicity thiamin tetrahydrofurfuryl disulfide (**8.113**, Scheme 8.30) was designed as a lipid-soluble prodrug of thiamin.[128] The prodrug permeates rapidly through red blood cell membranes (as a model for other membranes) where it reacts with glutathione to produce thiamin.[129]

To diminish the toxicity of the antimalarial drug primaquine (**8.115**, Primaquine) and target it for cells that contain the malaria parasite, a macromolecular drug delivery system was designed[130] (**8.116**). The lactose-linked albumin was used for improved uptake in the liver via the asialoglycoprotein receptor system. Because the concentration of free thiol in the blood is relatively low, but is high intracellularly, it was expected that thiol reduction of

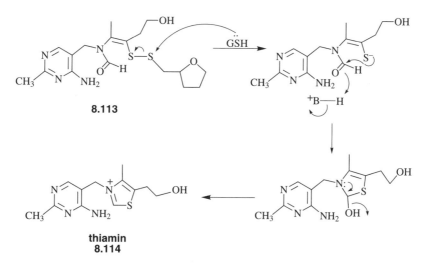

8.113

thiamin
8.114

Scheme 8.30 ▶ Conversion of thiamin tetrahydrofurfuryl disulfide to thiamin

the disulfide linkage would occur mostly inside the cell. It is not known if after disulfide reduction the cysteinyl residue is detached by hydrolysis or remains attached to primaquine. The therapeutic index of **8.116**, however, is 12 times higher than that of the free drug in *Plasmodium*-infected mice.

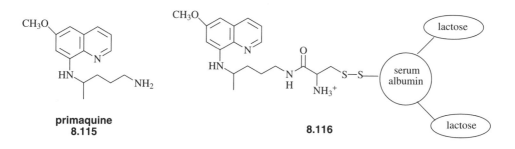

primaquine
8.115

8.116

e. Nitro Reduction

The mechanism of action of the antiprotozoal agent ronidazole (**8.117**, Dugro) is not known, but on the basis of metabolism studies using several radioactively labeled analogs, it was suggested that **8.117** is activated by initial four-electron reduction of the 5-nitro group to the corresponding hydroxylamine, which can react with protein thiols by one of two mechanisms (Scheme 8.31).[131]

A phosphoramidate-based prodrug was designed for intracellular delivery of nucleotides that requires initial nitro group reduction.[132] A prodrug for the delivery of the anticancer nucleotide 5-fluoro-2′-deoxyuridine 5′-monophosphate (**8.121**) is **8.118** (Scheme 8.32). Bioreduction of **8.118** gives **8.119**, which undergoes elimination to **8.120**, followed by spontaneous cyclization, elimination, and hydrolysis to **8.121**. Compound **8.118** gave excellent inhibition of cell proliferation and thymidylate synthase inhibition (see Chapter 5, Section 5.5.C.3.e, p. 298) via intracellular conversion to **8.121**.

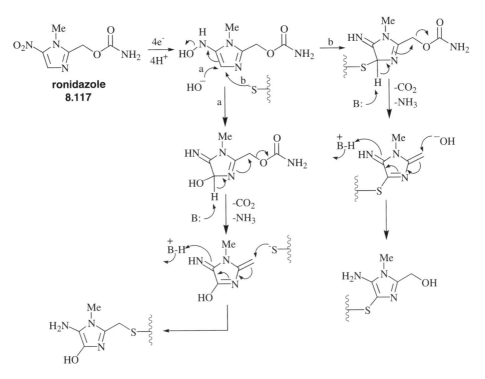

Scheme 8.31 ► Reductive activation of ronidazole

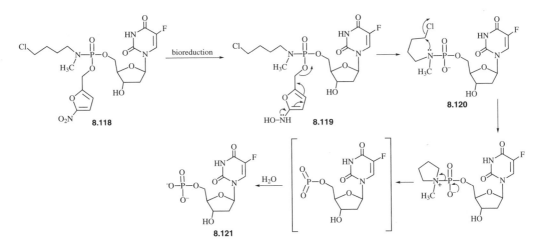

Scheme 8.32 ► Phosphoramidite-based prodrug for intracellular delivery of nucleotides

B.7 Nucleotide Activation

The antineoplastic agent 6-mercaptopurine (**8.122**, Purinethol) produces a 50% remission rate for acute childhood leukemias. Although **8.122** inhibits several enzyme systems, these inhibitions are irrelevant to its anticancer activity. Only tumors that convert the drug to its nucleotide are affected. 6-Mercaptopurine is activated by a hypoxanthine-guanine phosphoribosyltransferase-catalyzed reaction with 5-phosphoribosylpyrophosphate

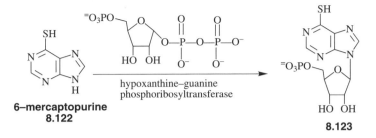

Scheme 8.33 ▶ Nucleotide activation of 6-mercaptopurine

(Scheme 8.33). The nucleotide (**8.123**) inhibits several enzymes in the purine nucleotide biosynthetic pathway, but the most prominent site is one of the early enzymes in the *de novo* pathway, namely, phosphoribosylpyrophosphate amidotransferase, which catalyzes the conversion of phosphoribosylpyrophosphate to phosphoribosylamine.[133] 5-Fluorouracil (see Chapter 5, Section 5.5.C.3.e, p. 298) is similar to 6-mercaptopurine in the sense that it must first be converted to the corresponding deoxyribonucleotide in order for it to be active.

B.8 Phosphorylation Activation

The antiviral drug acyclovir (**8.124**, R = H; Valtrex), one of the drugs for which Gertrude Elion and George Hitchings received the Nobel prize in 1988, is highly effective against genital herpes simplex virus and varicella zoster virus infections.[134] Its structure can be drawn so that it closely resembles the structure of 2′-deoxyguanosine (**8.125**), the nucleoside that is metabolically converted into 2′-deoxyguanosine triphosphate and is incorporated into the viral DNA. Acyclovir itself is inactive, but it is selectively phosphorylated by a viral thymidine kinase to the corresponding monophosphate (**8.124**, R = $PO_3^=$).[135] Uninfected cells do not phosphorylate acyclovir, and this accounts for the selective toxicity of acyclovir toward viral cells. The second step in the activation of acyclovir is the conversion of the monophosphate (**8.124**, R = $PO_3^=$) to the diphosphate (**8.124**, R = $P_2O_6^{3-}$), catalyzed by guanylate kinase.[136] The final activation step is the conversion of the diphosphate to the triphosphate (**8.124**, R = $P_3O_9^{4-}$), which could be accomplished by a variety of enzymes, particularly phosphoglycerate kinase.[137] Further selective toxicity is derived from the fact that acyclovir triphosphate is selectively taken up by viral α-DNA polymerases because its structure resembles that of the essential DNA precursor, 2′-deoxyguanosine triphosphate. The K_i for viral α-DNA polymerase is up to 40 times lower than that for normal cellular α-DNA polymerase.[138] Acyclovir triphosphate is a substrate for the viral α-DNA polymerase but not for the normal cellular α-DNA polymerase; however, incorporation of acyclovir triphosphate into the viral DNA leads to the formation of a dead-end complex (an enzyme–substrate complex that is no longer active) after the next deoxynucleotide triphosphate unit is incorporated.[139] This disrupts the replication cycle of the virus and destroys it. Even if the phosphorylated acyclovir were released from the virus cell, it would be too polar to be taken up by normal cells, and, as indicated above, the triphosphate is a poor substrate for normal human α-DNA polymerase anyway. Therefore, this drug exhibits a high degree of selective toxicity against viral cells.

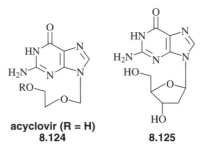

acyclovir (R = H)
8.124   8.125

As might be predicted from knowledge of the mechanism of acyclovir, acquired resistance to the drug can occur by three different mechanisms. Because of the importance of the thymidine kinase to the activation of acyclovir, resistance arises from a deletion of this enzyme or by a change in its substrate specificity.[140] The third mechanism is an altered viral α-DNA polymerase.[141] The degree of inhibition by acyclovir triphosphate of several different α-DNA polymerase mutants encoded by drug-resistant viruses correlated with the degree of resistance conferred by the mutation *in vivo*.[142] In some cases the enzyme mutation resulted in a decrease in binding of acyclovir triphosphate (higher K_m), and in others the mutation caused a reduction in catalytic activity for incorporation of acyclovir triphosphate into DNA (lower k_{cat}).

The largest shortcoming to the use of acyclovir is the fact that only 15–20% of acyclovir is absorbed after oral administration. Consequently, prodrugs for the prodrug acyclovir have been designed to improve gastrointestinal absorption and to protect acyclovir against biotransformations to inactive metabolites. 2,6-Diamino-9-(2-hydroxyethoxymethyl)purine (**8.126**) is converted to acyclovir by the enzyme adenosine deaminase[143] (catalyzes the hydrolysis of adenosine to inosine) and 6-amino-9-(2-hydroxyethoxymethyl)purine (**8.127**, 6-deoxyacyclovir) is oxidized to acyclovir by xanthine oxidase.[144] The latter compound is 18 times more water soluble than acyclovir. In humans urinary excretion of acyclovir is five to six times greater when **8.127** is given than an equivalent dose of acyclovir. Valaciclovir (**8.128**, Zovirax), the L-valyl ester of acyclovir, is a bipartate carrier-linked prodrug of acyclovir that has a three- to fivefold higher oral bioavailability while retaining the excellent safety profile.[145] The enzyme that catalyzes the hydrolysis of valaciclovir has been isolated and characterized.[146]

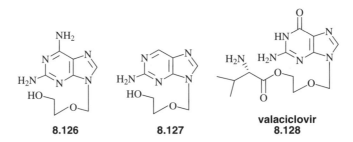

8.126   8.127   valaciclovir
                                  8.128

Ganciclovir (**8.129**, one trade name is Cytovene)[147] is an analog of acyclovir that has a structure and conformation resembling those of 2′-deoxyguanosine even closer than does acyclovir. This compound is about as potent as acyclovir against herpes simplex viruses and varicella zoster virus, but is much more inhibitory than acyclovir against human cytomegalovirus,[148] an important pathogen in immunocompromised and acquired immune deficiency syndrome (AIDS) patients.

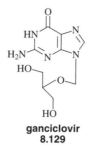

**ganciclovir
8.129**

The carbon isostere of ganciclovir, penciclovir (**8.130**, Denavir), is more potent and longer acting than acyclovir.[149] However, it is still poorly absorbed when given orally.[150] To increase absorption, a prodrug of the prodrug was designed called famciclovir (**8.131**, Famvir),[151] the diacetyl and 6-deoxy analog of penciclovir, which is used orally.[152] Greater than 75% absorption is attained with rapid conversion to penciclovir.[153] Metabolism studies showed that the diacetyl groups are hydrolyzed off prior to oxidation of the purine.[154] The *pro-S* acetoxyl group is hydrolyzed before the *pro-R* acetoxyl group.[155]

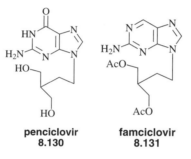

**penciclovir   famciclovir
8.130   8.131**

B.9 Sulfation Activation

The hypotensive[156] and hair growth[157] activity of minoxidil (**8.132**, R = −; Rogaine) requires a sulfotransferase-catalyzed sulfation to minoxidil sulfate (**8.132**, R = SO_3^-).[158] Because minoxidil sulfate is more potent than minoxidil, and inhibitors of the sulfotransferase inhibit minoxidil activity but not minoxidil sulfate activity, it is apparent that minoxidil is a prodrug for minoxidil sulfate. As described in Chapter 7, Section 7.4.C.3, p. 460, sulfation is a common mechanism in drug metabolism for the deactivation and excretion of drugs, so it is quite unusual for sulfation to be involved in prodrug activation.

**minoxidil (R = ⁻)
minoxidil sulfate (R = SO₃⁻)
8.132**

B.10 Decarboxylation Activation

The striatal tracts in the brain, which are important for the control of voluntary movements, contain a balance of the inhibitory neurotransmitter dopamine and the excitatory neurotransmitter acetylcholine. An imbalance in the dopaminergic and cholinergic components produces disorders of movement. In Parkinson's disease there is a marked deficiency in the dopaminergic component, which is attributed to the loss of dopaminergic neurons and a low concentration of dopamine in the substantia nigra. The obvious treatment for Parkinson's disease would be to give high doses of dopamine (**8.133**, R = H), but this does not work because dopamine does not cross the blood–brain barrier. However, there is an active transport system for *L*-amino acids; consequently, *L*-dopa (levodopa) (**8.133**, R = COOH, Larodopa) is transported into the brain where it is decarboxylated by the pyridoxal 5′-phosphate-dependent enzyme (see Chapter 4, Section 4.3.A.2, p. 195) aromatic *L*-amino acid decarboxylase (also called dopa decarboxylase) to dopamine. Because the *D*,*L*-mixture produces unwanted side effects, *L*-dopa (levodopa; Sinemet is the combination of levodopa and carbidopa; see below) is used as a prodrug for dopamine. Unfortunately, because dopaminergic neurons cannot be rejuvenated, levodopa does not reverse the course of the disease, it merely halts (actually only slows) its progression.[159]

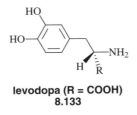

levodopa (R = COOH)
8.133

As discussed in Chapter 5, Section 5.5.C.3.d, p. 295, dopamine is a substrate for monoamine oxidase B; consequently, as levodopa is being converted to dopamine in the brain, monoamine oxidase B is degrading the dopamine. An inactivator of monoamine oxidase B, selegiline (Eldepryl), is now used in combination with levodopa to minimize the degradation of the dopamine generated by levodopa.[160]

One major complication with the use of levodopa therapy arises from the fact that aromatic *L*-amino acid decarboxylase also exists in the periphery (outside of the central nervous system), and greater than 95% of the orally administered levodopa is decarboxylated in its first pass through the liver and kidneys. Possibly only 1% of the levodopa taken actually penetrates into the central nervous system. If the peripheral aromatic *L*-amino acid decarboxylase could be inhibited without inhibition of the same enzyme in the brain, the levodopa would be protected from this undesired metabolism. This, in fact, is possible because inhibitors of aromatic *L*-amino acid decarboxylase are charged molecules, and unless they are actively transported, they will not cross the blood–brain barrier. Carbidopa (**8.134**, Lodosyn) is used in the United States and benserazide (**8.135**, one trade name is Prolopa) is used in Europe and Canada in combination with levodopa (the combination with carbidopa is Sinemet) for the treatment of Parkinson's disease.[161] With the combined use of a peripheral aromatic *L*-amino acid decarboxylase inhibitor, the optimal effective dose of levodopa can be reduced by greater than 75%.

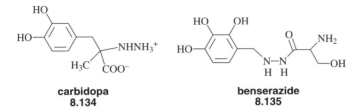

carbidopa
8.134

benserazide
8.135

In Chapter 5, Section 5.5.C.3.c, p. 292, the application of inactivators of monoamine oxidase A (MAO A) as antidepressant agents was discussed. Although MAO inactivators are used in the treatment of depression, a severe cardiovascular side effect can result unless the diet is controlled to minimize the intake of tyramine-containing foods. This side effect results from the concurrent inactivation of the peripheral MAO A along with brain MAO A. A brain-specific MAO A inactivator would give the desired antidepressant effect without the undesirable cardiovascular effect. A prodrug approach for the brain-selective delivery of a MAO A-selective inactivator was developed at formerly Marion Merrell Dow (now Aventis).[162] This particular type of prodrug was termed a *dual enzyme-activated inhibitor* because the activating enzyme is, by design, part of the same metabolic pathway as the enzyme that is targeted for inhibition. In this case the activating enzyme is aromatic *L*-amino acid decarboxylase, and the target enzyme is MAO A. (*E*)-β-Fluoromethylene-*m*-tyramine (**8.136**, R = H) is a mechanism-based inactivator (see Chapter 5, Section 5.5.C, p. 285) of monoamine oxidase with selectivity for MAO A.[163] The corresponding amino acid, (*E*)-β-fluoromethylene-*m*-tyrosine (**8.136**, R = COOH) is not an inhibitor of MAO, but it is a good substrate for aromatic-*L*-amino acid decarboxylase, which converts **8.136** (R = COOH) to **8.136** (R = H). The amino acid (**8.136**, R = COOH) is actively transported into the central nervous system and is concentrated in the synaptosomes. Because brain aromatic *L*-amino acid decarboxylase is located predominantly in monoamine nerve endings, **8.136** (R = COOH) is decarboxylated to **8.136** (R = H) at the desired site of action. To prevent inactivation of peripheral MAO A, **8.136** (R = COOH) is administered with carbidopa, which blocks peripheral *L*-aromatic amino acid decarboxylase-catalyzed decarboxylation of **8.136** (R = COOH). This results in brain-selective MAO A inactivation with little or no peripheral MAO A inhibition and only a minimal tyramine effect.

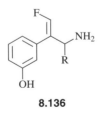

8.136

Not only is dopamine a major inhibitory neurotransmitter, but it also plays an important role in the kidneys. Dopamine increases systolic and pulse blood pressure and renal blood flow. If it is desired to have selective delivery of dopamine to the kidneys to attain renal vasodilation without a blood pressure effect, a prodrug for dopamine can be used. There is a high concentration of *L*-γ-glutamyltranspeptidase, the enzyme that catalyzes the transfer of the *L*-glutamyl group from the *N* terminus of one peptide to another, in kidney cells. Consequently, an *L*-γ-glutamyl derivative of an amino acid or amine drug could be cleaved selectively in the kidneys.[164] *L*-γ-Glutamyl-*L*-dopa (**8.137**) is selectively accumulated in the kidneys, and the *L*-dopa released by *L*-glutamyltranspeptidase is decarboxylated to dopamine by aromatic

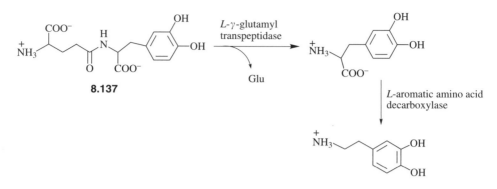

Scheme 8.34 ▶ Metabolic activation of *L*-γ-glutamyl-*L*-dopa to dopamine

L-amino acid decarboxylase, which also is abundant in kidneys (Scheme 8.34).[165] Even at high concentrations of this compound little central nervous system effect is apparent. This, then, is an example of a site-selective carrier-linked prodrug of a bioprecursor prodrug for dopamine.

Drug design is typically initiated with approaches to maximize the pharmacodynamic properties of molecules (increased binding to a receptor). A compound may be found that has the desired *in vitro* properties, but has unfavorable *in vivo* properties. It should be apparent, then, from the discussion in this chapter, that it may be possible to alter the structure of this compound to improve its pharmacokinetic properties and, thereby, transform it into a promising drug candidate.

8.3   General References

Prodrugs

Anderson, B. D. Prodrugs for improved CNS delivery. *Adv. Drug Deliv. Rev.* **1996**, *19*, 171–202.

Bundgaard, H. (Ed.) *Design of Prodrugs*, Elsevier, Amsterdam, 1985.

Melton, R. G.; Knox, R. J. (Eds.) *Enzyme-Prodrug Strategies for Cancer Therapy*, Plenum, New York, 1999.

Stella, V. J.; Charman, W. N.; Naringrekar, V. H. *Drugs* **1985**, *29*, 455.

Macromolecular Drug Carrier Systems

Friend, D. R.; Pangburn, S. *Med. Res. Rev.* **1987**, *7*, 53.

Goldberg, E. P. (Ed.) *Targeted Drugs*, Wiley, New York, 1983.

Gregoriadis, G.; Senior, J.; Trouet, A. (Eds.) *Targeting of Drugs*, Plenum, New York, 1982.

Luo, Y.; Prestwich, G. D. *Curr. Cancer Drug Targets* **2002**, *2*, 209.

Naughton, D. P. *Adv. Drug Deliv. Rev.* **2001**, *53*, 229.

Poznansky, M. J.; Juliano, K. L. *Pharmacol. Rev.* **1984**, *36*, 277.

Roerdink, F. H. D.; Kroon, A. M. (Eds.) *Drug Carrier Systems*, Wiley, Chichester, 1989.

Takakura, Y.; Hashida, M. *Pharm. Res.* **1996**, *13*, 820.

Takakura, Y.; Mahato, R. I.; Hashida, M. *Adv. Drug Deliv. Rev.* **1998**, *34*, 93.

8.4 PROBLEMS

(Answers can be found in Appendix at the end of the book.)

1. The anti-inflammatory agent fluocinolone (**1**, Synalar) is too hydrophilic for topical application. Suggest a prodrug (other than one used for a steroid in the text).

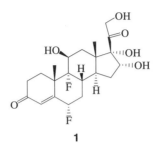

1

2. The cornea has significant esterase activity. Epinephrine (**2**) is an antiglaucoma agent that does not penetrate the cornea well. Suggest a prodrug (other than one used for epinephrine in the text).

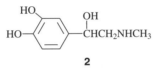

2

3. You have discovered a highly potent new drug (**3**) to treat West Nile virus, but it is soluble only to the extent of 50 mg/l, and it must be administered by injection. The prescribed dose is 400 mg. Design two prodrugs (one at each functional group) to get around this problem.

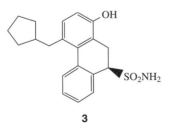

3

4. Resistance develops to your company's new antiviral drug (**4**) as a result of the encoding of a new viral urease that catalyzes the hydrolysis of the urea in your drug. As senior group leader, you need to devise a plan to rectify this catastrophe. Briefly describe two strategies that involve different prodrug approaches.

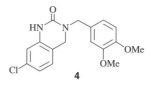

4

5. Design a prodrug to get **5** into the brain using a strategy similar to (but not the same as) the one used by Bodor (see Scheme 8.11, p. 523). In your prodrug strategy the prodrug activation should require monoamine oxidase for activation (see Chapter 5, Section 5.5.C.3.d, p. 275, regarding MPTP activation).

5

6. Design a copolymer-linked prodrug of ampicillin using a copolymer of poly(vinyl alcohol) and poly(vinyl amine).

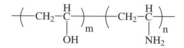

7. Compound **6** is a type of a tripartate antitumor prodrug which is activated by the enzyme β-glucuronidase. β-Glucuronidase catalyzes the hydrolysis of β-glucuronic acid acetals (**7**) to glucuronic acid (**8**). Give a reasonable mechanism for the activation of **6**.

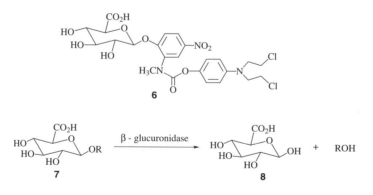

8. Draw a mutual prodrug for antitumor agents having different mechanisms of action.

9. Design bioprecursor prodrugs for the following drugs based on whatever hypothetical pathway you want. Show the enzymes involved in the conversions of your prodrugs to the drugs.

   A. acetaminophen

   B. cimetidine

   C. captopril

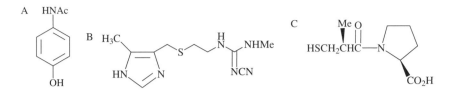

10. An antibody was raised against a tumor cell line and was conjugated to a β-lactamase. A nitrogen mustard was conjugated to a cephalosporin (**9**) for use in ADEPT. Draw a mechanism for the activation of the prodrug by the ADEPT conjugate.

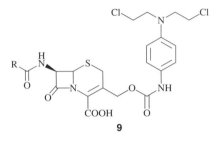

9

8.5 References

1. (a) Albert, A. *Selective Toxicity*, Chapman and Hall, London, 1951. (b) Albert, A. *Nature (London)* **1958**, *182*, 421.

2. Harper, N. J. *J. Med. Pharm. Chem.* **1959**, *1*, 467.

3. Wermuth, C. G. In *Drug Metabolism and Drug Design: Quo Vadis?*, Briot, M.; Cautreels, W.; Roncucci, R. (Eds.), Sanofi-Clin-Midy, Montpellier, 1983, p. 253.

4. Greene, T. W.; Wuts, P. G. M. *Protective Groups in Organic Synthesis*, 3rd ed., Wiley, New York, 1999.

5. Long, L., Jr. *Chem. Rev.* **1940**, *27*, 437.

6. Bundgaard, H. In *Design of Prodrugs*, Bundgaard, H. (Ed.), Elsevier, Amsterdam, 1985, p. 1.

7. Reynolds, W. F. *Prog. Phys. Org. Chem.* **1983**, *14*, 165.

8. Robinson, R.P.; Reiter, L. A.; Barth, W. E., Campeta, A. M.; Cooper, K.; Cronin, B. J.; Destito, R.; Donahue, K. M.; Falkner, F. C.; Fiese, E. F.; Johnson, D. L.; Kuperman, A. V.; Liston, T. E.; Malloy, D.; Martin, J. J.; Mitchell, D. Y.; Rusek, F. W.; Shamblin, S. L.; Wright, C. F. *J. Med. Chem.* **1996**, *39*, 10.

9. Dang, Q.; Brown, B. S.; van Poelje, P. D.; Colby, T. J.; Erion, M. D. *Bioorg. Med. Chem. Lett.* **1999**, *9*, 1505.

10. Bundgaard, H. In *Bioreversible Carriers in Drug Design*, Roche, E. B. (Ed.), Pergamon Press, New York, 1987, p. 13.

11. Johansen, M.; Bundgaard, H. *Arch. Pharm. Chem. Sci. Ed.* **1982**, *10*, 111.

12. (a) Johansen, M.; Bundgaard, H. *Int. J. Pharm.* **1980**, *7*, 119. (b) Bundgaard, H.; Johansen, M. *Int. J. Pharm.* **1981**, *8*, 183.

13. Bergmann, K. J. *Clin. Neuropharmacol.* **1985**, *8*, 13.

14. Kaplan, J.-P.; Raizon, B. M.; Desarmenien, M.; Feltz, P.; Headley, P. M.; Worms, P.; Lloyd, K. G.; Bartholini, G. *J. Med. Chem.* **1980**, *23*, 702.

15. Vaupel, P.; Kallinowski, F.; Okunieff, P. *Cancer Res.* **1989**, *49*, 6449.

16. Parveen, E.; Naughton, D. P.; Whish, W. J. D.; Threadgill, M. D. *Bioorg. Med. Chem. Lett.* **1999**, *9*, 2031.

17. Talley, J. J.; Bertenshaw, S. R.; Brown, D. L.; Carter, J. S.; Graneto, M. J.; Kellogg, M. S.; Koboldt, C. M.; Yuan, J.; Zhang, Y. Y.; Seibert, K. *J. Med. Chem.* **2000**, *43*, 1661.

18. Bundgaard, H. In *Design of Prodrugs*, Bundgaard, H. (Ed.), Elsevier, Amsterdam, 1985, p. 1.

19. (a) Anderson, B. D.; Conradi, R. A.; Knuth, K. E. *J. Pharm. Sci.* **1985**, *74*, 365. (b) Anderson, B. D.; Conradi, R. A.; Knuth, K. E.; Nail, S. L. *J. Pharm. Sci.* **1985**, *74*, 375.

20. Anderson, B. D.; Conradi, R. A.; Spilman, C. H.; Forbes, A. D. *J. Pharm. Sci.* **1985**, *74*, 382.

21. Masini, E.; Planchenault, J.; Pezziardi, F.; Gautier, P.; Gangnol, J. P. *Agents Actions* **1985**, *16*, 470.

22. Saulnier, M. G.; Langley, D. R.; Kadow, J. F.; Senter, S. D.; Knipe, J. O.; Tun, M. M.; Vyas, D. M.; Doyle, T. W. *Bioorg. Med. Chem. Lett.* **1994**, *4*, 2567.

23. Slojkowska, Z.; Krakuska, H. J.; Pachecka, J. *Xenobiotica* **1982**, *12*, 359.

24. Hadgraft, J. In *Design of Prodrugs*, Bundgaard, H. (Ed.), Elsevier, Amsterdam, 1985, p. 271.

25. Mandell, A. I.; Stentz, F.; Kitabuchi, A. E. *Ophthalmology* **1978**, *85*, 268.

26. (a) Pardridge, W. M. *Brain Drug Targeting: The Future of Brain Drug Development*, Cambridge University Press, Cambridge, UK, 2001. (b) Begley, D. J.; Bradbury, M. W.; Kreuter, J. (Eds.), *The Blood–Brain Barrier and Drug Delivery to the CNS*, Mercel Dekker, New York, 2000.

27. Bodor, N.; Brewster, M. *Pharmacol. Ther.* **1983**, *19*, 337.

28. Worms, P.; Depoortere, H.; Durand, A.; Morselli, P. L.; Lloyd, K. G.; Bartholini, G. *J. Pharmacol. Exp. Ther.* **1982**, *220*, 660.

29. Jacob, J. N.; Hesse, G. W.; Shashoua, V. E. *J. Med. Chem.* **1990**, *33*, 733.

30. (a) Harper, N. J. *J. Med. Pharm. Chem.* **1959**, *1*, 467. (b) Brandes, D.; Bourne, G. H. *Lancet* **1955**, *1*, 481.

31. (a) Niculescu-Duvaz, I.; Friedlos, F.; Niculescu-Duvaz, D.; Davies, L.; Springer, C. J. *Anti-Cancer Drug Design* **1999**, *14*, 517. (b) Denny, W. A. *Curr. Pharm. Design* **1996**, *2*, 281.

32. Xu, G.; McLeod, H. L. *Clin. Cancer Res.* **2001**, *7*, 3314.

33. (a) Sharma, S. K.; Bagshawe, K. D.; Burke, P. J.; Boden, J. A.; Rogers, G. T.; Springer, C. J.; Melton, R. G.; Sherwood, R. F. *Cancer* **1994**, *73*, 1114. (b) Rogers, G. T.; Burke, P. J.; Sharma, S. K.; Koodie, R.; Boden, J. A. *Br. J. Cancer* **1995**, *72*, 1357.

34. Vaswani, S. K.; Hamilton, R. G. *Ann. Allergy Asthma Immunol.* **1998**, *81*, 105.

35. Springer, C. J.; Dowell, R.; Burke, P. J.; Hadley, E.; Davies, D. H.; Blakey, D. C.; Melton, R. G.; Niculescu-Duvaz, I. *J. Med. Chem.* **1995**, *38*, 5051.

36. (a) Wentworth, P., Jr.; Janda, K. D. *Cell Biochem. Biophys.* **2001**, *35*, 63. (b) Hilvert, D. *Annu. Rev. Biochem.* **2000**, *69*, 751.

37. *Abzyme* is the generic term for a catalytic antibody.

38. (a) Wentworth, P.; Datta, A.; Blakey, D.; Boyle, T.; Partridge, L. J.; Blackburn, G. M. *Proc. Natl. Acad. Sci. USA* **1996**, *93*, 799. (b) Bagshawe, K. D. *Br. J. Cancer* **1989**, *60*, 275.

39. (a) Barbas, C. F., III; Heine, A.; Zhong, G.; Hoffmann, T.; Gramatikova, S.; Bjrnstedt, R.; List, B.; Anderson, J.; Stura, E. A.; Wilson, I. A.; Lerner, R. A. *Science* **1997**, *278*, 2085. (b) List, B.; Barbas, C. F. III; Lerner, R. A. *Proc. Natl. Acad. Sci. USA* **1998**, *95*, 15351.

40. Shabat, D.; Rader, C.; List, B.; Lerner, R. A.; Barbas, C. F. III *Proc. Natl. Acad. Sci. USA* **1999**, *96*, 6925.

41. (a) Deonarain, M.P.; Spooner, R. A.; Epenetos, A.A. *Gene Ther.* **1995**, *2*, 235. (b) Niculescu-Duvaz, I.; Spooner, R. A.; Marais, R.; Springer, C. J. *Bioconjugate Chem.* **1998**, *9*, 4.

42. Denny, W. A. *Curr. Pharm. Des.* **2002**, *8*, 1349.

43. Hay, M. P.; Sykes, B. M.; Denny, W. A.; Wilson, W. R. *Bioorg. Med. Chem. Lett.* **1999**, *9*, 2237.

44. (a) Grove, J. I.; Searle, P. F.; Weedon, S. J.; Green, N. K.; McNeish, I. A.; Kerr, D. J. *Anticancer Drug Des.* **1999**, *14*, 461. (b) Huber, B. E.; Richards, C. A.; Austin, E. A. *Adv. Drug Del. Rev.* **1995**, *17*, 279.

45. Weedon, S. J.; Green, N. K.; McNeish, I. A.; Gilligan, M. G.; Mautner, V.; Wrighton, C. J.; Mountain, A.; Young L. S.; Kerr, D. J.; Searle, P. F. *Int. J. Cancer* **2000**, *86*, 848.

46. Rigg, A.; Sikora, K. *Mol. Med. Today* **1997**, *3*, 359.

47. Muller, P.; Jesnowski, R.; Karle, P.; Renz, R.; Saller, R.; Stein, H.; Puschel, K.; Rombs, K.; Nizze, H.; Liebe, S.; Wagner, T.; Gunzburg, W. H.; Salmons, B.; Lohr, M. *Ann. N.Y. Acad. Sci.* **1999**, *880*, 337.

48. Panhda, H.; Martin, L. A.; Rigg, A.; Hurst, H. C.; Stamp, G. W. H.; Sikora, K.; Lemoine, N. R. *J. Clin. Oncol.* **1999**, *17*, 2180.

49. Denny, W. A.; Wilson, W. R. *J. Pharm. Pharmacol.* **1998**, *50*, 387.

50. Garceau, Y.; Davis, I.; Hasegawa, J. *J. Pharm. Sci.* **1978**, *67*, 1360.

51. Hussain, M. A.; Koval, C. A.; Myers, M. J.; Shami, E. G.; Shefter, E. *J. Pharm. Sci.* **1987**, *76*, 356.

52. Deberdt, R.; Elens, P.; Berghmans, W.; Heykants, J.; Woestenborghs, R.; Driesens, F.; Reyntjens, A.; Van Wijngaarden, I. *Acta Psychiat. Scand.* **1980**, *62*, 356.

53. Chouinard, G.; Annable, L.; Ross-Chouinard, A. *Am. J. Psychiat.* **1982**, *139*, 312.

54. Persico, F. J.; Pritchard, J. F.; Fischer, M. C.; Yorgey, K.; Wong, S.; Carson, J. *J. Pharmacol. Exp. Ther.* **1988**, *247*, 889.

55. Mandell, A. I.; Stentz, F.; Kitabuchi, A. E. *Ophthalmology* **1978**, *85*, 268.

56. Nielsen, N. M.; Bundgaard, H. *J. Med. Chem.* **1989**, *32*, 727.

57. De Haan, R. M.; Metzler, C. M.; Schellenberg, D.; Vanderbosch, W. D. *J. Clin. Pharmacol.* **1973**, *13*, 190.

58. Sinkula, A. A.; Morozowich, W.; Rowe, E. L. *J. Pharm. Sci.* **1973**, *62*, 1106.

59. Notari, R. E. *J. Pharm. Sci.* **1973**, *62*, 865.

60. de Duve, C.; de Barsy, T.; Poole, B.; Trouet, A.; Tulkens, P.; Van Hoof, F. *Biochem. Pharmacol.* **1974**, *23*, 2495.

61. (a) Wang, X.-W.; Xie, H. *Drugs Future* **1999**, *24*, 847. (b) Okuno, Y.; Iwashita, T.; Sugiura, Y. *J. Am. Chem. Soc.* **2000**, 122, 6848.

62. Cecchi, R.; Rusconi, L.; Tanzi, M. C.; Danusso, F.; Ferruti, P. *J. Med. Chem.* **1981**, *24*, 622.

63. Shen, W.-C.; Ryser, H. J.-P. *Mol. Pharmacol.* **1979**, *16*, 614.

64. Chu, B. C. F.; Howell, S. B. *Biochem. Pharmacol.* **1981**, *30*, 2545.

65. Zupon, M. A.; Fang, S. M.; Christensen, J. M.; Petersen, R. V. *J. Pharm. Sci.* **1983**, *72*, 1323.

66. Ringsdorf, H. *J. Polym. Sci., Polym. Symp.* **1975**, *51*, 135.

67. Rowland, G. F.; O'Neill, G. J.; Davies, D. A. L. *Nature* **1975**, *255*, 487.

68. Rowland, G. F. *Eur. J. Cancer* **1977**, *13*, 593.

69. Balboni, P. G.; Minia, A.; Grossi, M. P.; Barbarti-Brodano, G.; Mattioli, A.; Fiume, L. *Nature* **1976**, *264*, 181.

70. Voutsadakis, I. A. *Anti-Cancer Drugs* **2002**, *13*, 685.

71. Carl, P. L.; Chakravarty, P. K.; Katzenellenbogen, J. A. *J. Med. Chem.* **1981**, *24*, 479.

72. Bundgaard, H. In *Bioreversible Carriers in Drug Design*, Roche, E. B. (Ed.), Pergamon Press, New York, 1987, p. 13.

73. (a) Metges, C. C. *J. Nutrition* **2000**, *130*, 1857S. (b) White, A.; Bardocz, S. In *Polyamines in Health and Nutrition*, Kluwer Academic, Hingham, MA, 1999, pp. 117–122. (c) Roth, J. R.; Lawrence, J. G.; Bobik, T. A. *Annu. Rev. Microbiol.* **1996**, *50*, 137. (d) Conly, J. M.; Stein, K. *Prog. Food Nutr. Sci.* **1992**, *16*, 307.

74. Jansen, A. B. A.; Russell, T. J. *J. Chem. Soc.* **1965**, *2127*.

75. Bodin, N. D.; Ekström, B.; Forsgren, U.; Jalar, L. P.; Magni, L.; Ramsey, C. H.; Sjöberg, B. *Antimicrob. Agents Chemother.* **1975**, *9*, 518.

76. Daehne, W. V.; Frederiksen, E.; Gundersen, E.; Lund, F.; March, P.; Petersen, H. J.; Roholt, K.; Tybring, L.; Godtfredsen, W. O. *J. Med. Chem.* **1970**, *13*, 607.

77. Bodor, N. *Ann. N.Y. Acad. Sci.* **1987**, *507*, 289.

78. (a) Pop, E.; Wu, W.-M.; Shek, E.; Bodor, N. *J. Med. Chem.* **1989**, *32*, 1774. (b) Wu, W.-M.; Pop, E.; Shek, E.; Bodor, N. *J. Med. Chem.* **1989**, *32*, 1782.

79. Mllgaard, B.; Hoelgaard, A.; Bundgaard, H. *Int. J. Pharm.* **1982**, *12*, 153.

80. Kingsbury, W. D.; Boehm, J. C.; Mehta, R. J.; Grappel, S. F.; Gilvarg, C. *J. Med. Chem.* **1984**, *27*, 1447.

81. Greenwald, R. B.; Choe, Y. H.; Conover, C. D.; Shum, K.; Wu, D.; Royzen, M. *J. Med. Chem.* **2000**, *43*, 475.

82. (a) Shan, D.; Nicholaou, M. G.; Borchardt, R. T.; Wang, B. *J. Pharm. Sci.* **1997**, *30*, 787. (b) Testa, B.; Mayer, J. M. *Drug Metab. Rev.* **1998**, *30*, 787. (c) Wang, W.; Jiang, J.; Ballard, C. E.; Wang, B. *Curr. Pharm. Des.* **1999**, *5*, 265.

83. (a) Hartley, S.; Wise, R. *J. Antimicrob. Chemother.* **1982**, *10*, 49. (b) Baltzer, B.; Binderup, E.; Von Daehne, W.; Godtfredsen, W. O.; Hansen, K.; Nielsen, B.; Sørensen, H.; Vangedal, S. *J. Antibiot.* **1980**, *33*, 1183.

84. Sachs, G.; Chang, H. M.; Rabon, E.; Schackmann, R.; Lewin, M.; Saccomani, G. *J. Biol. Chem.* **1974**, *251*, 7690.

85. (a) Lee, Y.-H.; Phillips, E.; Sause, S. W. *Arch. Inf. Pharmacodyn.* **1972**, *195*, 402. (b) Brändström, A.; Lindberg, P.; Junggren, U. *Scand. J. Gastroenterol.* **1985**, *108* (*Suppl.*), 15.

86. Lindberg, P.; Brändström, A.; Wallmark, B.; Mattsson, H.; Rikner, L.; Hoffmann, K.-J. *Med. Res. Rev.* **1990**, *10*, 1.

87. Helander, H. F.; Ramsay, C.-H.; Regardh, C.-G. *Scand. J. Gastroenterol.* **1985**, *108* (*Suppl.*), 95.

88. Fryklund, J.; Gedda, K.; Wallmark, B. *Biochem. Pharmacol.* **1988**, *37*, 2543.

89. (a) Brändström, A.; Lindberg, P.; Bergman, N.-Å.; Alminger, T.; Ankner, K.; Junggren, U.; Lamm, B.; Nordberg, P.; Erickson, M.; Grundevik, I.; Hagin, I.; Hoffmann, K. J.; Johansson, S.; Larsson, S.; Löftberg, I.; Ohlson, K.; Persson, B.; Skånberg, I.; Tekenbergs-Hjelte, L. *Acta Chem. Scand.* **1989**, *43*, 536. (b) Brändström, A.; Bergman, N.-Å.; Lindberg, P.; Grundevik, I.; Johansson, S.; Tekenbergs-Hjelte, L.; Ohlson, K. *Acta Chem. Scand.* **1989**, *43*, 549. (c) Brändström, A.; Bergman, N.; Grundevik, I.; Johansson, S.; Tekenbergs-Hjelte, L.; Ohlson, K. *Acta Chem. Scand.* **1989**, *43*, 569. (d) Brändström, A.; Lindberg, P.; Bergman, N. Å.; Tekenbergs-Hjelte, L.; Ohlson, K. *Acta Chem. Scand.* **1989**, *43*, 577. (e) Brändström, A.; Lindberg, P.; Bergman, N. Å.; Tekenbergs-Hjelte, L.; Ohlson, K.; Grundevik, I.; Nordberg, P.; Alminger, T. *Acta Chem. Scand.* **1989**, *43*, 587. (f) Brändström, A.; Lindberg, P.; Bergman, N. Å.; Grundevik, I.; Tekenbergs-Hjelte, L.; Ohlson, K. *Acta Chem. Scand.* **1989**, *43*, 595.

90. Puscas, I.; Coltau, M.; Baican, M.; Domuta, G. *J. Pharmacol. Exp. Therap.* **1999**, *290*, 530.

91. (a) Puscas, I. In *Carbonic Anhydrase and Modulation of Physiologic and Pathologic Processes in the Organism*, Puscas, I. (Ed.), Helicon Publishing House, Timisoara, Romania, 1994, pp. 373–530. (b) Puscas, i. In *New Pharmacology of Ulcer Disease*, Szabo, S.; Mozsik, G. (Eds.), Elsevier, New York, 1987, pp. 164–179.

92. (a) Gremse, D. A. *Exp. Opin. Pharmacother.* **2001**, *2*, 1663. (b) Matheson, A. J.; Jarvis, B. *Drugs* **2001**, *61*, 1801.

93. Carswell, C. I.; Goa, K. L. *Drugs* **2001**, *61*, 2327.

94. Jungnickel, P. W. *Clin. Therap.* **2000**, *22*, 1268.

95. Kanda, Y.; Ashizawa, T.; Kakita, S.; Takahashi, Y.; Kono, M.; Yoshida, M.; Saitoh, Y. Okabe, M. *J. Med. Chem.* **1999**, *42*, 1330.

96. (a) Rückemann, K.; Fairbanks, L. D.; Carrey, E. A.; Hawrylowicz, C. M; Richards, D. F.; Kirschbaum, B.; Simmonds, H. A. *J. Biol. Chem.* **1998**, *273*, 21682. (b) Bruneau, J.-M.;

Yea, C. M.; Spinella-Jaegle, S.; Fudali, C.; Woodward, K.; Robson, P. A.; Sautes, C.; Westwood, R.; Kuo, E. A.; Williamson, R. A.; Ruuth, E. *Biochem. J.* **1998**, *336*, 299.

97. Davis, J. P.; Cain, G. A.; Pitts, W. J.; Magolda, R. L.; Copeland, R. A. *Biochemistry* **1996**, *35*, 1270.

98. Sutharchanadevi, M.; Murugan, R. In *Comprehensive Heterocyclic Chemistry II*, Shinkai, I. (Ed.), Elsevier, Oxford, UK, 1996, Vol. 3, pp. 221–260.

99. (a) Lahti, R. A.; Gall, M. *J. Med. Chem.* **1976**, *19*, 1064. (b) Gall, M.; Hester, J. B., Jr.; Rudzik, A. D.; Lahti, R. A. *J. Med. Chem.* **1976**, *19*, 1057.

100. Brodie, B. B.; Axelrod, J. *J. Pharmacol. Exp. Ther.* **1949**, *97*, 58.

101. (a) Colvin, M.; Chabner, B. A. In *Cancer Chemotherapy: Principles and Practice*, Chabner, B. A.; Collins, J. M. (Eds.), J. B. Lippincott, Philadelphia, 1990, p. 276. (b) Cox, P. J.; Farmer, P. B.; Jarman, M. *Cancer Treat. Rep.* **1976**, *60*, 299. (c) Hill, D. L. *A Review of Cyclophosphamide*, Thomas, Springfield, IL, 1975.

102. Benson, A. J.; Martin, C. N.; Garner, R. C. *Biochem. Pharmacol.* **1988**, *37*, 2979.

103. Connors, T. A.; Cox, P. J.; Farmer, P. B.; Foster, A. B.; Jarman, M. *Biochem. Pharmacol.* **1974**, *23*, 115.

104. (a) Oliverio, V. T. In *Cancer Medicine*, 2nd ed., Holland, J. F.; Frei, E., III (Eds.), Lea & Febiger, Philadelphia, 1982, p. 850. (b) Weinkam, R. J.; Shiba, D. A.; Chabner, B. A. In *Pharmacologic Principles of Cancer Treatment*, Chabner, B. E. (Ed.), W. B. Saunders, Philadelphia, 1982, p. 340.

105. Raaflaub, J.; Schwartz, D. E. *Experientia* **1965**, *21*, 44.

106. Kreis, W.; Piepho, S. B.; Bernhard, H. V. *Experientia* **1966**, *22*, 431.

107. Tsuji, T.; Kosower, E. M. *J. Am. Chem. Soc.* **1971**, *93*, 1992.

108. Froede, H. C.; Wilson, I. B. In *The Enzymes*, 3rd ed., Boyer, P. (Ed.), Academic Press, New York, 1971, Vol. 5, p. 87.

109. Sussman, J. L.; Harel, M.; Frolow, F.; Oefner, C.; Goldman, A.; Toker, L.; Silman, I. *Science* **1991**, *253*, 872.

110. Dougherty, D. A. *Chem. Rev.* **1997**, *97*, 1303.

111. Jansen, E. F.; Nutting, M.-D. F.; Balls, A. K. *J. Biol. Chem.* **1949**, *179*, 201.

112. Wilson, I. B. *Biochim. Biophys. Acta* **1958**, *27*, 196.

113. Shek, E.; Higuchi, T.; Bodor, N. *J. Med. Chem.* **1976**, *19*, 113.

114. Shigeta, M.; Homma, A. *CNS Drug Rev.* **2001**, *7*, 353.

115. (a) Summers, W. K. *J. Alzheimer's Dis.* **2000**, *2*, 85. (b) Kurz, A. *J. Neural Trans. (Suppl.)* **1998**, *54*, 295.

116. Argade, A. B.; Devraj, R.; Vroman, J. A.; Haugwitz, R. D.; Hollingshead, M.; Cushman, M. *J. Med. Chem.* **1998**, *41*, 3337.

117. Venhuis, BG. J.; Wikström, H. V.; Rodenhuis, N.; Sundell, S.; Dijkstra, D. *J. Med. Chem.* **2002**, *45*, 2349.

118. (a) Frigerio, A.; Fanelli, R.; Biandrate, P.; Passerini, G.; Morselli, P. L.; Garattini, S. *J. Pharm. Sci.* **1972**, *61*, 1144. (b) Johannessen, S. I.; Gerna, N. M.; Bakke, J.; Strandjord, R. E.; Morselli, P. L. *Br. J. Clin. Pharmacol.* **1976**, *3*, 575.

119. Neely, J. R.; Morgan, H. E. *Ann. Rev. Physiol.* **1974**, *36*, 413.

120. Barnish, I. T.; Cross, P. E.; Danilewicz, J. C.; Dickinson, R. P.; Stopher, D. A. *J. Med. Chem.* **1981**, *24*, 399.

121. (a) Kirsner, J. B. *J. Am. Med. Assoc.* **1980**, *243*, 557. (b) Eastwood, M. A. *Ther. Drug Monit.* **1980**, *2*, 149.

122. Brown, J. P.; McGarraugh, G. V.; Parkinson, T. M.; Wingard, R. E., Jr.; Onderdonk, A. B. *J. Med. Chem.* **1983**, *26*, 1300.

123. Reist, E. J.; Benitez, A.; Goodman, L.; Baker, B. L.; Lee, W. W. *J. Org. Chem.* **1962**, *27*, 3274.

124. Andrei, G.; Snoeck, R.; Goubou, P.; Desmyter, J.; DeClercq, E. *Eur. J. Clin. Microbiol. Infect. Diseases* **1992**, *11*, 143.

125. Whitley, R.; Alford, C.; Hess, F.; Buchanan, R. *Drugs* **1980**, *20*, 267.

126. Kotra, L. P.; Manouilof, K. K.; Cretton-Scott, E.; Sommadossi, J.-P.; Boridinot, F. D.; Schinazi, R. F.; Chu, C. K. *J. Med. Chem.* **1996**, *39*, 5202.

127. Shen, T. Y.; Winter, C. A. *Adv. Drug Res.* **1977**, *12*, 90.

128. Matsukawa, T.; Yurugi, S.; Oka, Y. *Ann. N.Y. Acad. Sci.* **1962**, *98*, 430.

129. Stella, V. J.; Himmelstein, K. J. In *Design of Prodrugs*, Bundgaard, H. (Ed.), Elsevier, Amsterdam, 1985, p. 177.

130. Hofsteenge, J.; Capuano, A.; Altszuler, R.; Moore, S. *J. Med. Chem.* **1986**, *29*, 1765.

131. Miwa, G. T.; Wang, R.; Alvaro, R.; Walsh, J. S.; Lu, A. Y. H. *Biochem. Pharmacol.* **1986**, *35*, 33.

132. Tobias, S. C.; Borch, R. F. *J. Med. Chem.* **2001**, *44*, 4475.

133. (a) McCollister, R. J.; Gilbert, W. R., Jr.; Ashton, D. M.; Wyngaarden, J. B. *J. Biol. Chem.* **1964**, *239*, 1560. (b) Caskey, C. T.; Ashton, D. M.; Wyngaarden, J. B. *J. Biol. Chem.* **1964**, *239*, 2570. (c) Henderson, J. F.; Khoo, M. K. Y. *J. Biol. Chem.* **1965**, *240*, 3104.

134. (a) Elion, G. B. *J. Med. Virol.* **1993**, (*Suppl 1*), 2. (b) Richards, D. M.; Carmine, A. A.; Brogden, R. N.; Heel, R. C.; Speight, T. M.; Avery, G. S. *Drugs* **1983**, *26*, 378. (c) Elion, G. B. *J. Antimicrob. Chemother.* **1983**, *12* (*Suppl. B*), 9.

135. Furman, P. A.; McGuirt, P. V.; Keller, P. M.; Fyfe, J. A.; Elion, G. B. *Virology* **1980**, *102*, 420.

136. Miller, W. H.; Miller, R. L. *J. Biol. Chem.* **1980**, *255*, 7204.

137. Miller, W. H.; Miller, R. L. *Biochem. Pharmacol.* **1982**, *31*, 3879.

138. Furman, P. A.; St. Clair, M. H.; Fyfe, J. A.; Rideout, J. L.; Keller, P. M.; Elion, G. B. *J. Virol.* **1979**, *32*, 72.

139. Reardon, J. E.; Spector, T. *J. Biol. Chem.* **1989**, *264*, 7405.

140. Larder, B. A.; Cheng, Y.-C.; Darby, G. *J. Gen. Virol.* **1983**, *64*, 523.

141. (a) Coen, D. M.; Schaffer, P. A.; Furman, P. A.; Keller, P. M.; St. Clair, M. H. *Am. J. Med.* **1982**, *73* (*1A*), 351. (b) Schnipper, L. E.; Crumpacker, C. S. *Proc. Natl. Acad. Sci. USA* **1980**, *77*, 2270.

142. Huang, L.; Ishii, K. K.; Zuccola, H.; Gehring, A. M.; Hwang, C. B. C.; Hogle, J.; Coen, D. M. *Proc. Natl. Acad. Sci. USA* **1999**, *96*, 447.

143. Good, S. S.; Krasny, H. C.; Elion, G. B.; de Miranda, P. *J. Pharmacol. Exp. Ther.* **1983**, *227*, 644.

144. Krenitsky, T. A.; Hall, W. W.; de Miranda, P.; Beauchamp, L. M.; Schaeffer, H. J.; Whiteman, P. D. *Proc. Natl. Acad. Sci. USA* **1984**, *81*, 3209.

145. Weller, S.; Blum, M. R.; Doucette, M.; Burnette, T.; Cederberg, D. M.; de Miranda, P.; Smiley, M. L. *Clin. Pharmacol. Ther.* **1993**, *54*, 595.

146. Burnette, T. C.; Harrington, J. A.; Reardon, J. E.; Merrill, B. M.; de Miranda, P. *J. Biol. Chem.* **1995**, *270*, 15827.

147. Bailey, S. M.; Hart, I.; Lohmeyer, M. *Drugs Future* **1998**, *23*, 401.

148. (a) McGavin, J. K.; Goa, K. L. *Drugs* **2001**, *61*, 1153. (b) Spector, S. A. *Adv. Exp. Med. Biol.* **1999**, *458*, 121. (b) Elion, G. B. In *Antiviral Chemotherapy: New Directions for Clinical Application and Research*, Mills, J.; Corey, L. (Eds.), Elsevier, New York, 1986, p. 118.

149. (a) Sutton, D.; Kern, E. R. *Antiviral Chem. Chemother.* **1993**, *4*, 37. (b) Vere Hodge, R. A.; Cheng, Y. C. *Antiviral Chem. Chemother.* **1993**, *4*, 13. (c) Earnshaw, D. L.; Bacon, T. H.; Darlison, S. J.; Edmonds, K.; Perkins, R. M.; Vere Hodge, R. A. *Antimicrob. Ag. Chemother.* **1992**, *36*, 2747.

150. Boyd, M. R.; Bacon, T. H.; Sutton, D. *Antimicrob. Ag. Chemother.* **1988**, *32*, 358.

151. (a) Bacon, T. H. *Int. J. Antimicrob. Agents* **1996**, *7*, 119. (b) Jarvest, R. L. *Drugs Today* **1994**, *30*, 575.

152. Vere Hodge, R. A. *Antiviral Chem. Chemoth.* **1993**, *4*, 67.

153. (a) Cirelli, R.; Herne, K.; McCrary, M.; Lee, P.; Tyring, S. K. *Antiviral Res.* **1996**, *29*, 141. (b) Vere Hodge, R. A.; Sutton, D.; Boyd, M. R.; Harnden, M. R.; Jarvest, R. L. *Antimicrob. Agents Chemother.* **1989**, *33*, 1765.

154. (a) Winton, C. F.; Fowles, S. E.; Pierce, D. M.; Vere Hodge, R. A. *Anal. Proc.* **1990**, *27*, 181. (b) Winton, C. F.; Fowles, S. E.; Vere Hodge, R. A.; Pierce, D. M. In *Analysis of Drugs and Metabolites*, Reid, E.; Wilson, I. D. (Eds.), Royal Society of Chemistry, Cambridge, UK, pp. 163–171.

155. Vere Hodge, R. A.; Earnshaw, D. L.; Jarvest, R. L.; Readshaw, S. A. *Antiviral Res.* **1990** (*Suppl.*), *1*, 87.

156. McCall, J. M.; Aiken, J. W.; Chidester, C. G.; DuCharme, D. W.; Wending, M. G. *J. Med. Chem.* **1983**, *26*, 1791.

157. Buhl, A. E.; Waldon, D. J.; Baker, C. A.; Johnson, G. A. *J. Invest. Dermatol.* **1990**, *95*, 553.

158. Meisheri, K. D.; Johnson, G. A.; Puddington, L. *Biochem. Pharmacol.* **1993**, *45*, 271.

159. Bernheimer, H.; Birkmayer, W.; Hornykiewicz, O.; Jellinger, K.; Seitelberger, F. *J. Neurol. Sci.* **1973**, *20*, 415.

160. (a) Deleu, D.; Northway, M. G.; Hanssens, Y. *Clin. Pharmokin.* **2002**, *41*, 261. (b) Myllyla, V. V.; Sotaniemi, K. A.; Hakulinen, P.; Mki-Ikola, O.; Heinonen, E. H. *Acta Neurol. Scand.* **1997**, *95*, 211.

161. (a) Galler, R. M.; Hallas, B. H.; Fazzini, E. *J. Am. Osteop. Assoc.* **1996**, *96*, 228. (b) Lieberman, A. *Curr. Opin. Neurol. Neurosurg.* **1993**, *6*, 339.

162. (a) Palfreyman, M. G.; McDonald, I. A.; Fozard, J. R.; Mely, Y.; Sleight, A. J.; Zreika, M.; Wagner, J.; Bey, P.; Lewis, P. J. *J. Neurochem.* **1985**, *45*, 1850. (b) McDonald, I. A.; Lacoste, J. M.; Bey, P.; Wagner, J.; Zreika, M.; Palfreyman, M. G. *Bioorg. Chem.* **1986**, *14*, 103. (c) Fagervall, I.; Ross, S. B. *J. Neurochem.* **1989**, *52*, 467. (d) Huang, S.-C.; Quintana, J.; Satyamurthy, N.; Lacan, G.; Yu, D.-C.; Phelps, M. E.; Barrio, J. R. *Nucl. Med. Biol.* **1999**, *26*, 365.

163. McDonald, I. A.; Lacoste, J. M.; Bey, P.; Palfreyman, M. G.; Zreika, M. *J. Med. Chem.* **1985**, *28*, 186.

164. Magnan, S. D. J.; Shirota, F. N.; Nagasawa, H. T. *J. Med. Chem.* **1982**, *25*, 1018.

165. (a) Wilk, S.; Mizoguchi, H.; Orlowski, M. *J. Pharmacol. Exp. Ther.* **1978**, *206*, 227. (b) Kyncl, J. J.; Minard, F. N.; Jones, P. H. *Adv. Biosci.* **1979**, *20*, 369.

Answers to Chapter Problems

Chapter 2

1. **Advantages**

   ▶ You don't have to be creative.

   ▶ You don't have to understand what causes the disease state.

   ▶ You can test whatever becomes available.

   ▶ You may get hits with compounds that would not have been tested in a rational approach because structures are unrelated to what you *think* should be important.

   Disadvantages

   ▶ There are an almost infinite number of possibilities.

   ▶ You may be testing compounds completely unrelated to the biological system that is important.

   ▶ Time and expense may be excessive.

2. Start with analogs of the substrate, arginine, and make small changes.

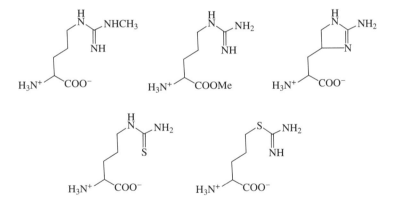

3. A. Cut pieces of **1** away and determine the effect on binding. If binding is decreased, then the group may have been in the pharmacophore. If binding is increased, then the group may have been preventing binding. If no binding effect, then the group may not be involved.

   B. **2**—one or both of the CH_3 groups interferes with binding

   　　3—only the top CH_3 interferes

4—phenyl is in the pharmacophore

5—carbonyl is important; maybe because without it the amine is protonated, and a cation interferes with binding or maybe F interferes with binding

6—F does not interfere

7—less conformational flexibility is better

4. A random screen that includes privileged structures would give you the ability to find compounds without having to know the target.

5. TI $= LD_{50}/ED_{50} = 5$. Because this is not a potentially lethal disease, the therapeutic index would not be acceptable; this is not a safe compound.

6. There are many good answers; below is a possibility.

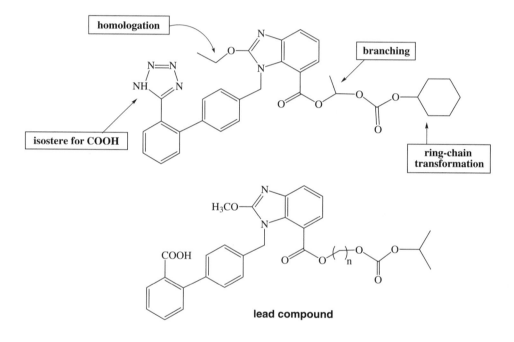

lead compound

7. Most bioisosteric changes result in a change in size, shape, electronic distribution, lipid solubility, water solubility, pK_a, chemical reactivity, and/or hydrogen bonding capacity. These modifications can affect both the pharmacokinetics and pharmacodynamics, so a simple bioisosteric replacement may destroy activity.

8.

▶ It is assumed that the synthetic reactions all proceed with high yields and at about the same rates for all members of the library.

▶ In testing, it is assumed that the presence of all other members of the library does not interfere with the binding of the best member to the test receptor.

▶ It is assumed that by-products (impurities) will not be effective.

9. A.

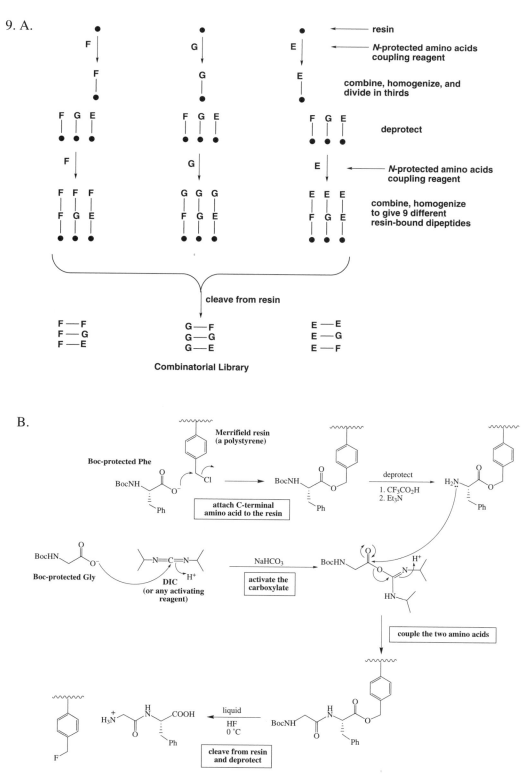

Combinatorial Library

B.

This looks like a lot of work, but peptide synthesizers make it simple.

10.

Tag   1 2 3 ¦ 4 5 6 ¦ 7 8 9 ¦ 10 11 12   read right to left for Merrifield resin
       0 1 0 ¦ 1 0 0 ¦ 0 0 1 ¦ 0 1 1   011 001 100 010
                                                 Glu - Lys - His - Ser

11. You need to discuss this with regard to the principles of SAR by NMR.

   A. The carboxylic acid group may be essential for binding to the receptor. Therefore, compounds **8** and **9** are not oriented on the receptor as shown in the problem or the two compounds were connected in the wrong place.

   B. There is more than one correct answer. The structures below are possibilities.

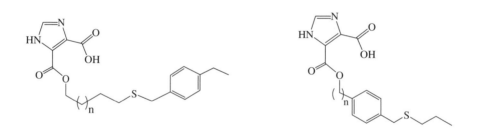

   These structures allow the carboxylate to remain free to bind to the receptor.

12. There are *many* correct answers. Although all of the compounds below are peptidomimetics, none may work because the bioactive conformation is not known, and none of these may be in the correct conformation for binding to the receptor.

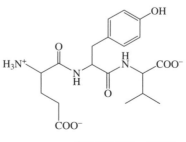

Glu-Tyr-Val   (EYV)

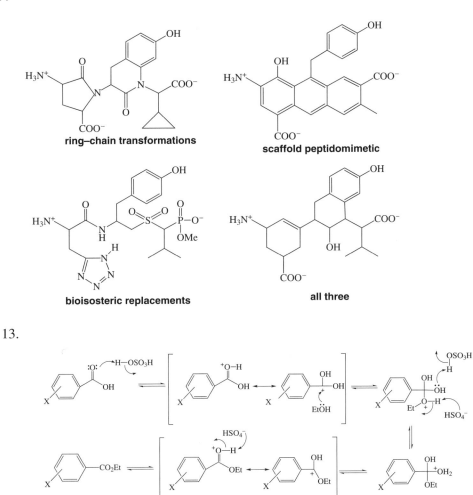

ring–chain transformations

scaffold peptidomimetic

bioisosteric replacements

all three

13.

The intermediates in brackets are highly electron deficient. Therefore, X = electron donating would increase rate, but X = electron withdrawing would decrease rate.

14. A.

$$\log P_{\text{(cinnamyl chromone)}} = \log P_{\text{(chromone)}} + \pi_{\text{CH=CH}} + \pi_{\text{PhCH}_2}$$

$$\pi_{\text{CH=CH}} = 1/3 \log P_{\text{benzene}} = 1/3(2.13) = 0.71$$
$$\pi_{\text{PhCH}_2} = \pi_{\text{Ph}} + \pi_{\text{CH}_2} = 2.13 + 0.50 = 2.63$$
(*Note:* $\pi_{\text{PhCH}_2} = \log P_{\text{PhCH}_2\text{NH}_2} - \pi_{\text{NH}_2}$
$$= \log P_{\text{PhCH}_2\text{NH}_2} - (\log P_{\text{MeNH}_2} - \pi_{\text{Me}}) = 1.09 - (-0.57 - 0.50)$$
$$= 2.16)$$

$$\log P = 1.39 + 0.71 + 2.63$$
$$= 4.73$$

B.

$$\log P \text{(structure)} = \pi_{CH_3CH_2CH_2} + \pi_{CH_3I} + \pi_{OCH=CH_2} - 0.2 \text{ (for branching)}$$

$$\pi_{OCH=CH_2} = \log P_{CH_2=CHOCH_2CH_3} - \pi_{CH_2CH_3} = 1.04 - 1.00 = 0.04$$
$$\pi_{CH_3CH_2CH_2} = 3\pi_{CH_3} = 1.50$$

$$\log P \text{(structure)} = 1.50 + 1.69 + 0.04 - 0.20 = 3.03$$

C. The log P and π values are constitutive, so the log P values obtained will depend on which log P or π values of pieces of the molecule were used to do the calculation for the whole molecule. To demonstrate the constitutive properties of lipophilicity, consider the following examples:

$$\log P_{ICH_2COOH} - \log P_{CH_3I} \text{ should give } \pi_{COOH} = 0.87 - 1.69$$
$$= -0.82$$

But $\log P_{NCCH_2COOH} - \log P_{CH_3CN}$ also should give $\pi_{COOH} = -0.33 - (-0.34)$
$$= 0.01$$

Therefore, use log P values of molecules as close in structure as possible to the one for which you need a log P or π value.

Also, $\pi_{CH_3O} \neq \log P_{CH_3OH}$. However, π_{CH_2OH} does equal $\log P_{CH_3OH}$. Because $\pi_{Me_2N} = \log P_{Me_2NH}$ ($\pi_{Me_2N} = \log P_{Me_3N} - \pi_{Me} = 0.27 - 0.50 = -0.23 = \log P_{Me_2NH}$) and the p$K_a$ of amine is much greater than alcohol, the ionization factor may be important.

Therefore, there will be different correct answers depending on which molecules you choose to determine your π values.

15. A. If the molecule does not ionize, then pH will not have an effect, and the log P will not change with pH.

  B. At low pH ($<$pH 5), the pyridine ring is protonated, producing an equilibrium between the charged form and the neutral form; the more in the charged form, the larger the negative for the log D value. As the pH increases, the equilibrium changes more and more to favor the neutral form, thereby making the log D less negative.

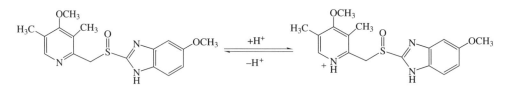

Between pH 5 and 9 the molecule is completely in the neutral form.
At higher pH values ($>$pH 9) deprotonation of the imidazole ring occurs, giving an equilibrium between the neutral form and the stabilized anion, which again lowers the log D.

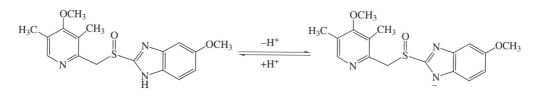

16. No. The molecular weight is too high (663) and there are too many hydrogen bond donors (7) and acceptors (13) and too many rotatable bonds (21).

17. A. This molecule is basic, so it is in equilibrium with the protonated form, which cannot cross membranes and therefore exhibits low activity.

    B. Change the pK_a of the pyridine by adding electron-withdrawing groups to it or using fewer electron-donating groups that are isosteric:

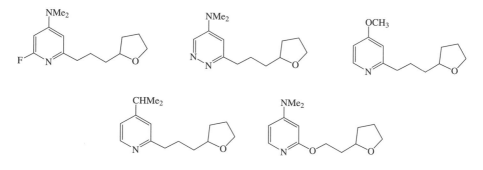

18. A. Either there is a steric problem at position 4 or the optimum π and/or σ value has been exceeded.

    Try the 3-Br compound to check for steric hindrance

    Try the 4-CN compound $(-\pi)$

    Try the 4-CH$_3$ compound $(-\sigma)$

    Try the 4-NH$_2$ compound $(-\pi, -\sigma)$

    B. Two variables have been introduced: a change in position and a change in π. Therefore, you cannot determine if it is a steric effect or a π effect. Try the 3-Br compound to see if steric.

19. DOCK is an algorithm for novel drug discovery that determines the best fit of a large number of random compounds into a known receptor structure. Each molecule is docked into the receptor in a number of geometrically allowable orientations to determine shape complementarity. First, the Connolly molecular surface (Connolly, M. L. *Science* **1983**, *221*, 709) for the receptor is developed, then a space-filling negative image of the receptor site is created, then a database of compounds (potential ligands for the receptor) is used to match the molecular structure to the negative image of the receptor. See Kuntz *et al.*, *Acc. Chem. Res.* **1994**, *27*, 117 for applications.

    CoMFA is a 3D QSAR methodology that involves the use of partial least-squares data analysis to compute separately the contributions of steric (shape) and electrostatic (electronic) molecular mechanic force fields of a set of molecules. These results provide parameters that can be correlated with noncovalent receptor binding and the biological properties of the molecules.

20. When a ligand binds to a receptor, it can change the receptor conformation. If you use the unliganded crystal structure, you could be docking compounds into the wrong binding site conformation. The liganded structure also gives you an idea of which interactions are important.

21.

▶ Steric effects affect the surface area and volume.
▶ Electronic effects affect the pK_a, the charge, and the electrostatic potential.
▶ Lipophilic effects affect the $\log P$.
▶ H-bonding affects the number of H-bond acceptors, donors, $\log P$, and conformation.

◻ Chapter 3

1. **a** electrostatic interaction/ hydrogen bonding
    b hydrogen bonding
    c hydrophobic effect
    d dipole–dipole interaction
    e ion–dipole interaction

2. There may be another lysine close to the lysine, lowering its pK_a.

$$\text{LysH}^+ \qquad \text{LysH}^+$$

Disfavored so one lysine becomes more acidic (lower pK_a).

There may be a compensating aspartate near the histidine, raising its pK_a.

$$\text{HisH}^+ \qquad \text{Asp}^-$$

This is favorable, so the pK_a of histidine increases (to keep it protonated).

3.

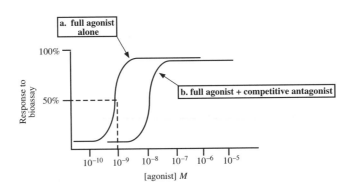

4.

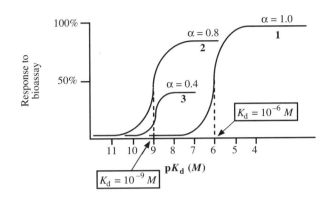

5. A. Affinities are measured by the K_d. The order of increasing affinity is **1** < **2** < **4** < dopamine < **3**

   B. Efficacies are measured by the increase in activity. The order of increasing efficacy is **2** < **1** ≤ **4** < **3** = dopamine

   C. **1** partial inverse agonist
      2 antagonist
      3 full agonist
      4 partial agonist

6. A. Generally, only one enantiomer is active or, at least, much more potent (the eutomer) than the other enantiomer (the distomer). The distomer may be responsible for side effects or toxicities. The distomer may be the eutomer for a different activity. The distomer may block the eutomer from binding to the receptor. One enantiomer may be an agonist, and the other an antagonist for the same receptor.

   B. The best way to increase the eudismic ratio is to administer only the eutomer. Otherwise, incorporate the chiral center in the pharmacophore.

7. A.

B.

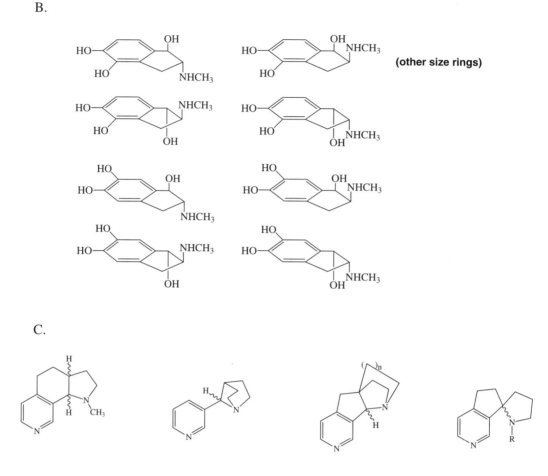

(other size rings)

C.

Ulrich, T.; Krich, S.; Binder, D.; Mereiter, K.; Anderson, D. J.; Meyer, M. D.; Pyerin, M. *J. Med. Chem.* **2002**, *45*, 4047.

8. It should be a pure antidepressant. It has an α-angle because all of the central atoms have sp^3 hybridization. It has a β-angle because the center ring is seven membered and the other two are six membered. It has a γ-angle because the top two center atoms are sp^3.

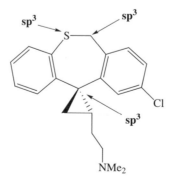

9.

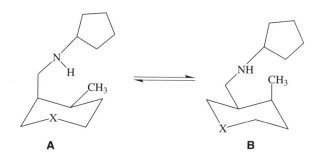

A   **B**

B is more stable than **A** when X = CH₂ because of smaller 1,3-diaxial interactions. When X = S, the ring size is larger (atomic size of S > CH₂), so the equilibrium does not favor B as strongly. When X = O, H-bonding to NH changes equilibrium to favor A.

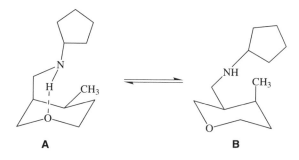

A   **B**

When X = NH, the H bond is even stronger, favoring **A** more. The bioactive conformation is **A**.

10. This is supposed to be related to the discussion of the discovery of cimetidine. Without knowledge of the structure of the receptor or the structure of any lead compounds, a lead must be discovered. A random lead discovery approach could be tried. A more rational approach would be to prepare analogs of tyramine. The phenolic OH and the protonated amino groups may be important to agonist activity. Therefore, you may want to modify these groups to avoid agonism, for example, by replacing the OH with other groups, such as OCH_3, CH_3, SH, NH_2. CH_3 and NH_2 are isoteres, SH would be more ionized, OCH_3 cannot ionize. The phenyl can be replaced by $-(CH_2)_3-$ or by another ring system or be further substituted. The $-CH_2-CH_2-$ can be lengthened or shortened or substituted. The amino group can be replaced by amidino, guanidino, ureido, etc. At each stage, antagonist and agonist activities need to be evaluated by bioassay.

☐ Chapter 4

1. Substrate recognition and catalysis of a reaction or specificity and rate acceleration.

2.

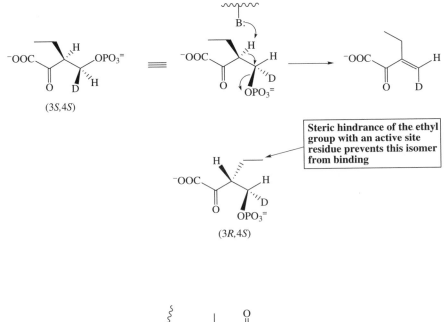

(3S,4S)

Steric hindrance of the ethyl group with an active site residue prevents this isomer from binding

(3R,4S)

3.

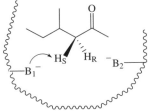

B_1 is the active site base responsible for deprotonation of H_S. B_2 is not a strong enough base and/or not close enough to H_R to remove it.

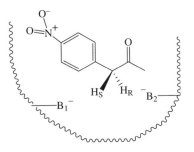

The *p*-nitrophenyl group lowers the pK_a of both H_S and H_R. Now, maybe B_2 is a strong enough base to remove H_R some of the time. Or maybe there is a small conformational change when it binds that moves B_2 closer to H_R.

4.

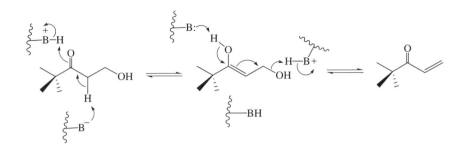

Simultaneous protonation of the ketone carbonyl and deprotonation of the α-proton gives the enol, which can be coupled with protonation of the β-hydroxyl group to effect β-elimination. Because the pK_a values of residues inside an enzyme can be quite different from those in solution, these protonations and deprotonations may be more facile than expected.

5.

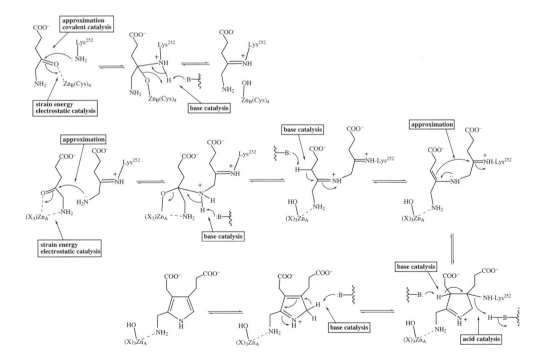

6.

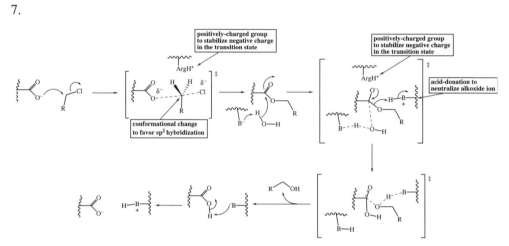

7.

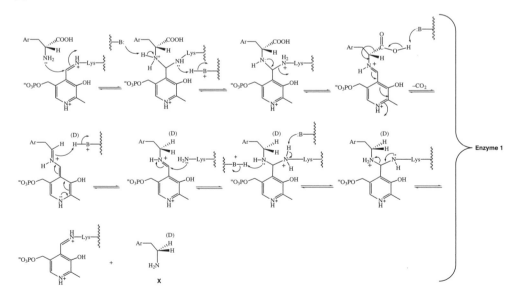

8. Both the substrate (aspartate) concentration and the α-keto acid (pyruvate) concentration. One of them may have been completely consumed.

9.

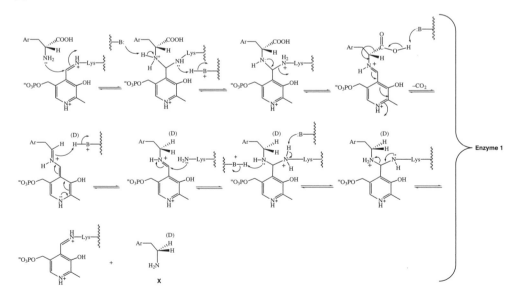

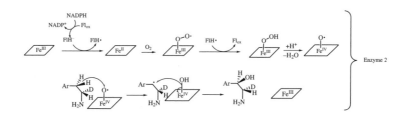

10.

A.

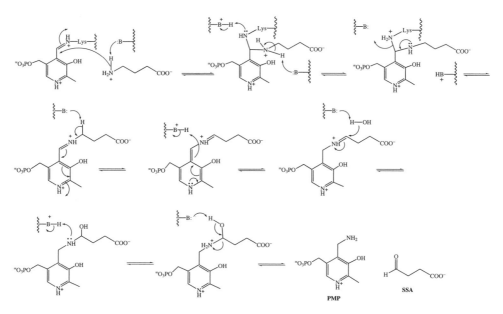

In order for the enzyme to continue catalysis, the PMP must be converted back to PLP. α-Ketoglutarate is required for that half of the reaction.

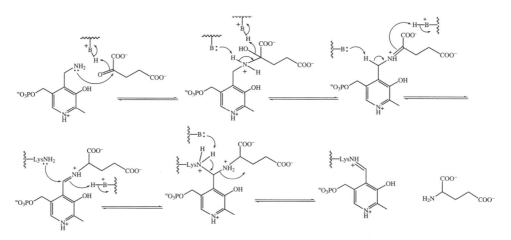

B. Without α-ketoglutarate, the coenzyme cannot get back into the active PLP form, so one turnover is all you get.

C. See mechanism in part A. The ^{18}O is incorporated into the carbonyl oxygen of succinic semialdehyde. Because of the acidity of protons adjacent to imines and carbonyls, there could be exchange with deuterium, but the answer based on the mechanism shown is that the succinic semialdehyde would have ^{18}O in the aldehyde carbonyl, and the glutamic acid would have a deuterium at the α-carbon.

11.

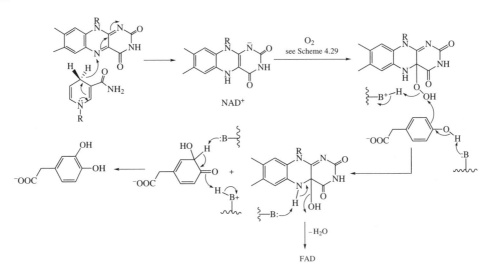

12.

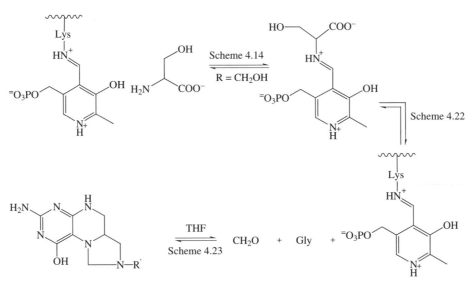

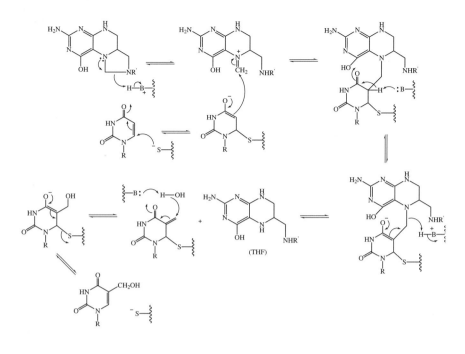

13.

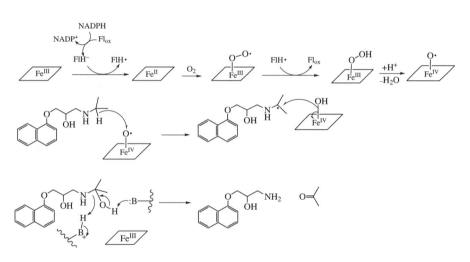

14.

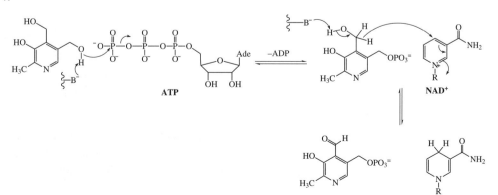

15.

Enzyme reaction (1):
Requires *L*-serine and tetrahydrofolate (or folate and NADPH)

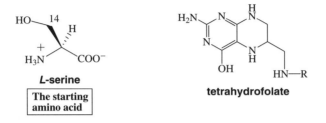

L-serine

| The starting |
| amino acid |

tetrahydrofolate

These give [^{14}C]methylenetetrahydrofolate

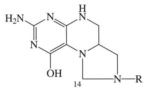

methylenetetrahydrofolate

Enzyme reaction (2):
Requires NAD$^+$ to oxidize [^{14}C]methylenetetrahydrofolate to [^{14}C]formyltetrahydrofolate

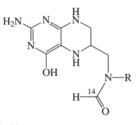

N^{10}-[^{14}C]formyltetrahydrofolate

Enzyme reaction (3):
Transfers the [^{14}C]-formyl group from [^{14}C] formyltetrahydrofolate to methionine

◻ Chapter 5

1. To block the metabolism of a substrate, if the substrate produces the desired therapeutic effect; to block the production of a product, if the product causes the problem; to block a metabolic pathway in a foreign organism or tumor cell.

2. Try to determine thermodynamic or kinetic differences in the enzymes from the two sources so that selective inhibition might be possible. Synthesize substrate analogs and measure binding constants for the two enzymes to see if there is a difference in binding affinities.

3. There are many correct answers for both parts.

A.

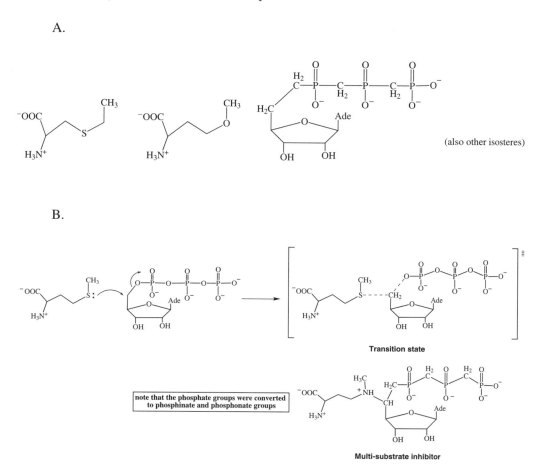

(also other isosteres)

B.

note that the phosphate groups were converted to phosphinate and phosphonate groups

Transition state

Multi-substrate inhibitor

4. In a slow, tight-binding inhibitor the E•I complex is more stable. The enzyme has changed conformation so it has to change back to release the inhibitor.

5. Convert the carboxylic acid, which presumably is deprotonated and forms an electrostatic interaction with the active site lysine residue, into an amino group, which will be protonated and could form an electrostatic interaction with the aspartate of the resistant enzyme.

6. A. Design an inhibitor of TXA_2 synthase, an antagonist of the TXA_2 receptor, or a dual-acting enzyme inhibitor/receptor antagonist.

   B. Design a dual-acting drug that both antagonizes the TXA_2 receptor and inhibits TXA_2 synthase.

   C. The idea is to combine parts of each molecule into a single compound that could bind to both TXA_2 synthase and to the TXA_2 receptor.

Examples would be:

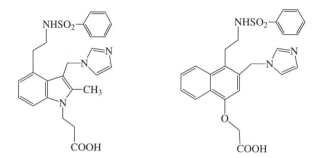

The following compound was shown to be active:

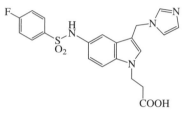

Dickinson, R. P., Dack, K. N., Steele, J., and Tute, M. S. *Bioorg. Med. Chem. Lett.*
1996, *6*, 1691. See also, Dickinson, R. P.; Dack, K. N., Long, C. J., and Steele, J. *J.
Med. Chem.* **1997**, *40*, 3442.

7.

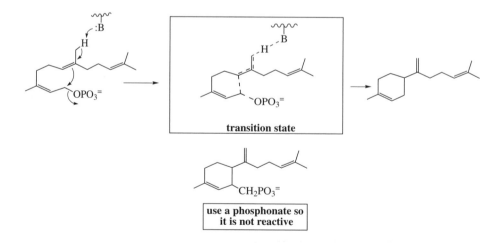

8. A dihydrofolate reductase inhibitor, because dihydrofolate reductase is needed to reduce
the dihydrofolate to tetrahydrofolate produced from methylene tetrahydrofolate in the
thymidylate synthase reaction.

9. It coordinates to the ferric ion.

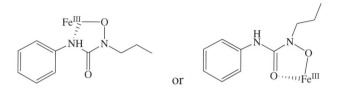

10.

▶ A—water soluble isostere of phenyl; the methyl group may impart a hydrophobic effect and expand the pharmacophore

▶ B—ketomethylene peptidomimetic for an amide

▶ C—hydrophobic group; the fluorine prevents metabolism (see Chapter 7)

▶ D—ring-chain glutamine mimic; more lipophilic

▶ E—Michael acceptor for affinity labeling with an active site nucleophile. This is an example of a quiescent affinity labeling agent. The Michael acceptor does not react with thiols, such as glutathione under physiological conditions, but the active site cysteine residue in rhinovirus protease is exceptionally nucleophilic and undergoes Michael addition with irreversible inactivation of the enzyme.

Dragovich, P. S. *et al., J. Med. Chem.*, **1999**, *42*, 1213.

11. A. Design a selective isoform-1 inhibitor.

B. Try to make a selective reversible inhibitor that has a group large enough to bind at the cysteine bonding site, but not in the phenylalanine binding site. Design an affinity labeling agent that reacts with the cysteine residue.

12. A.

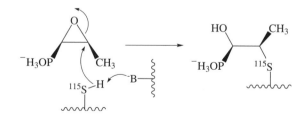

B. 1. mutation of the Cys
2. production of a new enzyme (FosA) that catalyzes the glutathione-dependent opening of the epoxide (see Armstrong and coworkers, *J. Am. Chem. Soc.* **2002**, *124*, 11001)

13. A. This is a mechanism-based inactivator.

Substituents on the PLP omitted for clarity

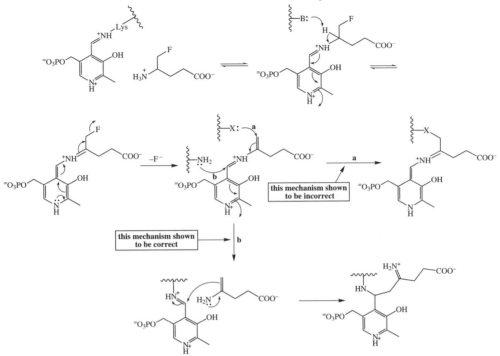

Nanavati, S.M.; Silverman, R.B. *J. Am. Chem. Soc.* **1991**, *113*, 9341–9349.

B. Give glutamine, which can cross the blood–brain barrier and get hydrolyzed to Glu and may shift the Glu-GABA imbalance. Give a glutamic acid decarboxylase inhibitor to prevent degradation of the glutamic acid and block GABA formation.

14.

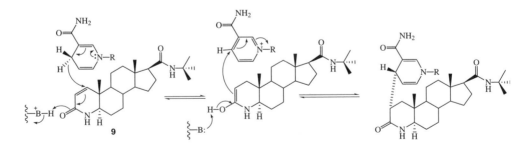

Bull, H. G. *et al. J. Am. Chem. Soc.* **1996**, *118*, 2359–2365.

Chapter 6

1. See Scheme 6.1 in the text.

2.

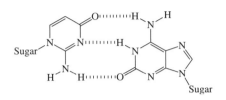

3. Incorporate a flat, aromatic, or heteroaromatic portion to stack between DNA base pairs.

4. Cancer cells replicate much faster than most normal cells. Therefore, they have a greater need for DNA precursors. Their uptake systems for these metabolites, and also antimetabolites, are very efficient. Therefore, there is selective toxicity for the cancer cells. Because of rapid cell division, cancer cell mitosis can be stopped preferentially to normal cells, where there is sufficient time for repair mechanisms to be effective.

5.

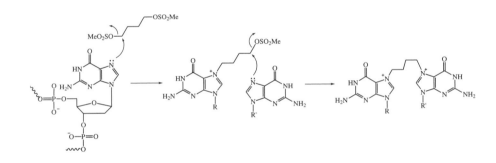

6.

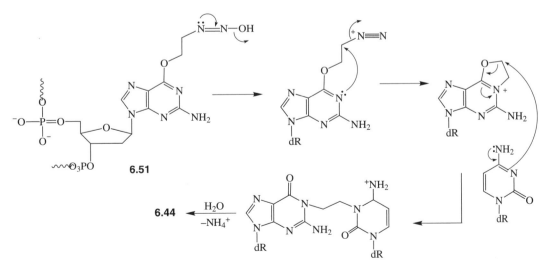

7. Design a compound to inhibit the new enzyme, and give both the DNA alkylating agent and the new inhibitor in combination.

8.

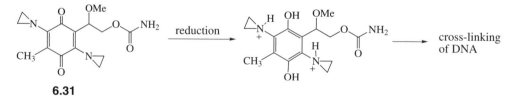

9.

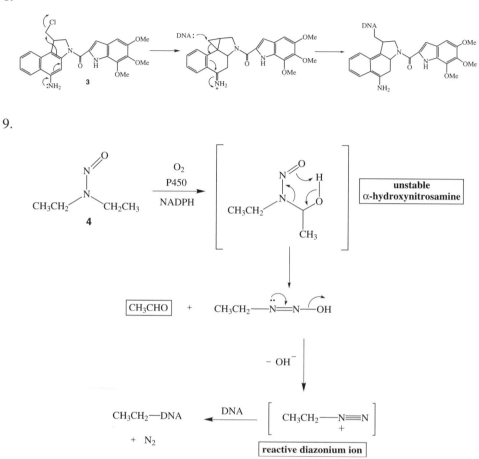

Leoppky, R. N. *Drug Metab. Rev.* **1999**, *31*, 175–193.

10. The quinone lowers the pK_a of the aziridine nitrogens by resonance, so it is not protonated. Reduction of the quinone to the hydroquinone prevents the delocalization of the nonbonded electrons of the aziridine nitrogen, so it becomes more basic, is protonated, and is activated for DNA attack.

6.31

11. The geometry of the molecule is such that the enediyne orbitals do not interact until reduction of an adjacent moiety takes place, which leads to the enediyne orbitals coming close enough to react, giving the 1,4-dehydrobenzene diradical (Bergman rearrangement).

12.

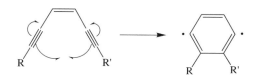

13. Any structure with a flat, aromatic part and an enediyne.

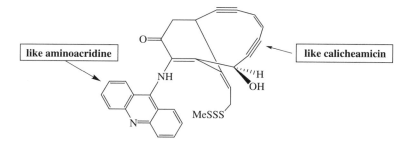

like aminoacridine

like calicheamicin

14.

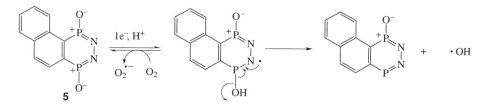

15.

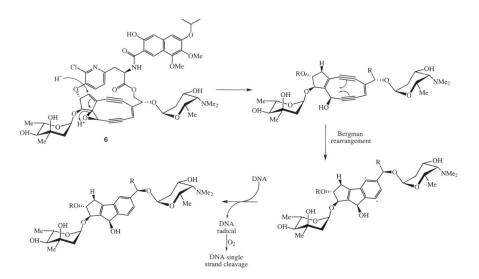

☐ Chapter 7

1. Metabolism is enzyme catalyzed and, therefore, will be dependent on chirality. The enantiomers can be substrates for two different enzymes.

2. By changing the stereochemistry at one site in the molecule there will be an effect on how the molecules bind at the active site of P450. The oxygenation occurs at the heme-oxo species, so whatever part of the substrate molecule is closest to this reactive species will be oxygenated.

3.

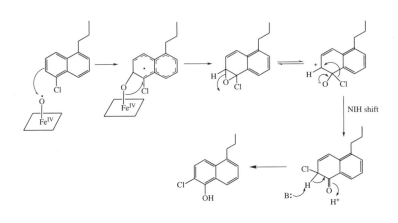

4.

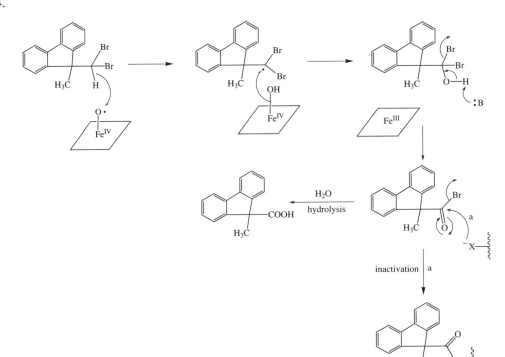

5.

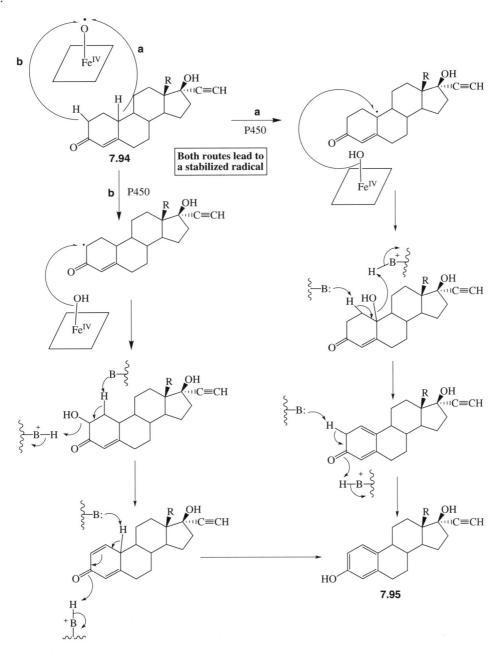

Both routes lead to
a stabilized radical

7.94

7.95

6.

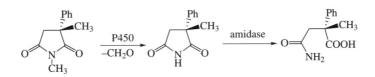

7.

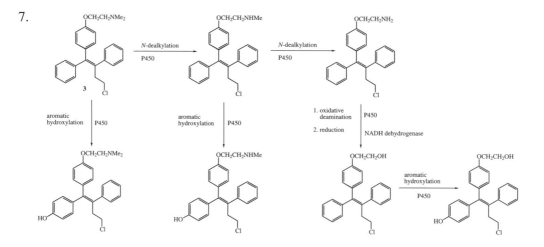

Bethou, F. *et al., Biochem. Pharmacol.* **1994**, *47*, 1883–1895.

8.

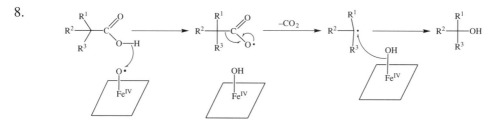

Komuro, M.; Higuchi, T.; Hirobe, M. *Bioorg. Med. Chem.* **1995**, *3*, 55–65.

9.

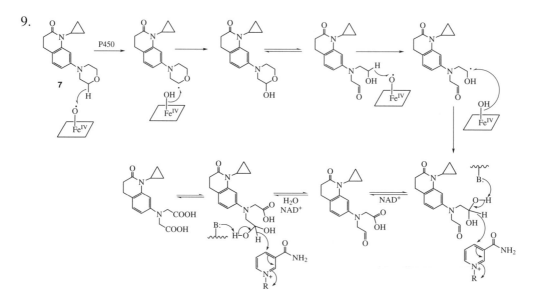

The aldehydes also could be oxidized by P450.

10. Glutathione protects cells from potent electrophiles by a variety of nucleophilic reactions.

11.

Reaction	Phase	Type of Reaction	Cofactor
a	I	*N*-dealkylation	Heme
b	I	Oxidative deamination	Heme
c	I	Oxidation	NAD$^+$
d	II	Amino acid conjugation	ATP/CoASH
e	I	Aromatic hydroxylation	Heme
f	I	Aromatic hydroxylation	Heme
g	II	Methylation	*S*-adenosylmethionine
h	I	Oxidation	Heme
i	II	Sulfate conjugation	PAPS
j	I	Reduction	NADPH
k	II	Glucuronidation	UDP-glucuronic acid
l	II	Acetylation	Acetyl CoA
m	I	*S*-oxidation	Heme

12.

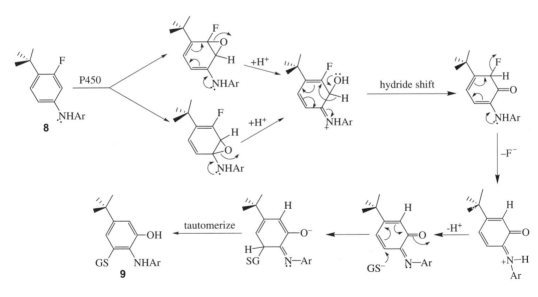

13.   A.

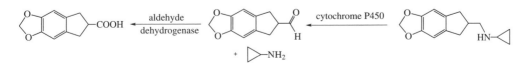

B.

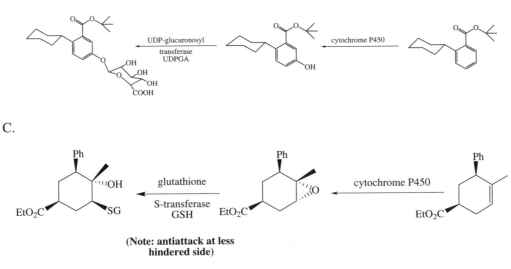

C.

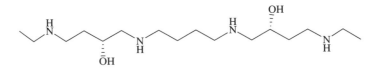

(Note: antiattack at less
hindered side)

14. A. Soft drug approach

   B. A compound found effective is shown below, but there are many correct answers
   where the compound is modified to induce metabolism. In this case, the hydroxyl
   groups could be conjugated, but the actual metabolites were not determined.

Bergeron, R. J. *et al.*, *J. Med. Chem.* **1996**, *39*, 2461.

⬛ Chapter 8

1.

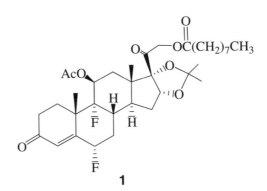

1

Any *one* of these changes might suffice, but possibly more than one would be needed.

2.

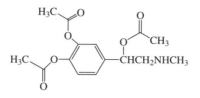

Or any ester at these positions

3.

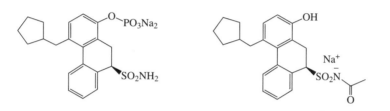

4. Make a prodrug that prevents hydrolysis by the new enzyme, but can be converted to the drug after activation. In this case, the additional RCO group may not be a substrate for the resistant enzyme or may inhibit the resistant enzyme. This may give the prodrug the ability to get to a different site for activation (hydrolysis of the RCO group).

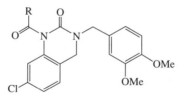

Design a molecule to inhibit the new enzyme (XH) and then make a mutual prodrug that can undergo hydrolysis to the two drugs. A possible example is shown below.

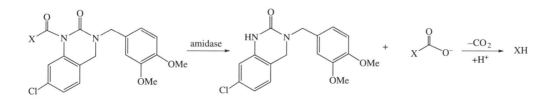

5.

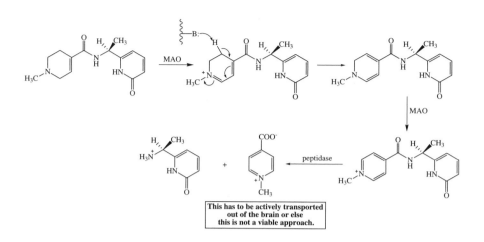

6.

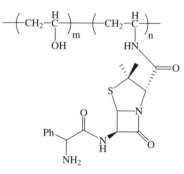

The ampicillin will be released by an amidase. If not, a spacer has to be inserted between the polymer and ampicillin.

7.

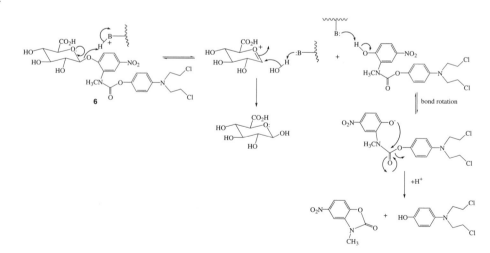

Schmidt, F. *et al. Bioorg. Med. Chem. Lett.* **1997**, *7*, 1071–1076.

8.

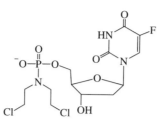

There are many correct answers. Here I show floxuridine, an inactivator of thymidylate synthase, linked to a nitrogen mustard through a phosphoramidate bond (tumor cells have high phosphoamidase activity).

9. A.

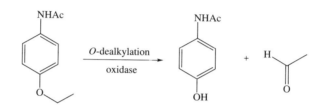

B.

C.

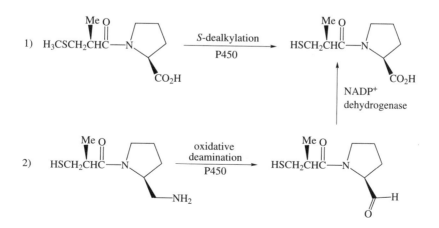

10.

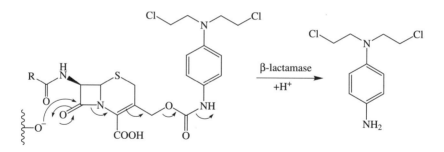

Svensson, H.P., Frank, I. S., Berry, K. K., and Senter, P. D. *J. Med. Chem.* **1998**, *41*, 1507–1512.

Index

593